Natural Resources
Technology, Economics and Policy

Natural Resources Technology, Economics and Policy

Editor

U. Aswathanarayana

Mahadevan International Centre for Water Resources Management, Hyderabad, India

CRC Press is an imprint of the Taylor & Francis Group, an **informa** business

A BALKEMA BOOK

CRC Press
Taylor & Francis Group
6000 Broken Sound Parkway NW, Suite 300
Boca Raton, FL 33487-2742

First issued in paperback 2019

CRC Press/Balkema is an imprint of the Taylor & Francis Group, an informa business

Typeset by MPS Limited, Chennai, India

ISBN-13: 978-0-415-89791-4 (hbk)
ISNB-13: 978-0-367-38158-5 (pbk)

Library of Congress Cataloging-in-Publication Data

Natural resources : technology, economics and policy / editor,
U. Aswathanarayana.
p. cm.
Includes bibliographical references and indexes.
ISBN 978-0-415-89791-4 (hbk. : alk. paper) — ISBN 978-0-203-12399-7 (ebook)
1. Natural resources. 2. Natural resources—Management—Technological innovations. 3. Natural resources—Government policy.
4. Environmental policy. I. Aswathanarayana, U.
HC85.N354 2012
333.7—dc23

2011053273

Published by: CRC Press/Balkema
P.O. Box 447, 2300 AK Leiden, The Netherlands
e-mail: Pub.NL@taylorandfrancis.com
www.crcpress.com – www.taylorandfrancis.com

Visit the Taylor & Francis Web site at
http://www.taylorandfrancis.com

and the CRC Press Web site at
http://www.crcpress.com

POINTS TO PONDER

"The only source of sustainable growth is technology"

– Anonymous

"Listen to the technology, find out what it is telling you"

– Carver Mead (Caltech)

"Tracking vectors in technology over time to judge when an intriguing innovation is ready for a market place. Technical progress, affordable pricing and consumer demand – all must jell to produce a blockbuster product"

– Steve Jobs of Apple Inc.

"Real innovation in technology involves a leap ahead, anticipating needs that no one really knew they had, and then delivering capabilities that define product categories"

– David B. Yoffie of Harvard Business School

"Technology changes. Economic laws do not"

– Hal Varian (Berkeley, USA)

An Ode to Africa
(where the editor served for 21 years)

It is morning in Africa,
The impala gets up and starts running,
It knows that unless it runs faster than the fastest lion, it will not survive

It is morning in Africa,
The lion gets up and starts running,
It knows that unless it runs faster than the slowest impala, it will starve.

It does not make a difference whether you are an impala or a lion.
When the sun is up, you will have to start running.

Contents

Preface

The purpose of the volume is two-fold: to provide knowledge and skills which would enable candidates to obtain gainful employment in managing natural resources (water, soils, minerals, and ecosystem) in economically-viable, ecologically-sustainable, and socially-equitable ways, and to draw attention to technologies which would enable Man to live within the resilience limits of Nature while utilizing natural resources. *Technology has to be continuously adapted depending upon the emerging biophysical and socioeconomic realities.*

Section 1 introduces the ethos of the book.

Section 2 devoted to water is the longest of all sections – it actually covers about half of the book. Among the natural resources, water undoubtedly is the most critical, as food security, health and hygiene and environmental security are all dependent upon water security. Optimal use of water resources should be at the heart of developmental planning in any country. Since about 80% of the water is used in agriculture, it is this sector that should achieve a low water footprint at the earliest. Notwithstanding the advances in capture and culture fisheries, man is dependent upon the soil for about 90% of his food requirements. Also, if the soils are deficient in micro-nutrients, the crops grown in such soils will also be deficient in micronutrients. The future of food security will therefore depend upon the sustainable management and conservation of soil resources, in conjunction with water resources. The section covers the applications of earth, space, atmospheric and information technologies in water and soil resources management.

There is general public abhorrence of the mining industry because it pollutes and disfigures the environment. Section 3 on Mineral Resources Management deals at length with control technologies which can minimize the adverse impact on the environment on the one hand, and put mineral wastes to beneficial use on the other.

Section 4 on Energy Resources Management deals with the technology and economics of fossil fuels, nuclear fuels and renewable fuels, and strategies for low carbon footprint.

Section 5 on Bioresources and Bioenergy deals with role of biodiversity in medicine, agriculture and forestry, among others, and ways of maintaining ecological services to support human livelihoods.

Section 6 on Disaster Management deals at length with Integrated Preparedness systems for disasters, natural or man-made.

The last section 7 deals with possible ways of living in harmony with nature.

The theme song of India's freedom movement against the British, "*Vande Mataram*" – "(I salute Mother)", invokes a Motherland of clean water, fruitful groves, cool breezes and verdant crops. This could be the universal theme song of environmentalists all over the world.

The knee replacement surgery performed by the eminent surgeon, Dr. A.V. Guravareddy, and his associates, Drs. T. Chiranjeevi, P. Kavitha, Lt. Col. K. Bhaskar Reddy, and facilitated by my friend, Shri S.N. Jayasimhan, made it possible for me to stand on my own legs – literally and figuratively. My wife, Vijayalakshmi, supported by our kind neighbour, Satya, has kept me going through her 24-7 ministrations.

I am grateful to Drs. K.S. Murti, B.B.G. Sarma, T. Harikrishnan, and R. Shankar for kindly reviewing the manuscript, and to Shri P. Lakshminarayana for computer assistance. The copyright permission of the Kerala Tourism Development Corporation, Tiruvananthapuram, India, for the image (paddy fields) on the cover page, is thankfully acknowledged.

Dr. C.N.R. Rao, F.R.S., the doyen of Indian science, has kindly contributed a perceptive Foreword, which greatly enhances the value of the book. I am grateful to the distinguished chapter authors for their valuable contributions. The volume is meant to be an international, university-level textbook-cum-reference book in natural resources management. It will be useful to students, professionals and administrators in the areas of natural resource management, environmental engineering, irrigation engineering, energy engineering, mineral engineering, biodiversity, and disaster management.

U. Aswathanarayana
Hyderabad, India

Foreword

Technology has been growing by leaps and bounds, but quite often, technological advances do not reach the classroom at the same pace. The present volume is not only up-to-date technology-wise, but also has the merit of integrating technology with economics and policy, which is as it should be.

Dr. Aswathanarayana's pioneering researches in nuclear geology under the supervision of the Late Prof. C. Mahadevan, his post-doctoral studies in Caltech (USA), and Oxford (England), his long stint of over two decades in Africa as teacher-researcher-consultant, and his authorship of books dealing with the management of soils, waters and minerals, have equipped him to bring out this textbook on natural resource management. He has also drawn upon the expertise of subject-specialists, whose contributions have been woven into a coherent and relevant book.

Among the natural resources (water, soils, minerals and ecosystems, etc.), the book gives primacy to water, particularly drinking water (~200 out of ~400 pages of the book are devoted to water), a consideration, which I strongly endorse. The wellness of a community is closely linked to water security. Availability of safe drinking water is crucial for the maintenance of health. Water scarcity, which most countries face, has a strong adverse affect on food productivity, industrial activity where water is a part of the chain, ecosystem integrity and productivity. In the interest of ecology and economy, it is necessary to place water at the heart of the development planning. Just as we have embraced a low-carbon economy to meet the challenges of climate change, we need to go in for a low-water economy to meet the challenges of water scarcity. We should observe a low water footprint in all our activities. How this is to be achieved is covered in the chapters on economic frameworks for informed decision-making regarding water use, sequential use of water resources, protection of groundwater and wastewater reuse systems.

Mining cannot be avoided, but mining industry has a poor public image as a spoiler of the environment. The section on mineral resources management gives an account of the technologies, which could minimize the adverse environmental impact of mining and put the mining wastes to beneficial use.

The section on energy resources management gives a detailed comparative account of the technology and economics of fossil fuels, nuclear fuels and renewable fuels, and ways of reducing the carbon footprint.

The section on bioresources and biodiversity deals with the role of biodiversity in medicine, agriculture and forestry, and the use of biotechnology in conserving biodiversity.

Natural hazards such as earthquakes and tsunamis cannot be prevented, but by being prepared, damage to life and property can be minimized. The section on disaster management gives an account of science and people-based preparedness systems. It provides information on the Fukushima – Daiichi nuclear disaster.

I compliment Dr. Aswathanarayana on preparing this volume which will be useful to university students and professionals.

C.N.R. Rao, F.R.S.
Chairman, Science Advisory
Council to Prime Minister
and
Hon. President, Jawaharlal Nehru Centre
for Advanced Scientific Research
Bangalore, India

List of Figures

List of Tables

Units, Abbreviations and Acronyms, Definitions and Conversion Constants

Units

bbl – barrel
bcm – billion cubic metres
tmc – thousand million cubic feet
boe – barrels of oil equivalent (1 BOE = 159 L)
kW_{el} – kilowatt electric capacity
kWh – kilowatt hour
kW_{th} – kilowatt thermal capacity
lge – litre gasoline equivalent
mbd – million barrels per day
Mbtu – million British thermal units
Mha – million hectares
MJ – megajoule = 10^6 joules
Mpg – miles per gallon
MW_e – megawatt electrical
MWh – megawatt hour
Mt – millions of tonnes
PJ – Petajoule = 10^{15} joules
EJ – Exajoule = 10^{18} joules
ZJ – Zettajoule = 10^{21} joules
toe – tonne of oil equivalent
TW – terawatt = 10^{12} watts
TWh – terawatt hour
Wp – watt-peak

Abbreviations and Acronyms

(source: Principally from IEA's *Energy Technology Perspectives*, 2008. pp. 602–613)

AFC – Alkaline Fuel Cell
AIRS – Advanced Infra Red Sounder
AMSR – Advanced Microwave Scanning Radiometer
API – American Petroleum Institute
AVHRR – Advanced Very High Resolution Radiometer
B2B – Business-to-Business
B2C – Business-to-Consumer
B2G – Business-to-Government
BEMS – Building Energy Management System
BFB – Bubbling Fluidised Bed
BIGCC – Biomass Integrated Gasification with Combined Cycle
BtL – Biomass to Liquids
CAES – Compressed Air Energy Storage System
CAT – Carbon Abatement Technologies
CBM – Coal-Bed Methane
CCS – CO_2 Capture and Storage
CDM – Clean Development Mechanism
CdTe – Cadmium Telluride
CFB – Circulating Fluidised Beds
CFL – Compact Fluorescent Light-bulb
CHP – Combined Heat and Power
CIS – Copper-Indium-Diselenide
CIGS – Gallium-doped Copper – Indium – Diselenide
CNG – Compressed Natural Gas
CSP – Concentrating Solar Power
CTL – Coal To Liquids
DME – Dimethyl Ether
EGR – Enhanced Gas Recovery
EIA –Environment Impact Assessment
EPR – European Pressurised Reactor
FBC – Fluidised Bed Combustion
FDI – Foreign Direct Investment
FGD – Flue Gas Desulphurisation
GRACE – Gravity Recovery and Climate Experiment
HIRS – High Resolution Infra Red Sounder
HTGR – High Temperature Gas Cooled Reactor
IAEA – International Atomic Energy Agency
IEA – International Energy Agency
IET – International Emissions Trading
IGCC – Integrated Gasification Combined Cycle
IGFC – Integrated Gasification Fuel cell combined Cycle
ITER – International Thermonuclear Experimental Reactor
LAI – Leaf Area Index

LED – Light Emitting Diode
LNG – Liquified Natural Gas
LPG – Liquid Petroleum Gases
MODIS – Moderate Resolution Imaging Spectroradiometer
MSU – Microwave Sounding Unit
NDVI – Normalized Vegetation Difference Index
NEA – Nuclear Energy Agency
NGL – Natural Gas Liquids
NOAA – National Oceanographic and Atmospheric Administration
NSG – Nuclear Suppliers Group
O&M – Operating and Maintenance
OECD – Organization for Economic Cooperation and Development
OPEC – Organization of Petroleum Exporting Countries
PFBC – Pressurised Fluidised Bed Combustion
PM-10 – Particulate matter of less than ten microns in diameter
PPP – Purchasing Power Parity
P&T – Partitioning and Transmutation
PV – Photovoltaics
PWR – Pressurised water Reactor
RDD&D – Research, Development, Demonstration and Deployment
RETs – Renewable Energy Technologies
SACS – Saline Aquifer CO_2 Storage
SCSC – Supercritical Steam Cycle
SMR – Small and Medium-sized Reactor
SSM/I – Special Sensor Microwave Imager
T&D – Transmission and Distribution
USCSC – Ultra Super Critical Steam Cycle
VHTR – Very High Temperature Reactor

Definitions

Source A: *Energy Technology Perspectives*, 2008, pp. 601–605
Source B: *International Energy Markets*. 2004. Carol A. Dahl, Pennwell Corporation, pp. 475–533.

Ad valorem tax: A tax that is a percentage of the price of a good or a service. (B)
Amortization: Allocating the cost of intangible assets over their legal life as specified in the tax code. (B).
Avoided cost: The amount avoided for the incremental purchase or the production of a good. (B)
Benefits of Pollution: Any costs that you forego by being able to pollute rather than to abate. Benefits of pollution are then equal to the costs of abatement. (B)
Biodiesel: Biodiesel is a diesel-equivalent, processed fuel made from the transesterification (a chemical process which removes the glycerine from the oil) of vegetable oils or animal fats. (A)
Biogas: A mixture of methane and carbon dioxide produced by bacterial degradation of organic matter and used as fuel. (A)

Blackouts: A non-isolated power loss over an extended period of time due to capacity shortage. It may result from peak loads higher than available capacity or from equipment failure (B).
Black liquor: A by-product from chemical pulping processes which consists of lignin residue combined with water and the chemicals used for the extraction of lignin. (A)
Breakeven Pricing: Charging a price for which revenues exactly equal all costs including opportunity costs. (B)
Brent Forward Market: The over-the-counter market for buying Brent Crude oil at some future date. (B)
Clean Coal Technologies (CCT) Technologies designed to enhance the efficiency and the environmental acceptability of coal extraction, separation and use. (A)
Clearinghouse: An institution that is a part of an organized exchange that guarantees each transaction and matches buyers to sellers when contracts come due. (B)
Coases Theorem on Externalities: In the absence of transaction costs and market power, that private markets will arrive at an optimal allocation in the presence of market externalities no matter how property rights are originally distributed. (B)
Cross Price Elasticity: The percentage change in quantity of one good that results from the percentage change in price of another good. (B)
Data Mining: Techniques for extracting information from large databases. (B)
Deregulation: Removing government regulations. (B)
Discounted Cash Flow (DCF): The present value of future flows of income. (B)
Discount Rate: The interest rate for converting or discounting future cash values to present values. (B)
Energy Futures: A standardized contract offered and guaranteed on an organized exchange to buy or sell an energy product in the future. (B)
Enhanced Coal-bed Methane Recovery (ECBM): A technology for the recovery of methane through CO_2 injection into uneconomic coal seams. (A)
Enhanced Gas Recovery (EGR): A speculative technology in which CO_2 is injected into a gas reservoir in order to increase the pressure in the reservoir, so that more gas can be extracted. (A)
Financial Derivatives: Financial assets that derive their value from an underlying asset upon which they are based. (B)
Hydrocracking: Refinery process that heats heavy oil products under pressure in the presence of hydrogen to remove sulphur and increase lighter product yields. (B)
Marginal Production Cost: The cost of the last unit of production. (B)
Marketable Permits: Permits to pollute that can be bought and sold in the market place. (B)
Metcalfe's Law: A network's value increases as the square of the number of connections. (B)
Multivariate Time Series: A statistical forecasting technique in which a variable is forecast by using historical values of itself and other related variables. (B)
Negative Externalities: An externality is an effect from an economic activity that involves some one not directly involved in the economic activity. (B)
Opportunity Cost: What you forego by undertaking an economic activity (B).
Outage: A temporary loss of power from isolated electricity transmission, generation or distribution failure. (B)
Peak load pricing: Charging higher prices during peak hours than off-peak hours. (B)

Pollution tax: A payment of tax to the government for the right to pollute. (B)
Price Elasticity of Demand: Percentage change in the quantity demanded of a good divided by the percentage change in its own price. (B)
Public good: A good that no one is excluded from using. (B)
Purchasing Power Parity (PPP): The rate of currency conversion that equalizes the purchasing power of different currencies, by making allowances for the differences in price levels and spending patterns between different countries. (A)
Renewables: Includes biomass and waste, geothermal, solar PV, solar thermal, wind, tide, and wave energy for electricity and heat generation. (A)
Straight-line Depreciation: Using an annual depreciation charge for tax purposes equal to the value of the asset divided by its allowed depreciable life. (B)
Strike Price: The price at which a put or a call entitles you to sell or buy the underlying asset. (B)
Total Final Consumption (TFC): It is the sum of the consumption by the different end-use sectors: industry (including manufacturing and mining), transport, other (including residential, commercial and public services, agriculture/forestry and fishing), non-energy use (including petrochemical feedstocks), and non-specified. (A)
Total Primary Energy Demand: Total Primary Energy Demand represents domestic demand only, including power generation, other energy sector, and total final consumption. It excludes international marine bunkers, except for world energy demand where it is included. (A)

Conversion constants

Length, area, volume

1 micron (μm) = 10^{-6} m = 10^{-4} cm = 10^{-3} mm = 10^{4} A
1 Ångstrom (Å) = 10^{-4} μm = 10^{-8} cm = 10^{-10} m
1 hectare (ha) = 100 m × 100 m = 10^{4} m^{2} = 2.47 acres
1 sq.km (km^{2}) = 100 ha = 247 acres ; 1 acre = 4840 sq. yds = 4046.8 m^{2}
1 cu.km (km^{3}) = 10^{3} ha m; 1 ha m = 8.1 acre-ft
1 US gallon = 3.875 L; 1 Imperial gallon = 4.546 L
1 barrel (crude oil) = 42 US gallons = 35.80 Imp. Gallons
1 acre-ft = 1235 m^{3} = 326,000 gallons
1 tmc (thousand million cu.ft) = 28.317 million m^{3}

Weight

1 tonne (t) = 10^{3} kg = 10^{6} g; 1 kg = 2.2046 lbs = 32.150 oz.
1 troy oz = 31.10348 g = 20 pennyweights (dwt) = 480 grains = 1.0971 av. Oz
1 pennyweight (dwt) = 1.5517 g = 24 grains
1 part per million (ppm) = 10^{-6} g/g = l g/tonne = 0.032 oz/t = 0.644 dwt/t 0.644
1 part per billion (ppb) = 10−9 g/g = 1 mg/tonne

Time

1 year = 365.25 days = 8766 hours = 5.26 × 10^{5} mins = 2.156 × 10^{7} sec

Radiation units

Measure of radiation	Old unit	S.I. Unit	Conversion
Activity	Ci (curie)	Bq (becquerel)	1 Bq = 2.7×10^{-11} Ci; 1 Ci = 3.7×10^{10} Bq
Exposure	R (Röntgen) = 89 ergs/g	C/kg (Coloumb/kg)	1 R = 2.58×10^{-4} C/kg; 1 C/kg = 3876 R
Exposure rate	μR/h	C/kg/s	
Absorbed dose	rad; 1 rad = 100 ergs/g	Gy (Gray)	1 Gy = 100 rads
Absorbed dose rate	100 rads/s	Gray/s	
Dose equivalent	100 rem	1Sv (sievert)	

The prefixes, m = milli = $10^{-}3$; μ = micro = 10^{-6}; and p = pico = 10^{-12}, are used with radiation units as necessary.
Legal consumption limit of radioactive isotopes in food, Cs-134 & 137 and I-131: 500 Bq/kg.
Permissible Radiation environment levels: 100 microsieverts/hr.
Source: "*World Energy Outlook 2007*", International Energy Agency, Paris, 2007, pp. 633–641.

General Conversion factors for Energy

To From	TJ Multiply by	Gcal	Mtoe	MBtu	GWh
TJ	1	238.8	2.388×10^{-5}	947.8	0.2778
Gcal	4.1868×10^{-3}	1	10^{-7}	3.968	1.163×10^{-3}
Mtoe	4.1868×10^{4}	10^{7}	1	3.968×10^{7}	11630
MBtu	1.0551×10^{-3}	0.252	2.52×10^{-8}	1	2.931×10^{-4}
GWh	3.6	860	8.6×10^{-5}	3.412	1

TJ = Tera Joules; Gcal = Gigacalories; Mtoe = Million tonnes of oil equivalent
MBtu = Million British Thermal Units; GWh = Gigawatt hours
1 million tonnes of oil equivalent = 1.9814 million tonnes of coal
= 0.0209 million barrels of oil/day
= 1.2117 billion cubic metres of gas

Conversion factors for mass

To: From:	kg Multiply by	t	lt	st	lb
kilogramme (kg)	1	0.001	9.84×10^{-4}	1.102×10^{-3}	2.2046
tonne (t)	1000	1	0.984	1.1023	2204.6
long ton (lt)	1016	1.016	1	1.120	2240.0
short ton (st)	907.2	0.9072	0.893	1	2000.0
pound (lb)	0.454	4.54×10^{-4}	4.46×10^{-4}	5.0×10^{-4}	1

Conversion factors for volume

To: From:	gal U.S. Multiply by	gal U.K.	bbl	ft^3	L	m^3
U.S. gallon (gal)	1	0.8327	0.02381	0.1337	3.785	0.0038
U.K. gallon (gal)	1.201	1	0.02859	0.1605	4.546	0.0045
Barrel (bbl)	42.0	34.97	1	5.615	159.0	0.159
Cubic foot (ft^3)	7.48	6.229	0.1781	1	28.3	0.0283
Litre (L)	0.2642	0.220	0.0063	0.0353	1	0.001
Cubic metre (m^3)	264.2	220.0	6.289	35.3147	1000.0	1

About the Editor

U. Aswathanarayana has extensive international teaching, R&D and institutional capacity-building experience in a number of countries. He is the author of several books on natural resources and renewable energy published by CRC Press/Balkema. He was elected to the Fellowship of the Third World Academy of Sciences (TWAS) in "recognition of his outstanding contributions to science, and to the development of science in the South". He is the recipient of, amongst others, the Excellence in Geophysical Education and of the International Award of the American Geophysical Union, and of the Certificate of Recognition of the International Association of Geo-Chemistry (IAGC).

Section I

Introduction

Chapter 1.1

Symbiotic relationships between mangroves and coastal communities

Dr. M.S. Swaminathan, the eminent agricultural scientist of India, described a case where science-illuminated resource management brought about profound improvement in the quality of life of a community (*The Hindu*, May 22, 2010).

The mangroves of Pichavaram, Tamilnadu, India, came into the public eye when it was realized that the mangroves saved the fishermen community of Pichavaram from the fury of the Indian Ocean tsunami of Dec. 2004, by serving as speed-breakers. The fishermen were extremely poor and did not have proper nets. The children do not go to school, for the simple reason that there is no school in the neighbourhood. They were not receiving government benefits that are given to Scheduled Castes and Tribes, as the bureaucracy was not sure where they stand in the caste hierarchy. Swaminathan got these questions sorted out. A primary school was established with the help of some donors. Swaminathan makes a profound observation: "*Saving mangrove forests without saving the children for whose well-being the forests are being saved, makes no sense*". After the 2004 tsunami, the government constructed brick-built houses for the community. There is now a secondary school in Pichavaram. The fishermen now know the value of the mangroves – how the root exudates enriches water with nutrients, and thereby help in augmenting the fisheries, and how the mangroves protect the coastal waters from salinization. Mangroves are known to have a genetic element which enables them to grow in salt water. MSSRF, the research institute built by Swaminathan, uses this genetic element to create new varieties of salt-tolerant paddy and other crops. Such salt-tolerant crops will become very useful when the coastal soils get salinized due to sealevel rise consequent upon global warming.

As against this, wrong practices caused by the ignorance of the importance of symbiosis between mangroves and man led to horrendous degradation of the environment in Quelimane, Mozambique. The uncontrolled cutting down of mangrove trees for fuelwood, timber and coffins, led to extensive, *irreversible* salinization of coastal waters and soils. Mangroves grow slowly – it may take about 20 years for mangroves to grow to a height of 10 m. Though copious groundwater is available at depth of one metre, it is brackish and is not potable. Coconut trees seem to have a beneficial effect on the quality of groundwater. When groundwater is drawn from ponds within clusters of coconut trees, the water is less brackish. The soils have been so degraded that a large area of about a thousand hectares has become a saline waste. People defecate in the mangroves as they do not have latrines, and the faeces end up in fish, which are eaten by people. When there is a cholera epidemic, the cholera-contaminated faeces end up in fish, which are consumed by the people, and a cycle gets triggered with tragic consequences.

After a study of the situation, the author came up with a science-illuminated strategy to reverse the degradation, whereby the cutting down of mangroves is avoided. Salt-tolerant paddy varieties are to be grown in the salinized soils. Fish spawn is put in the paddy fields. The fish eat the dead paddy leaves, and the droppings of fish fertilize the soil. When the paddy is harvested, the fish are harvested along with it. Fast-growing, salt-tolerant casuarinas trees are planted along the bunds of the paddy fields (casuarina trees grow to a height of 10 m in five years). The renovated ecosystem will then be able to provide food (paddy and fish) and fuelwood and timber for house construction.

Chapter 1.2

Earth system science for global sustainability

The International Council of Scientific Unions (ICSU) in cooperation with the International Science Council (ISSC) came up with a holistic strategy on Earth System research that integrates our understanding of the functioning of the Earth system – and its critical thresholds – with global environmental change and socio-economic development (Reid et al., 2010).

The five Grand Challenges identified are summarized below:

1. Forecasting – to improve the usefulness of forecasts of future environmental conditions and their consequences for people.
2. Observing – to develop, enhance and integrate the observation systems needed to manage global and regional environmental change.
3. Confining – to determine how to anticipate, recognize, avoid and manage disruptive global environmental change.
4. Responding – to determine what institutional, economic and behavioural changes can enable effective steps towards global sustainability.
5. Innovating – to encourage innovation in developing technological, policy and social responses to achieve global sustainability.

Chapter 1.3

"Virtual" natural resources

The economics of natural resources management can be understood in terms of "virtual" natural resources. It is generally difficult and prohibitively expensive to transfer large quantities of water, soils, and low unit cost minerls and rocks (such as limestones) to some other places where they are required. It is, however, possible to share the *benefits* of water, instead of physically sharing the water. "Virtual" water is the amount of water needed to produce goods – for instance, 1000 L (one m^3) of water are needed to produce 1 kg of grain, and 13,000 L of water is needed to produce 1 kg of beef. Thus, a country which exports one tonne of grain is in effect exporting 1000 m^3 of "virtual" Water.

While the world's population has doubled in the past half century, the consumption of meat has quadrupled. About 2 kg of grain are needed to produce a weight gain of one kg of chicken. For pork, it is 3 kg, and for beef, 8 kg. For the annual production of about 200 million tonnes of meat, livestock are now fed about 40% of all grain harvested. At a given level of nutrition (say, a minimum amount of 2200 daily calories), the grain and soya bean needed to feed a pig to produce pork, will feed a man ten times the number of days, if consumed directly instead of as pork (source: *National Geographic*, Oct. 1958).

About 70% of all water is used in food production. It is possible to convert the per capita consumption of various items of food and other uses in terms of "virtual" water. The consumption of "virtual" water is 1400 L per capita per day in Asia, and about 4000 L in the case of Europe and North America. The following countries are among biggest net exporters of "virtual" water: USA, Canada, Thailand, Argentina, India, Vietnam, France and Brazil. The largest net importers of "virtual" water are Sri Lanka, Japan, South Korea, Germany, Italy, etc.

Chapter 1.4

Natural resources and globalization

1.4.1 GENERAL CONSIDERATIONS

Globalization describes a process by which regional economies, societies and cultures have become integrated. The process is driven by a combination of economic, technological, sociocultural, political and biological factors. Economic globalisation is the most important part of globalisation, whereby national economies get integrated into international economy through trade, foreign direct investment, capital flows, migration and spread of technology.

An important consequence of globalisation is the modernisation and industrialisation of previously agrarian countries like India and China. These countries are gaining more knowledge and wealth. As more and more people move from Below-Poverty-Line to middle class status, they consume superior food, purchase more cars (Tata's Nano car costing about USD 2200/- is the cheapest car in the world, and is highly affordable), consume more fuel and more consumer goods. This necessitated a massive development of natural resources, including mineral resources within these countries. Besides, these countries are also making large investments in developing mineral deposits in Africa and elsewhere.

China produces 92% of the world's Light Rare Earth Elements, like cerium and lanthanum, and 99% of the world's Heavy Rare Earth Elements like Dysprosium. These rare earth elements are used in smart phones, clean energy industries, entertainment electronics, etc. There was an uproar from countries like Japan and USA which use the REE in their high-tech industries, when China restricted the export of the rare earth elements. USA had to intervene strongly to ease the situation.

Steel which is used for the construction of the wind turbines, accounts for 90% of the cost of the turbine. Turbine fabrication costs are being brought down by replacing steel with lighter and more reliable material, and by improving the fatigue resistance of the gear boxes. During the last five years, there has been a phenomenal growth in the use of rare-earth elements in the energy industries. Tiny quantities of dysprosium can make magnets in electrical motors lighter by 90%, thereby allowing larger and more powerful wind turbines to be mounted. Use of terbium can help cut the electricity use of lights by 80%. Dysprosium prices have gone up sevenfold since 2003, with the current market price being USD 116/kg. Terbium prices have quadrupled during the period 2003–2008 to USD 895/kg. The recession brought down the price to USD 451/kg.

Mineral resources are finite, and when once used, are gone forever. Technology has a great potential to address the depletion problem. Since globalization connects the world, more scientists from around the world would be able to work to find solutions to a global problem. For instance, since oil production is peaking, and oil plays a critical role in the world economy, attempts are being made to substitute natural gas in its place.

1.4.2 DIFFERENT ASPECTS OF GLOBALISATION

- *Industrial* – Globalisation has led to the emergence of worldwide production markets and broader access to a range of foreign products for consumers and companies. There has been extensive movement of materials and goods within and outside countries. International trade in manufactured goods increased more than 100 times (from $95 billion to $12 trillion) during the 50-year period (1955–2005).
- *Financial* – The worldwide emergence of financial markets greatly facilitated external financing by borrowers. So much so that by the early part of the 21st century, more than $1.5 trillion in national currencies were being traded daily. The unregulated growth of these worldwide structures had disastrous consequences. The global financial crisis of 2007–2010 is a consequence of this lack of regulation. Countries like India and China which have strong regulatory system recovered quickly, but the Industrialised countries is stll struggling to get out of the recession, despite various stimuli.
- *Economic* – Free exchange of goods and capital led to the emergence of global common market. This has a corollary. As the markets are interconnected, economic collapse in one area has an impact in other areas. Lehman Brothers, a US bank with assets of over USD 600 billion, was considered too large to fail. But fail it did most spectacularly – it declared bankruptcy on Sept. 15, 2008. This shook the banking industry around the world to its roots. Many banks and companies around the world collapsed. The objective of a company is to make profit. As a consequence of globalisation, companies find it more profitable to produce goods and services in the lowest cost locations. For instance, readymade garments for the US market are produced in China because of low labour costs. As the labour costs are increasing in China, the readymade garments industry is moving to countries like Vietnam. In the case of industrial activities, production is moved to areas with least pollution restrictions and worker safety regulations. For instance, aluminium industry has been shifted from South Africa to Mozambique. More and more persons in the former developing countries are becoming richer. Four Indians are among the top richest persons in the world in 2008. In 2007, China had 415,000 millionaires and India 123,000.

- *Job Market* – Till recently, a worker in a country has to confine himself to a job in his country. This is no longer so. Advances in Information and other technologies and improvements in communications have changed the situation drastically. For instance, in the Silicon Valley in California, Indian technologists and entrapreneurs dominate the Information Technology industry. They earn fabulous salaries and many of them are millionaires. Thus, profound changes in income distribution have taken place as a consequence of globalisation.
- *Political* – Political power goes with economic power. USA with GDP of USD 14 trillion, has been considered the most powerful country in the world. During the last ten years, GDP of China grew 16 times to USD 5 trillion, while during the same time, India's GDP grew seven times to USD 1.2 trillion. China has replaced Japan to become the second richest country in the world, after USA. There is a good possibility that in the next twenty years, the GDP of China may exceed that of USA, with corresponding increase in its political clout. Though theoretically all countries, rich and poor, have the same political rights in the international organisations, poor contries rarely figure in global decisions.
- *Informational* – A profound revolution in communications has taken place with the advent of fibre optic communications, satellites, and increased availability of telephone and Internet. Individuals, institutions, corporations, and governments can get the information promply, and react to it.
- *Competition* – Corporations have no option but adapt their production and marketing strategies to the emerging situation in the global business market. They have to upgrade their products and reduce costs continuously in order to stay in business. For instance, as a consequence of technological changes, considerations of economy, and government policy, vapour lamps are rapidly replacing filament lamps.
- *Ecological* – Global environmental challenges such as climate change, ozone hole, cross-boundary water and air pollution, depletion of global fish stocks, and the spread of invasive species, can only be solved with global cooperation. Developed countries locate industries, such as paper and pulp, which are not only highly polluting, but would need lots of freshwater, in the developing countries, thus passing on the ecological damage.
- *Cultural* – Globalisation has greatly accelerated cross-cultural contacts, cultural diffusion, and desire to live better, etc. Multiculturalism promotes peace and understanding between people. Globalisation is manifested in the form of greater international travel and tourism (in 2008, there were 922 million international tourist arrivals), high remittance flows to developing countries ($328 billion in 2008), worldwide fads and pop culture, such as YouTube, and widespread use of some dishes, such as Swedish meatballs, Indian curry, Japanese noodles, etc. McDonald fast food is a good example of coroporitisation and globalisation of food (since Hindus in India do not eat beef).

1.4.3 NATURAL RESOURCES AND VIOLENT CONFLICTS

When South Africa was being ruled by the apartheid regime, they played havoc with peace in the neighbouring countries. For instance, in Angola, the rebel leader, Jonas Savimbi, financed his campaign from the illegal sale of diamonds marketed through

South African companies. As against this, the Govt. of Angola got their funds from the oil industry, which is high-tech and not accessible to Savimbi. Millions of people died in the civil war, which came to an end after the dealth of Savimbi.

At one time, Mozambique had 1700 elephants, the largest number in Africa. Rebel leader, Dhakama, slaughtered the elephants and sold their tusks to South Africans, who then exported them to China and Japan. Thus, the elephant paid for the war in Mozambique.

Sierra Leone in west Africa has been notorious for the production and illegal sale of diamonds (the inhumanity involved is touchingly shown in the movie, "*The Blood Diamond*"). Even government machinery was mobilised to transport illegal diamonds – it was said that the illegal diamonds were sent to Europe in diplomatic bags of the neighbouring Government of Liberia (the President of Liberia, Charles Taylor, was later tried for crimes against humanity, for collusion in the sale of illicit diamonds in exchange for arms for Sierra Leone).

Chapter 1.5

Innovation chain and economic growth

It is Innovation that makes it possible to produce more products and services from less material and at less cost, and thus holds the key for economic development (excerpts of this chapter have been drawn from the author's work, "*Green Energy: Technology, Economics and Policy*", 2010, chap. 17).

Technology development goes hand in hand with the innovation process. The framework conditions necessary for the successful prosecution of the innovation are macroeconomic stability, education and skills development, favourable business climate, protection of Intellectual Property (IP) rights, etc. Innovation process is not necessarily linear, and may not proceed smoothly – there can be many impediments enroute. RD&D is only part of the Innovation scheme – it needs to be adapted depending upon the feedback from the markets and technology users.

Innovation Chain has five phases: Basic Research → Research and Development → Demonstration → Deployment → Commercialisation (diffusion). The schematic working of the Innovation System is depicted in Fig. 1.5.1.

Governments and private sector have roles to play in all the five phases of the Innovation Chain (Figs. 1.5.2 and 1.5.3). Generally, governments are expected to play a greater role in the early part of the chain (such as, basic science), though some large industrial houses (such as, SONY, GE) have extensive basic research programmes. Down the line, private sector alone would be involved in the last phase, i.e. commercialization. Roles overlap in the case of Applied R&D, Demonstration and Deployment. Sometimes there may be difficult technical problems that markets fail to address. Through specifying technology standards and participation in full-scale "in the field" demonstration projects, governments may induce private companies to achieve higher technological performance.

Governments could play an effective role in the progress of the innovation chain in three ways: (i) direct funding of basic research in universities and national laboratories,

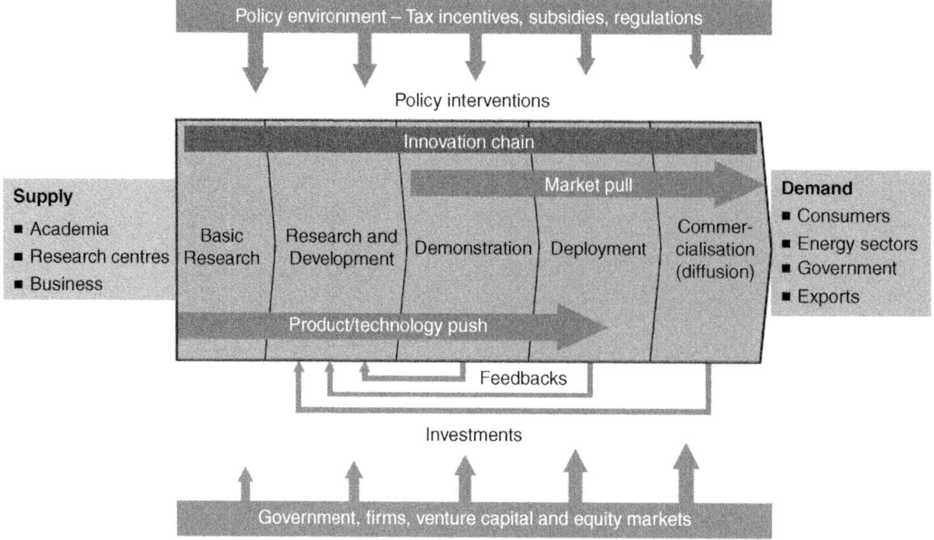

Figure 1.5.1 The schematic working of the Innovation system
Source: Energy Technology Perspectives, 2008, p. 170, © IEA – OECD

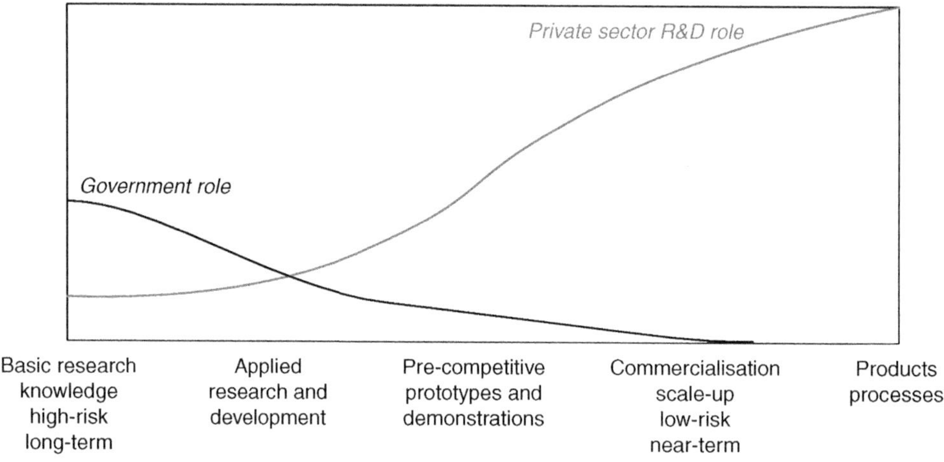

Figure 1.5.2 Illustration of the respective government and public sector RD&D roles in phases of research over time
Source: "*Deploying Renewables: Principles of Effective Policies*", 2008, p. 170, © *IEA – OECD*

(ii) granting protection under IP rights, to enable the innovators to make money from their findings, and (iii) market measures that can indirectly stimulate private sector investment.

While it is generally agreed that the present level of RD&D funding is inadequate, estimates of what constitutes the "right level" of funding, varies widely, from two

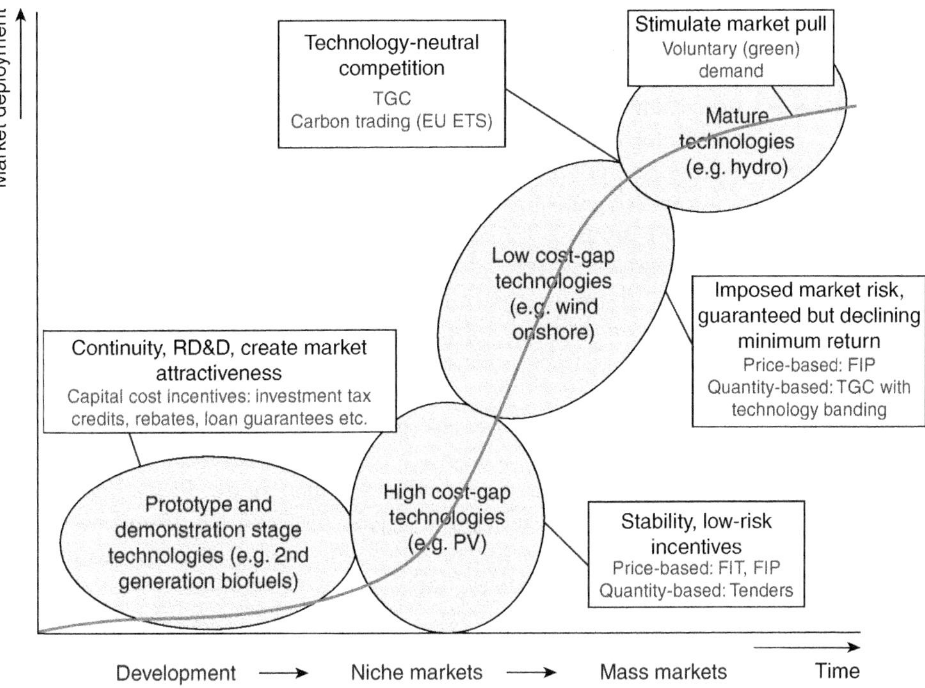

Figure 1.5.3 Combination of framework of policy incentives in function of technology maturity
Source: "*Deploying Renewables: Principles for Effective Policies*", 2008, p. 25. © IEA – OECD

to ten times the present level. Stern et al. (2006) recommended the doubling of the present level of funding. Anderson (2006) suggests that the necessary investment may be estimated as the difference between the average incremental costs of investment in new technologies and that of mature technologies.

Innovation becomes progressively more expensive, as we move along the chain, from basic research to demonstration. RD&D costs are much lower than deployment costs. Also, synergies need to be explored whereby publicly funded RD&D can stimulate privately funded RD&D.

By its very nature, innovation is a risky business. There is no guarantee that greater level of RD&D spending will automatically lead to higher success rates in the commercialization of technologies. Some RD&D projects may yield "above-cost" returns, while some may not give any returns. Under the circumstances, a practical approach will be to develop a portfolio of projects in order to hedge risks.

Basic research and applied R & D should proceed in tandem, feeding each other.

Many projects do not survive the transition from the publicly-funded demonstration to commercial viability. Murphy and Edwards (2003) call it a "Valley of Death". At this point, the investment costs and risks are very high. Neither the public sector nor the private sector considers it their duty to fund this phase. Neither "*technology-push*" nor the "*market- pull*" may be sufficiently powerful to smoothen the transition.

Many technologies need long lead times to come to fruition, and may require extensive applied research and testing, before an invention gets commercialized.

There are a number of ways in which governments could help in navigating the "Valley of Death": Economic incentives like tax credits, production subsidies, and guaranteed procurement, and knowledge incentives, such as, codification and diffusion of generated technical knowledge. Public–private research consortia can play a valuable role in technology transfer and commercialization. In some countries, technology parks are established to facilitate technology transfer. In these parks, governments give support to individuals or groups of scientists and technologists to perform basic research and applied R&D. When something viable emerges from this effort, the same group is helped to commercialize the invention. Governments can also create demand to new technologies through the promulgation of regulatory requirements. This would induce the supply side to respond to new regulations.

Sometimes, the same RD&D activities are performed simultaneously in different countries. This redundancy can be avoided by the countries pooling their RD&D budgets to perform pre-competitive research that will benefit all. Individual industry players can then draw on this common pool of R&D knowledge to build their own enterprises. Such a process strengthens technology development – for instance, the technology strengths of an industrialized country can be combined with the lower manufacturing costs in a developing country.

A private sector investment in RD&D is constrained by the following market realities (Stiglitz and Wallsten, 1999):

- A private company has no incentive in pursuing RD&D which brings about society-wide benefits, since the benefits RD&D do not uniquely accrue to it but could be made use of by several other companies. For instance, RD & D in regard to artificial photosynthesis (which involves the production of hydrogen and zero carbon dioxide) has been estimated to cost USD 20–30 billion. Despite the profound importance of this research, no private company would undertake this research. Such research is best sponsored by governments.
- Private sector would not be interested in innovations that do not bring financial benefits to the company concerned, even though they may bring great environmental benefits to the whole community (for instance, clean air).
- As the performance of companies is judged by the share holders on a short-term basis, the private sector generally goes in for RD&D that can bring benefits quickly. Thus, areas of RD&D that by their very nature necessarily takes a long time to come to fruition, tend to be ignored by the private sector. Japanese corporations are better than US corporations in taking a long-term view of RD&D expenditures.

Basic research for achieving public good (say, carbon dioxide abatement technologies) is best undertaken by public agencies. Governments may support the research by private agencies nearer to market level, from where the private sector would find it attractive to take it over wholly on their own. Alternately, governments may design a framework to value the public benefits of research funded by private agencies.

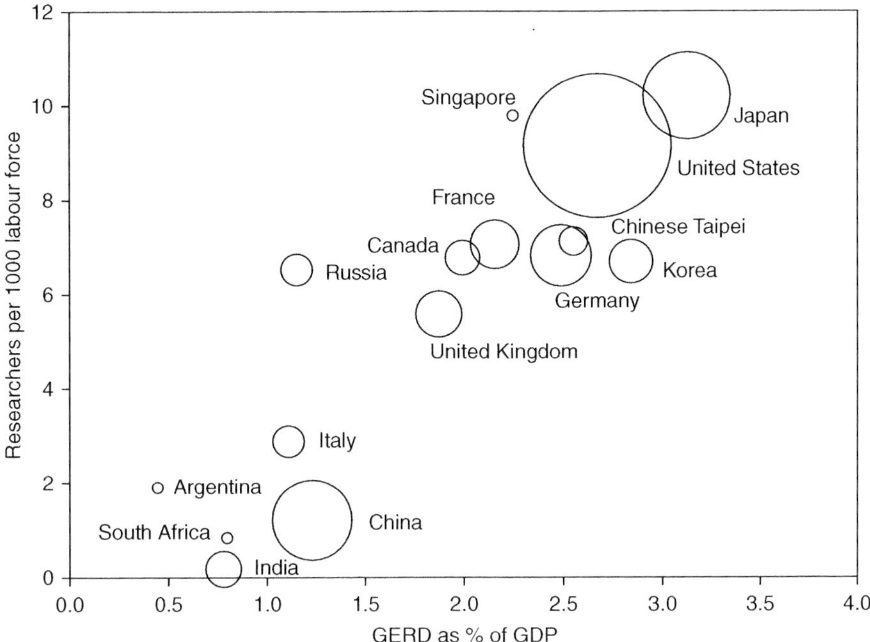

Figure 1.5.4 Relationship between the government expenditure and the number of researchers
Source: *Energy Technology Perspectives*, 2008, p. 172, © IEA-OECD

Energy RD&D budgets as a percentage of GDP is 0.08% in the case of Japan, and 0.03% in other OECD countries. There is greater emphasis in science and innovation in rapidly industrializing non-OECD countries like China and India. Good part of energy RD&D in China and India is devoted to adapting and improving technology from the OECD countries. It is, however, a matter of time before they will be able to undertake more sophisticated RD&D activities on their own.

Figure 1.5.4 plot of the relationship between the Government Expenditure in R&D (GERD) as a % of GDP versus the number of researchers per thousand people in the labour force. The circles represent the size of the expenditure in USD billion. The figure shows that the developing countries need to strengthen RD&D investments in hardware and human capital.

References

Anderson, D. (2006) *Costs and Financing of Abating carbon emissions in the Energy Sector.* London: Imperial College.

Deploying Renewables: Principles of Effective Policies (2008) Paris: IEA – OECD.

Energy Technology Perspectives (2008) Paris: IEA – OECD.

Murphy, I., and Edwards, P. (2003) Bridging the Valley of Death: Transitioning from public to private sector financing. Golden, CO: National Renewable Energy Laboratory.

Reid, W.V., Chen, D., Goldfarb, L., Hackman, H., Lee, Y.T., Mokhele, K. and Ostrom, E. (2010) Earth System Science for Global Sustainability. *Science*, v. 330, pp. 916–917, 12 Nov. 2010.

Stern, N. et al. (2006) *The Stern Review of the Economics of Climate Change*. 2007. Cambridge, U.K.: Cambridge University Press.

Stiglitz, J., and Wallsten, S. (1999) Public – Private Techology Partnership: Promises and Pitfalls. *American Behavioural Scientist*, **43** (**11**), 52–77.

Section 2

Water resources management

Chapter 2.1

Holistic water resources management, based on the hydrological cycle

U. Aswathanarayana (India)

2.1.1 INTRODUCTION – WATER AND CULTURE

"Highest good is like water ... because water excels in benefiting the myriad creatures without contending with them and settles where none would like to be. In the world, there is nothing more submissive and weak than water, Yet for attacking that which is hard and strong, nothing can surpass it ... Tasteless, it accepts all tastes, colourless, all colours, reflecting the sky, refracting the white stones of its bed, dissolving or suspending the soils and minerals over which it flows. The pulse of the bodies is liquid, as indeed all living pulses are ...

Sayings of Lao Tzu, as quoted and annotated by Vikram Seth, in his book, *From Heaven Lake*, 1983, p. 15.

Every human being on earth is a stakeholder in water resources management. And so, for that matter, is every plant and animal, domesticated and wild – only they do not have a constituency. We need to protect other living things, not for any altruistic reasons, but in our own self-interest. We should realize that our well-being is inseparable from the well-being of the ecosystem. If frogs are dying, we would surely be next in line.

All major civilizations of the world arose on the banks of rivers (Nile Valley, Euphrates – Tigris, Indus Valley).

Egypt has been rightly called the "Gift of Nile". The Nile waters are so crucial to the life of Egypt that virtually all the institutions – governments, religion, gods and goddesses, priesthood, rituals, etc. – arose from the need to measure the flood waters of the Nile. Egypt is thus a most extraordinary example of intertwining between governments and irrigation, for millennia. The earliest village of Marimda in the Nile Delta is said to be 7,000 years old. The earliest irrigation ditches were dug about 5,000 years ago (presently, there are 16,000 km of irrigation canals in the Nile Delta). Observations on the water level in the Nile started as long ago as 2000 B.C. (ancient nilometers are preserved in Aswan). On the basis of the measurement of flood waters in the Nile, priests would figure out the size of the flood water and hence the kind of harvest that could be expected. Higher flood waters meant bigger harvests, and lower flood waters meant poor harvests. The need to measure the dimensions of land under cultivation led to the development of geometry.

The physical wellbeing of man (drinking water, food, sanitation, industry, etc.) is critically dependent upon the water availability. There is also a spiritual dimension to water in some cultures. Water had and continues to have, a central role in the rituals, worship and prayers of the Hindus (*abhishekam*). Ganga means water in Sanskrit. Hindus believe that all water in India – surface and underground – is ultimately connected to Ganga. So much so, even when a man is taking bath with water from the well in his backyard, he ritually invokes the connection of that water with Ganga. Hindus generally cremate the dead and ceremonially immerse the ashes in the perennial rivers. *Life originated in water, and is returned to water.*

Sir Arthur Cotton (1803–1899) was an engineer in the employ of the East India Company. He was stationed at Rajahmundry in the banks of the mighty Godavari river. It deeply saddened him to see people migrating to Rangoon in Burma (now called Mynmar) to do menial jobs, despite the water wealth of the region. Braving the admonitions of the East India Company, he built during 1851–55 the Dhowleswaram *anicut* (barrage) and canal system on Godavari. The irrigation availability profoundly improved the economy of the region. One day when he was inspecting a canal on horseback, Sir Arthur found a brahmin offering obeisance to sungod (*arghyam)* standing in the waters of the canal. Sir Arthur was intrigued to find that the brahmin was mentioning the name of Cotton in his prayers. He went to the brahmin and asked him why he was mentioning the name of Cotton. The brahmin explained to him that he is invoking the blessings of gods on Cotton for bringing the sacred waters of the Godavari to his village, and thereby enabling him to acquire merit and spiritual happiness. Till then, Cotton was happy that what he did improved the material well-being of people. But what the brahmin told him about the spiritual happiness, deeply touched Cotton, and he had tears in his eyes.

2.1.2 WATER BALANCE

There are four kinds of waters – Rainwater, Surface Water, Groundwater and Soil Water (Soil Moisture). We have no control over precipitation – it has to be treated as a given.

Surface water and groundwater form a continuum – most rivers start as groundwater springs. Human interventions are possible to apportion the precipitation between surface water and groundwater. The position about soil water is more complex.

Table 2.1.1 gives the volumes and rates of water exchange in the various parts of the hydrosphere.

2.1.2.1 Residence times

The rate of movement of water in the various parts of the hydrosphere varies very widely. The movement of the material between the pools is called the flux. The residence time (t) of an element in a reservoir, is given by the equation:

$$t = m\left(\frac{dt}{dm}\right)$$

Table 2.1.1 Rates of water exchange in the various parts of the hydrosphere

Parts of the hydrosphere	*Volume (10^3 km^3)*	*Elements of balance (10^3 km^3)*	*Rate of water exchange*
Oceans	1,370,000	452	3000 yr
Groundwater	60,000	12	5000 yr*
Groundwater in the zone of active water exchange	(4000)	12	330 yr**
Ice sheets	24,000	33	8000 yr
Surface water on land	280	39	7 yr
Rivers	1.2	39	11 d
Soil moisture	80	80	1 yr
Atmospheric vapor	14	525	10 d
Hydrosphere as a whole	1,454,000	525	2800 yr

*The figure will be 4,200 years if we take into account groundwater runoff directly into the oceans.
**The figure will be 280 years if we take into account the groundwater runoff directly into the oceans.
Source: L'vovich 1979, p. 58

where m = mass of the element in the reservoir or the pool, and dt/dm = rate of input (or output) to the pool.

The term, rate of water exchange, used by L'vovich (1979, p. 58) refers to the time required for "hypothetical replacement of the entire volume of a given part of the hydrosphere in the process of the water cycle". He defines the rate of water exchange (A) as the ratio of the volume of a given part of the hydrosphere (γ) to the input or output elements of its balance in the process of the water cycle (w). Thus,

$$A = w/\gamma$$

where A would correspond to the number of years required for the complete renewal of the water supply. Hence, the rate of water exchange of L'vovich corresponds to the residence time. The faster the turn over, the shorter the residence time.

The data given in Table 2.1.1 leads us to several very significant conclusions:

- The rate of exchange of groundwater in the zone of active exchange (about 300 yr) is sharply different from that of the deep groundwater (5,000 yr), which is mineralized and moves very slowly.
- Water moves fastest in rivers (11 d) and in the atmosphere (10 d). It is not an accident that the two residence times are comparable, as runoff in rivers follows precipitation from the atmosphere. The short residence time of river water and precipitation means that these waters are fresh.

2.1.2.2 *Water balance of an area*

The water balance of an area is computed from the following equations:

$$P = S + U + E$$
$$S + U = R$$
$$W = P - S = U + E$$
$$K_U = U/W$$
$$K_E = E/W$$

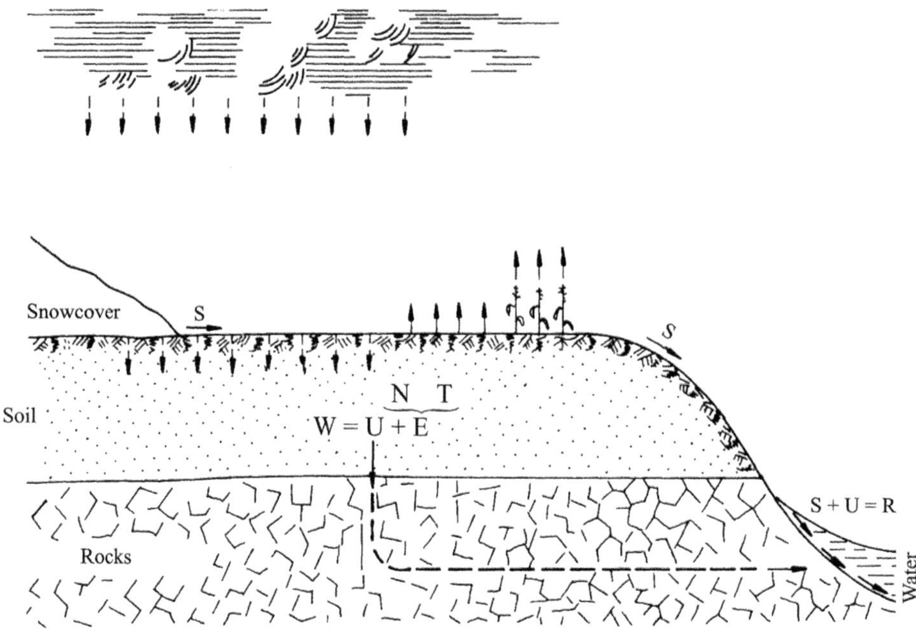

Figure 2.1.1 Diagram of water balance in a given area
Source: L'vovich, 1979, p. 65
P – Precipitation; R – Total runoff; U – Groundwater runoff; S – Surface runoff; W – total wetting of the area (annual infiltration) including surface retention; N – Unproductive evaporation (evaporation proper); T – Transpiration plants; and E – Evapotranspiration

where P = precipitation, S = surface (flood) stream flow, U = underground flow into the rivers (the stable part of the stream flow), E = evapotranspiration, R = total runoff, W = total wetting of an area, K_U and K_E = groundwater runoff and evaporation coefficients respectively, which show what parts of annual infiltration end up as groundwater runoff and evapotranspiration.

It is not always possible to round off the water balance with a single river basin, for the simple reason that the boundaries of the surface catchment of a river may not be co-terminous with the boundaries of the underground basin. For instance, artesian water received in the basin of the Don river crops out in the basin of the Dnieper river in Russia. The scale of the phenomena is, however, not large. Similar phenomena are also observed in the karst regions. A comparison of the water balance of the river basins could reveal the possible existence of a passage of water by underground routes from one river basin to another (L'vovich 1979, p. 64).

The water balance of a land area is schematically shown in Figure 2.1.1.

2.1.3 GREEN AND BLUE WATERS

When rainfall occurs, a part of it infiltrates into the soil and becomes the soil moisture (*green water resource*). This is the water that is available to plants through root uptake, and constitutes the main water source of the rainfed agriculture. The surface runoff, which may be stored in dams, lakes and aquifers, is called the *blue water resource*. It is the main source of water for irrigated agriculture. In the course of the hydrological cycle, green water which constitutes about 65% of the global precipitation, flows back as vapour to the atmosphere (evaporation, interception and transpiration), whereas blue water which amounts to about 35% of the global precipitation, flows as liquid recharging groundwater, and flowing in rivers to lakes, and ultimately to the ocean. The total green water flow from croplands globally is around 6,800 km^3/yr which corresponds to around 6% of global precipitation. It has been estimated that 5,000 km^3/yr of green water flow originates from rainfed agriculture, and the rest 1,800 km^3/yr from irrigated agriculture (Rockström et al., 1999).

The distinction between irrigated and rainfed agriculture does not actually have any conceptual basis in hydrology. Green water in rainfed agriculture originates from naturally infiltrated rainfall, whereas irrigated agriculture involves supplementation with blue water.

Increased water use in one sector (say, agriculture) will affect water availability in other sectors, such as direct human use (water supply) and ecosystem use (terrestrial and aquatic ecosystems). Experience has shown that rainwater harvesting and building of small-scale water storage systems have hardly any adverse impacts on downstream flows. Investing in water management in rainfed agriculture has several beneficial effects, such as reduction in land degradation, improvement in water availability (to enable more food to be produced) and water quality downstream.

Table 2.1.2 summarises the various ways of improving crop yields and water productivity. The plant water availability in the root zone and hence the crop productivity, can be enhanced by mulch practices, drip irrigation techniques and crop management to enhance canopy cover. Table 2.1.3 gives the major water management strategies, involving and blue and green water resources.

If steps are taken to minimize non-productive green water losses, more green water will be available for crop production. This would lead to higher yields for the same amount of green water use. Expansion of irrigation in effect involves augmenting of green water.

Green water can also be augmented by *in-situ* rainwater harvesting, before the rainfall generates runoff. This would reduce the blue water source.

Saline water, municipal and industrial wastewater can also be used in irrigation if combined with proper management (Karlberg, 2005). If not properly controlled, irrigation with contaminated water may introduce pathogens and toxic chemicals in vegetables and crops.

Some kinds of water management involve only demand management. For instance, mulching and growing of larger canopy plants would result in lower soil evaporation, thereby enhancing the productivity of green water.

When *ex-situ* water harvesting expands irrigation, the amount of water available to sustain the downstream ecosystems and downstream industrial, domestic use, will be reduced. Trade-offs with other land uses, such forestry and pasture will have impacts on biodiversity.

Yields are very low in rainfed agriculture. If investments in *in-situ* or *ex-situ* water harvesting in rainfed agriculture could increase yields from 1 t/ha to 2 t/ha, it would amount to increase in crop productivity in terms of water consumption from approximately 3,500 m^3 of water per tonne of grain to less than 2,000 m^3/tonne. Water

Table 2.1.2 Rainwater management strategies and corresponding management options to improve yields and water productivity

Rainwater management strategy		*Purpose*	*Management options*
Increase plant water availability	External water harvesting systems	Dry spell mitigation, Protective irrigation, Spring protection, Groundwater recharge, Enable off-season irrigation, Multiple water use	Surface micro dams, Sub-surface tanks, Farm ponds, Percolation dams/tanks, Diversion and recharging structures
	In-situ water harvesting systems, soil and water conservation	Concentrate rainfall through runoff to cropped area and/or other use	Bunds, Ridges, Broad-beds and furrows, Micro-basins, Runoff strips
		Maximise rainfall infiltration	Terracing, Contour cultivation, Conservation agriculture, Dead furrows, Staggered trenches
	Evaporation management	Reduce non-productive evaporation	Dry planting, Mulching, Conservation agriculture, Intercropping, Windbreaks, Agroforestry, Early plant vigour, Vegetative bunds
Increase plant water uptake capacity	Integrated soil, crop and water management transpiration	Increase proportion of water balance flowing as productive	Conservation agriculture, Dry planting (early), Improved crop varieties, Optimum crop geometry, Soil fertility management, Optimum crop rotation, Intercropping, Pest control, Organic matter management

Source: Chritchley and Siegert, 1991

collected *ex-situ* can also be used to raise a cash crop in winter. As the atmospheric demand would be low at that time, this would result in higher water productivity.

If flood water can be captured upstream and used in agriculture, there will be less runoff and fewer problems with erosion. When agriculture is developed in degraded lands with low infiltration capacity, more groundwater is formed and there will be reduced floods and erosion during heavy storms.

2.1.4 CONJUNCTIVE USE OF WATER RESOURCES

The purpose of the conjunctive use of water resources is simple: surface water surplus during the wet periods is temporarily stored in groundwater reservoirs, to be withdrawn for use during the dry periods. Conjunctive use is particularly necessary in the monsoon regions. During the rainy period lasting for a few months, the rivers are full and there is more water than could be used. During the dry period when water is desperately needed, the rivers are dry or have very little flow. Conjunctive use of

Table 2.1.3 Implications of water management strategies on blue and green water resources

Water management strategy	*Implications on blue and green water resources*
Improving water productivity (demand management) e.g. evaporation management, integrated soil, crop and water management, deficit irrigation	Reduce green water losses
Expanding irrigation (supply augmentation) e.g. *ex-situ* rainwater harvesting and supplemental irrigation	Adding blue water to the field, blue to green redirection
Improving local use of rainfall (supply augmentation) e.g. *in-situ* rainwater harvesting such as conservation agriculture	Reduce blue water losses, increase green water resources
Agricultural area expansion (supply augmentation)	Convert green water use in natural ecosystems, to green water use in agriculture. Possible effects on blue water generation
Use of non-conventional water sources (supply augmentation) e.g. desalinisation of seawater, use of marginal quality water, reuse of drainage water from cities and industries	Adding more water to the hydrological cycle, generating more blue and green water

water is a sensible strategy to address the mismatch in time and space between water demand and water availability (for details, see IAHS publication no. 156, "Conjunctive Water Use").

Buras & Hall (1961) used dynamic programming and economic considerations for optimizing the conjunctive use of water. The latter involves the determination of the necessary volume of the groundwater reservoirs to be used conjunctively with the surface reservoirs, in the long term.

2.1.4.1 Design and management of groundwater reservoirs

A groundwater reservoir needs to have two facilities: areas where water can be recharged naturally or artificially, and pumping installations. The availability of surface water in a given project is first determined on the basis of runoff records for the last 50 years. That data is then used to estimate the volume of the groundwater reservoir needed for conjunctive use. Geological considerations determine the maximum storage capacity that a groundwater reservoir could have. The extent to which the stored water could be withdrawn depends upon the availability of finances (e.g. pumping and recharge costs), and environmental considerations (e.g. saline intrusion, subsidence of land). Models should therefore simulate parameters such as storage coefficient (S), transmissivity (T) and the contour conditions of the aquifers.

The water balance of the groundwater reservoir is given by the following formula (Correa, 1987):

$$V_{(t+1)} = V_{(t)} + I_{(t)} - O_{(t)} + R_{(t)} - P_{(t)} \tag{2.1.1}$$

where $V_{(t)}$ = groundwater volume at the end of period, t, $I_{(t)}$ = underground inflow to groundwater reservoir in period of time, t, $O_{(t)}$ = Underground outflow from groundwater reservoir in period of time, t, $R_{(t)}$ = Groundwater recharge (natural and artificial) in period of time, t, $P_{(t)}$ = pumping volume from groundwater reservoir in period of time, t.

The variables, $I_{(t)}$ and $O_{(t)}$, are dependent upon the piezometric surfaces, but their yearly mean values can be made use of in the calculations. The volume of water that needs to be pumped will be the deficit between the water demand and the quantity of surface water available. Evidently, the maximum pumping capacity of the system limits the quantity of groundwater that could be extracted.

The groundwater recharge involves the following terms:

- Groundwater recharge from precipitation: the rate of infiltration is controlled by the nature and intensity of precipitation on one hand, and the permeability of the rocks and soils in the zone of aeration, on the other. Groundwater recharge from precipitation may be insignificant in arid areas,
- Artificial recharge: the maximum possible volume of artificial recharge would depend upon the geological conditions (recharge area), physical characteristics (permeability) and economic considerations (costs),
- Channel seepage, deep percolation of unconsumed irrigation water on unconfined aquifers: this may be taken as constant for every year, if the water demand is constant,
- Percolation from streams and ponds: this depends upon the surface water surplus that flows over the recharge area (unconfined aquifer).

The long-term equilibrium of the groundwater reservoirs or safe yield is given by the following equation. It corresponds to the condition that the total inputs and outputs to the groundwater reservoir are equal (Correa, 1987).

$$\sum_{t=1}^{N} (<F_{(t)}> + Rj_{(t)} + I_{(t)} - P_{(t)} - O_{(t)}) = 0 \tag{2.1.2}$$

where N = Years of observed runoff sums, $<F_{(t)}>$ = function of artificial recharge + percolation from streams and ponds, $Rj_{(t)}$ = percolation from irrigation losses, $I_{(i)}$ = underground inflow to groundwater reservoir in period of time, t, $P_{(t)}$ = pumping volume from groundwater reservoir in period of time, t, $O_{(t)}$ = Underground outflow from groundwater reservoir in period of time, t.

The above equation has a solution if the mean value of water demand in N years (Wd_{m}) falls within the range of WD_{min} and WD_{max}, corresponding to the coefficient of recharge (k) being within the range of k_{min} and k_{max}. The system would evidently not be sustainable if $Wd_{\mathrm{m}} > WD_{\mathrm{max}}$. This point can be illustrated with a simple example. If the monthly earnings of a person vary from (say) a minimum of USD 3,000 per month to a maximum USD 6,000/m, with an annual average of USD 48,000, the "safe" amount that such a person could spend (without getting into problems) is USD 4,000 per month. If such a person starts spending USD 5,000 per month, he will land himself in troubles.

The coefficient of recharge (k) can also be used to determine the recharge policy. For water demand with $k_{max} > k > 1$, it is necessary to have a recharge policy to ensure the long-term sustainability of the reservoir system. On the other hand, for water demand with $k_{min} < k < 1$, no artificial recharge policy would be needed as the system can be sustained through controlling the natural percolation from the river.

The objective function to be applied for the optimization of the system is based on the difference between the groundwater actually stored and the target groundwater volume at the beginning and at the end of the hydrological year.

2.1.4.2 Case history of conjunctive use through over-irrigation of paddy fields

Tsao (1986) gave a case history from Taiwan of an innovative conjunctive use through over-irrigation of paddy fields.

Wetlands serve as "kidneys" (purifiers) of the contaminated water, besides playing a useful role in the hydrological cycle. The Cho-Shui river basin in Taiwan with an area of 5,335 km^2, has fertile soils, average annual temperature of 22°, and average annual precipitation of 2,017 mm. It is thus ideally suited for paddy cultivation round the year. But the catch is, that 80% of the precipitation occurs during May to September. It is dry during October to April. Since the river is heavily loaded with sediment, and since a single-purpose hydropower reservoir already exists on the river, it has not been found to be feasible to construct surface reservoirs. Under the circumstances, recourse has been taken to the withdrawal of groundwater for irrigation during the dry period. Consequently, more than 100,000 wells have been dug. To complicate the matters further, large quantities of groundwater have been withdrawn to raise eels, shrimp, fish and clams. As is to be expected, the groundwater levels and yields declined sharply, land subsided, and incursions of seawater occurred (in some places, the zero groundwater level is 25 m below the mean sea level). The tidal wave due to typhoon Wyne in 1980 overflowed the sea dykes erected near the coast, and caused great damage to the paddy fields.

Two types of conjunctive use of water through artificial recharge were attempted. Injection recharge was technically feasible, but as the water is heavily loaded with sediment, the experiment proved to be too expensive, and was therefore abandoned.

Artificial recharge through the over-irrigation of the paddy fields proved to be techno-economically feasible. Since the Cho-Shui river basin is a granary of Taiwan, any modification of the irrigation system has profound socio-economic implications. For this reasons, detailed simulation studies were made before decisions were taken, and elaborate monitoring systems have been set up to keep track of the environmental and socio-economic implications of over-irrigation.

An orientation study was initially made with a 195 ha plot, where the shallow top soil was underlain by a 30 to 50 m deposit of alluvial coarse sand and gravel which serves as a natural recharge area. Since the average intake rate of paddy fields is 150 mm d^{-1}, the groundwater responded well to over-irrigation. A salt balance study was also made in order to determine the optimal rate of fertilizer application, so as to minimize the loss of fertilizers due to over-irrigation. Weirs were constructed to provide ponds for the supply of irrigation water. As the drainage water from the paddy fields is invariably contaminated, it could not be used for aquaculture which is extremely sensitive to contamination.

Experience has shown that an over-irrigation of 50 mm d^{-1} is entirely feasible. About 0.77 Bt (billion tonnes) of groundwater per year can be safely extracted from the system, while satisfying all the constraints. Mathematical models predicted that with an artificial recharge of 86.8 Mt (million tonnes) per year by over-irrigation, it is possible to prevent the intrusion of saline water inland.

2.1.5 WATER RESOURCES ENDOWMENTS OF COUNTRIES

The total global runoff is more than adequate to meet the demands of water for the humankind for many decades to come. But the catch in it is that the distribution of freshwater resources in the world is extremely uneven. Anthropogenic activities tend to degrade the freshwater resources, thus complicating the problem even further. Under the circumstances, the transfer of waters between one country to another and between different parts of a country are unavoidable, and every effort should be made to minimize the problem.

The total runoff from the earth's rivers is estimated to be 42,700 $km^3 yr^{-1}$. The Amazon alone accounts for about one-sixth of the total runoff. The total sustainable yield of freshwater in the world has been estimated to be 12,500 km^3. According to Shiklomanov (1998), the total global withdrawal by 2025 is projected to be 5,100 $km^3 yr^{-1}$, with a consumption of 2,860 $km^3 yr^{-1}$.

There are several major rivers (i.e. those whose long-term runoff is more than 100 $km^3 yr^{-1}$) in the world (source: Shiklomanov 1998, p. 14). Though some rivers like the Danube and the Rhine in Europe, Nile in Egypt, Colorado in USA and Kaveri (Cauvery) in India, are intensively used, most of the rivers in the world are used only to a limited extent. Despite enormous techno-socio-economic problems, massive transfer of waters (e.g. Three Gorges dam on River Yangtze in China which transfers about 70 $km^3 yr^{-1}$ of water from south to north) has to be undertaken in some situations. For example, transfer of water from the Congo River could provide water supplies to several parts of Africa. Similarly, transfer of waters from the Ganga-Brahmaputra system within India can virtually alleviate water supply problems in good part of the country. It should be emphasized that high dams are not the only means of bringing about water transfers – a number of other techniques are available (e.g. groundwater recharge).

The water resources endowments of the countries are given in Table 2.1.4.

IGBP jointly with its human resources analogue, IHDP, came up with an Integrated Land and Water Management paradigm for Africa (IGBP Newsletter, Dec. 1999), which is valid for most of the developing countries. The paradigm has the following components:

1. *Horizontal integration*: integration among adjacent land users and land uses within catchments; between upstream and downstream users; among domestic, industrial, urban and other users, and among governments sharing river systems,

Table 2.1.4 Water resources endowment, and annual water withdrawal of countries

Country	*1*	*2*	*3*	*4*
Mozambique	0.8	1	13.0	13 + 40 = 53
Tanzania	0.5	1	3.0	8 + 28 = 36
Nepal	2.7	2	7.8	6 + 149 = 155
Bangladesh	22.5	1	19.6	6 + 205 = 211
Madagascar	16.3	41	22.8	17 + 1658 = 1675
Nigeria	3.6	1	2.5	14 + 30 = 44
India	380.0	18	2.2	18 + 594 = 612
China	460.0	16	2.3	28 + 434 = 462
Kenya	1.1	7	1.1	13 + 35 = 48
Pakistan	153.4	33	3.3	21 + 2032 = 2053
Indonesia	82.0	3	12.8	9 + 443 = 452
Egypt	56.4	97	0.9	84 + 1118 = 1202
Zimbabwe	1.2	5	1.8	18 + 111 = 129
Thailand	31.9	18	3.0	24 + 575 = 599
Tunisia	2.3	53	0.4	42 + 283 = 325
Malaysia	9.4	2	22.6	176 + 589 = 765
Iran	45.4	39	1.7	54 + 1308 = 1362
Mexico	54.2	15	3.8	54 + 847 = 901
South Africa	9.2	18	1.2	65 + 339 = 404
Brazil	35.0	1	43.0	91 + 121 = 212
Ireland	0.8	2	14.1	43 + 224 = 267
Israel	1.9	88	0.4	72 + 375 = 447
Spain	45.3	41	2.8	141 + 1033 = 1174
Belgium	9.0	72	1.2	101 + 816 = 917
UK	28.4	24	1.2	101 + 406 = 507
Australia	17.8	5	19.0	849 + 457 = 1306
France	40.0	22	3.4	116 + 612 = 728
Canada	42.2	1	98.5	193 + 1559 = 1752
USA	467.0	19	9.4	259 + 1903 = 2162
Germany*	41.2	26	2.1	67 + 601 = 668
Sweden	4.0	2	20.5	172 + 307 = 479
Japan	107.8	20	4.4	157 + 766 = 923
Switzerland	3.2	6	6.9	115 + 387 = 502

*Before reunification.

1. Total internal renewable water resources (km^3), 2. Percentage of annual withdrawal out of total water resources (%), 3. Internal renewable water resources ($\times$ 1,000 m^3 per capita per year), 4. Annual per capital withdrawal of water in m^3 (domestic + industrial and agricultural = total). The countries are arranged in order of increasing income. (source of 1, 3 and 4. World Development Bank; source of 3. Human Development Report, 1997, of UNDP).

2. *Vertical integration*: integration among the range of organizations and institutions functioning at different scales and strives to achieve (a) maintenance of adequate amounts and quality of water to all water users, (b) prevention of soil degradation, (c) food security, and (d) prevention and resolution of conflicts between water users'.

Ways of integrating biophysical and socioeconomic approaches of land and water management, are shown in Fig. 2.1.2.

Numerous hydrological, biophysical and socioeconomic issues of water resources management of an area (agroclimatic zone/watershed/river basin) in a country need

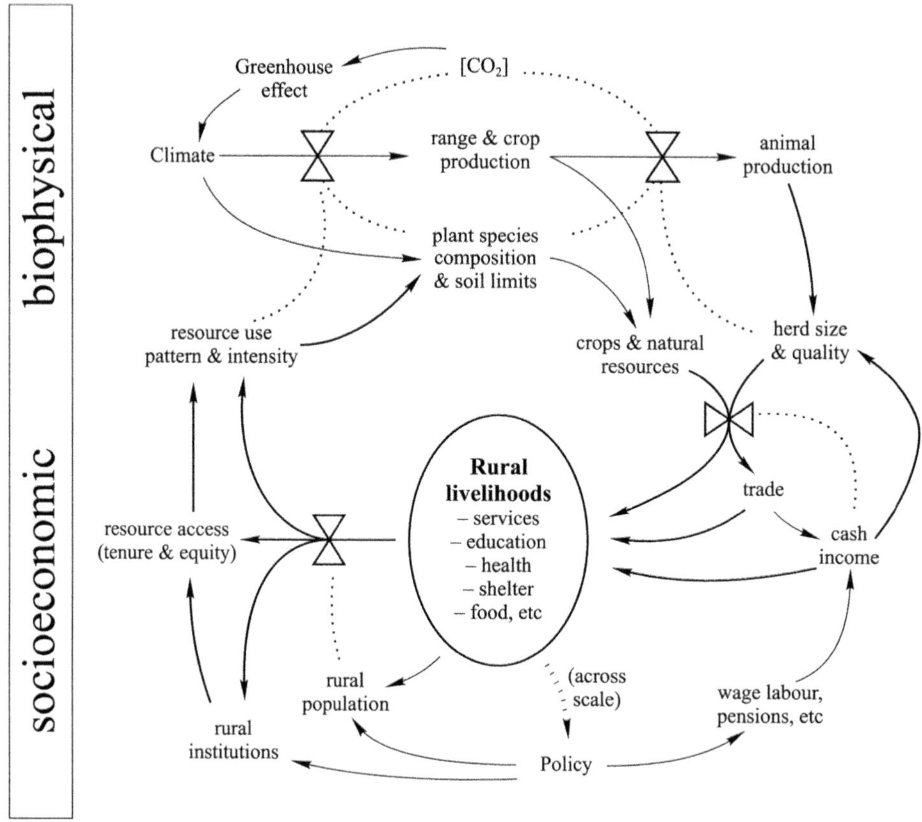

Figure 2.1.2 Integrating biophysical and socioeconomic approaches in land and water management
Source: IGBP Newsletter, Dec. 99

to be integrated in order to optimize the use of water resources. Vörösmarty et al. (1999) designed key biogeophysical datasets that can be used to build typology models (Table 2.1.5).

2.1.6 DECISION – SUPPORT SYSTEM FOR WATER RESOURCES MANAGEMENT

The water resource issues have so many dimensions and complexities that decision-support is necessary for water managers. Kaden et al. (1989) describe a simple computer-aided water management system, to assist operators, managers and planners (Fig. 2.1.3). It has three components:

1. Measuring and information systems (data acquisition, transmission and storage),
2. Software system (user software), and
3. Organizational system (organizational structure, legal and economic regulations).

Table 2.1.5 Key datasets for use in typology models

Geophysical parameter	*Space/time scale*	*Source/archive*
Surface hydrology attributes		
– River networks and Basin	30 minute	Vörösmarty et al. (1999)
– Boundaries (STN30 v. 5.12)	30 minute	Fekete & Vörösmarty
– Discharge	Station data	RivDis, R-HydroNET, R-ArcticNET Vörösmarty et al.
– Runoff	30 minute	Fekete & Vorosmarty
– Digital topography	1 km	Hydro 1K EDC
Surface attributes		
– Land cover	1 km	EDC (IGBP 1996)
– Soils	Vector	FAO (1996)
Geology/lithology	Vector	UNESCO (1999)
N deposition (NOy and NHx)	10°	Dentener et al.
N equivalent fertilizer use/input	30 minute	CSRC-UNH
Livestock	1°	Lerner et al.
Population	30 minute	CSRC-UNH
Climatological Data		
– Air temperature	30 minute	Wilmott & Matsuura
– Precipitation	30 minute	Wilmott & Matsuura
Evapotranspiration	30 minute	Fekete & Vorosmarty
Lake density and volume	Vector	ESRI
Reservoir induced aging	Basin scale	Vörösmarty et al.
Constituent data (various forms of N, P, C, Si, susp. solids and discharge)	Station Data Mean Annual (530) Mean monthly (52)	GEMS/GLORI GEMS/WATER

Source: Vörösmarty et al. (1999)

Haagsma (1995) describes a more sophisticated, Internet-linked Decision Support System (Fig. 2.1.4). We should realize that decisions in regard to water resources are not necessarily taken on the basis of objective information alone, and that political considerations do come into the picture. Large amounts of data have to be transformed into useful information. The aim of the decision support system is to make available the relevant information to both the experts and water authorities, to enable them to have informed consultation.

We should distinguish between coordination and integration. In the case of coordination, there is hierarchy whereby one model delivers the data to another in a specified manner, but there may be no feed-back (for instance, a water technician may make hundreds of piezometric measurements and then hand over the information to his boss who may be a local water engineer, who in his turn may send a summary of it to his superior, and so on; the important point is that there is no feed-back to the technician).

On the other hand, integration has no such built-in hierarchy. Models, information (including non-machine-readable information) and databases can be integrated into a decision-support system. Integration is thus analogous to different ministries setting up a joint task force to solve a problem. Different ministries bring in their own view-point to bear on the issue. There is no hierarchy here. Haagsma (1995) describes

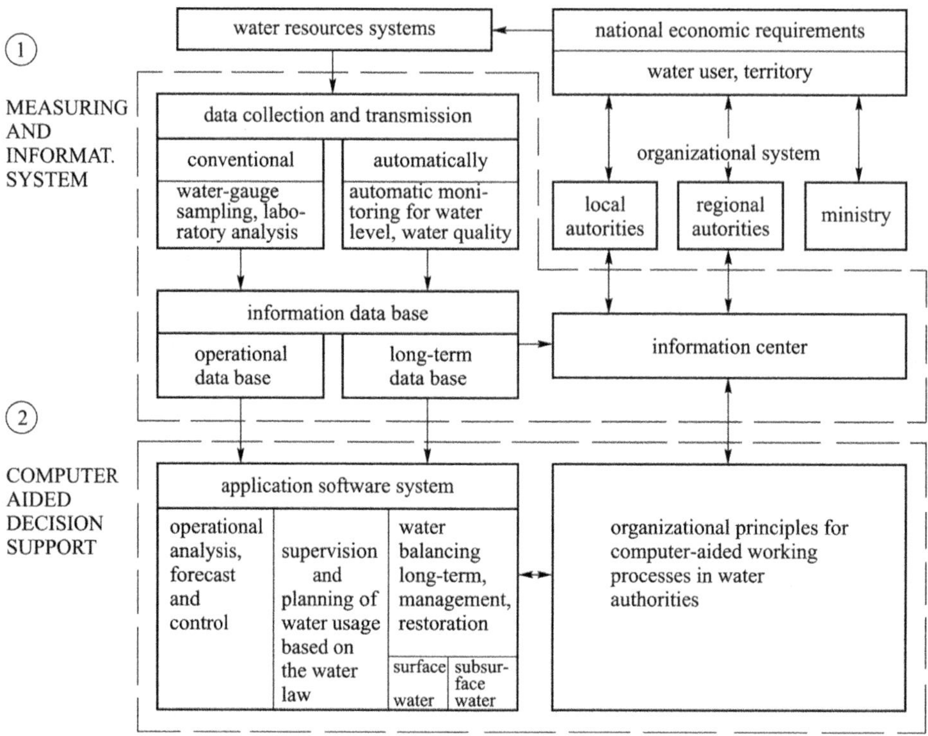

Figure 2.1.3 Computer-aided water management system
Source: Kaden et al., 1989

a recursively linked approach whereby different models can be used for different parts of the system (e.g. study of groundwater – surface water interaction, according to the time scales of the underlying hydrological processes) and run simultaneously, communicating through a network. Communication is facilitated by a communications server.

2.1.7 PARADIGM OF GLOBAL WATER RESOURCES MANAGEMENT

There is little doubt that in the twenty-first century, water issues are going to emerge as the most serious problem facing humankind.

The visionary paradigm for water resources management that L'vovich proposed way back in 1979 continues to be valid to this day. It presupposes:

- Cessation of the discharge of the sewage into rivers and lakes,
- Reuse of waste water,
- Closed circuit recycling of industrial water,

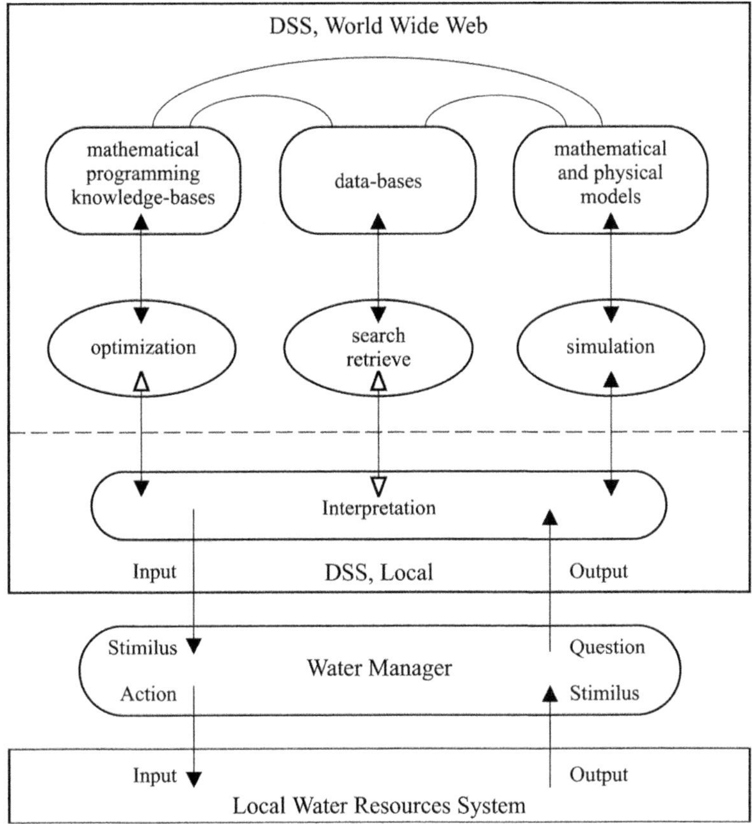

Figure 2.1.4 Decision-support system for water resources management
Source: Haagsma, 1995

- Underground reservoirs of the same order of storage capacity as surface reservoirs,
- Protection of forest cover to reduce runoff, and increase percolation and soil moisture,
- Regulation of evaporation,
- Water transfers (Fig. 2.1.5).

Shiklomanov (1998) made the following recommendations to address the situation:

1. Protection of water quality. Groundwater is sought to be protected through an understanding of its vulnerability.
2. Drastic decrease in specific water consumption, particularly in irrigation and industry: Shiklomanov 1998, p. 25) gives a forecast of the increase of population and increase in irrigated land (in M ha) upto the year 2025. In 1995, the agriculture sector accounted for 67% of the total water withdrawal and 86%

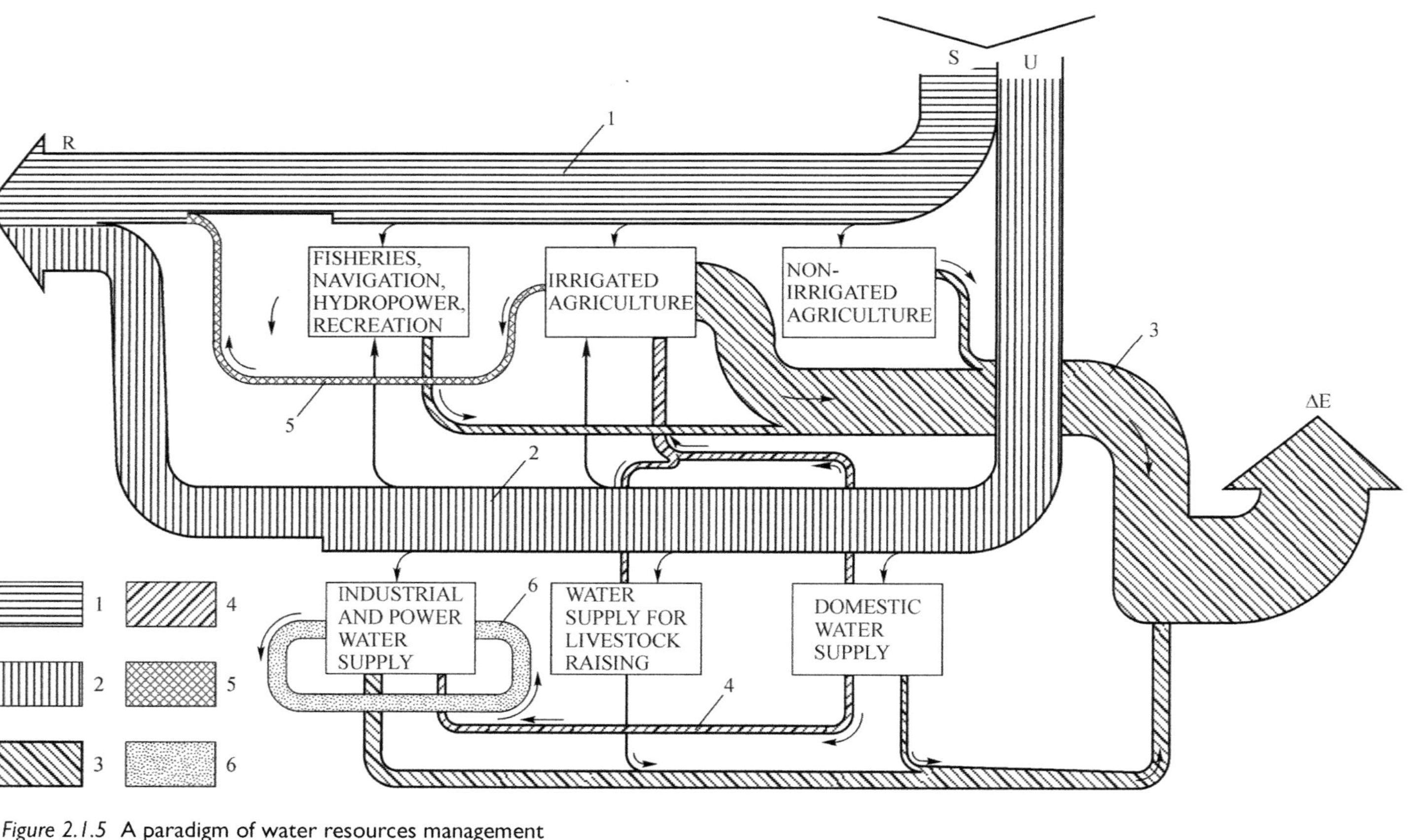

Figure 2.1.5 A paradigm of water resources management
1. Unstable surface runoff, 2. Stable runoff resources, 3. Unrecoverable water consumption, 4. Sewage and polluted river water, 5. Water returning following irrigation, 6. Closed water recycling system
Source: L'vovich 1979, p. 373

of its consumption. This is not sustainable. In future, the efficiency of irrigation has to be improved greatly (through measures such as drip irrigation and cutting down of conveyance losses), so that with lesser total withdrawal (about 60%), more food could be grown.

3. Ways and means of reducing the water consumption in industries by extensive recycling.
4. Complete cessation of the practice of discharging waste water into the hydrological systems (such as, rivers and lakes): As pointed out by Shiklomanov (1998), every cubic metre of contaminated water discharged into rivers and lakes spoils up to 8–10 m^3 of good water. This is a monstrous situation, and every effort should be made to stop this practice.
5. Harvesting of precipitation and making a more efficient use of runoff: rainwater harvesting from rooftops, harvesting of surface runoff, groundwater recharge, etc.
6. Greater use of salt and brackish waters.
7. Influencing the precipitation-forming processes.
8. Use of water stored in lakes, underground aquifers and glaciers: Conjunctive use of water resources, and the design and management of groundwater reservoirs.

2.1.8 HOW BEST TO USE WATER RESOURCES – INDIA AS A CASE

The problems of water resource use and possible solutions are described with India as a case. The ideas proposed here are applicable to most countries in the world.

The Chinese character for "political order" is based on the symbol for "water", symbolizing that those who control water, control people (In the menu cards in the Chinese restaurants, the character for a pork dish not only indicates the ingredients of the dish, but also whether the pig has been raised in the range or in the yard!).

2.1.8.1 India's water budget

India's land area is 3.28 million sq.km, and average annual rainfall is 1,170 mm. So the total rainfall input is 3,838 cu.km, rounded as 4,000 cu.km.

Table 2.1.6 gives the principal annual components of India's water budget.

Gupta and Deshpande (2004) gave the figure of 40% on the basis of the above calculation, and the Planning Commission uses the same figure. Narasimhan (2008) favours the higher figure of 69.5% for India's evapotranspiration given by Jain et al. (2007), as it is of the same magnitude as the evapotranspiration (ET + infiltration) figure of 60.5 – 66.4% (three estimates) given for world's land by Shiklomanov (1997), evapotranspiration estimate (59.4 – 82.1% – four estimates) for the Amazon basin (Marengo, 2006), and 69.8% for California (Dept. of Water Resources (2005), etc. Evapotranspiration, just like precipitation, varies from year to year. For instance, in California, it was 63.4% in 1998 (wet year), 69.8% in 2000 (normal year), and 90% in 2001 (dry year). It is hence even possible that India's Evapotranspiration may actually be even higher than the estimate (69.5%) of Jain et al. (2007).

Table 2.1.6 Principal annual components of India's water budget

Component	Volume (km^3)	Precipitation (%)
Precipitation	3838	100
Potential flow in rivers	1800	48.7
Natural recharge	432	11.3
Available water	1869 + 432 = 2301	60.0
Evapotranspiration	3838 – (1869 + 432) = 1537	100 – (48.7 + 11.3) = 40.0

Source: Narasimhan, 2008

The estimate of available water in India varies sharply depending upon the figure of evapotranspiration used – it will be 2,301 km^3 if the 40% figure of Gupta and Deshpande (2004) is used, and 1,460 km^3 (1,260 km^3 of surface flow and 200 km^3 of groundwater recharge) if the evapotranspiration figure of Jain et al. (2007) is used. Assuming 48.8% extractability, then the "estimated utilizable water resources" will be 1122 km^3, as per Gupta and Deshpande model (2004) and 712 km^3 as per Jain et al. model (2007). Since the current use is 634 km^3, it should be realized that shortage of water is an imminent challenge, rather than a long-range problem.

There are reasons to believe that the Ministry of Water Resources is using the overestimated figures. It is urgently necessary for surface water hydrologists, meteorologists and climate scientists to undertake appropriate studies to derive credible estimates.

2.1.8.2 Paradigm change needed

The series of articles in *The Hindu* of 2010 by R.R. Ayer, T.N. Narasimhan and V.K. Gaur and the article by Rohini Nilekani (*Times of India*, May 28, 2011) recommend that the past approaches need to be reversed in the following ways in order to address the water resources problems of India:

(i) Shifting from large, capital-intensive, centralized projects involving dams and canals, to small, decentralized, local, community-led, water harvesting and watershed development programmers. The two systems need not be considered as mutually exclusive. Since India already has a vast network of water reservoirs and canals, there has to be a transition from this to the proposed decentralized systems.

(ii) Avoidance of massive extraction of water from the rivers, and "killing" of the rivers.

(iii) Considering the State as trustee holding natural resources in public trust for the community.

Water has traditionally been assumed to be freely available. This led people to make use of water from the most convenient source, say, irrigation with groundwater. During the last 30 years, the number of tube wells in India went up from 2 million to 23 million, with the consequence that the water table has been going down from 1 to 3 m per year. This led to extensive depletion of groundwater and soil salinization. GRACE (Gravity Recovery And Climate Experiment) satellite technology has made it possible to monitor the depletion of groundwater. Tiwari, Wahr, and Swenson (2009) found that northern India and its surroundings which is the home of 600 million people and the most heavily irrigated region in the world, has been losing groundwater at the rate of $54 \pm 9\,km^3/yr$. This is probably the largest groundwater loss in any comparable region in the world. Evidently this is not a sustainable situation. Drastic steps need to be taken to reverse this trend.

Excessive use of fertilizers and pesticides to improve crop yields have severely contaminated the groundwater. For instance, the nitrate levels in groundwater in Haryana (114–1,800 mg/l) are much higher than the permissible levels of 50 mg/l. The toxicity of groundwater in Ferozepur district of Punjab has reached such high levels that children are falling victim to crippling disorders (e.g. loss of eyesight, psoriasis, skeletal disorders, organ failures, etc.). In one Government school, 110 out of 180 students are found to be physically challenged.

The availability of fresh water is finite. It is therefore imperative to manage the water use in the context of the quantum of water availability rather than convenience in the use of water. There should be a change from a supply-demand mindset to holistic, science-based management. This could mean painful decisions, but that cannot be avoided.

With the water availability constraint as a given, we need to fit in the plethora of existing water use and distribution structures to optimize its use among the stakeholders. This is a stupendous task involving all Indians, ranging from lay citizens, farmers, corporations, state functionaries and learned academies.

Water resources management structures

The emphasis in the proposed setup is on local management. A good example of this is the Silicon Valley Watershed management in California, USA, described by Narasimhan in *The Hindu*. The citizens of the Valley are the owners of the water in the watershed, which is a collection of 23 smaller watersheds, and covering an area of 3,400 sq.km They operate the system involving integration of surface water, groundwater, artificial recharge, imported water, water reuse, water treatment and public education. The watershed management is assisted by qualified scientists, engineers and economists. There is thus good synergy between science and public policy.

Our water resources are finite (around 4000 BCM). With increasing population, our per capita availability of water has been declining steadily since 1947. Within this decade, our per capita availability of water may fall below the accepted 1,700 cubic meter per capita per year. Hence the urgent need to achieve a low water footprint.

Technical measures for the Minimization of water use

Rice which is a staple food in India and China, is highly water-intensive crop, requiring 3,000–5,000 L of water to grow one kg of rice. Sashidhar and his associates (2008) developed a new kind of aerobic rice which does not require transplantation and can be cultivated like maize. Most importantly, it requires 45% less water than normal rice. China developed new strains of drought-tolerant rice. "On the same amount of land that Chinese farmers grow 4,000 kg of rice each year, Indians grow no more than 1,600 kg, and they use ten times more water than is necessary" (*Michael Specter*, New Yorker, Oct. 23, 2006).

Advances in remote sensing and information technologies have made it possible on one hand to monitor the aquifer depletion, and on the other hand, to enforce compliance with water rights of individual farmers. There is a clear correlation between evapotranspiration (ET) and groundwater pumpage in an area. Landsat data is processed through software called METRIC to develop seasonal evapotranspirational levels for an area, and thereby monitor the aquifer depletion. The water rights entitlement for an individual farmer is determined depending upon the area of the farm, rainfall, aquifer characteristics, etc. The compliance with water rights entitlement is monitored remotely for each farmer through the GPS location of a well, capacity of the pump, measurement of power consumption and well flow, and the extent of groundwater pumpage. If it has been found that a particular farmer is withdrawing groundwater beyond his entitlement, he is warned and heavily penalized. If he persists in drawing water which really belong to his neighbour, or the community, the electricity connection to the water pump could be cut off.

Numerous hydrological, biophysical and socioeconomic issues of water resources management of an area (agroclimatic zone/watershed/river basin) need to be integrated in order to optimize the use of water resources. In order to bring about the transition from the silos approach to hydrological approach, we need to have discussions among engineers, earth scientists, social scientists and others, and build institutions and training facilities to organize databases, monitor the systems, disseminate the information and come up with techno-socio-economic options to aid the policy makers.

Vörösmarty et al. (1999) designed key biogeophysical datasets that can be used to build typology models for a country (Table 2.1.5 in this chapter), which can be applied to India.

2.1.8.3 Tools to enhance agricultural productivity

(i) Soil Health Cards: A soil may have inherent fertility arising out of its mineralogy, humus content, and ability to hold moisture. Fertility may be induced in the soil by the addition of suitable fertilizers. The use of manure for fertilizing the soil is probably as old as the agriculture itself (the Romans even had a particular god, Sterculius, to preside over the protection of the fertility of the soil!). The increase in harvest due to the use of fertlisers varies greatly depending upon the nature of the soil, agroclimatic conditions, and crop management. It is now generally agreed that manure and fertilizer are complementary but not competitive. For purposes of management of soil nutrients and conditioning, it is necessary to keep track of the soil health in a farm. The health of a soil is characterized by the following soil properties: (i) Texture and structure, (ii) Water-soluble salts (electrical conductity), Keen box data (density, porosity, Maximum water holding capacity, field capacity), soil pH, CEC (Cation Exchange Capacity – CEC) of soil/roots, Organic carbon, and Assay of biological processes. Farmers are provided with the Soil Health Cards for their farms, on the basis of which they are advised on fertilization, depending upon the crop that is proposed to be grown, and agroclimatic calendar. The common field-scale deficiencies are those of micronutrients, such Zn, Mn, Fe, S, etc. (vide *Curr. Sci.*, v. 82, no. 7, pp. 797–807, 2002).

(ii) Precision farming: Global Positioning System (GPS) is at the heart of this exercise. Tractor-mounted, GPS-controlled system collects soil samples of a farm according to a grid system. The samples are analysed for about 17 physical and chemical characteristics, and the resultant data are made use of to generate composite colour-grams of the various parameters. The stencils thus generated can be put to many uses, such

as the balancing of the soil fertility of the field with respect to major, secondary and micronutrients needed to grow the desired crop. Instead of uniformly applying one particular fertilizer all through the farm, a tractor-mounted, computer-controlled spreader applies the necessary fertilizers at various points as needed (by comparing the ambient fertility at a given point to the fertility that is needed to be present at that point, for crop growth). This kind of customized fertilizer application not only effects great savings in fertilizer costs, but also leads to maxium economic yields of crops. The system can also be used to optimize the spacing of seeds, and the application of herbicides and pesticides, the application of irrigation water depending on the moisture content of the soil, and the crop that is proposed to be grown. At harvest, crop yields are recorded along the same grid pattern. This will enable the farmer to fine-tune the various variables to maximize the output at minimal costs of the inputs.

More importantly, this robotic system works round the clock and is unaffected by weather (say, rains or fog). In the Indian context, the most cost-effective arrangement will be for the farmer to have the precision farming as a service which can be offered by private companies or farmers' cooperatives (vide *The Hindu*, Mar. 25, 2008).

(iii) How to make use of Soil microorganisms to improve soil productivity (Johri et al., 2008): The Below Ground Bio-Diversity (BGBD) of the soil microorganisms has a significant bearing on the soil productivity because of the involvement of microorganisms in nutrient cycling, sequestration of atmospheric carbon, modification of the soil physical structure and water regimes, enhancement of plant health by interacting with pathogens and pests, predators and parasites, and enhancement of plant defense through induced systematic resistance and other mechanisms. The erosion of the biological component of the soils led to the general decline in the soil fertility of the Indo-Gangetic Plains (IGP) of India which produces about 50% of the food grains of the country. It is possible to reverse the productivity decline through the management of the biodiversity of the soils which needs to be implemented at three different scales: (i) Keystone biota level management – biological nitrogen fixation, mycorrhizas, earthworms, etc. (ii) Soil level management – organic matter inputs, balanced mineral fertilizers and amendments, tillage and irrigation, and (iii) Cropping system level management – cropping system design in space and time, choice of crops and varieties, genetic manipulation, microsymbioses, rhizosphere microbial dynamics, etc.

2.1.8.4 Conclusion

In order to bring about the transition from the silos approach to holistic hydrological approach, we need to have discussions among engineers, earth scientists, social scientists and others, and build institutions and training facilities to organize databases, monitor the systems, disseminate the information and come up with techno-socio-economic options to aid the policy makers.

In the interests of ecology and economy, it is necessary to place water at the heart of the development planning. Just as we embraced a low-carbon economy to meet the challenge of climate change, we need to go in for a low-water economy. We should observe a low water footprint in all our activities – agriculture, industry, domestic. Since agriculture accounts for 80% of water use, it is this sector that we should seek low water footprint at the earliest. How this is to be achieved is covered in the chapters on sequential use of water resources and wastewater reuse systems.

In the Africa – India Forum Summit held in Addis Ababa, Ethiopia, on May 24, 2011, President Abdoulaye Wade of Senegal is said to have stated that the Indian technical assistance helped his country to emerge from a rice-importing country to a rice-exporting country in a matter of four years (*The Hindu*, May 25, 2011). This demonstrates the power of technology in improving the quality of life of the people.

The International Council of Scientific Unions (ICSU) in cooperation with the International Science Council (ISSC) came up with a holistic strategy on Earth System research that integrates our understanding of the functioning of the Earth system – and its critical thresholds – with global environmental change and socio-economic development (Reid et al., 2010). Water resources management constitutes a critically important part of the earth system science. By studying the matter and energy flows between water, land and biota, Indian science could contribute to the preservation and optimal use of water resources in India and elsewhere. Here is a great opportunity for Indian scientists to contribute to global advancement of science.

REFERENCES

Buras, N. and Hall, W.A. (1961) An analysis of reservoir capacity requirements for conjunctive use of surface and groundwater storage. In *Groundwater in Arid Zones*. IAHS Publ. No. 56.

Chritchley, W. and Siegert, K. (1991) *Water Harvesting, A Manual for the Design and Construction of Water Harvesting Schemes for Plant Production*. Food and Agricultural Organization of the United Nations (FAO), Rome, pp. 129.

Correa, N.R. 1987. Determination of necessary volume of the groundwater reservoirs for optimal conjunctive use for irrigation in arid region. In T.H. Anstey & U. Shamir (eds.), *Irrigation and Water Allocation*. IAHS Publ. 169, 221–228.

Department of water Resources, State of Calfornia 2005. *Water Plan Update*. Public Review Draft. Vol. 3.

Gupta, S.K., and Deshpande, R.D. (2004) Water for India in 2050: First-order assessment of available options. *Curr. Sci.*, 93, 932–941.

Haagsma, U.G. (1995). The integration of computer models and databases into a decision-support system for water resources management. In S.P. Isonomic et al. (eds.) *Modeling and Management of sustainable basin-scale water resources systems*. IAHS Publ. 231: 253–261.

Jain, S.K, Agarwal, P.K. and Singh, V.P. (2007) *Hydrology and Water Resources of India*. The Netherlands: Springer-Verlag, p. 1258.

Johri, B.N., Chowdhary, D.K. and Chaudhuri, S. (2008) How to use soil microorganisms to optimize soil productivity. In "*Food and Water Security*" (Ed. U. Aswathanarayana), pp. 51–62, London: Taylor and Francis Group.

Kaden, S., Becker, A. and Ganuck, A. (1989) Decision-support system for water management. In D.P. Loucks & U. Shamir (eds.) *Closing the Gap between Theory and Practice*. IAHS Publ. 180: 11–21.

Karlberg, L. (2005) Irrigation with saline water using low-cost drip-irrigation systems in sub-Saharan Africa. PhD thesis, KTH Architecture and the Built Environment, Stockholm, Sweden, p. 26.

L'vovich, M.L. (1979) *World Water Resources, and their future*. Washington, D.C.: American Geophysical Union.

Marengo, J.A. (2006) On the hydrological cycle of the Amazon Basin: A historical Review and current state-of-art. Revista Brasileira de Meteorologia. 21, 1–19.

Narasimhan, T.N. (2008) A note on India's water budget and evapotranspiration. *J. Earth System Science*, 117, no. 3, 237–240.

Reid. W.V., Chen, D., Goldfarb, L., Hackman, H., Lee, Y.T., Mokhele, K. and Ostrom E. (2010). Earth System Science for Global Sustainability: Grand Challenges. *Science*, v. 330, pp. 916–917, 12 Nov. 2010.

Rockström, J., Gordon, L., Folke, C., Falkenmark, M. and Engwall, M. (1999) Linkages among water vapor flows, food production and terrestrial ecosystem services. *Conservation Ecology* 3(2), 5 p.

Sashidhar, H.E. (2008) Aerobic Rice – An efficient water management strategy for rice production. In "*Food and Water Security*" (Ed. U. Aswathanarayana). pp. 131–139, London: Taylor and Francis Group.

Shiklomanov. I.A. (1997) *Comprehensive assessment of the fresh water resources of the world*. World Meteorological Organization, 88 p.

Shiklomanov, I.A. (1998) *World Water Resources: A new appraisal and assessment for the twenty-first century*. Paris: UNESCO.

Tiwari, V.M., Wahr, J., and Svenson, S. (2009) Dwindling groundwater resources in northern India from satellite gravity observations. *Geophy. Res. Lett.*, v. 36. L 1840, 17 Sept. 2009.

Tsao, Y.-S. (1987) Over-irrigation of paddy fields for the purpose of artificially recharging groundwater in the conjunctive use scheme of Cho-Shui River Basin. In T.H. Anstey & U. Shamir (eds.) *Irrigation and Water Allocation*. IAHS Publ. 169.

Vörösmarty, C. et al. (1999) The Global Hydrological Archive and Analysis System (GHAAS). Abstract no. HW1/25/A5, IUGG, Birmingham.

Chapter 2.2

Economic frameworks to inform decision-making

U. Aswathanarayana (India)

Water scarcity is crippling the economies in many countries. It is impeding the growth of industries whose chains depend upon water availability, and is adversely affecting the sustainability of communities and ecosystems on which they depend. "Business – as – usual" is not a viable option, for the simple reason that by 2030 the global demand for water would be more than 40% of the supply.

McKinsey Consultants in a highly thought-provoking document entitled, "*Charting our water future: Economic frameworks to inform decision-making*" (2009), came up with a strategy of solutions-driven dialogue among the stakeholders to address water scarcity issues during the period upto 2030, on the basis of an integrated fact base about technical levers and their costs. Chapter 2.2 is built on these concepts.

The important point that McKinsey report makes is that, though it is a formidable problem, it is entirely possible to close the growing gap between water supply and demand in an affordable and sustainable manner.

The report identifies supply- and demand-side measures that could constitute a more cost effective approach to closing the water gap and achieve saving for BASIC countries (Brazil, South Africa, India and China). These countries collectively account for 40 percent of the world's population, 30 percent of global GDP and 42 percent of projected water demand in 2030.

The report makes the important point that solutions to these challenges are in principle possible and need not be prohibitively expensive. For a given basin, integrated solutions for closing the demand-supply gap in a cost-effective way are sought on the basis of three approaches: (i) technical improvements, (ii) improving water productivity under a constant set of economic activities, and (iii) ways of actively reducing withdrawals by changing the set of underlying economic activities, through an understanding of how the economic activities in a given country are affected by the reduction of withdrawals.

2.2.1 AN INTEGRATED ECONOMIC APPROACH TO WATER SCARCITY

The Report specially addresses the water scarcity issues of BASIC countries.

The cost of closing the gap in the case of the BASIC countries is estimated by McKinsey to be \$19 billion annually by 2030, which works out to be just 0.06% of the combined GDP of the four countries for 2030. Globally, the cost of the closing the demand-supply gap in the least costly way available is approximately \$50–60 billion annually, which is 75% less than supply-only solution.

In the case of India, two developments are driving up the demand for water – Indian population is growing, and a large portion of them is moving towards middle- class diet, with increasing requirements of rice, wheat and sugar. Consequently, the demand of water in India is expected to grow about 1500 billion m^3, almost double the present current water supply of 740 billion m^3. Unless drastic action is taken, India's most populous river basins, notably the Ganga, Krishna and Indian portion of the Indus basin, face the biggest absolute gap.

The Report developed an elegant "water marginal cost curve" which is a decision-making tool based on a "microeconomic analysis of the cost and potential of a range of existing technical measures to close the projected gap between demand and supply in a basin" (Fig. 2.2.1).

The technical options available for bridging the demand-supply gap in a given sector are evaluated on a like-for-like basis, efficiency and productivity measures, and the most-effective technical measure is then chosen. Each of these technical measures is represented with a block on the curve. The height of the of the block represents the

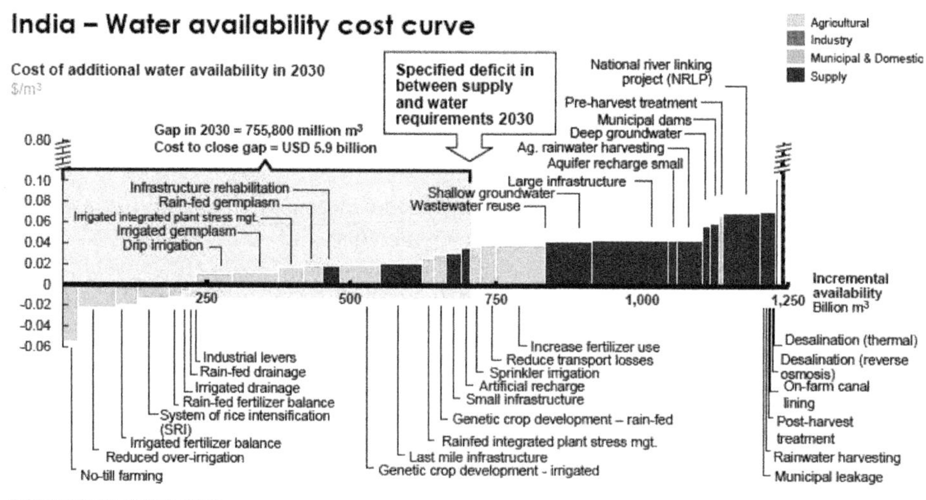

Figure 2.2.1 India – Water availability cost curve
© *Charting Our Water Future*, 2009, p. 19

unit cost, while the width of the block represents the amount of additional water that becomes available consequent upon the adoption of the measure.

An analysis of the water marginal cost curve leads to the following conclusions:

1. The Agricultural sector is of overwhelming importance in bridging the supply-demand gap.

   The bridging in the case of the agricultural sector is to be accomplished through

   (i) Improved application of well-established technologies such as, drip and sprinkler irrigation.

   (ii) Increasing crop productivity ("More crop per drop") through adopting measures such as, "no-till farming", improved drainage, utilization of the best available germplasm or other seed development, optimizing fertilizer use, and application of crop stress management, including both improved practices (such as integrated pest management) and innovative crop protection technologies.

In the case of India, the profound importance of the agricultural measures (vide the left-hand side of the cost curve) can be judged from the fact that their proper implementation in respect of both irrigated and rainfed crops production, can collectively close 80% of the gap. The remaining 20% of the gap can be bridged by the avoidance of conveyance losses through the rehabilitation of the existing irrigation and completion of the lining of canals.

McKinsey estimates that the total cost of supply and agricultural levers works out to about $6 billion per annum, which is just 0.1 of the projected GDP of India 2030.

The above analysis does not take in to account unforeseen adverse developments, such as institutional barriers, labour markets and GDP levels.

The above measures are already known and have been implemented in various parts of the country to a lesser and greater extent. The Water Marginal Cost Curve for India has the great merit of demonstrating the feasibility and affordability of the proposed measures, beyond any doubt.

1. How to make efficient use of water in industry and municipal systems

India can benefit from the Chinese experience in this regard. In China, industrial and municipal water use has been growing at a fast rate of about 3% per annum. Under the circumstances, China brought down the allocation of water for agriculture to 50% (as against 70% in the case of India) without affecting productivity, through enacting water-saving regulatory reforms aggressively. The water saving potential of power, wastewater reuse, pulp and paper, textile and steel industries is expected to be about 25% of the demand-supply gap and results in net saving of about $24 billion. McKinsey estimates that the net capital expenditure to close the remainder of the gap amounts to $8 billion, or less than 0.06 percent of projected 2030 GDP.

2. There is a strong linkage between the quality and quantity of water

India can learn from the experience of São Paulo state (Brazil) in this regard. In the case of São Paulo, resolving water quality issues is crucial for practical usage for environmental reasons. A supply infrastructure option would cost $530 million per year or 0.07% of GDP). As against this, a least-cost option based on efficiency and productivity measures, is available at a net annual cost of $285 million (0.04% percent of the state's projected 2030 GDP). Industries can improve their finances by reducing water use by using levers in spring-valve installations and sensitivity sensors.

Reduction in leakage of utility water supplies can save nearly 300 million m^3 of water. Wastewater reuse in industrial processes and public works uses is equivalent to roughly 80 million m^3 in new water.

3. *Most solutions imply cross-sectoral trade-offs*

India can learn from the experience of South Africa in this regard. South Africa developed a multi-sector solutions, based on improvements in available water supply which can close 50 percent of the country's projected supply-demand gap to 2030: agricultural efficiency and productivity improvements (30%) in seven river sub-basins, and industrial and domestic levers in industrial centres such as Johannesburg and Cape Town (20%). There can be a annual saving of about $418 million by improving efficiency and smaller use of water in processes such as paste-thickening and water-recycling in mining, and dry-cooling, and pulverized beds in power. This is a win-win situation and saves not only water but also significant savings in input costs. This makes the solution almost 50% cost-negative.

Each country has to prepare its own water resources management strategy based on "Water Marginal Cost Curve".

2.2.2 ROLE OF THE PRIVATE SECTOR IN THE WATER RESOURCES MANAGEMENT

Private sector has a major role in the water resources management. McKinsey has identified the following five sectors in this regard.

(i) Agriculture, horticulture, food industries and agricultural value chain players

Food security is closely linked with water security as food cannot be grown without water. Presently, agriculture accounts for 70% of the water use. It is critically important that this percentage has to be brought down to at least 50%, as China has accomplished.

Agriculture has to go hand-in-hand with animal husbandry. Animals can survive drought much better than plants. Animals need large amounts of drinking water – 40 L/d for cattle, horses and camels, 15 L/d for hogs, 10 L/d for sheep. Animals can drink semi-potable water without any ill effects. Thus, while 500 ppm of Total Dissolved Solids (TDS) is the potable limit for humans, animals can safely drink 700 ppm TDS water. In water-scarce South Africa, waste bathwater is given to pigs.

Impala can survive on just 1.5 L/d (one sixth that of sheep). So impala is raised in ranches as a meat animal instead of sheep. Ostrich does not need water at all. It also provides valuable meat, and feathers, and one ostrich egg suffices for sixteen omelettes.

No wonder ostrich farms have become popular in South Africa, Australia and Arizona (USA).

In India, much food goes waste not only because of rodents and pests, but also because there are not adequate facilities for food preservation. For instance, perishable vegetables such as tomatoes, which have been grown after much expenditure of water, are thrown away as cattle-feed or sold at throw-away prices, for want of refrigeration and food processing facilities. Also, every part of paddy, wheat and sugarcane should be put to proper use, without being wasted (e.g. rice brawn oil, boards from paddy husk and straw, etc.). In some cases, fermenting could increase the nutritional value of food.

McKinsey estimates that aggregate agricultural income in India could increase by $83 billion by 2030 if the full potential agricultural measures is mobilized.

(ii) Financial institutions

That the agricultural sector in India is starved of investment should be evident from two statistics – the agricultural growth has been just one-fourth of the industrial growth, and about 17,000 farmers committed suicide in 2009. Here is a heart-rending case of a farmer's death. A farmer borrowed Rs. 200,000 (eq. $4,500) to raise paddy and chillies (mirch) in his rented farm. As luck had it, unseasonal rains destroyed his crops, and he became destitute. Two years later, he got a notice from a bank that the Government granted him Rs. 900 ($20) for the chillies crop that he lost. When he went to the bank to collect his Rs. 900 he was just given Rs. 400. The bank stated that they used Rs. 500 to open an account in his name, as required by regulations. The farmer collapsed and died on the spot.

There is little doubt that the farmer indebtedness and suicides are traceable to credit problems. Banks in Maharashtra give loans at 7% interest to buy a Mercedes, but the same bank charges 12% interest for a tractor – probably on considerations of recoverability of loans. The issue gets even more serious due to unwillingness of farmer to change crop. In Vidarbha part of Maharashtra which is notorious for farmer suicides, some farmers faced with scarcity of water, switched to soya beans for which there is good market, and survived. Most often, for traditional reasons, land is never in the name of a woman – she can neither raise a loan, nor her suicide is counted as a farmer suicide. Andhra Pradesh circumvented the problem by providing 3% interest to women's self-help groups.

Financial institutions should make use of the cost curves to identify appropriate technical measures and financial costs to close the water supply-demand gap. For instance, drip irrigation has great potential for lending and equity investments alike. It is expected that this technology penetration will grow at the rate of 11% per year through 2030.

This would be a win-win situation, involving increased manufacturing capacity and credit for farmers.

In the case of China, investment is needed in the sector of municipal leakage reduction. São Paulo and South Africa require investment in water transfer schemes.

(iii) Large industrial water users

Large industrial users of water, such as, metals, mining, petroleum, paper and pulp and energy industries, face both water and energy challenges. An innovative way-out is to use waste water to the maximum extent possible. For instance, municipal waste water with pathogens is unsuitable for use in food industry, but it can be safely used in thermal power cooling. In Arizona, USA, the cooling water in the nuclear power stations, is recycled municipal waste water. The demand and supply analysis and the cost curves are a transparent way for decision-making for investing in water efficiency solutions.

In South Africa, for example, the basins with the largest gaps are those with the largest industrial water demand. Under the circumstances, technologies have to be chosen in the context of water scarcity such as dry cooling and fluidized-bed combustion in power generation, and paste tailings in mining.

(iv) Technology providers

Technologies for augmenting the supply of water to close the gap are described below:

(i) Recycled wastewater: This is by far the largest source. It has been estimated in 1995 that the volume of waste water is 326 km^3/yr in Europe, 431 km^3/yr in North America, 590 km^3/yr in Asia, and 55 km^3/yr in Africa. Depending upon the compositional characteristics of recycle waste water, it is used for agricultural irrigation for edible and non-edible crops, pasture irrigation and livestock farming (e.g. as is being practiced in Tel Aviv, Israel). Waste water may be used in spray irrigation, overland runoff and rapid infiltration.

    In the case of industries, waste water can be used in power plant cooling, processing plants and construction. While municipal wastewater with pathogens is unsuitable for use in food industry, there is no problem whatsoever in using such water for thermal cooling. For instance, a 3,810 MW nuclear power plant in Phoenix, Arizona, USA, is cooled with sewage water at a design flow rate of 276,000 m^3/d.

(ii) Deasalinization: Desalinisation of seawater is energy-intensive and is practiced mostly by the energy-rich gulf countries. The cost of desalinization varies from \$1,500 to 4,500 per 1,000 m^3, with the cost at Jubail, South Arabia being \$2,700 per 1,000 m^3.

(iii) Reverse Osmosis (RO) is the preferred process for the desalinization of brackish water. The process is based on the principle that when water is forced through semi-permeable membrane under pressure, the dissolved salts are held back. The Central Salt and Marine Chemicals Research Institute (CSMCRI), Bhavnagar, India, designed both stationary and mobile units, which produces clean water at the rate of 300–400 L per m^2 of membrane surface.

(iv) Desalinisation coupled to heavy water production. A plant for this purpose is in operation in Tuticorin, South India.

(v) Concentrated solar power (CSP) systems concentrate direct sunlight to reach high temperatures. This heat is used to power a steam turbine which drives a generator. CSP plants could be used to produce fresh water in the process.

(vi) Saline agriculture: *Salicornia* is a salt-tolerant bush which grows well in salinised coastal soils. Its leaves could be used as fodder, and its seeds could be used to produce industrially usable oils.

Technology providers may use the cost curves to evaluate the market potential and cost competitiveness of different technology options in socioeconomic and biophysical setting of a given country.

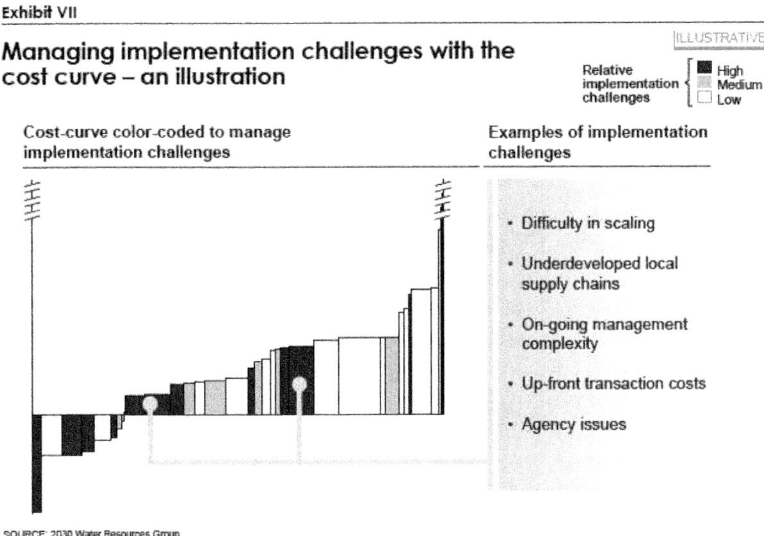

Figure 2.2.2 Managing implementation challenges in the cost curve
© *Charting our water future*, 2009, p. 24

In the case of China, membrane technology is still 2–3 times more expensive than conventional water treatments. Membrane technology has the market potential of 85 billion m^3 by 2030, which is 56 times the figure in 2005.

(v) Construction sector

Construction sector will need to provide the large scale infrastructure (buildings, roads, ports, airports, etc.). In the case of South Africa and Brazil, there is a gap of about 50% in the infrastructure. In India, the position is better with 14% gap, though this would require an investment of $1.4 billion.

2.2.3 TOOLS FOR POLICY MAKERS

Policy makers would like to know the issues that would arise in implementing a technical decision suggested by the cost curve – whether there are any difficulties in implementing it, whether there are any secondary impacts and whether the implementation would affect other water policies.

The cost curve may be refined in three ways:

(i) *Ease of implementation*: In the case of India and China, levers are grouped independent of the economic sector. This may throw up two options: smaller number of central decision-makers which makes for easier implementation, or a very large number of stakeholders such as farmers and water users at the grassroots level, making implementation a cumbersome process. Involving a

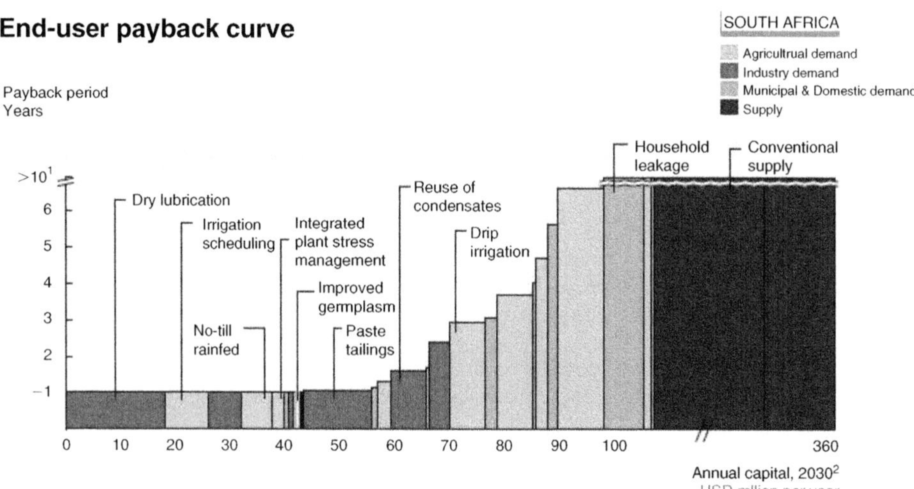

Figure 2.2.3 End-user payback curves
© *Charting our water future*, 2009, p. 26

large number of stakeholders would have the potential to change the behaviour of such people. Experience has shown that where the implementation required the action of a few central decision-makers, the implementation costs are not only greater (typically \$17 billion per year), but also would not able to cover all the issues. The second option is therefore recommended.

(ii) ***Construction of scenarios to assess the impact of policy decisions on water demand:*** A policy maker would like to know how a country's water supply-demand gap would be if free electricity is provided to pump groundwater. Some states provide subsidized electricity. If this benefit is withdrawn, production of irrigation crop will decrease by 5% and water demand would be reduced by 8%. The actual cost of closing the gap is expected to be 10%. This has to be weighed against reduced crop production and reduced economic activity.

An ethanol boom in Brazil would increase the demand for water from 2.6 billion m^3 to 6.7 billion m^3. The cost of closing the water demand-supply gap would double, or increasing even more if priority is given to the supply decision.

(iii) ***Construction of a "Pay Back" curve to quantify the economics of adoption for endusers.***

The way an enduser would look at a supply-demand gap may be different from that of the government. For instance, a low-income farmer may need to recover his investment in a short period – say, three years. As against this, an industrial water user could afford to have greater flexibility. In China and India, 75% of the gap could be closed through measures with a payback time of three years or less. In São Paulo state,

Brazil, 86% of the programmes have a payback time of 5 years. The programmes need to be modified to reduce the pay back period in order to make them more attractive to endusers.

2.2.4 QUO VADIS?

The McKinsey Report is based on the policy formulations: (i) Business-as-usual is no longer an option for most countries. (ii) The challenges of bridging the demand-supply gap are formidable, but not insurmountable, (iii) Every country needs to develop a *fact-based* vision of water resources at state and country level, which will link cost and economic data to water resources data, including environmental requirements. Such a database would help policy makers, governmental agencies and private sector to design and implement various water supply programmes. By identifying measures which have the greatest impact on water resources, investment can be channeled to those measures.

The situation is not static – political, social, economic and technical developments may necessitate new approaches. Here innovation could play a crucial role.

Technology hubs, research and education need to be developed for the purpose.

The cost curve and the end-user payback curve may be useful to banks and private financiers to increase their investments, and small-scale users to get microfinance support (say, drip irrigation). Large industrial water users can play a valuable role to reduce water use through improved technologies. Governments may facilitate water supply programmes through defining clear ownership rights, appropriate tariffs and subsidies pricing and standards.

Since water concerns every body, it is necessary that we should all work together to develop a sustainable water economy.

REFERENCE

Charting our Water Future (2009) McKinsey Consultants.

Chapter 2.3

Multiple perspectives on water: A synthesis

Ramaswamy R. Iyer
Former Secretary to the Govt. of India, Ministry of Water Resources, New Delhi
Now Honorary Research Professor, Centre for Policy Research, New Delhi

2.3.1 NATURE OF WATER

Water is many things in one: a basic life-need and right; an amenity; a cleaning agent; a social good (e.g. for firefighting, hospital use, use in schools and public institutions, etc.); a requirement for economic activity (agriculture, industry, commerce); a means of transportation; an occasional manifestation as floods; a part of our social, political and cultural life; and a substance that plays a role in rituals (Iyer, 2003, Iyer, 2007, pp. 12–14, 224–227; Iyer, 2009. pp. 568–569; Iyer, 2010, Iyer, 2011, and Bandyopadhyay 2009). It is therefore not surprising that there are many different perceptions of water, and correspondingly, there are multiple perspectives on water.

2.3.2 PERSPECTIVES ON WATER

2.3.2.1 Traditional perspective

In earlier times people not only met their water needs from a river or lake but lived with it in a close relationship: it was part of the landscape, lives, culture and religion of a people. The people's relationship with the river or lake was part of their relationship with nature as a whole. They did not consciously use the word "ecology" – that word came into use much later – but ecology in fact governed their lives. One is not romanticising the past or imagining a golden age; there were inequities, injustices and exclusions in pre-colonial India. The point is that the numbers of people and their needs were not such as to generate acute conflicts (though conflicts over water were not unknown: the Buddha is said to have resolved a water-related conflict between two communities); and human beings had not yet developed the capacity to subject their habitats or water sources to stress. Water was then regarded as nature's bounty, and rivers were worshipped as goddesses (with the exception of the Brahmaputra, which was masculine and therefore a god). The aspect of the god or goddess was not always benign; occasionally life-giving water became fearsome in the form of floods. People accepted that aspect too and evolved ways of living with it and minimising damage, except on those rare occasions when floods became catastrophic: people then called that phenomenon an Act of God. There were diverse local water management systems in different parts of the country – tanks and anicuts in the south, johads in Rajasthan, stepwells in Gujarat, ahars and pynes in Bihar, canals in the north – with some structures built by princes or chieftains, but mostly managed by the community (Agarwal and Narain, 1997).

2.3.2.2 Riparian perspective

That perspective, which we may call *the old or pre-colonial or traditional perspective*, underwent some changes with the advent of colonial rule. With the British common law traditions came *the riparian perspective*. Even earlier, that perspective must have been implicit in a rudimentary fashion. Proximity to rivers must have conferred advantages on some homes, farms and settlements. Relations between upstream and downstream people must have been of a "riparian" kind, though that concept might not have been in existence. However, riparianism as a legal principle is no doubt of common law origin. Under this perspective, proximity to a river brings with it a right to use the waters of that river, but with an obligation not to cause harm or injury to a lower riparian. That is a greatly over-simplified statement of a very complex matter: complex because of a vast body of case law (See Puthucherrill in Iyer, 2009, pp. 97–133; and Iyer *ibid*, pp. 592–3). With the growth of technology and the possibility of transporting water over long distances, proximity to a river has lost some of its old importance. At the level of households, farms and villages, the riparian perspective is perhaps not as significant as it once was, but the doctrine of riparian rights is not without value as a moderator of the assertion of the state's sovereign power, often described as "eminent domain". (Strictly speaking, the term "eminent domain" refers to the state's sovereign right to take over private property on the payment of just compensation. The term, in use in American law, has gained currency in general discussion in this country, and is loosely used in the sense of the overriding power of the state, without specific reference to the acquisition of property). Moreover, inter-State and even inter-country river water disputes are essentially riparian disputes, and riparian doctrine plays a part in their resolution by tribunals within the country or by treaties between countries. Both the old Helsinki Rules (Helsinki, 1966) and the more recent UN Convention on International Water Courses (UN, 1997) are riparian documents.

2.3.2.3 State perspective

With colonial rule came not only British common law but also control by the colonial state. This is what we may call *the state perspective*. The colonial state asserted overriding sovereign control and even ownership over all natural resources, though it did recognize customary law and traditional practice to some extent. That perspective is enshrined in the Acts, variously known as Irrigation Acts or Irrigation and Drainage Acts or Canal Acts, enacted by most provinces. In substance water was regarded as state property. That perspective (along with old colonial laws) continued into the post-colonial period. State Governments in independent India, like their colonial forbears, assert state control over water. Undoubtedly the state has a number of functions to perform in relation to water: legislation, policy formulation, planning, administration, "governance" at all three levels, ensuring/enforcing rights, resolving water-related disputes, and so on, and of course it is concerned with the international aspects. It needs powers to perform these functions, and draws those powers from the Constitution and the laws, and its authority to use force. It tends to assert sovereign powers, and is generally inclined towards command and control. This (combined with other factors) has had the effect, over the years, of weakening old systems, traditions and practices of local community-management of water.

2.3.2.4 Engineering perspective

Colonial rule meant not only a colonial bureaucracy but also colonial engineers. The advent of colonial rule was accompanied by the advent of 'modernity' which meant among other things western engineering. The colonial government undertook irrigation works extensively, largely for the purpose of raising more revenue from land, and thus began the era of dam-building. Here again, the post-colonial state has not only continued that activity but greatly enlarged and accelerated it. Water came to be largely a matter for engineers. *The engineering perspective* became very important. At the heart of that perspective lies the idea of subjugation and control of nature for human purposes. This in fact goes back to ancient times and inspired the Promethean legend, but the human capacity to subdue nature became marked only after the industrial revolution. It was then that the Promethean view came into its own. With that world view, the engineer became or tried to become the master of rivers. It is interesting to contrast the Promethean legend of the west with our own Bhagiratha legend. Prometheus is said to have brought fire to earth in defiance of the gods, whereas Bhagiratha brought the Ganga through prayer. However, the Promethean attitude to nature came into India with western engineering, and was ardently embraced by our own engineers and administrators, and by our intelligentsia as a whole. The modern urban view of rivers is very well described in the following lines from T. S. Eliot's *Four Quartets*:

> I do not know much about gods; but I think that the river
> Is a strong brown god—sullen, untamed and intractable,
> Patient to some degree, at first recognised as a frontier;
> Useful, untrustworthy, as a conveyor of commerce;
> Then only a problem confronting the builder of bridges.
> The problem once solved, the brown god is almost forgotten
> By the dwellers in cities—ever, however, implacable.
> Keeping his seasons and rages, destroyer, reminder
> Of what men choose to forget. Unhonoured, unpropitiated
> By worshippers of the machine, but waiting, watching and waiting.

We should perhaps substitute the word "dams" for "bridges" in that quotation to reflect Indian thinking. In this country dams became symbols of development in the popular mind. Dam-building was an important part of economic planning in India. In the year 2000, there were over 4,500 large dams by ICOLD definition in India. That number must have gone up since. (The International Commission on Large Dams (ICOLD), established in 1928, defines a large dam as a dam with a height of 15 m or more from the foundation. If dams are between 5–15 m high and have a reservoir volume of more than 3 million cubic metres, they are also classified as large dams. Using this definition, the World Commission on Dams estimated that there were over 45,000 large dams around the world).

The principal concerns of the engineering perspective are the design, construction, operation and maintenance of dams, reservoirs, canal systems; public tubewells; water supply, drainage and sanitation systems; etc. It has authority in technical matters, and is generally driven by a professional belief in the primacy of technology. The engineering perspective is neatly captured in a catchy light-hearted statement reported to have been made by an American water engineer: "I love pushing rivers around" (quoted in Conca, 2005). That probably represents the attitude of our own water engineers and water bureaucrats, though they may not explicitly say so. The apotheosis of that kind of thinking, that cavalier attitude of manipulating rivers, was reached in the Inter-Linking of Rivers Project announced with fanfare in 2002. Curiously enough, the advocates of that project invoked the name of Bhagiratha, though their approach was really Promethean. "Inter-linking" was a bad enough expression, but our former President Dr. Abdul Kalam was fond of using an even more incongruous term: "Networking of Rivers". Fortunately the project is now in the doldrums, but it has not been formally abandoned, and may get revived at any time. (On the Inter-Linking of Rivers Project, see Bandyopadhyay et al., 2004 and Alagh et al., 2006.)

2.3.2.5 Economic perspective

A bit later than the engineering approach to rivers came the economic approach, and the two converged. The Dublin Statement of 1992 (Dublin, 1992) contained some good principles (1, 2 and 3) but principle 4 declared water to be an economic good, and that laid the foundation for the treatment of water as a commodity, subject to market forces like any other commodity. That *economic perspective* has grown in strength and is widely held now. This perspective sees water as economic good subject to market forces. It believes in water markets, full economic pricing of water, cost-benefit analyses as the basis for investment decisions on projects, the privatisation of water services, private sector participation in water resource projects, etc. It is essentially an ideological perspective. Its strong advocates are the World Bank, IMF, ADB, and economists of a certain persuasion. Allied to this is the *"growth" perspective,* which focuses almost exclusively on economic growth at a certain desired rate, and tends to be impatient with social, community, rights, equity, environmental or other perspectives. This in turn has connections with the *business or corporate perspective* which is concerned with a supply response to demand, the objective being profits. It is associated with the corporate search for profitable investment, sensing profits in scarcity. It advocates the privatisation of water services. Corporate social responsibility is professed but tends to be subordinated to the imperative of profits. The economic, growth and corporate perspectives may be distinct, but they form a cluster; and they have an affinity with the engineering perspective.

It is that engineering-economic combination that has been dominant for a long time. However, it has been challenged by other perspectives, notably the ecological perspective and the social justice perspective. Both these had their origins in a growing unease with big projects, particularly big dam projects. The environmental impacts of big projects were strikingly noticeable in the case of big dams; similarly, the problem of large-scale displacement of people and the difficulties of their resettlement and rehabilitation were particularly acute in the context of big-dam projects.

2.3.2.6 Ecological perspective

Taking the *ecological perspective* first, it must be noted that it was in fact a modest "environmental impact" perspective to begin with. Big dams, once considered wholly beneficial and celebrated as the temples of modern India, began to cause some concern when their environmental impacts – for instance, disruption of the routes and habitats of wildlife, submergence of agricultural land or forests, and so on – became evident. As more and more such impacts came to light, and the larger and wider implications of such interventions were better understood, the relatively simplistic "environmental impact" perspective became a deeper and more sophisticated "ecological perspective". Stages in that evolution were the Stockholm Conference of 1972 (Stockholm 1972); the Brundtland Commission Report of 1987 (Brundtland, 1987) and the emergence of the idea of sustainable development; the Dublin Principles (Dublin, 1992) and the Rio "Earth Summit" of 1992 (Rio, 1992); the report of the World Commission on Dams 2000 (WCD, 2000); and so on.

In its fully developed form, the ecological perspective is concerned with the environmental/ecological footprint of "developmental" activity and the pursuit of growth; the impacts of water-resource development projects; the heavy infliction of pollution and contamination on soil, water, air; the death of rivers; the reduction of biodiversity; global warming, climate change; the destruction of our habitat; and the urgent need to pull back from the brink of disaster.

The ecological perspective was making slow headway with considerable difficulty in the face of resistance by the engineering-economic perspective, but has gained some strength during the last few decades from the growing realisation of the phenomena of climate change and global warming. It is tempting to think that the ecological perspective has now widened and deepened into a "Planet Earth perspective", but it is perhaps premature to come to that conclusion, because resistance from the advocates of "economic growth" and "development" continues to persist and gather strength.

2.3.2.7 Social justice perspective

We come next to *the social justice perspective*. This had its origins in certain commonly experienced injustices and inequities; for instance, serious disparities between the rich and the poor in urban and rural water and sanitation services; denials of service to or exclusions of the poor; injustices to Scheduled Castes or Tribes; the inequitable incidence of the costs and benefits of major projects (the social costs being borne by one group and the benefits going to another); forced displacement by projects and deficiencies or failures in resettlement/rehabilitation; inequities in access to irrigation water in the command areas of projects (head-reach and tail-end farms); and so on. Apart from seeking specific remedies, the empowerment of the powerless and voiceless is among the objectives of this perspective. The champions of the perspective are NGOs and other civil society institutions and movements. The victims of injustice themselves (the poor, the dalits, the Scheduled Castes, and Scheduled Tribes) also take up their own cause through movements.

2.3.2.8 Rights perspective

Closely allied to the social justice perspective is *the rights perspective*. This has figured in the context of displacement for dams, land acquisition for "development" etc., but did not originate in them: it goes back to the fight for civil liberties during the colonial period. Internationally, the concept of human rights has a long history. In relation to water, the rights perspective focuses on the fundamental or human right to water; right to sanitation; traditional rights of access of communities (tribal or other) to rivers, lakes, forests, and other sources of sustenance and livelihoods; and so on. The perspective is espoused by human rights and tribal rights activists, legal experts, NGOs, etc. (Incidentally, in the UN context, though water was recognised as a basic human need, there was some resistance to its recognition as a human right, but on 26 July 2010 the UN General Assembly declared safe and clean drinking water and sanitation to be a human right). In India, the right to water (as essential for life) is a fundamental right by judicial interpretation of the right to life (See Baxi in Iyer, 2009, pp. 149–166; and Iyer *ibid*, pp. 588–592).

It will be seen at once that those who regard water as a fundamental right or a human right cannot accept the economic perception of water as commodity, the idea of tradable water entitlements or water markets, and the pricing of water on the basis of "full cost recovery". These are fundamentally opposed perspectives (Barlow and Clarke 2002, Shiva, 2002, Shiva, et al., 2002, Dharmadhikari, 2002, and Barlow, 2007).

2.3.2.9 Women's perspective

Related to the social justice and rights perspectives is the *women's perspective*. It is useful not to subsume the latter under the former but recognise it as a distinctive perspective that stresses the burden on women of fetching water from long distances as well as managing water in the home, with no voice in water-planning or water-management institutions. Some changes are beginning to take place: share in PRIs (women sarpanches), representation in water users' associations under PIM, etc. However, the empowerment of women has to go much further. Their role as economic actors has to be recognised (See Kuntala Lahiri-Dutt in Iyer, 2009, pp. 275–307).

2.3.2.10 Community perspective

Another distinctive perspective is the *community perspective*. The issues here are the relationship between state and civil society, the empowerment of the people *vis-à-vis* the state (or the corporates), the community management of common pool resources, mobilisation of people for local water augmentation and management, social control of water use and sanctions against misuse, voice in water policy formulation and water management, etc. This perspective invokes customs and traditions, customary law, informal institutions, and legal pluralism. It also has recourse to formal law where possible. There can be constructive cooperation between the state and community, but that is not always the case. The assertion of the community perspective often runs counter to the state perspective, and the invocation of customary law can fall foul of formal law. For instance, the mobilisation of the community for local water-harvesting and use backed by social control may be regarded with disfavour by the state, as in fact happened at one stage in Rajasthan (Iyer, 2009, p. 581). The state-community relationship tends to be an uneasy one.

However, the relationship can be harmonised by *the public trust doctrine* which may be regarded as yet another perspective. According to this perspective the state holds water and other natural resources in public trust for the community. This enables the state to legislate and take executive action in relation to water, but as a trustee for the community. It thus empowers both the state and the community. The public trust doctrine is explicitly recognised in the laws of some States in the USA (See Narasimhan, pp. 544–548, in Iyer, 2009). In India, the Supreme Court has held that the public trust doctrine is part of Indian law (cf. the SC's observation "These resources meant for public use cannot be converted into private ownership. Thus the public trust doctrine is a part of the law of the land" 1997, I SCC P-388/389).

2.3.2.11 Private property perspective

Unfortunately, the *private property perspective* goes against this. This difficulty arises in the case of groundwater, because the ownership of land carries with it the ownership of the water under the land. There is thus a difference in the legal regimes applicable to surface water and groundwater respectively. While the state tends to assert its control over surface water, a private property regime prevails in the case of groundwater. This anomaly needs to be rectified by bringing all forms of water under one legal regime, namely that the state holds all of them in public trust for the community.

2.3.2.12 Citizen/water-user perspective

A part of the community perspective that can be isolated is *the citizen/water-user perspective*. Users tend to assert requirements for various uses (drinking, domestic, commercial, industrial, agricultural, etc.) quite strongly, but show poor recognition of the obligations of economical and efficient use, avoidance of waste and conflict, conservation of the resource, and protection of the environment. The constitutional/legal imposition of responsibility on the users (mainly through the Water Pollution Control Act) is inadequate and ineffective. The Central Groundwater Authority established under the Environment Protection Act has also been a weak instrument.

2.3.2.13 Water quality perspective

That brings us to *the water quality perspective*, which is concerned with the enforcement of water quality standards, and the prevention and control of pollution and contamination of water. It works through the Water (Prevention and Control of Pollution) Act 1974, the related Water (Prevention and Control of Pollution) Cess Act 1977, Amendment Acts, Rules under the Acts, etc, and other related legislation. It has an indirect relationship with the Environment (Protection) Act 1986. As mentioned above, this perspective has been quite weak in this country.

2.3.2.14 Historical-cultural perspective

Water is an inseparable part of the landscape, society, culture, history and religion of a people, and we could identify a *social-cultural-historical-sacred perspective or cluster of perspectives*, but this is part of the traditional perspective mentioned right at the beginning. The traditional perspective is not a thing of the past: it continues into the present and remains important.

2.3.2.15 Legal perspective

Finally, there is the *legal perspective*. This is in fact not a separate perspective. Legal issues arise in all perspectives. Issues that are specifically and essentially legal could include the constitutional division of legislative powers; Centre-State and inter-State relations on water; inter-State river-water disputes; riparian law, international water law; questions of ownership and/or control of water; right to water and water rights; groundwater regulation; water quality legislation; etc. However, these too are not merely legal but also socio-political questions.

2.3.2.16 An overarching ethical or *Dharma* perspective

While this account of perspectives on water began in a sequential manner, it was found difficult to maintain that approach. It is not the case that one perspective is followed by another; several perspectives are prevalent simultaneously, and some are in conflict with others. All the perspectives listed above are partial perspectives, with varying degrees of validity (from strong to moderate to weak to negative), and some are incompatible with certain others. If these perspectives are to be integrated and harmonised into a coherent whole, some will have to be regarded as the overarching, governing perspectives, and all others subsumed under them. In the author's view, the ecological and social justice perspectives will have to be the overarching perspectives, and all other perspectives subordinated to them. While many will probably agree that several of the perspectives listed above should be so subordinated, there may be some discomfort at the subordination of the "growth" and rights perspectives. Rights are indeed important, but they cannot be absolute; they have to be part of the idea of justice. Elevating "rights" to the position of an absolute principle is the surest way of promoting disharmony and conflict. The author sees nothing wrong in subsuming the rights perspective under the social justice perspective. As for growth, it is clear enough that it will have to be brought within the discipline of ecological imperatives. If the pursuit of growth at x% results in an ecological footprint that Planet Earth cannot take, then x% growth simply cannot be pursued. That is the meaning of sustainability.

Keeping in mind Gandhiji's position that rights flow from responsibilities, we can consider combining the ecological and social justice perspectives into *a Moral Responsibility perspective, or in other words, an Ethical or Dharma perspective on water.* Let us think in terms of our responsibility or dharma in relation to:

- the poor, deprived, disadvantaged, or disempowered;
- other humans sharing the resource with us, including those in our State or other States, our country or other countries, our generation or future generations;

- other species or forms of life;
- rivers, lakes, aquifers, forests, nature in general, Planet Earth itself.

That is the overarching perspective that this writer would like to propose in relation to water, which is the subject of this paper, with the suggestion that it has perhaps an application beyond water.

REFERENCES

Agarwal, Anil and Sunita Narain (eds). 1997. *Dying Wisdom: Rise, Fall and Potential of India's Traditional Water Harvesting Systems*, New Delhi Centre for Science and Environment.

Alagh, Yoginder K., Ganesh Pangare and Biksham Gujja (eds). 2006. *Interlinking of Rivers in India: Overview and Ken-Betwa Link*, New Delhi: Academic Foundation.

Bandyopadhyay Jayanta, 2009. *Water, Ecosystems and Society: A Confluence of Disciplines*, Sage Publications, New Delhi, 2009.

Bandyopadhyay, Jayanta and Shama Perveen. 2004. "Interlinking of Rivers in India: Assessing the Justifications", *Economic and Political Weekly*, 11 December.

Barlow, Maude, 2007. *Blue Covenant: The Global Water Crisis and the Coming Battle for the Right to Water*, The New Press, New York, Books for Change, Bangalore, 2008.

Barlow, Maude and Tony Clarke. 2002. *Blue Gold: The Fight to Stop the Corporate Theft of the World's Water*, published in India in 2003, New Delhi: LeftWord Books.

Conca, Ken, *Governing Water: Contentious Transnational Politics and Global Institution Building*, MIT Press, 2005.

Dharmadhikari, Shripad. 2002. *Water: Private Limited*, Barwani: Manthan Adhyayan Kendra.

Dublin Statement on Water and Sustainable Development, 1992.

Helsinki Rules on the Uses of the Waters of International Rivers, adopted by the International Law Association at Helsinki in August 1966.

Iyer, Ramaswamy R. 2003. *WATER: Perspectives, Issues, Concerns*. New Delhi: Sage Publications, New Delhi 2003.

—— 2007. *Towards Water Wisdom: Limits, Justice, Harmony*, Sage, New Delhi 2007.

—— 2009. (ed.) *Water and the Laws in India*, Sage, New Delhi 2009.

—— 2010. "Approach to a New National Water Policy", *The Hindu*, 29 October 2010, Leader Page main.

—— 2011, "National Water Policy: An Alternative Draft for Consideration", *Economic and Political Weekly*, supplement, Vol XLVI, Nos 26 & 27, 25 June 2011, pp. 201–214.

Shiva, Vandana 2002. *Water Wars: Privatization, Pollution and Profit*, New Delhi: India Research Press.

Shiva, Vandana, Radha Holla Bhar, Afsar H. Jafri and Kunwar Jalees. 2002. *Corporate Theft of the World's Water*, New Delhi: Navdanya.

United Nations. 1997. UN Convention on the Non-Navigable Uses of International Water Courses (passed by the General Assembly, 1997).

World Commission on Dams, 2000. Report: *Dams and Development: A New Framework for Decision-Making*, Earthscan Publications Ltd, London and Sterling, VA, November 2000.

Chapter 2.4

Water pollution*

U. Aswathanarayana (India)

2.4.1 PATHWAYS OF POLLUTION

Geoenvironment comprises rocks, soils, fluids, gases and organisms. It is linked to and influenced by, climate, terrain, and vegetal cover. Human activities affect the geological, physical, chemical and biochemical processes taking place in rocks, soils, hydrological systems and associated media. Surface water and groundwater are the components of the geoenvironment that by far the most affected by man. Protection of surface water from depletion and pollution (chemical, organic, thermal, mechanical, etc. contamination) constitute the most important task of the community, for the simple reason, when once the water (particularly groundwater) gets badly polluted, it is horrendously difficult, sometimes even impossible, to detoxify it. Microorganisms affect the hydrochemical processes and water quality. For instance, thionine bacteria oxidize hydrogen sulphide to sulphuric acid, and render the water more corrosive. Atmospheric pollution can penetrate the hydrosphere and biosphere, and can bring about undesirable changes in climate, soils, vegetation and water quality.

The issues of water pollution are discussed under different headings: sources (contaminants generated by activities, such as, agriculture, mining, industrial and households, etc.,) and types (inorganic and organic, biological and radioactive contamination).

Pollutants may enter the surface water directly, and the groundwater through fractures/pores in rocks and soils. Dissolved pollutants in the surface water move much faster than the dissolved constituents in the groundwater. The hydraulic connection between the groundwater and the surface water has a profound implication – for

*This chapter draws the basic data and concepts from Chapter 8 of the author's works, "*Water Resources Management and the Environment*", 2001.

instance, steps taken to protect the groundwater quality by reducing infiltration into the groundwater may lead to greater runoff, and consequent effect of river flow and quality.

2.4.2 ACTIVITIES THAT CAN CAUSE GROUNDWATER POLLUTION

The type of activities that can cause groundwater contamination at various levels are summarized as follows (source: US Environmental Protection Agency's Citizen's Guide to Groundwater Protection, April 1990):

Surface of the ground: Infiltration of polluted surface water, Land disposal of wastes, Stockpiles, Dumps, Sewage sludge disposal, Use of salt in de-icing of roads (in cold countries), Animal feedlots (and manure heaps), Fertilizers and pesticides. Accidental spills. Airborne source particulates. Human and animal faeces, etc.

In the ground, above the water-table: Septic tanks, cesspools and privies, Holding ponds and lagoons, Sanitary landfills, Waste disposal in excavations, Leaks from underground storage tanks, Leaks from underground pipelines, Artificial recharge, Sumps and dry wells, Graveyards, etc.

In the ground, below the water-table: Waste disposal in wells, Drainage wells and canals, Underground storage, Mines, Exploratory wells, Abandoned wells, water-supply wells, groundwater withdrawal, etc.

In the public mind, chemical pollution is almost always associated with industrial effluents. It is not adequately realized that household activities (particularly, in high income homes) could introduce potentially harmful compounds into the groundwater. The toxic and hazardous components contained in some of the common household products (such as, motor oils, paints, refrigerants, toilet cleaners, swimming pool chemicals, etc.) include hydrocarbons, heavy metals, toluene, xylene, benzene, trichloroethylene, acetone, 1,2,2 trifluoroethane, chlorinated phenols, sulfinates, pentachlorophenols, sodium hypochlorite, etc. Thus, households can play a significant role in limiting groundwater contamination by conserving water, maintaining the septic system in good condition, using fertilizers and pesticides for the home gardens to the minimum extent possible, and taking care of the disposal of the toxic and hazardous chemicals, etc.

2.4.3 LEACHATES FROM SOLID WASTES, SOURCE-WISE

Contamination of groundwater can occur when the solid wastes are leached. The leachate may reach the groundwater through either natural filtration or the deliberate disposal of liquids in association with solids (paper 1.1, Groundwater contamination, in "Water Resources and Land use planning", Laconte & Haimes, 1982).

2.4.3.1 Solid wastes from agriculture

While solid and liquid agricultural wastes do sometimes occur in concentrated form, the more common situation is the low-level groundwater contamination in rural areas as a consequence of leaching of excess nutrients from the inorganic and organic fertilizers applied to both arable and pastoral land, and the residues of pesticides and herbicides that may be leached from the soil.

Nitrate which is the principal contaminant arising from agricultural activities, arises from two sources:

- Use of fertilizers: The consumption of commercial fertilizers has been increasing rapidly in all parts of the world – for instance, the fertilizer consumption in 1981/1982 was 10 times more than it was in 1950's. It ranges from about 120 kg ha^{-1} in the Developed countries to less than 40 kg ha^{-1} in the Developing countries. The application of ammonium sulfate and potassium chloride fertilizers manifest themselves in the form of increased concentrations of sulfate and chloride in the drainage water,
- Mineralization: This is the process whereby the bacteria transform the organically-bound nitrogen in the soil to inorganic forms, following the ploughing of the established or temporary grasslands. Studies show that the potential quantities of nitrate released by ploughing may exceed the total, annual quantities of nitrogen normally applied.

Where the livestock is free ranging, the faeces and urine of the animals get deposited in a disseminated manner. In such a situation, the soil bacteria can readily degrade the animal wastes, and there is little risk of contamination of groundwater due to faecal bacteria. But serious problems can arise in the case of stock yards and dairy farms, where large number of animals are kept. In USA, livestock produces 130 times more waste than humans. It has been reported that one large hog farm in Utah produces more sewage than the City of Los Angeles (California, USA)!

2.4.3.2 Solid wastes from mining

Mining industry produces more solid wastes than any other industry. Presently, there are more than 40,000 mines in the world, which produce an aggregate volume of $33 \times 10^9\ m^3\ yr^{-1}$ of rock (Vartanyan, 1989). Certain kinds of solid wastes, such as, waste products of quarrying for building stone, lime for cement and agricultural use, filters (e.g. gypsum, barytes), and roadstone, are generally inert. Consequently, water percolating through them does not undergo any significant chemical changes. If the waste contains crushed material, the Total Suspended Solids (TSS) content of the percolating water could increase.

When the soil or sediment cover is removed in the process of quarrying, their filtering and attenuation capabilities would have been lost, thus exposing the groundwater to greater risks of pollution. Groundwater could also be contaminated due to some ancillary activities associated with quarrying, such as, accidental spillages of fuel oil, leakages from storage tanks or toilets for workers, or draining of water from the surrounding areas into the quarry, etc.

Solid wastes arising from the mining of coal, lignite, metallic sulfides, uranium, etc. tend to contain pyrite (FeS_2). Under oxidizing conditions, and in the presence of catalytic bacteria, such as *Thiobacillus ferrooxidans*, pyrite gets oxidized into sulfuric acid and iron sulfate. Thus, surface runoff and groundwater seepages associated with waste piles tend to be highly acidic, and corrosive, and contain high concentrations of iron, aluminum, manganese, copper, lead, nickel and zinc, etc. in solution and suspension. The discharge of such waters (known as Acid Mine Drainage or AMD) into streams destroys the aquatic life, and the stream water is rendered non-potable. The ubiquitous gangue minerals, such as calcite and quartz, are less soluble and reactive (ways and means of ameliorating AMD have been discussed in detail under Section 3).

2.4.3.3 Solid wastes from household, commercial and industrial sources

Household solid wastes are largely composed of biodegradable, putrescible matter. The process of biodegradation is accompanied by a rise in the initial temperature within the waste mass, and the generation of carbon dioxide and methane gases (methane spontaneously catches fire, and billowing smoke is a common sight in the municipal waste piles; instances are known of people dying due to methane poisoning, when they enter chambers where organic matter is undergoing putrefaction). The content of Total Organic Carbon (TOC) tends to be very high in the leachates from the domestic wastes. More than 80% of TOC is in the form of volatile fatty acids (acetic, butyric, etc.) which are responsible for the characteristic odor of municipal garbage dumps. In due course, the organic carbon content will change over into higher molecular weight substances, such as carbohydrates. The rate of change depends upon temperature and moisture content. The change may take place in a matter of 5–10 years, in the case of humid, temperate climates, and probably in a shorter period in the case of humid, tropical regions. In arid regions, bacterial degradation may be impeded due to lack of moisture.

The generally high sulfate, chloride and ammonia concentrations in the leachates from the household wastes, reach their peak concentrations within a year or two of disposal of wastes, and then steadily decrease over a long period of time (of the order of decades). There may be seasonal fluctuations in the concentrations of these components as a consequence of the changes in ambient temperature and infiltration rates.

Heavy metals (such as, lead, cadmium, mercury and arsenic) may be immobilized, either by getting adsorbed on the organic matter within the fill, or precipitated as sulfides. Municipal garbage invariably contains plenty of cellulose derived from waste paper. The halogenated hydrocarbons may get adsorbed on cellulose and then evaporated or degraded. Most municipalities co-dispose domestic wastes, with commercial wastes. The two kinds of wastes may not differ greatly in content, except that the commercial wastes may contains oils, phenols and hydrocarbon solvents which may not be biodegradable.

By virtue of their dissolved constituents and high BOD, the leachates from domestic and commercial wastes could degrade the aquifers by producing anoxic conditions.

In February 2000, leachates from the cyanide wastes of an Australian-operated gold mining company in Romania entered the Danube river through the Tisza (a tributary), and caused an ecological disaster. For several tens of km of the stretch of the Danube river in Hungary, and Yugoslavia, hundreds of tonnes of dead fish were found floating in the waters, and the birds which ate the fish also died.

Fly ash is reactive because of its high surface area: volume ratio. Leaching of the fly ash may produce effluents containing toxic elements, such as, Mo, F, Se, B and As. Low pH leachates from fly ash may give rise to problems of iron floc formation in surface waters. When the flue gases are scrubbed, the resulting sludge will typically contain cyanide and heavy metals. Its pH will be low, unless neutralized by lime. It has been reported that mixtures of sludge, lime and fly ash will set rapidly to a load-bearing, low-permeability, solid which is not easily leachable. Two benefits accrue from this process – on one hand we will have a useful construction material, and on the other, we would have minimized the pollution risk.

An environmentally-sound, and technoeconomically viable approach to minimize the contamination potential of the wastes, such as fly ash and red mud, is to put them to some useful purpose soon after they are produced (vide Sec. 3).

Practitioners of most of the religions in the world bury their dead (with the exception of the Hindus who cremate their dead and immerse the ashes in flowing waters – by an extraordinary coincidence, the aboriginal Yanomami Indians in northeastern Brazil cremate their dead, and promptly disperse the ashes, though not necessarily in waters!). Drainage water from the graveyards may contain organic residues and pathogens, and could contaminate the surface water and groundwater with which they may come into contact. The realization of such a risk led to legislation in mid-nineteenth century England, prohibiting the construction of shallow wells within 200 m of the graveyards. It can happen that due to rapid growth of a town, a cemetery which was initially located in the outskirts of the town, may now be located right in the heart of the town, and cannot be shifted due to religious sensibilities. It is critically important that no shallow borewells be allowed to be constructed within 500 m of a cemetery. If such wells already exist and their use is unavoidable, the water should be regularly be monitored for pathogens and organic residues, before being allowed to be used.

The solid wastes produced in household, commercial and industrial activities which could contaminate water resources through leaching and effluent production, are listed in Table 2.4.1 (source: Laconte & Haimes 1982).

2.4.4 POLLUTION FROM LIQUID WASTES, SOURCE-WISE

2.4.4.1 Agriculture and animal husbandry

The large volumes of liquid and semi-solid faecal matter produced by the animals could readily infiltrate to the water-table, and seriously impair the quality of groundwater. Where domestic water supplies are drawn from shallow wells, or boreholes located near the animal farms, the contaminated water could cause serious health problems.

Because of their extremely high BOD (Biological Oxygen Demand) values and the presence of organic contaminants arising from fermentation, the farmyard slurries and silage effluents are capable of causing intense groundwater contamination, including the onset of anoxic conditions. Consequently, special precautions need to be taken to protect the water resources from contamination, such as:

> Covering the wells to protect the ingress of contaminants from above, and provision of casings and linings, or sills and impermeable aprons around the wellhead. If these steps are found to be inadequate, recourse has to be taken to either collect and treat the wastes, or to confine the wastes hermetically through the use of water-impermeable geomembranes.

Table 2.4.2 (source: Laconte & Haimes, 1982) describes the contaminant effluents from agriculture and animal husbandry activities.

Table 2.4.1 Solid waste production from household, commercial and industrial sectors

Source	*Potential characteristics of leachate*	*Rate of solid waste production*
Household wastes	High sulfate, chloride, ammonia, BOD, TOC, and suspended solids from fresh wastes. Bacterial impurities. In humid climates leachate composition changes with time. Initial TOC mainly volatile, fatty acids (acetic, butyric, proprionic), subsequently changing to high molecular weight organics (humic substances, carbohydrates). Period of change, ~5 to 10 yrs after deposition of wastes in humid, temperate regions.	0.2–0.4 t yr^{-1} $person^{-1}$. Characteristic landfill size: 10^4–10^8 m^3. Rate of leachate production dependent on climate.
Commercial wastes	Similar to domestic wastes. may also include phenols, mineral oil, and hydrocarbon solvent wastes.	Co-disposal with domestic wastes.
Industrial wastes	Variable. May contain toxic substances – heavy metals, oils, phenols, solvents, pesticide/herbicide residues.	0.3 t yr^{-1} $person^{-1}$ in industrialized societies; Industrial landfills: 10^4–10^6 m^3. May be co-deposited with domestic, commercial wastes.
Power generation (thermal)	Pulverized fuel ash. Upto 2% by weight of soluble constituents, sulfate. May contain concentrations of Germanium and Selenium. Fly ash and flue gas scrubber sludges. Finely particulate, containing disseminated heavy metals. Sludges of low pH unless neutralized by lime addition.	10^4–10^6 t yr^{-1}.

Source: "Water Resources and Land Use Planning" 1982

Table 2.4.2 Contaminants resulting from effluents related to agriculture and animal husbandry

Source	*Potential characteristics of leachate/effluent*	*Rate of effluent or solid waste production*
Intensive units, livestock yards	Effluent formed by washing down, diluting faeces and urine, 3 to 10 times. SS, BOD and N reduced. Chloride – 200–400 mg l^{-1}	Cattle units: 10^3–10^5 m^3 yr^{-1} Pig units: 10^3–10^4 m^3 yr^{-1} Poultry units: 10^4–10^5 m^3 yr^{-1}
Washing/drainage from farm buildings and yards	High suspended solids, high organic content, high BOD, Mineral oil from machinery. Fecal bacteria	Variable quality (10^3 $m^3 yr^{-1}$) washing wastes/head of livestock
Silage	High suspended solids, BOD $1–6 \times 10^4$ mg l^{-1}. Organic components – carbohydrates, phenols	0.3 m^3 yr^{-1} t^{-1} 10^2–10^3 m^3 yr^{-1} $livestock^{-1}$

Source: "Water Resources and Land Use Planning"

2.4.4.2 Liquid wastes from mining

Drainage waters from coal collieries tend to have high suspended and dissolved solids of iron and sulfate (derived from the oxidation of sulfates), and chlorides (derived from connate water trapped within the sedimentary rocks). Discharge of such waters on the surface and their subsequent percolation could contaminate the groundwater seriously (in the early part of the last century, the discharge of mine drainage water severely contaminated about 13 km^2 of Chalk aquifer in southern England). The drainage waters from metallic mines tend to be acidic, and have higher concentrations of dissolved metals. The drainage waters may also contain organic flocculents used in the screening and dressing of metallic ores.

Oil deposits are often associated with hot brines carrying traces of hydrocarbons. In the early phases of development of the hydrocarbons, it is not uncommon for the hydrocarbon/water systems to be under artesian conditions. In such a situation, the contaminated water may spill on the ground and percolate into shallow aquifers, or it could leak upwards into incompletely grouted production well. When hydraulic mining is employed to mine evaporite deposits, care should be taken to ensure that the brines do not contaminate the groundwater through surface spills and pipeline leakages.

The heavy withdrawal of groundwater in the coastal regions and estuaries could lead to the incursion of saline water into the coastal freshwater aquifers, thereby degrading them. There have been several instances of giant tidal waves generated by tropical cyclones, salinizing the arable land, surface water and the groundwater (On Oct. 29, 1999, a 10 m high tidal wave generated by a super-cyclone swept a stretch of 150 km. along the coast of Orissa province in eastern India, destroying every thing in its path, and rendering the surface and groundwaters saline – vide Chap. 6.3).

Mineralized waters may sometimes occur at depth – in the form of connate water trapped below the zone of natural groundwater circulation, or they may arise from the leaching of evaporite beds terminating against an aquifer. Salinization of fresh groundwater can occur in inland areas due to upconing of such mineralized waters.

Overpumping of the groundwater may result in the lowering of the water-table below the stream bed levels. If the river concerned is perennial or seasonally influent, and if the river water is already contaminated, this would inevitably induce undesirable recharge of the aquifer with the contaminated water of the river.

Table 2.4.3 (source: Laconte & Haimes, 1982) describes the contaminant effluents from mining activities.

2.4.4.3 Liquid wastes from household, commercial and industrial sources

The potential of the human sewage to contaminate the groundwater arises out of its high BOD, suspended solids, faecal bacteria and viruses, chloride and ammonia. Most of these are removed when the sewage is subjected to Primary (mechanical) treatment, and Secondary (biological treatment). But some anions persist in the final effluent (typically in the ranges of 100–200 $mg\,l^{-1}$ for Cl, and 30–40 $mg\,l^{-1}$ for NO_3^-–N). Where an aquifer is contaminated by such treated effluents, the movement of the contaminant plume could be readily monitored on the basis of increased chloride concentrations.

Table 2.4.3 Contaminants resulting from liquid wastes from mining

Source	*Potential characteristics of leachate/effluent*	*Rate of effluent/leachate production*
Oil and gas well brines	High total solids (10^3–10^5 mg l^{-1}), High Ca^{2+} and Mg^{2+} (10^3–10^5 mg l^{-1}), High Na^+ and K^+ (~10^4 mg l^{-1}), High Cl^- (10^4–10^5 mg l^{-1}), High SO_4^{2-} (10–10^3 mg l^{-1}), Oil, upto 10^3 mg l^{-1}, Possibly high temp.	10^3–10^4 m^3 d^{-1} per well
Saline intrusions, due to overpumping close to coastlines	Na^+(10^3–10^4 mg l^{-1}), Mg^{2+} (10^2–10^3 mg l^{-1}), Ca^{2+} (10^2 mg l^{-1}), K^+ (10-10^2 mg l^{-1}), Cl^- (10^3–10^4 mg l^{-1}), SO_4^{2-} (10^2–10^3 mg l^{-1}), Alkalinity (as $CaCO_3$) 10^2 mg l^{-1}	Rate of landward movement of saline incursion varies with pumping regime and aquifer type (example: 4 km in 40 yr along the estuary of River Thames in England)

Source: "Water Resources and Land Use Planning"

In some countries, the mains which carry the sewage water also carry storm water washed from paved surfaces. The storm water generally contains considerable particulate matter of vegetable, animal and mineral origin. In the cold countries, common salt is used for de-icing of roads, and consequently the storm water may contain seasonally high concentrations of sodium chloride. Contamination due to automobile exhausts and accidental oil spills may lead to immiscible layers and emulsions containing hydrocarbons and organo-metallic complexes, such as tetraethyl lead. Bacterial contamination may be high initially, but intense flushing due to large quantities of storm water will quickly dissipate the pollution.

When wastewater is treated in the sewage works, some quantity of sludge is produced. The sludge typically contains 4 to 7% solids, with organic residues constituting more than half of the solids. Sewage sludge is used in agriculture as a fertilizer and soil conditioner. It has been observed that sludge application reduces the surface runoff, and gives some protection against soil erosion.

The composition of sewage sludge varies greatly, depending on the source of sewage (domestic, industrial, etc.). The mean concentration level and range (percent content of dry matter) of the fertilizer content are as follows: N – 2.2 (1.5–3.5), P – 1.7 (1.0–2.5), K – 0.15 (0.05–0.3). The undesirable consequences of sludge application arise from the heavy metal content of the sludge, which may be taken up by the plants.

The metals present in sewage sludge may be divided into three categories on the basis of their availability to plants:

- Unavailable forms, such as insoluble compounds (oxides or sulfides),
- Potentially available forms, such as insoluble complexes, metals linked to ligands, or forms attached to clays and organic matter, and
- Mobile and available forms, such as hydrated ions or soluble complexes (Peter O'Neill 1985, p. 207). A soil which has high pH (i.e. alkaline) and high cation exchange capacity, will be able to immobilize the metals added to it via the sewage

sludge. A soil which is presently alkaline may not always remain that way. Thus, if at a later stage the pH drops (i.e. becomes acidic), the metals may get released.

Elements such as iron, zinc and copper are essential elements needed by man. The same elements may become toxic at high concentrations, however. Cadmium is a highly toxic element. Because of its geochemical affinity with zinc, it enters the plants along with zinc.

In a study made in Norway, sludge application at rates of 60–120 t (dry matter) per ha increased the Ni and Zn contents of plants, but had no significant effect on the concentration levels of Cd and Pb. Excessive application of sludge could increase the content of heavy metals in the crops to levels which render them unfit for human consumption.

Liming (1.5–6 t of $CaCO_3$ ha^{-1}) reduces the heavy metal uptake by plants *in the short run*. The heavy metals will continue to persist in the top soil, however.

It has been reported that sludge-amended soils almost invariably contain higher levels of Cd than garden soils. Potential increase in the cadmium content of plants due to the application of sewage sludge constitutes the most important health hazard, which can be mitigated by increasing the available content of zinc in the soil. Sludge should not be applied to soils which are used for growing vegetables.

Toxic, liquid wastes are some times disposed by deep well injection (In 1962, in the Denver Basin, USA, the forceful injection of toxic, fluid wastes into a borehole triggered low-magnitude earthquakes, but that is a different story). The following precautions need to be taken to ensure that the injection of toxic wastes does not contaminate the overlying aquifer:

- Impermeable, confining layers exist above and/or beneath the layer into which the wastes are injected,
- The injection borehole should be so constructed that there is no possibility of leakage of wastes around the casings,
- The injection zone and local aquifers should be regularly monitored,
- Any abandoned oil or gas or water supply or exploration boreholes present in the vicinity of the injection wells, should be sealed, and
- The injection pressures should not be so high as to lead to hydraulic fracturing of the confining beds.

There may be accidental spills when liquid wastes stored in tanks are transported by road or rail. Also, there may be leakages when the wastes are transported by pipeline. Such spills or leakages could contaminate the groundwater, particularly in the case of shallow water table aquifers. The magnitude of such contamination may have an enormous range – while the leakage of a few cubic metres of oil from a domestic tank may contaminate a water well nearby, an undetected leak of several thousand cubic metres of oil from a pipeline could jeopardize a whole aquifer.

2.4.5 CONTAMINANTS, TYPE-WISE

The presence of certain contaminants in water supplies can cause health risks to humans and animals. The effects of contaminants vary greatly depending upon the general

Table 2.4.4 Characteristics of the waste effluents from different industries

Industries	*Characteristics of the waste effluents*
Food and drink	High in BOD and suspended matter; taste and odor problems
Textile	Characteristically alkaline
Tannery	High concentration of dissolved chlorides, sulfides and chromium
Petrochemicals	High BOD, toxic sulfur compounds and phenols
Metallurgical & metal finishing	Characteristically acid, with high suspended solids; metal finishing wastes additionally contain heavy metals, phenols and oils
Thermal power production (cooling water)	Suspended matter in cooling water will not increase as it does not come into contact with any particulate matter during circulation. It may, however, get reduced in volume due to evaporation, and raised in temperature by about 10°C. Consequently, the concentration of dissolved constituents will increase four-fold, and carbonate minerals may be partly precipitated due to loss of dissolved carbon oxide because of increasing temperature. Loss of dissolved oxygen may also occur. The disposal of such oxygen-depleted, heated water to the ground may have the following consequences: – soil or aquifer material may be leached, – when the warm water cools down as a result of contact with cooler, less mineralized groundwater, some components may be reprecipitated within the rock pores, thus reducing the permeability of the rock, and sequestering the recharged waters within an oxygen depleted zone. If the heated water is recharged via boreholes, microbial growth may take place, and clog the wall screens
Production of gas by coal distillation (by, say, horizontal retort method)	Crude tar oils with significant content of phenolic compounds are produced. When hydrated ferric oxide is used to purify the raw gas, it may be contaminated with sulfides, free sulfur and cyanides. Impurities in the gas itself may dissolve in water. Leaching of these wastes could contaminate the groundwater with phenols and cyanide

Source: Laconte & Haimes 1982

state of health, age, diet, body weight, genetic factors, etc. of the individuals. Consequently, it can happen that different members of a community drinking water from (say) the same contaminated source, may be affected differently. For convenience, the contaminants may be classified type-wise, as chemical, biological and radioactive.

As should be expected, the composition of the industrial effluents depends upon the industrial process involved (Table 2.4.4).

2.4.5.1 Chemical contaminants

In 1984, the World Health Organization (WHO) has prescribed norms for the quality of potable (i.e. drinking) water, and these have been updated in 1993. The Federal Drinking Water Standards, prescribed by the US Environmental Protection Agency (EPA) in 1988), are widely used all over the world. US EPA has elaborate procedures for continuously updating the figures for the Maximum Contaminant Levels (MCLs) and their enforceability for various components (inorganic, pesticides, and radioactive contaminants) on the basis of laboratory investigations, etiological and epidemiological studies and consultations with the authorities, stakeholders and interest groups (for instance, US EPA has organized a consultation in San Diego in June, 2000, to examine

Table 2.4.5 Quantum of production of contaminants from industrial and miscellaneous sources

Food and drink manufacturing	High BOD. Suspended solids often high. Colloidal and dissolved organic substances, odors	10^3–10^7 m^3/yr
Textiles and clothing	High Suspended solids. High BOD. Alkaline effluent	10^4–10^6 m^3/yr
Tanneries	High BOD, total solids, hardness, chlorides, sulfides, chromium	10^3–10^6 m^3/yr
Chemicals		
Acids	Low pH	10^5–10^9 m^3/yr
Detergents	High BOD, saponified residues, low pH, high organic acids	
Explosives	Low pH, High organic acids, alcohols, oils	
Insecticides/ Herbicides	High TOC, toxic benzene derivatives, low pH	
Synthetic resins and fibers	High BOD	
Petroleum and petrochemical	– refining High BOD, chloride, phenols, sulfur comp. – process High BOD, Suspended solids, chlorides, variable pH	10^6–10^8 m^3/yr
Thermal power	Increased water temperature. Slight increase in dissolved solids by evaporation of cooling wastes	10^3–10^4 m^3/yr/ megawatt
Engineering works	High suspended solids, soluble cutting oils, trace heavy metals, variable BOD, pH	10^4–10^7 m^3/yr
Foundries	Low pH, High suspended solids, phenols, oil	10^7–10^9 m^3/yr
Plating and metal finishing	Low pH, High content of toxic heavy metals, sometimes as sludge	10^7–10^9 m^3/yr
Deep well injection	Various concentrated liquid wastes, often toxic. Brines. Acid and alkaline wastes. Organic wastes	10^4–10^6 m^3/yr
Leakage from storage tanks and pipelines	Aqueous solutions, hydrocarbons, petrochemicals, sewage	
Accidental spillages	Various liquids in transit, hydrocarbons, petro-chemicals, acids, alkalis, solvents. Liquids may enter surface drains or soakaways	Generally 10 m^3 per incident

Source: Paper 1.1 in Laconte & Haimes 1982

the need for, and implications of, the downward revision of the MCL for arsenic from its current level of 50 μg l $^{-1}$).

Pesticides (e.g. herbicides, insecticides, fungicides, nematocides, etc.) are an important group of chemicals which could contaminate groundwater.

Pesticides are widely used in agriculture, public health, and several other purposes. There are more than 10,000 formulations of pesticides, of which about 450 are widely used. The extensive use of pesticides during the last fifty years not only led to their becoming ubiquitous pollutants in soils, crops, groundwater, human and animal tissues, etc. in industrialized countries, but also led to their dispersal to far-off places; pesticides residues have been found in Antarctic penguins. Phreatic or unconfined aquifers (i.e. aquifers extending to the earth's surface) are most vulnerable to pesticide pollution.

The main types of compounds used as pesticides are as follows (Manahan 1991):

Insecticides (insect killers):
- Organochlorines (e.g. DDT, Lindane, Aldrin, Heptachlor)
- Organophosphates (e.g. Parathion, Malathion)
- Carbamates (e.g. Carbaryl, Carbofuran)

Herbicides (weed killers):
- Phenoxyacetic acids (e.g. 2,4-D; 2,4,5-T; MCPA)
- Toluidines (e.g. Trifluralin)
- Triazines (e.g. Simazine, Atrazine)
- Phenylureas (e.g. Fenuron, Isoproturon)
- Bipyridyls (e.g. Diquat, Paraquat)
- Glycines (e.g. Glyphosate "Tumbleweed")
- Phenoxypropionates (e.g. "Mecoprop")
- Translocated carbamates (e.g. Barban, Asulam)
- Hydroxy nitriles (e.g. Ioxynil, Bromoxydynil)

Fungicides (fungus killers):
- Nonsystemic fungicides
  - Inorganic and heavy metal compounds (e.g. Bordeaux Mixture – Cu)
  - Dithiocarbamates (e.g. Maneb, Zineb, Mancozeb)
  - Pthalimides (e.g. Captan, Captafol, Dichofluanid)
- Systemic fungicides
  - Antibiotics (e.g. Cycloheximide, Blasticidin S, Kasugamycin)
  - Benzimidazoles (e.g. Carbendazim, Benomyl, Thiabendazole)
  - Pyrimidines (e.g. Ethirimol, Triforine)

As evidence mounted on the toxic effects of pesticides on human beings and animals, some pesticides, such as DDT, Aldrin and Dieldrin were completely banned in the Industrialized countries. Though the ban was effected in the 1980s, these pesticides continue to persist in the environment to varying extents:

Pesticide	*% decay*	*Time taken (years)*
DDT	44	8
BHC	50	1
Dieldrin	49–53	3

In the place of the banned pesticides, substitutes such as pyrethroids and organophosphorus compounds, have been introduced in developed countries. These substitutes are environmentally more acceptable, though more expensive. For instance, pyrethrins (derived from the flower of the plant, *Chrysanthemum cinerariaefolium*), undergo rapid photolytically induced oxidation, and are therefore short-lived. Several synthetic analogues of pyrethrin have been developed. Intensive work is underway to develop a whole range of pesticides, based on the seeds of the Indian tree, *neem* (*Azirdirachta Indica*). Some of the organophosphorus pesticides, such as Chlorpyrifos, are highly effective against pests, but less toxic to mammals. They are applied in relatively small quantities (200–1,200 g ha^{-1}). They break down by hydrolysis and their persistence in the soil is in the range of 60–120 days.

Soil processes have great ability to decompose organic matter. The enzymatic activities of the microorganisms are capable of degrading and detoxifying a variety of substances added to the soil. However, the chemical structures of some pesticides are such as to preclude their enzymatic degradation. Several synthetic pesticides (such as 2,4,5-T or 2,4,5-trichlorophenoxyacetic acid) are heavily substituted with chlorine, bromine, fluorine or nitro or sulfonate groups which are not found in biological tissues. Microorganisms cannot degrade such pesticides, which therefore persist in the soil for long periods (say, about 15 years).

Those pesticides characterized by low vapor pressure and low solubility in water tend to persist in the soil. The extent of persistence depends upon the temperature, soil type, and soil microbiology. For instance, the herbicide, trifluralin, may persist in the soil between 15 to 30 weeks. Herbicides tend to be lost more readily from moist soil than from the same soil when it is dry. The molecules of herbicides that are strongly adsorbed on dry soils get dislodged under damp conditions.

Herbicides interact with soil components in complex ways. Urea derivatives, triazines, carbamates and nitrophenyl ethers, etc. tend to be strongly adsorbed on soil organic matter. Quaternary ammonium compounds are adsorbed on clays and metal hydroxides. The herbicides may also interact with the exudates from soil microorganisms.

The probable range of persistence (in weeks) of some important herbicides has been given by Hassal (1987, p. 253). Some herbicides, such as, Propham, Dalapon and Aminotriazole, have shorter persistence periods of 3–10 weeks, whereas some others, such as, Diuron and Simazine, have much longer persistence periods of 30–120 weeks. The persistence is greatly influenced by the rate of application, rainfall, soil type, temperature and microbial population. Besides, some crops are more tolerant than others for the same herbicide.

The soil type not only influences the period of persistence of a herbicide, but also affects the toxicity of a given dose to the weeds. For instance, carbamate and urea herbicides are more toxic to the weeds in light soils than to the same weeds in heavy organic soils. Consequently, these herbicides persist in organic soils for longer periods.

2.4.5.2 Biological contaminants

Waste waters from urban areas contain a variety of faecally-excreted human pathogens, such as helminths, protozoans, bacteria and viruses. The concentration of pathogens in human faeces is high (10^4–10^7 organisms per gm. of faeces), and they have persistence times ranging from weeks to months.

Bacteria and viruses transmissible by groundwater, may cause a variety of diseases, such as, typhoid/paratyphoid, enteric fever, cholera, bacterial dysentery, amoebic dysentery, non-specific gastroenteritis, leptospiral jaundice (Weil's disease), poliomyelitis, infective hepatitis, etc.

The biological contamination of groundwater may be caused by

- Leaching of deposits of solid faecal material,
- Spray irrigation with sewage water, and
- Migration of domestic sewage from septic tank or pit latrine.

Though aesthetically objectionable, the presence of live algae in water does not cause disease. But water with dead algae is unacceptable, as they release toxins. Bacteria are unicellular organisms, with sizes of the order of 0.5 to 10 μm. Thus, they are similar in size to clay particles. The mobility of bacteria in aquifers may be constrained by their ability to move through pore spaces in rocks. Viruses are smaller than bacteria, with sizes of the order of 20 to 200 nm. They are incapable of reproduction outside a host organisms, and they have no metabolic functions. Since viruses tend to be negatively charged, they can only be adsorbed at positively charged sites. The adsorption of viruses reaches its peak at pH 7, which is also the normal pH level of groundwater. At higher pH, the adsorption of viruses decreases because of the lower availability of positively charged sites.

The maximum length of travel of the biological pollutants in aquifer sediments is of the order of 20 to 30 m.

2.4.5.3 Radioactive contaminants

The following six nuclides are of particular hydrological concern because of their toxicity, environmental mobility and relatively long half-life: ^{3}H, ^{90}Sr, ^{129}I, ^{137}Cs, ^{226}Ra, and ^{239}Pu. Uraniferous rocks contain ^{226}Ra, and leaching of such rocks could contaminate the surface waters and groundwaters. ^{226}Ra is also released during the processing of the uranium ores. The rest of the nuclides are produced in the course of nuclear power generation, or during nuclear accidents (see Aswathanarayana 1995, pp. 162–165, for a summary account of the Chernobyl Reactor Accident of April 26, 1986). They may enter the groundwater systems due to accidental escape or intentional release. A variety of radioactive tracers (both natural and man-made) are used to estimate the transport and attenuation of inorganic contaminants in the groundwater flow systems and age-dating of groundwaters.

Groundwater is depleted in some isotopes because of adsorption in the soil. For instance, ^{137}Cs is strongly retained in the soil, and very little of it goes into the groundwater. In contrast, ^{90}Sr is not bound in the soil, and is therefore passed on into groundwater. As groundwater is an important source of drinking water in many communities, the extent of infiltration of radioisotopes in the soil assumes great importance. The extent of lime-lag between the time of emission of a radionuclide into the atmosphere, and the time it reaches the hydrosphere, could be illustrated with an example. For instance, ^{131}I in the hydrosphere due to Chernobyl has reached the peak level of 1.1 k Bq l^{-1} on May 3, 1986 (i.e. one week after the accident).

The organ in which a given radionuclide accumulates preferentially, is called the critical organ. For instance, bone is the critical organ for ^{90}Sr, because of the geochemical coherence of Sr with Ca in the bone. As ^{3}H and ^{137}Cs are transported by the blood stream, the whole body constitutes the critical organ for these isotopes.

2.4.6 ANTHROPOGENIC ACIDIFICATION OF WATERS

2.4.6.1 Acid rain

Combustion of fossil fuels (coal, oil, natural gas, etc.) for power generation, transport and other industries, produces the emissions of NO_x and SO_x. These gases get oxidized in the atmosphere and precipitate as dilute nitric and sulfuric acid solutions, generally known as Acid Rain.

Acid rain degrades the quality of air, soils, surface and groundwaters, biota, etc. It has emerged as a serious environmental problem in the industrialized countries. For instance, the pH of the rainwater in the Netherlands decreased by about one unit – from 5.4 in 1938 to 4.52 in 1980, due to increase in the concentrations of SO_4^{2-} and NO_3^- (Appelo 1985). As a consequence of evapotranspiration, the acidity of the solution that enters the soil goes down to a pH to about 3. Appelo et al. (1982) reported that the youngest groundwater got acidified down to a pH of about 4 in the carbonate-free sandy aquifer in the Veluwe area of the Netherlands.

The lowering of the pH down to 3, is a consequence of the oxidation of the higher concentration of ammonia in the 1980 rainwater, by the following reaction:

$$NH_4^+ + 2O_2 \rightarrow NO_3^- + 2H^+ + H_2O \qquad (2.4.1)$$

Ammonia and manure are used as fertilizers in large amounts. When ammonia is oxidized by oxygen, nitrate is produced by the following reaction:

$$NH_3 + 2O_2 \rightarrow NO_3^- + H^+ + H_2O \qquad (2.4.2)$$

Plants use the nitrate thus produced, and proton production is balanced by the production of HCO_3^- in the denitrification process, by the following reaction:

$$5CH_2O + 4NO_3^- \rightarrow 2N_2 + 4HCO_3^- + CO_2 + 3H_2O \qquad (2.4.3)$$

Another source of acidification of waters is the reduction of pyrites, FeS_2. Reducing sediments such as those with organic matter, invariably contain some amount of pyrite. When the watertable in a well goes down due to pumping, pyrite that might be present in the sediments may get oxidized to sulfate by the following reaction:

$$2FeS_2 + 15/2\,O_2 + 5H_2O \rightarrow 2FeOOH + 4SO_4^{2-} + 8H^+ \qquad (2.4.4)$$

The oxidation of pyrite is one of the most strongly acid-producing reactions in nature.

Pyrite is a ubiquitous mineral in mine dumps of not only coal but also of sulfide ores. Acid Mine Drainage also involves the oxidation of pyrite.

2.4.6.2 Susceptibility of rocks to acidification

Susceptibility of rocks to acidification is dependent upon the rate of dissolution of the constituent minerals. Aquifers of limestones, dolomites and other rocks containing carbonate minerals are unlikely to develop acid groundwater, because of the rapid dissolution of the carbonate minerals. Quartzose rocks like granites and granodiorites, clean sandstones and quartzites which contain slowly-dissolving minerals, are most susceptible to acidification. Basic and ultrabasic rocks which contain more quickly dissolving silicates are less susceptible to acidification.

Aluminium dominates the buffering processes in the aquifers. Al is an oxyphile element, and always occurs in nature as oxides and hydroxides, such as gibbsite,

$Al(OH)_3$ and clay minerals. When acid waters come into contact with a soil or an aquifer, they interact with gibbsite releasing Al as per the following equation:

$$Al(OH)_3 \rightarrow Al^{3+} + 3OH^- \qquad (2.4.5)$$

2.4.7 WATER POLLUTION ARISING FROM WASTE DISPOSAL

The relevance of waste disposal to water resources management arises from the possibility that leachates from landfill sites could contaminate water and resources.

2.4.7.1 What is waste disposal?

Some authorities differentiate between the terms, waste deposit and waste disposal. They hold that "Disposal" has a wider meaning and greater finality than "Deposit". In other words, every deposit on land does not constitute disposal. If the purpose of a deposit is not disposal, the operation will be simply a deposit and not a disposal.

Wastes arise from municipal, industrial, agricultural, mining, demolition, sewage sludge, etc. sources. Hazardous wastes are special wastes which are difficult to handle or cause harm to human health or the environment. They may be inflammable, corrosive, toxic, reactive, carcinogenic, infectious, irritant, harmful, or ecotoxic.

The issues of waste disposal have been comprehensively dealt with by Bagchi (1990) and Petts & Eduljee (1994). The ways in which the treatment and disposal of wastes have impacts on the flows and quality of surface water and groundwater are shown in Tables 2.4.6 and 2.4.7 (Petts & Eduljee, 1994, p. 167).

A desk study should precede field investigations. The sources of information are maps, archival data about the surveys made earlier, research studies and monitoring information available from relevant statutory authorities. The following account lists the relevant areas of information (Petts and Eduljee, 1994, p. 172):

1. Land-use: current land-uses including identification of possible polluting industries; aquatic habitats that need to be conserved; nature and extent of use of water in the catchment area,
2. Topography: relief of the area, based on topographic maps, aerial photographs or space photographs; patterns and flows of surface water drainage; kind and depth of soils,
3. Geological setting: stratigraphic sequence; thickness and lateral extent of layers; geological structures and their geometries; mineral resources, if any, and the extent of their present and projected utilization,
4. Hydrogeology and water supply: Aquifers – their nature, lithology, porosity and permeability, depth of the water-table, direction of flow; sources of recharge; water quality; present level of abstractions; site flood potential; availability of sewage treatment, etc.

Time was when the guiding philosophy of waste disposal was simply, "Out of sight, Out of Mind". A well-known case is that of Love Canal, near Niagara Falls, USA. The company which produced the chemical wastes, simply covered them up with soil, and

Table 2.4.6 Sources and effects of waste treatment and disposal upon surface water

Phase	*Source of effect on run-off and flows*	*Source of effect on water quality*
Construction	Soil compaction by heavy vehicles, Provision of temporary drainage, Earth works, Elimination of on-site depression storage, Diversion of surface waters.	Earthworks, Washing effluents from construction vehicles, Leakage or spillage of oils, hydraulic fluids, Accidental release of surface run-off.
Operation	Uncontrolled discharge of surface water, Leachate level break-out, Provision of artificial ground surface, Provision of engineered drainage systems, Raised landfill areas, Removal of vegetated areas.	Accidental release of surface run-off, Spillage from transport accidents, Discharge of effluents, Leachate level break-out to surface, Leachate leakage to groundwater, and then to river, Spillage of chemicals, etc. from unloading or filling, Leakage from storage tanks, Uncontrolled discharge from surface water, Deposition of pollutants from air.
Restoration/ closure	Provision of artificial ground surface, Engineered low permeability cap, Shape of the final landform – slope profile, Leachate level break-out.	Leachate break-out, Leachate leakage.

Source: Petts & Eduljee, 1994, p. 167

Table 2.4.7 Sources and effects of waste treatment and disposal upon groundwater

Phase	*Source of effect on groundwater level*	*Source of effect on groundwater quality*
Construction	Removal of topsoil and exposure of subsoil of lower permeability, Soil compaction	Leaching of contaminants in soils, Leakage of spillage of oils, hydraulic fluids, etc.
Operation	Structures/buildings, Provision of artificial surfaces, Provision of engineered drainage systems	Leachate leakage, Contaminated drainage from paved areas, Spillage of chemicals, etc. Leakage from storage tanks, Leakage from underground sumps or drains, Contaminated drainage from rainfall on unpaved areas
Restoration/ closure	Provision of artificial ground surface	Leachate leakage

Source: Petts & Eduljee, 1994, p. 167

left. Later, in 1978, land developers built houses on the disposal site. It was found that the community suffered bouts of ill health, and some children were born with birth defects. These were traced to leachates and toxic vapors from the chemical wastes, contaminating the groundwater and indoor air.

Till about 1950s, wastes were deemed to be unwanted, useless material, which had to be got rid of at the lowest cost. By 1990's, the position changed drastically. Waste management, involving the production, handling, storage, transport, processing, treatment and ultimate disposal of the wastes, has become a big industry (for instance, good part of the multi-billion dollar budget of US EPA is devoted to the clearance of Superfund sites). The objective of the waste disposal is to minimize the environmental impact and maximize the resource recovery.

Presently, the basic paradigm in regard to waste management involves the following approaches: Prevention of waste (closing of cycles), Maximum recycling and reuse of material, Safe disposal of any waste which cannot be reused through combustion as fuel, incineration and landfill. Landfill is the most preferred option for waste disposal. Landfill is described as "engineered deposit of wastes onto and into land, with deposit usually taking place predominantly below the ground surface in voids, which have often been formed by mineral extraction or quarrying" (Petts & Eduljee, 1994, p. 25).

Prior to disposal, wastes are treated to recover materials and/or energy content of the wastes, and to convert the wastes to a form which permit safe and efficient disposal. The treatments are custom-made, depending upon the nature of the waste, on the basis of the Best Practicable Treatment Option (BEPO) (summarized from Salcedo et al., 1989):

- *Chemical treatment*: Neutralization, precipitation, ion exchange, oxidation/reduction, solidification,
- *Physical treatment*: Screening, sedimentation, flotation, filtration, centrifugation, reverse osmosis, ultrafiltration, distillation/stream stripping, adsorption,
- *Biological treatment*: Activated sludge, aerated lagoons, composting, anaerobic digestion,
- *Thermal treatment*: Wet air oxidation, incineration, vitrification.

Such treatments benefit waste disposal through steps, such as reduction in volume (e.g. wastes to be landfilled can be reduced by 80–90% through incineration), reduction in toxicity (e.g. conversion of cyanide wastes to non-toxic carbon dioxide and water), change in the physical form (e.g. removal of heavy metals in liquid wastes through precipitation), etc.

2.4.7.2 Leachate management

Leachate is produced when water (rainfall, snow, surface or groundwater intrusion, or water in the waste itself) percolates through the landfill material. As should be expected, the composition of the leachate would depend upon the extent of water infiltration, the nature of the waste and the rate of its degradation, the method of operation of the waste site and the procedures of leachate management. The production of leachate can be minimized by good landfill practices, but cannot be avoided altogether. The leachate may be brown to nearly black in color, and its polluting potential can be 10–100 times that of raw sewage.

The leachate may contain the following components:

- Major ions: Ca, Mg, K, Fe, Na, ammonium, bicarbonate, sulfate and chloride,
- Trace metals: Mn, Zn, Cu, Cr, Ni, Pb, Cd,

- Organic compounds: These are usually expressed in terms of TOC, COD, or BOD. Besides, hazardous substances, such as pesticides, benzene, phenols, may be present,
- Microbiological components.

Farquhar & Rovers (1973) described the decomposition processes that occur at the waste site. The *acetogenic* leachates produced in the early stages are characterized by high organic strength, whereas in the subsequent *methanogenic* phase, landfill gases such as methane, are produced, leaving behind a poorly degradable residue.

The cardinal principle in the disposal of wastes, particularly hazardous wastes, is that they should be effectively isolated from the biosphere for a sufficient length of time, until they no longer present a risk to the biosphere. Two kinds of barriers are envisaged: the *natural* barrier of soil or rock which should prevent or keep within acceptable limits, both the flow of water into the waste or seepage of contaminated water from it, and the *technical* barrier composed of man-made material whose purpose is to seal the waste in order to reduce leaching to a minimum, and to hinder chemical reaction of the waste material with the soil or rock which could adversely affect the capacity of the natural barrier to prevent contact with the biosphere (Archer et al., 1987).

Composite impermeable barriers, involving HDPE (High-density Polyethylene) liners, and natural materials, such as bentonite, are fairly effective in preventing leachate migration. The manufacturers quote a design life of 25–40 years for the geomembranes, but it should be kept in mind that the period of leachate generation on site could extend to 100 years or more.

US EPA's HELP (Hydraulic Evaluation of Landfill Performance) model can be used to model the landfill water balance.

The operational period for a landfill barrier system cannot be precisely determined in advance. However, assuming that it is possible to eliminate groundwater intrusion and external surface drainage, an order of magnitude figure for the operational period could be estimated from the following equation (Petts & Eduljee, 1994, p. 175):

$$Q = I - E - aW \qquad (2.4.6)$$

where, Q = free leachate generated ($m^3 yr^{-1}$), I = total liquid input including liquid waste ($m^3 yr^{-1}$), E = actual evaporative losses ($m^3 yr^{-1}$), a = absorptive capacity of waste ($m^3 t^{-1}$), and W = weight of waste deposited ($t\, yr^{-1}$).

Mitigation of water impacts may be achieved through the following measures (Petts & Eduljee, 1994, p. 185):

- Minimization of leachate generation,
- Containment of leachate within the landfill,
- Control over leachate quality,
- Collection and disposal of leachate as it is generated,
- Monitoring, and
- Contingency plans, in case the groundwater has been contaminated.

2.4.8 TRANSPORT OF CONTAMINANT SOLUTES IN AQUIFERS

2.4.8.1 Movement of contaminant solutes in groundwater

The chemistry of groundwater is affected by a large number of factors, such as, physiographical, geological, physico-chemical, physical, biological, anthropogenic, etc. The processes affecting the migration of solutes in an aquifer are summarized in Table 2.4.8 (after Schvarcev, 1983, quoted by Sytchev, 1988, p. 85).

The movement of the groundwater is generally slower and less turbulent than that of the surface water. Though chemical contamination of groundwater may sometimes be caused by natural sources (e.g. calcium-rich waters in a limestone country), most of the contamination occurs due to substances produced and introduced by man. These include synthetic chemicals (such as, solvents, pesticides and hydrocarbons), landfill leachates (i.e. liquids that have leached landfill dumps, and carry dissolved substances from them), organic wastes (e.g. bacteria and viruses from night soil dumps, hospital garbage), etc.

The issues concerned with the movement of contaminants in the groundwater have been dealt with by Petts & Eduljee (1994, p. 159), Reichard et al. (1990) and Devinny et al. (1990).

When contaminated water moves through the soil column in the unsaturated zone, the contaminants present in the water tend to be removed through processes such as anaerobic deposition, filtration, ion exchange, adsorption, etc. On the other hand, the content of dissolved solids may increase as a consequence of the reaction between soil and water. The velocity of movement of the contaminant through and into the groundwater depends upon the physical and chemical characteristics of the soil, the nature of the contaminants and the flow system of the zone. It may be highly variable – ranging from nearly instantaneous to hundreds of years. If the aquifer concerned is composed of coarse-grained material like gravels, water may travel quickly, with residence times of a few days or weeks, before it is abstracted or reaches a river. On the other hand, if

Table 2.4.8 Migration of solutes in an aquifer system

Action	*Main process*
Transfer of matter	Molecular diffusion; Convective diffusion mass transfer
Removal of solutes from groundwater	Hydrolysis; Sorption and ion exchange; Precipitation of compounds with low solubility; Filtration of coarsely dispersed matter; Bioaccumulation
Mobilization of material from rock matrix	Solution and leaching; Desorption
Combination of removal and mobilization processes	Formation of complexes; Oxidation – reduction and biogeochemical reactions; Radioactive decay
Interaction with water molecules	Hydration and dehydration of material; Underground evaporation and freezing; Dilution and concentration; Membrane effects

the aquifer is composed of fine-grained material such as chalk, water will move slowly. Consequently, water may reside in such aquifers for years, or even centuries, before it is abstracted or discharged.

The manner of spreading of a pollutant in an aquifer depends upon the nature of the pollutant and the aquifer. If the pollutant concerned is soluble (e.g. chloride or sulfate salts), it will get dissolved in the groundwater, and will move as a plume along the same path and with the same velocity as groundwater. On the other hand, if the pollutant is insoluble or poorly soluble (e.g. chlorinated solvents), its behaviour will be markedly different from that of the soluble pollutant. The insoluble pollutant will tend to remain as a separate phase, and "may sink below the water table, and flow separately along low permeability layers encountered at depth in the aquifer, or be trapped by capillary forces and act as a long-term contaminant source" (Petts & Eduljee, 1994, p. 159).

In order to model the sources of groundwater pollution and to recommend remedial action, we need to know the spatial, chemical, and physical characteristics of the source, and its temporal behavior. The spatial characteristics include information on the location, depth, areal extent, and whether it is a point source (e.g. septic tank), or a line source (e.g. sewage channel). The model could be used to predict the behavior of the contaminant plume. Models are increasingly being used to trace the source(s) of the existing pollution, by the technique of backward pathline tracking. Van der Heijde (1992) gave some case histories in this regard.

Appelo & Postma (1996) gave a detailed account of how to use the geochemical model, PHREEQE/PHREEQM, developed by Parkhurst et al. (1980), to find solutions to the problems that are commonly encountered, such as, Nitrate reduction in groundwater from agricultural land by organic matter, Recovery of freshwater injected in a brackish aquifer for storage purposes, Acidification of groundwater, Salinization of freshwater by seawater incursion, etc.

REFERENCES

Appelo, C.A.J., et al. 1982. Controls of groundwater quality in the NW Veluwe catchment (in Dutch), Soil Protection Series 11, Staatssuitgeverji, Den Haag 140 p.

Appelo, C.A.J. & D. Postma 1996. *Geochemistry, Groundwater and Pollution.* Rotterdam: A.A. Balkema.

Archer, A.A., G.W. Luttig & I.I. Snezhko (eds.) 1987. *Man's dependence on the earth.* Nairobi – Paris: UNEP – UNESCO.

Aswathanarayana, U. 1995. *Geoenvironment: An Introduction.* Rotterdam: A.A. Balkema

Aswathanarayana, U. 2001. *Water Resources Management and the Environment.* Lisse: A.A. Balkema

Devinny, J.S., L.G. Everett, J.C.S. Lu & R.L. Stollar 1990. *Subsurface Migration of Hazardous Wastes.* New York: Van Nostrand Reinhold.

Farquhar, G.J. & F.A. Rovers 1973. Gas production during refuse decomposition. *Water, Air and Soil Pollution* 2: 483–495.

Hassal, K.A. 1987. *The Chemistry of Pesticides.* ELBS Ed. Basingstoke, UK: Macmillan.

Laconte, P. & Y.Y. Haimes (eds) 1982. *Water Resources and Land-use Planning: A Systems Approach.* The Hague: Martinus Nijhoff Publishers.

Manahan, S.E. 1991. *Environmental Chemistry* (5th ed.), Chelsea, Michigan USA: Lewis Publishers.

O'Neill, P. 1985. *Environmental Chemistry*. London: George Allen & Unwin.
Parkhurst, D.L., D.C. Thorstenson & L.N. Plummer 1980. *PHREEQE – A computer program for geochemical calculations*. US Geol. Surv. Water Resour. Inv. 80–96: 210 p.
Petts, J. & G. Eduljee 1994. *Environmental Impact Assessment for waste treatment and disposal facilities*. Chichester: John Wiley.
Reichard, E., C. Cranor, R. Raucher & G. Zapponi 1990. *Groundwater Contamination Risk Assessment: A Guide to understanding and managing uncertainties*. Wallingford, UK: Int. Assn. Hydrol. Sci.
Salcedo, R.N., F.L. Cross & R.L. Chrismon 1989. *Environmental impacts of Hazardous waste treatment, Storage and Disposal facilities*. Lancaster, Penn., USA: Technomic Publ.
Sytchev, K.I. 1988. *Water Management and Geoenvironment*. Paris – Nairobi: UNESCO – UNEP.
Van der Heijde, P.K.M. 1992. Computer modeling in groundwater protection and remediation. In P. Melli & P. Zannetti (eds), *Environmental Modeling*: 1–21. Amsterdam: Elsevier.
Vartanyan, G.S. (ed.) 1989. *Mining and Geoenvironment*. Paris – Nairobi: UNESCO – UNEP.

Chapter 2.5

Sequential use of water resources*

U. Aswathanarayana (India)

2.5.1 WATER QUALITY IN RELATION TO WATER USE

Water is a mobile resource. Unlike gasoline which when once used is gone forever, water can be used and reused many times (e.g. municipal water supply → waste water → bio-ponds → irrigation water → groundwater recharge → withdrawal from wells, etc.). The kind of changes that water undergoes in terms of its quality, quantity, location and timing because of a given use, would determine its value for the succeeding uses.

Water quality for certain purposes (say, drinking) may be impaired by natural causes (e.g. highly fluorous spring and river waters in northern Tanzania) or anthropogenic causes (e.g. effluent discharges or acid rain).

Generally, water in the natural state is fit for most human uses, but this is not always so. The waters of the rivers in arid zones (e.g. Ethiopia) are unfit for drinking purposes, as high evaporation make them salty. The *As* and *F* contents of the hydrothermally influenced rivers (e.g. Firehole River in the Yellowstone National Park, Wyoming, USA) are three orders of magnitude higher than the most common natural concentrations. Some river waters are unsuitable for irrigation because of their high salt content, and high sodium to calcium ratio. High Dissolved Organic Carbon (DOC) values found in the plain rivers in the tropical regions render them unsuitable for use for some industrial purposes. Similarly, high level of hardness in carbonate-draining rivers, high content of silica in basalt-draining rivers, and high TSS in the arid zones with high slopes or in volcanic areas, impair the use of the river waters for particular purposes.

*Basic data and concepts in this chapter have largely been drawn from the author's own work, "*Water Resources Management and the Environment*", 2001, pp. 283–309.

Table 2.5.1 Water quality constraints in water use

Pollutant	*A*	*B*	*C*	*D*	*E*	*F*	*G*
Pathogens	xx	0	xx	0 to xx (1)	x	na	na
Suspended solids	xx	xx	xx	x	x	x (2)	xx (3)
Organic matter	xx	x	xx	x to xx (4)	+	x (5)	na
Nitrate	xx	x	na	0 to xx (1)	+	na	na
Salinity (9)	xx	xx	na	x to xx(10)	xx	na	na
Inorganic toxics (11)	xx	xx	x	x	x	na	na
Organic toxics	xx	xx	x	x to xx (1)	x	na	na
Protons (acidification)	x	xx	x	x	na	x	na
Aquatic biota	x (5, 6)	x (7)	xx (12)	x to xx (4)	+	x (5, 13)	x (8)

xx – Presence of the pollutant markedly impairs the use of water for the concerned purpose; can only be used for the purpose after major treatment; x – Minor impairment of use, needing minor treatment; 0 – No impairment of use is involved. na – Not applicable. + – Degradation of water quality in this way may be beneficial for this specific use.
1: Food industries; 2: Abrasion, 3: Sediment settling in channels, 4: Some industries (e.g. electronic industries) require low DOC, 5: phytoplankton may cause clogging of filters, 6: Bacterial activity may affect odor and taste, 7: Higher algal biomass is acceptable in fish ponds, 8: Development of water hyacinth (*Eichhornia crasspipes*), 9: Also includes B and F, 10: Ca, Fe, Mn in textile industries, 11: As, Cd, Cu, Hg, Pb, Se, Sb, Sn, Zn, etc. 12. Loss of transparency from algal development, 13. Development of zebra mussels in pipes (*Dreissena polymorpha*).

Impairment may be caused not only due to degradation in the chemical and physical attributes of water, but also due to aquatic organisms. In the recent decades, the zebra mussel (*Dreissena polymorpha*) which clogs water intakes, has spread all over Europe and North America. The thick and vast mats of water hyacinth (*Eichhornia crassipes*) in tropical waters are seriously impeding fluvial transport.

The various uses of water can be classified into three categories, depending upon the stringency of specifications for a particular use:

- Most demanding uses: A – Drinking water; B – Fisheries;
- Less demanding uses: C – Recreation, D – Industrial uses, E – Irrigation,
- Least demanding uses: F- Power and cooling; G – Transport.

Table 2.5.1 (source: Meybeck 1998) shows how the water quality contaminants constrain the use of water.

The information provided in Table 2.5.1 indicates the directions for proper use and reuse:

- The highest quality water (at the rate of, say, 2–4 l per capita per day) should be reserved for drinking purposes (as different from water for other domestic uses, such as bathing, washing clothes and utensils and sanitation, which can use slightly inferior water). In many Indian homes, the traditional practice has been to keep drinking water separately from water to be used for washing purposes. It is no doubt technologically feasible but prohibitively expensive to purify contaminated water (say, municipal waste water) to a level of purity acceptable for domestic purposes,

- Municipal waste water which contains (say) pathogens can be used in most industries, except food industries, and
- Sewage which contains organic matter and nitrates is not only acceptable but even be desirable for use in irrigation.

In the market economies of the Industrialized countries, the economic value of water is determined by the market. But the market prices do not reflect the true worth of water, as they do not take into account social and environmental factors and as they may get distorted due to income disparities, subsidies, etc. On the other hand, a water-supply system which ignores the market in deciding about water prices, will not be sustainable. Most public institutions use three sets of prices: market prices (based on demand and supply), administered prices (decided upon by the government for specific purposes, on socio-economic and political grounds, such as domestic water supply or irrigation) and accounting prices ("shadow" prices that reflect the true economic value of water).

There are different levels of considering the value of water. An individual user (e.g. householder/farmer/industry) would look at water from a narrow perspective of their particular net revenues or satisfaction. A regional perspective would additionally take into account some aspects which are not take into consideration by individual user, such as employment generation, new economic activities, poverty alleviation, improved sanitation and health of the community, tourism, etc. A national perspective would take a broader view of integrating water resources planning in the context of national income, national productivity, etc. Conflicts can arise between these perspectives, because of NIMBY (Not In My Back Yard) approach. A farmer knows that canals need to be dug to distribute water from the reservoir, but he wants the canal to be located in somebody else's farm. Similarly, the beneficial effects of water use may accrue to users in one part of the region (e.g. irrigated agriculture), whereas another part of the region is exposed to the detrimental effects (e.g. drainage water). Besides, in order to protect the natural water system, the State may have to enforce some limits (such, as maintaining minimum streamflow rates). The various conflicts have to be reconciled in a fair, equitable and ecologically-sound manner.

Based on the technique of decomposition of linear programs developed by Dantzig (1963), Stephenson (1989) developed a layered model for use in the development of water resources in a country at different levels (national, departmental, river basin and project levels). "Each successively lower model is optimized on its own, but uses shadow values on output imposed by the successively higher models. The lower models in turn feed back optimal plans and technical aspects, such as output, to the higher model which is further able to refine planning" (Stephenson, 1989, p. 63).

2.5.2 ESTIMATES OF WATER VALUE FOR DIFFERENT USES

2.5.2.1 Environment and Ecosystem

The complexity and importance of impact of water use on environment and ecosystems is increasing steadily. As Global Water Partnership puts it, "Until realistic values can be placed on ecological services, economic planning techniques will ignore or marginalize

Table 2.5.2 Economic valuation of non-market issues

Effects category	*Valuation method option*
Productivity	Market valuation via prices or surrogates; Preventive expenditure; Replacement cost/shadow projects/cost-effective analysis; Defensive expenditure
Health	Human capital or cost of illness; Contingent valuation; Preventive expenditure; Defensive expenditure
Amenity (e.g. lake)	Contingent valuation/ranking; Travel cost; Hedonic property method
Existence values (eco-systems; cultural assets)	Contingent valuation

Source: Turner & Adger 1996

the role of the environment as provider and user of water resources". For instance, the economic value of a tree is not confined to the market value of its timber – it should include the cost of growing such a tree, the environmental benefit that the tree has been providing (e.g. absorbing carbon dioxide, protecting against erosion, habitat for birds, biodiversity) and the adverse effects that would arise when the tree is felled.

Turner & Adger (1996) developed innovative methods of economic valuation of non-market issues involved in the assessment of economic value of water (Table 2.5.2).

2.5.2.2 Crop irrigation

Roughly half the water used in irrigation is lost through evaporation and transpiration. Whereas evaporation from the surface is a waste, transpiration by crop plants is necessary and productive. The value of water depends upon the climate, soil, crop grown (high-valued vegetables and fruits versus low-valued crops and forages), stage of growth of the crop and the techniques of application of water. The value of water may be estimated at the point of delivery to the farmer or for the watershed as a whole.

2.5.2.3 Municipal water use

Water for municipal use is given the highest priority, as water is essential for life. Two aspects of municipal water use need special mention: the water has to be of the highest quality, though the quantities involved are small.

The criteria used for the estimation of the economic value of water for productive use are not applied in the case of municipal use of water.

2.5.2.4 Industrial water use

Water is used in the industry for cooling, transportation and washing, and as a solvent. In some industrial processes, water enters the composition of the finished product. The principal water uses are in thermal power production, chemical and petroleum plants, ferrous and non-ferrous metallurgy, wood pulp and paper industry, etc. Cooling accounts for about 80% of the water use. In some countries, withdrawals of water for industry are more than 50% of all withdrawals. In thermal power generation, only about 0.5–3% of the water intake is actually consumed, and recycling is the norm. But some industrial processes (e.g. food industries) consume 30–40% of the water intake.

The value of water for industrial use is estimated on the basis of the "cost of alternative processes that will produce the same product while using less water" (Janusz, 1998, p. 413). The costs of internal recycling of water are usually low. But it is generally more expensive to remove the toxic dissolved materials which the industrial wastewater may be carrying. The costs vary greatly depending upon the extent of degradation in water quality occurring in the process. In general, the costs of water supply and wastewater treatment for industry are usually about 2–3% of the production costs. The total water consumption in the industrial sector is stabilizing or even decreasing, as more and more industries are switching over to water-free or dry technologies.

2.5.2.5 Waste assimilation

The economic value of water for waste assimilation is estimated on the basis of the alternative cost of treating the effluent. A given flow of water can only assimilate a certain quantity of effluent with certain characteristics (such as, TDS, SS, BOD, SAR, etc.) without exceeding the quality standards. The treatment costs needed to bring down the effluent characteristics in order to stay within the standards, gives a good estimate of the value of water.

2.5.2.6 Navigation

The economic value of water for navigation is the difference between the costs of water transport vis-á-vis the lowest cost of alternative mode of transportation. It is known that transport of goods through river barges (in terms of USD per ton-km) is cheaper than other modes, such as rail or truck (towed barges plying over the Mississippi river in USA routinely carry large quantities of ores, various kinds of construction materials, scrap, etc.). If, however, a navigational facility has to be constructed anew, the value of water for navigation may be zero or even negative.

2.5.2.7 Hydropower generation

The value of water for hydropower generation is the difference between the costs of hydropower (including transmission) and other lowest-cost means of power generation, say, coal-fired thermal power plants (including transmission).

2.5.2.8 Recreation

The value of water for recreation depends upon a number of factors such as location, accessibility, scenic setting, water quality, etc. The valuation is done on the basis of the behavior of tourists by means of a survey – how much extra travel costs, admission price, etc. they are willing to bear to visit a particular kind of recreational facility.

2.5.3 WATER VALUE IN SYSTEM CONTEXT

Since many combinations of use and reuse of water are possible simultaneously or in sequence, it is critically important to know how the various uses of water combine and interact in space and time. Consideration has therefore to be given to return flow and reuse. A given physical unit of water can be used over and over again if water is used fast and returns large quantities of clean or well-treated water. Even though the value addition for each use of water is small, a pattern of multiple use can generate large values. The location and timing of water use becomes important in the system context of water use. The system gains more value when the waste-generating users are located as far downstream as possible.

Wollman (1962) made a pioneering study of water value in the system context on the value of water in the San Juan and Rio Grande basins of New Mexico, USA. The demand of water far exceeded the supply for water use categories, such as, irrigation, municipal and industrial uses, recreation and others. The study evaluated the effects on the economy of New Mexico state as a consequence of the use of water in several different ways. A knowledge of the value addition through the use of an acre-foot (1 acre-ft = 1235 m^3) of water for alternative uses permitted informed decision about the directions in which the economy could move in order to make the best use of all its resources, as the scarcity of water increases.

For populist reasons, a government may take an ad hoc decision about the upstream use of water, and this may create serious problems and civil unrest in the country on the downstream side. Such conflicts could be avoided if informed decisions could be made on the basis of different of techno-socio-economic options that are available.

2.5.4 PRICE COORDINATION OF WATER SUPPLIES

Often, a given region may have a complex system of supplies and demands. Price coordination methods have to be developed to make the system techno-socio-economically efficient. Guariso et al. (1982) adapted the classical methods of price coordination to water resource systems. The objective of the exercise is to maximize the total regional net benefit. A disaggregated approach is used whereby "the marginal benefit at each demand point is equal to the marginal cost of delivering water to that point".

Traditionally, water use demands are projected in terms of fixed water requirements of the user (e.g. so many liters of drinking water per capita per day, so many ha cm of irrigation quality water for irrigating a crop, so many liters of water for refining one barrel of crude, etc.). Supplies of water are sought to be organized on the basis of the projected demand. But the picture is changing all over the world for two reasons. As population grows and industries and irrigated agriculture expands, several countries find that known supplies of good quality fresh water are grossly inadequate to provide all the requirements of the users. Water use has to be optimized, using pricing as a control. Also, research and development is leading to more efficient use of water, use of inferior quality of water and reuse and recycling of water. In other words, water demand cannot be assumed to be inelastic, but would depend in part on the availability and pricing of water of a particular quality.

As Zimmerman put it, "*Resources are not, they become*". Inferior quality water, drainage water and waste water which were unusable earlier, can be used now. For instance, acid mine water can be treated for use in agriculture and for drinking. Brackish water can be used for irrigation with gypsum amendment, or made use of to grow salt-tolerant plants. It therefore follows that the supplies and demands of water should be treated in dynamical terms. Thus, instead of investing more money in developing new supplies of water, it is generally cost-effective to use the same money to develop methods of efficient use of water and recycling and reuse of water.

There are two approaches for price coordination – aggregated and disaggregated. In the aggregated approach, a large mathematical model is formulated to represent all the components of supply and demand for the whole region. The equations are solved to maximize the net benefit from water use for the entire system. When the supplies and demands are complex and are managed centrally, the aggregated approach would, of course, the proper one to use, but the model would be too cumbersome and too expensive in terms of computer time to be practicable.

In the disaggregated approach, supply and demand is treated independently. They are coordinated sequentially by a supervisor. In effect, the supervisor would be engaged in a dialogue with the various supply and demand entities. The process will go on until it converges to produce an optimal balance or equilibrium between supply and demand.

The coordination algorithm is aimed at maximizing the total regional net benefit, but instead of the usual practice of computing the total benefits and costs, the algorithm of Guariso et al. (1982) seeks to achieve maximization of benefit on the basis of marginal benefits and costs ("... if a certain flow is to be transferred from a supply to a demand, the cost of delivering the final unit of water, which is the marginal cost of this flow, must be equal to the benefit generated by this final unit, or the marginal benefit"). This is analogous to determining the equilibrium price in a market.

The application of the price coordination method of the market to water resources management suffers from a short-coming, namely, that water supplies and demands are not independent. For instance, the upstream use of water affects the use of water downstream, and extensive withdrawal of groundwater leads to the reduction of surface water, and so on. The model of Guariso et al. (1982) does take this matter into account, but still involves the making of some assumptions, such as, that water of requisite quality is available, that all flows are made available at the same time, and that all supplies have the same reliability, and so on.

2.5.5 PRINCIPLES OF OPTIMIZATION

A demand unit may be a farm, a factory or a residence or a community or a region. The benefit (B) accruing to a demand unit is a function of the amount, Q, of water used. Thus, $B = B(Q)$. Where the demand unit is, say, a farm, it is critically important to know the timing and reliability (say, 95%, 90%) of supply of the irrigation water during the critical weeks of the growing season. Some irrigation models take into account the total amount of water needed for the whole season, rather than particular flow levels during particular periods. The assumption of a constant average flow in the present algorithm may be consistent with the output of these models.

If the water is paid for at a price p, the profit of the unit is $[B(Q) - pQ]$. Assuming that the unit is profit maximizing, the amount of water concerned can be obtained by solving the following optimization problem:

$$\operatorname*{Max}_{Q}[B(Q) - pQ] \tag{2.5.1}$$

If the Equation 2.5.1 is solved for all values of the parameter p, we can derive the following demand function, which gives the amount demanded by the unit as a function of the price of water:

$$Q = Q^{D}(p) \tag{2.5.2}$$

The necessary condition for optimality can be written as:

$$\frac{d}{dQ}B(Q) = p \tag{2.5.3}$$

The demand function (Eq. 2.5.2) may be interpreted as the marginal benefit of the unit.

A supply unit may be a reservoir, a pumping station, a runoff storage pond, or a desalination plant, etc. At a cost of $C(Q)$, the supply unit supplies Q amount of water at price p. The optimization problem for the unit is solved by

$$\operatorname*{Max}_{Q}[pQ - C(Q)] \tag{2.5.4}$$

The corresponding solution for the supply function is obtained by:

$$Q = Q^{S}(p) \tag{2.5.5}$$

We assume that the demand and supply structures have been designed for optimum performance. Thus, the benefit and cost functions, $B(Q)$ and $C(Q)$, are expected to have optimization built into them.

When a supply and demand unit is connected, the value of water exchanged can be simply obtained from Figure 2.5.1. The Equilibrium Point E in the plot corresponds to the following equation:

$$Q^{S}(p_E) = Q^{D}(p_E) \tag{2.5.6}$$

Q_E corresponds to the equilibrium flow which must be exchanged between the supply and demand units in order to maximize the total net benefit of the system, such that

$$\operatorname*{Max}_{Q}\ B(Q) = C(Q) \tag{2.5.7}$$

Similarly, pE corresponds to the equilibrium price which is the price "which leads the supply and demand units not only to exchange the same amount of water but also to select the particular value Q_E which maximizes the total net benefit of the system" (Guariso et al., 1982, p. 377). One way of solving the problem is to assume a particular

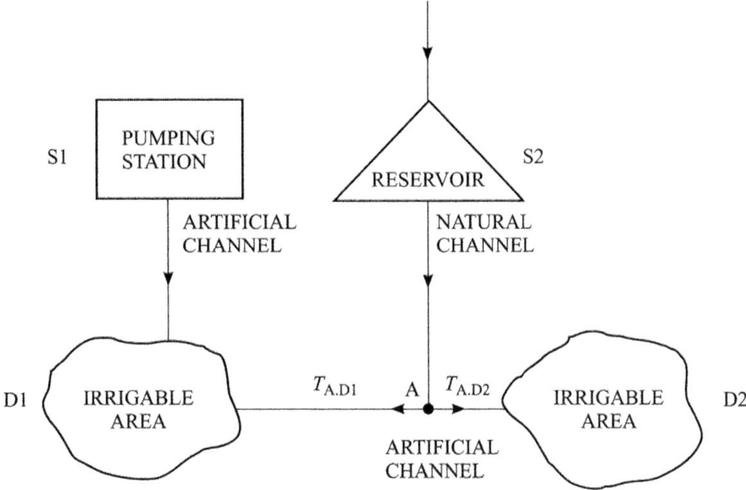

Figure 2.5.1 Example of the methodology of price coordination
Source: Guariso et al., 1982, p. 378

value for p and compute the corresponding imbalance, $Q^S - Q^D$ and the reiterating the price until $Q^S - Q^D$ is zero.

2.5.6 PRICE COORDINATION OF A TYPICAL IRRIGATION SYSTEM

The price coordination method is applied to a simple situation (Fig. 2.5.1).

An irrigation system has (say) two sources:

- A pumping station (S_1) where energy has to be spent for pumping groundwater, and from where water is transferred through an artificial channel to irrigate the area (D_1), and
- A reservoir (S_2) which has already been built, and from where water flows out by gravity through a natural channel (i.e. at no recurring expense) to an artificial channel to irrigate the area (D_2). Though the economics of capital and operating costs of the sources and transfer units are different, irrigation water has to be delivered at the same price to farmers in the areas D_1 and D_2.

The same irrigation system can be depicted in the form of interaction graph (Fig. 2.5.2).

Guariso et al. (1982) developed an algorithm for the Northwest Water Plan in Mexico. The system involved a groundwater supply, three surface water supplies from rivers, and four irrigation areas which would be interconnected through an interbasin water transfer. Convergence was achieved in 12–15 iterations. The model indicates a 6% increase in the total net benefit due to crop production. It also shows that the

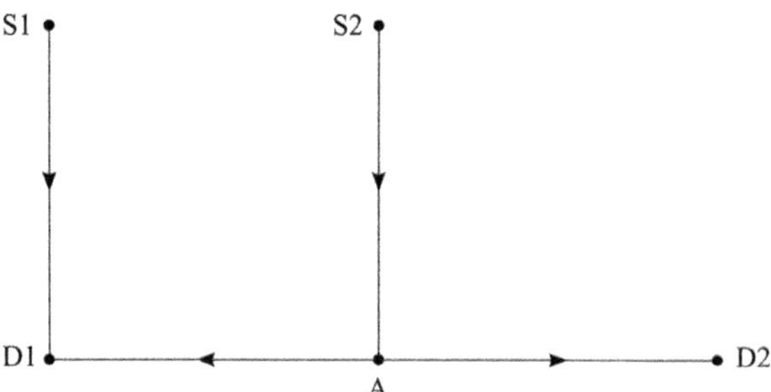

Figure 2.5.2 Interaction graph for price coordination
Source: Guariso et al., 1982, p. 381

market price of the water supplied is double the existing (probably subsidized) price if water is to be treated as a market commodity. The model can be further refined by bringing in parameters such as the quality of water, and the reliability of supplies.

2.5.7 OPTIMIZATION METHODS IN WATER MANAGEMENT

This section is largely drawn from the excellent and lucid summary by Holy (1982).

Systems engineering concepts are applied to the solution of the water management problems in relation to the human environment, namely, protection of water resources, multiplying the number of water resources, and improving the efficient utilization of water resources (Holy, 1982, p. 81). The objective of the water production process is to collect and store water having the required utility properties (quantity, quality, place and time) which may be required for final consumption (e.g. water for domestic use, irrigation water) or a means of production (e.g. water for transportation, water for power production). Water is thus an economic good. But unlike other economic goods like coal and oil, water cannot be treated in purely economic terms, since human life itself depends upon it. Water resources are sought to be protected and augmented by providing for well-head protection, protection around water springs, and wastewater reuse. The efficient use of water involves, among other things, the reduction in conveyance losses, and minimization of evaporation.

2.5.7.1 Water reservoirs

There may be many reasons for building water reservoirs, such as:

- Providing water for domestic purposes (potable water), industry (utility water), and agriculture (irrigation),
- Flood control,
- Power production,
- Inland water transport,
- Recreation, etc.

The construction of a reservoir perturbs the natural environment and affects the geographical, climatic and social conditions. There may be beneficial impacts on the economy and way of life of the people through the development of new economic activities (agriculture, power production, inland transport, recreation, etc.). There may be adverse effects on the health and productivity of men and animals (e.g. incidence of waterborne diseases, schistosomiasis, etc.), and there is the potential danger arising from the failure of the dam due to torrential rains or an earthquake.

Reservoirs affect the following components of the biophysical environment (Holy, 1982, p. 73):

- Solar radiation and the thermal balance of the accumulated water,
- Temperature of the air, and fog formation,
- Air currents (similar to those near the sea), and
- Air humidity. As the water level of the reservoir fluctuates, muddy areas with shallow water develop at the edges of the reservoirs. Water weeds and swamp plants grow in these areas, and form breeding grounds for mosquitoes.

Construction of reservoirs have the following adverse impacts:

- Large populations may have to be moved and resettled away from the areas that are going to be flooded,
- Silting reduces the life-span of the reservoir,
- Buildings, cultural monuments, soils, forests, mineral deposits, etc. may be irretrievably lost,
- Ecosystems in the impounded water may be severely disturbed (changes in thermal and chemical stratification, reduction in DO (Dissolved Oxygen) in water, movement of fish, etc.) and
- Problems of eutrophication, and the development of thick, and extensive mats of water hyacinth (*Eichhornia crassipes*) which seriously impede fluvial transport.

The favorable impacts of the reservoirs in the downstream side are the decrease in the flood flow, and the improvement in the low minimal rates in water courses. Reservoirs trap silt and insoluble pollutants. So the water quality improves in the discharge of the reservoir.

2.5.7.2 Water management systems

Water management systems are of crucial importance to the quality of life and national economy. They are often complex and costly. Most water projects are multipurpose. It is therefore necessary to identify the principal purpose of the project (water supply, power generation, recreation, etc.) and use the secondary purposes as normative limitations.

Economic activities which are dependent on water can be divided into two categories: water users and water exploiters. Water users remove water from its source (reservoir, pumping station, runoff pond, etc.) and use it for some purpose (agricultural, industrial, domestic, etc.) and then discharge it at a different place, reduced in quantity, and most often degraded in quality. On the other hand, water exploiters do not remove water from its source, but use it where it is and as it is, for production of hydropower, for river transport and fisheries, for recreation, etc. This division is, however, arbitrary. For instance, when large quantities of water are stored for purposes of power generation, there is considerable loss of water due to evaporation from the surface of water. Agriculture and forestry operations, draining of swamplands, construction of new townships, etc. have a great impact on the quality and quantity of water resources.

The objective of optimization is to attain the optimal relation between the purposes for which the system is being built and the means available for building it. The following summary is largely drawn from Holy (1982).

The standard function ϕ for the optimization of a water management system contains three groups of economic variables:

N – Unit costs of the individual equipments of the system for different purposes, P – Unit benefit of the system for water supply for individual purposes, Z – Unit losses that arise due to failure to supply water for individual purposes.

The model assumes that

- The system contains k reservoirs $(1, 2, 3, \ldots, n)$,
- N, P, and Z vary with time t. Value T corresponds to the assumed physical or economic lifespan of an installation or the whole system.

It therefore follows that

$$N = f(k, t) \tag{2.5.8}$$

$$P = f(k; i; t) \tag{2.5.9}$$

$$Z = F(i; t) \tag{2.5.10}$$

A general standard economic function which simulates a multi-purpose water management system, may be written as follows:

$$\phi_{\text{opt}} = \sum_{k=1}^{m} \sum_{i=1}^{n} \sum_{t=1}^{T} \phi(N_{kt}; P_{kit}; Z_{it}) = \max; \min; 0 \tag{2.5.11}$$

Optimization cannot be based on economic criteria alone. Many times, non-economic criteria, such as social priorities and environmental concerns, may have to be taken into account or may be given priority over purely economic criteria. In a multi-purpose water management system, this process is accomplished by the method of matrix of contradiction of purposes. We identify the principal purpose of water use, for which the system is to be optimized, and consider other purposes as limitations or contradictions.

The following criteria for contradiction may be considered:

1. Demand on reservoir volume,
2. Demand on water quality,
3. Demand on water level fluctuation.

There is bound to be contradiction in demands in most cases. There are a few exceptions – for instance, there may be no conflict involved if water is to be used for navigation and recreation. There can be movement of boats, and water sports in the water in a river or lake. The most frequently occurring contradiction is contradiction 1, followed by 3 and in some cases, 2.

Evidently, the priority of purpose would determine would determine the principal consideration that has to be taken into account.

Priority of purpose	*Principal consideration*
Drinking water	Water quality
Hydropower production	Water level in the reservoir
Irrigation	Reservoir volume

For instance, if priority is to be given for the use of water in irrigation, maintenance of adequate storage of water would be the prime concern. Limited importance will be given to other uses, such as, water supply, power, etc.

2.5.8 ALLOCATION OF WATER TO COMPETING USERS

2.5.8.1 General considerations

In most parts of the world, there is intense competition among the users for scarce water resources. The competitors may be individual users (e.g. factories) in an area or communities or provinces within a country (violent confrontation between the provinces of Tamilnadu and Karnataka in India about the sharing of the waters of the Kaveri (Cauvery) River, or even countries themselves (Syria, Turkey, Iraq about sharing of the waters of Euphrates – Tigris). While the value of the possible output should have primacy in allocating the water to the competing users, it is not possible to ignore some cultural, socio-economic, political and technical factors. Innovation and practicability should be the key words in addressing such problems.

How an innovative approach could provide a way-out of the water competition problems can be illustrated with two examples from the Indian sub-continent (Suresh, 1998).

The province of Tamilnadu in south India makes use of 95% of the surface water potential through 55 major reservoirs and more than 300 medium and small an cuts, barrages and diversion structures. On the other hand, the province of Kerala to the west has plenty of water (43 flowing rivers), but does not have enough land to construct the reservoirs. Thus, it makes economic sense to transfer the waters of Kerala towards the east, and construct the reservoirs in Tamilnadu where land is available. The Parambikulam Aliyar Project (PAP) seeks to divert 863 M m^3 of Kerala waters to Tamilnadu, in a way which will be mutually beneficial to both the states.

River Ganges ranks next only to Amazon in terms of the volume of water, with surface water availability of 446 million acre-ft (MAF) or about 551 km^3. The headwaters of the major tributaries of the Ganges lie in Nepal which contribute to about

40% of the annual flows and 71% of the dry season flows of the Ganges available at Farakka (in the West Bengal province of India). But Nepal does not have the resources to develop these water resources, and sell water and energy to the downstream countries, namely, India and Bangladesh. Techno-economic solutions do exist to solve these problems. What is needed is the political will and economic means to implement them.

2.5.8.2 Case histories

Two case histories are given to illustrate the use of linear programming methodology (Lee, 1976) to decide about the allocation of water to competing users.

Barandemaje (1988) tried to develop a policy aimed at optimizing the allocation of water to competing industrial users in Burundi (Africa). A finite amount of water (26,000 $m^3 d^{-1}$) has to be allocated to five competing industrial users (a cigarette company, a brewery, a textile mill, a paper mill and a dairy). A linear mathematical model was developed to conceptualize the five firms. It consists of an objective function and constraint equations. Each equation takes into account the water-use/net-profit relationship for each firm. The constraints are based upon the lower and upper water consumption capacities and the non-negativity conditions on the value of the decision variables. Optimization methods were applied to two kinds of situations: involving only competitive criteria (i.e. which kind of allocation would yield the maximum profits), and involving both social criteria (the needs of the dairy being given the highest priority and the cigarette company, the lowest) and economic competitive criteria. Even though the dairy may provide less profit per unit of water used relative to the cigarette company, the community needs milk more than it needs the cigarettes, and hence the higher priority for the dairy in water allocation. On the basis of this kind of analysis, the decision-makers could make their choices.

Singh et al. (1987) made use of goal programming (a case of linear decision problems which may have more than one objective) to allocate water to the winter crops (wheat, *ahu* paddy, pulses, oil seeds and potato) in Garufella catchment in Assam in northeast India. Garufella is a typical monsoon land – though the rainfall is very heavy (3,710 mm), 90% of it occurs in the monsoon period (June to October).

On the basis of data (water availability data, crop yield data, food requirement), the various parameters are maximized, taking into account various constraints and priorities.

Under this study, three objective functions are formulated for maximizing the net return, protein content, and the calorific value of crop, under various socio-economic constraints.

Maximization of net return

$$\text{Max } Z_1 = \sum_{i=1}^{n} A_i N_i \tag{2.5.12}$$

where A_i = area under i-th crop activity in hectares, N_i = net return from hectare from i-th crop activity in rupees, n = number of crops being considered.

Maximization of calorific value

$$\text{Max } Z_2 = \sum_{i=1}^{n} A_i Y_i C_i \tag{2.5.13}$$

where Y_i = yield of i-th crop activity (in kg ha^{-1}), C_i = calorie of i-th crop (calories kg^{-1}).

Maximization of net protein value

$$\text{Max } Z_3 = \sum_{i=1}^{n} A_i Y_i P_i \tag{2.5.14}$$

where P_i = nutrient value of i-th crop activity (in g kg^{-1}), C_i = calorie of i-th crop (calories kg^{-1}).

The three optimization models given above are subject to the following constraints.

Water availability constraint

This refers to the amount of surface water available for crop production.

$$\sum_{i=1}^{n_j} A_i R_{ij} \leq S_j \tag{2.5.15}$$

where S_j = surface water available in j-th month, R_{ij} = water requirement per unit area in excess of effective rainfall for the i-th crop in j-th month, n_j = total number of crops which are grown in j-th month.

Land availability constraint

The extent of land used for various crops cannot exceed the total available land. Also, the land allocated to a given crop has to remain unchanged from sowing to harvesting.

$$\sum_{i=1}^{n} A_i \leq TA \tag{2.5.16}$$

where TA = Total available land (in ha).

Minimum area constraint

In order to meet the minimum food requirements of the population, a given crop (say, wheat) needs to be grown in a minimum area.

$$A_i \geq T_i \tag{2.5.17}$$

where T_j = minimum area allocated to j-th crop.

Protein requirement constraint

$$\sum_{i=1}^{n} A_i Y_i P_i \geq PR \quad (2.5.18)$$

where PR = total protein requirement (in g).

Calorie requirement constraint

$$\sum_{i=1}^{n} A_i Y_i C_i \geq CR \quad (2.5.19)$$

where CR = total calorie requirement (in calorie units).

Each goal constraint (nutritional requirement constraint, Net return constraint, Production constraint based on the food habits of the people) may be assigned a positive or negative deviation variable or both.

The priorities assigned are as follows:

P_1: The highest priority is assigned to the maximization of net return.
P_2: The second priority is assigned to protein and calorific value. Between the two, a higher weight is given to the calorific value.
P_3: The production of different crops should be adequate to meet the actual food requirements of the population. Wheat is assigned a higher weight than paddy which is also cultivated during the monsoon season.

The existing situation (EXT) is compared with ten models, involving the following parameters: Total area (ha), Total water utilization (ha m), Net return (millions of rupees), Total protein ($\times 10^8$ g), and Total calories ($\times 10^{10}$ calories). From among these, the model which yielded the maximum net return, while providing adequate nutrition in tune with the food habits of the people, was chosen.

2.5.9 DECISION-MAKING PROCESS

Nijkamp & Rietveld (1982) gave a good account of the conceptual basis of the decision-making process. The process has the following elements:

1. DM (Decision Maker) has at his disposal I instruments, $x = (x_1, x_2, \ldots, x_I)$.
2. The vector of the instruments x is an element of the convex set K_I, being a subset of $\Re^I = (x \in K \subset \Re^I)$,
3. The DM considers J objectives: $w = (w_1, w_2, \ldots, w_j)$, which he wants to maximize,
4. For each combination of instruments, x, the effect on the set of objective w can be determined with certainty. Hence, a set of J concave objective functions w ($w = w_1, w_2, \ldots, w_j$) is assumed to exist, each mapping $x \in \Re^I$ to $w_j \in \Re^I$.

In the flow chart of the decision-making process, the analyst comes up with a variety of solutions. The Decision Maker (DM) evaluates them. By iteration, he will eliminate the least satisfactory solutions and in due course, arrives at the most satisfactory solution.

REFERENCES

Barandemaji, D. 1988. Optimum allocation of water to competing users. *Proc. Int. Training Course on Environment Management in Developing counties*. Leipzig.

Dantzig, G.B. 1963. *Linear Programming and Extensions*. Princeton, N.J.: Princeton Univ. Press.

Guariso, G., Maidment, D. Rinaldi, S. & Soncini-Sessa, R. 1982. In P. Laconte & Y.Y. Haimes (eds.), *Water Resources and Land-use Planning: A Systems Approach*. The Hague, Netherlands: Martinus Nijhoff Publishers. pp. 373–392.

Holy, M. 1982. Environmental aspects of water management. In P. Laconte & Y.Y. Haimes (eds), *Water Resources and Land-use Planning: A Systems Approach*. The Hague, Netherlands: Martinus Nijhoff Publishers. pp. 69–91.

Janusz, K. 1998. Economic value of water. In H. Zebidi (ed.), *Water: A Looming Crisis?* IHP-V, Tech. Doc. no. 18. Paris: UNESCO. pp. 407–416.

Lee, Sang M. 1976. *Linear optimization for management*. New York: Petrocelli/Charter.

Meybeck, M. 1998. Surface Water Quality: Global assessment and perspectives. In H. Zebidi (ed.) *Water: A Looming Crisis?* Paris: UNESCO, pp. 173–185.

Nijkamp, P. & Rietveld, R. 1982. Selecting a range of alternatives by individual or group decision-makers. In P. Laconte & Y.Y. Haimes (eds), *Water Resources and Land-use Planning: A Systems Approach*. The Hague, Netherlands: Martinus Nijhoff Publishers. pp. 41–45.

Singh, R., Soni, B. & Changkakoti, A.K. 1987. Optimal utilization of irrigation water in Garufella catchment in Assam, India. In T.H. Anstey & U. Shamir (eds), *Irrigation and Water Allocation*. IAHS Publ. no. 169.

Stephenson, D. 1989. Planning model for water resources development in developing countries. In D.P. Loucks & Uri Shamir (eds.), *Closing the Gap between Theory and Practice*, IAHS Pub. no. 180.

Suresh, S. 1998. Intersectoral competition for land and water policy between users and uses in Tamilnadu, India. In H. Zebidi (ed.), *Water: A Looming Crisis?* IHP-V, Tech. Doc. no. 18. Paris: UNESCO. pp. 441–445.

Turner, R.K. & Adger, W.N. 1996. *Coastal Zone Resources Assessment Guidelines*. LOICZ Reports and Studies. No. 4. Texel, The Netherlands: LOICZ.

Wollman, N. 1962. *The value of Water in Alternative Uses*. Albuquerque: The University of New Mexico Press.

Chapter 2.6

Wastewater reuse systems*

U. Aswathanarayana (India)

2.6.1 INTRODUCTION

Wastewater reuse has become an important part of the water resources management practice for two reasons: to protect the public from diseases, and to make proper use of an increasingly scarce resource, namely, water.

Wastewater means discarded water. Wastewater is just a euphemism for sewage. Wastewater is generally of two kinds:

1. Sanitary or foul sewage: composed of sanitary or domestic wastewater, liquid material collected from residences, buildings and institutions, industrial or trade wastes arising from manufacturing, municipal wastes, industrial effluents from dairies, bakeries, breweries, etc.
2. Storm water: Runoff of rainwater.

After a dry spell, the runoff from urban areas may contain organic carbon, pathogens, suspended solids, hydrocarbons, lead washed from the highways, acid rainfall, etc. Hence it is necessary to treat such water.

There are few places on earth where potable water is so plentiful and so freely available that there is no necessity to reuse the wastewater. There is little doubt that in future the reuse of wastewater will become routine all over the world, for the following reasons:

It has been estimated that in 1995 the volume of wastewater was 326 $km^3\,yr^{-1}$ in Europe, 431 $km^3\,yr^{-1}$ in North America, 590 $km^3\,yr^{-1}$ in Asia, and 55 $km^3\,yr^{-1}$ in Africa (Shiklomanov, 1998, p. 37).

---

*Basic data and concepts in this chapter have largely been drawn from the author's own work, "*Water Resources Management and the Environment*", 2001, pp.159–185.

Comprehensive treatment of wastewater before discharging into a water body is expensive, and is practiced only in Industrialized countries, and even there, the coverage is not 100%. Most countries tend to discharge the wastewater either untreated or partly treated, into the hydrological system. This has disastrous consequences. *Every cubic metre of waste water discharged into water bodies or water courses contaminates and degrades 8 to 10 cubic metres of good water.* Hence, prime waters get polluted and become non-potable. Thus most parts of the world are already facing the problem of degradation of quality of water because of the inherently wrong practice.

It is said that the Chinese term for Crisis consists of two ideograms representing Danger and Opportunity. The waste water problem has indeed a win-win solution. Ecological treatment of waste water has a number of merits: besides removing the pollutants effectively, it is low-cost, energy saving, resource recovery, easy operation.

2.6.2 BIO-POND TREATMENT OF WASTE WATER

The bio-pond treatment of wastewater is ideally suited for developing countries.

Wang (1991) gave a detailed analysis of the Chinese experience in the biological methods of treatment of waste water. In 1988, China had 86 conventional sewage treatment works with a total capacity of three million $m^3\ yr^{-1}$, accounting for about 11% of the municipal sewage flow in the whole country. As the country did not have the requisite financial and technical resources to build conventional sewage treatments for the rest 89% of the municipal sewage flow, they went in a big way for eco-pond technologies.

2.6.2.1 Land treatment

The acute water shortage in northern China necessitated the reuse of domestic and industrial waste water for irrigation and other purposes. Sewage irrigation increased from 42,000 ha in 1963, to 1.4 million ha in 1983. For instance, Tianjin city has 153,000 ha of sewage irrigation farmlands. Land treatment of wastewaters has been able to achieve high pollutant removal efficiencies as follows: SS, BOD, COD, total phosphorus and bacteria: 95%; trace metals, phenols and cyanide: 90%, COD, total nitrogen and potassium: 80%.

The effluent from the irrigated fields is used to charge the depleted aquifers. On an average, one m^3 of sewage applied on a farm increased the grain yield by 0.5 kg.

There have been failures also. In the case of some irrigation projects (less than 10% of the total area), the toxic and harmful substances in the industrial wastewater have resulted in decreasing yields and pollution of crops, soil mantle and groundwater. It is therefore crucial that toxic substances be removed from wastewater before it is used in irrigation.

2.6.2.2 Eco-ponds

The ecological and biochemical processes that go on in the eco-ponds under aerobic, facultative and anaerobic conditions, are shown in Figure 2.6.1. This diagram helps us to understand what kind of ecological conditions are necessary for what kind of ecological production. For instance, aerobic conditions have to be maintained in an

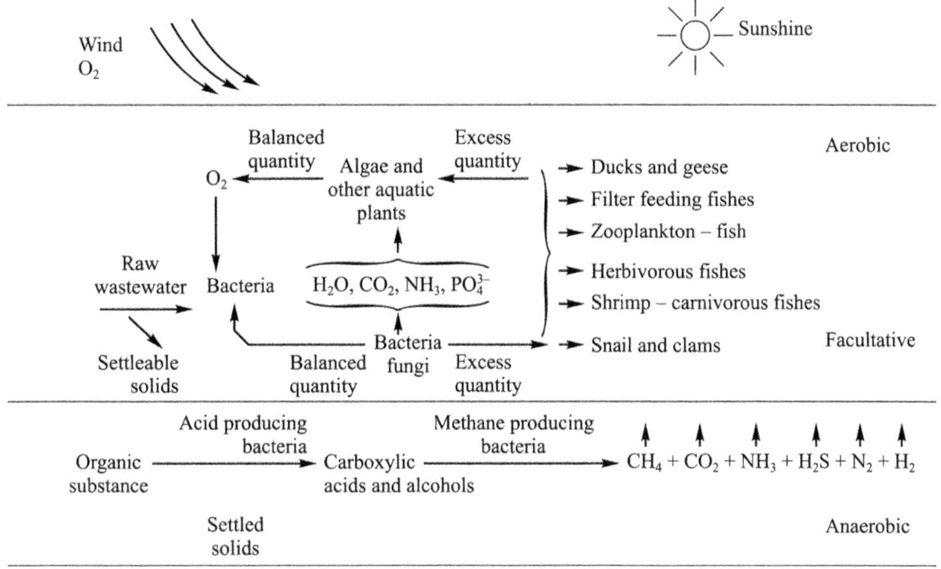

Figure 2.6.1 Ecological and biochemical processes in ecoponds
Source: Wang, 1991

eco-pond if the objective of developing a eco-pond is to grow aquatic plants for human and animal food. On the other hand, if the objective is the production of methane to be collected and used as fuel, anaerobic conditions are necessary.

The design of the flowsheet for treatment of waste water would take into consideration different local situations:

Where sewage is produced all the year round, whereas irrigational needs for reusing the waste water are seasonal, the following flowsheet is made use of:

Raw waste water → Preliminary treatment → Facultative ponds → Storage lagoons → Farmland irrigation

In areas where adequate land is not available and where the weather permits irrigation during most of the year, storage lagoons could be avoided and the ponds may directly supply water for irrigation. Then the following flowsheet can be adopted:

Raw waste water → Preliminary treatment → Facultative ponds → Year-round irrigation

The eco-pond systems are designed in the following ways, depending upon the climatic conditions and the composition of the sewage water (Wang 1991).

In arid or semi-arid, cooler areas

Domestic sewage → Preliminary treatment → Facultative ponds → Storage ponds or lagoons (fish farming/duck or geese raising in warm season) → Irrigation of farm land → Effluent recharging ground water

High strength organic wastewaters → Anaerobic ponds → Facultative ponds
→ Storage lagoons (fish farming/duck or geese raising in warm season)
→ Effluent recharging ground water

In humid, warmer areas

Organic wastewaters → Preliminary treatment → (Anaerobic ponds) → Facultative ponds (with growth of floating plants) → Fish farming in paddy fields → Lotus or reed ponds → Effluent to receiving waters

High strength organic wastewaters, such as those from sugar mills, wineries, food processing industries, are sent to anaerobic digesters after pre-treatment. Methane formed in the digesters is removed to be used as fuel in the houses of farmers (China produces 720 M m^3 of methane gas annually, and some 15 million farmers use methane gas as domestic fuel). The supernatant from the anaerobic digesters is then sent to hydrophyte-growing facultative ponds or duck/geese ponds or lotus/reed ponds. The sewage sludge may be applied as fertilizer on farmland, grassland or forest, or it can be used to grow mushrooms or earthworms (fish food).

2.6.2.3 Fish farming

Using the pretreated wastewater for fish farming has become lucrative in all parts of China. The high rate of fish production (4,000 to 16,000 $kg\,ha^{-1}$) with very low cultivation cost, makes fish farming more attractive financially than raising rice or wheat or vegetables. The case history of fish farming ponds in Changsha city, Hua-nan province of China, illustrates the benefits of ecological wastewater treatment: About 25,000 $m^3\,d^{-1}$ of municipal sewage and 50,000 $m^3\,d^{-1}$ of organic waste waters from industries and animal farms, are sent to the fish farms. The total area of sewage fish ponds in the city is 1,430 ha. Under conditions of sewage hydraulic loading of 400–500 $m^3\,ha^{-1}\,d^{-1}$, BOD_5 loadings of 20–30 $kg\,ha^{-1}\,d^{-1}$, and pond water temperature of 15–25°C, the ponds achieved the following removal efficiencies: SS 74–83%, BOD_5: 75–91%, Total nitrogen: 70%. The fish production ranged from 4500–6000 $kg\,ha^{-1}\,yr^{-1}$.

In macrohydrophyte ponds, water hyacinth (*Eichhornia crassipes*), water peanuts (*Alternathera phioloxeroides*) and water lotus (*pistia stratiotes*) grow fast for the following reasons:

- Their utilization of solar energy (3–5%) is much higher than crop plants (0.5–1.0%),
- The hydrophytes make use of nitrogen and phosphorus nutrients in the wastewater, and
- The bacterial number and biomass in a water hyacinth pond is 10–20 times greater than in an ordinary pond of equal volume. The hydrophytes are used as food by fish, poultry, pigs and other animals.

Relative to water hyacinth (*Eichhornia crassipes*) ponds, water peanut (*Alternathera philoxeroides*) ponds are more efficient in removing BOD_5, COD, TN and NH_3-N and less efficient in removing TP and PO_4^{3-}. The high pollutant reduction ability of water peanut pond is attributed to the bacteria, *Bacillus*, *Psecomonas*, and *Alcaligenes*. These adhere to the roots of the plants and bring about the degradation of material containing C, N and P into CO_2, NH_4-N, and PO_4-P which are then taken up by the roots of the plants. Thus, N and P in waste water are effectively removed, while benefiting the plant (Hang Xu et al., 1991).

Shi & Wang (1991) studied the purifying efficiency and the mechanism of the aquatic plants in the bio-ponds. They found that the increasing peroxidase activity in the plant body promoted the phytophysiological metabolic activity resulting in the acceleration of the pollutant removal rate.

2.6.3 TYPES OF WASTEWATER REUSE

The various issues involved in wastewater reuse have been discussed in a number of studies (National Academy of Sciences, 1972, Ayers, 1975, AWWA, 1985, Kramer, 1985, WHO 1989).

Figure 2.6.2 (source: *Future of Water Reuse* – Proc. of the Water Reuse Symposium, San Diego, CA., USA, Aug. 84, v.3, p. 1738) shows the various types of water reuse.

The following case histories have been drawn from Dean and Lund (1981, p. 237, 241, 243, 245).

A.1: Direct reuse: Highly treated, reclaimed waste water is piped directly into the drinking water system.

Case history: Windhoek, Namibia, SW Africa.

A.2: Indirect reuse: Highly treated waste water is discharged to the environment for dilution, natural purification and subsequent withdrawal for water supply (e.g. discharge to river upstream of water plant intake, or replenishment of groundwater supplies either through infiltration from the ground surface or by direct injection into the aquifers).

Case history: Thames River basin, London, England.

Non-potable reuse:

B.1: Agricultural irrigation, including irrigation for edible and non-edible crops, pasture irrigation, and livestock watering.

Case history: Dan Region project, Israel.

B.2: Industrial uses: Power plant cooling, processing plants and construction.

Case history: Contra Costa County, California, USA.

B.3: Landscape irrigation and recreation: including irrigation of turf and ornamental plants in golf courses, parks; playgrounds, landscaping, use of reclaimed water to fill artificial lakes for recreational and aesthetic purposes.

Case history: Santee, California, USA.

B.4: Industrial process water: water use in manufacturing.

B.5: Municipal non-potable use: use of water for toilet flushing, fire fighting, and air-conditioning.

B-6: Environmental applications: wildlife refuges, or in-stream benefits.

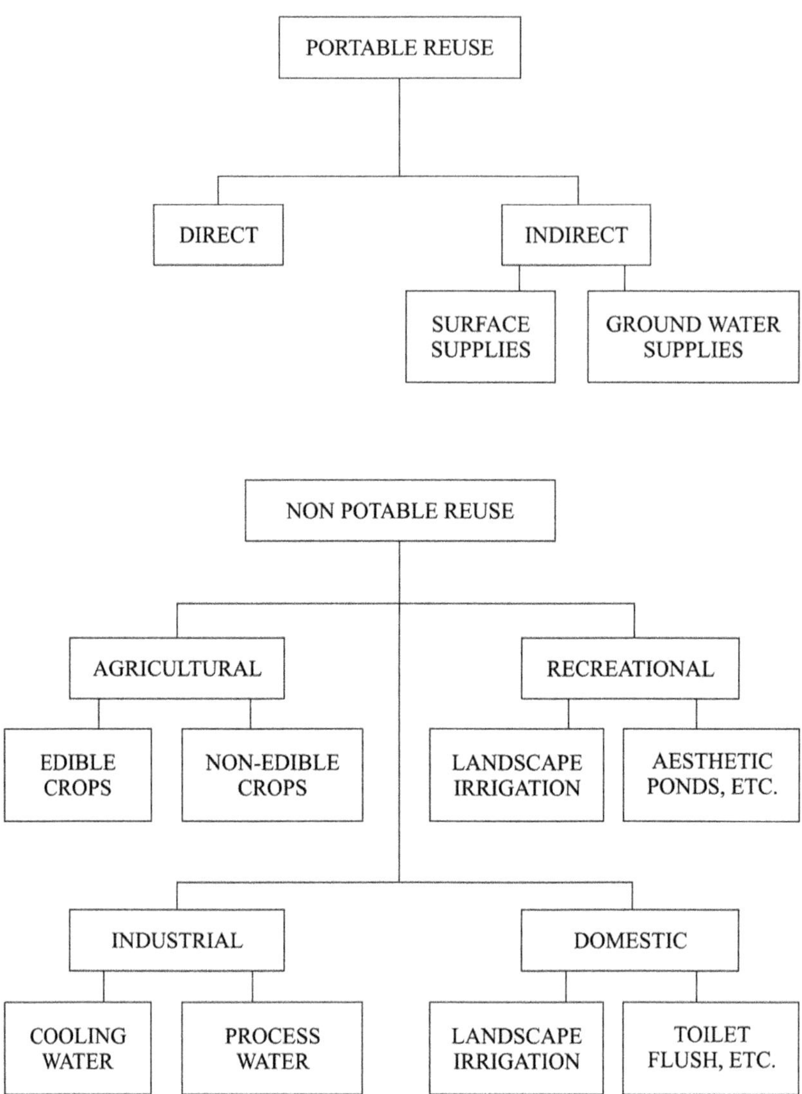

Figure 2.6.2 Types of wastewater reuse
Source: AWWA, 1985, p. 1788

2.6.4 USE OF WASTEWATER IN IRRIGATION

The standards of water quality for the use of wastewater in irrigation have been given by Matters (1981). Kramer (1985, p. 1669) listed the allowable limits for wastewater for specific uses (Table 2.6.1). He gave a summary of the regulations regarding the treatment requirements required for the use of wastewater in the state of Arizona, USA.

The land application of waste water has to take into consideration the removal of pathogenic microorganisms. The bacterial survival depends upon moisture, temperature, organic matter and the presence of antagonistic soil microflora. The removal of the viruses by adsorption depends upon the salt concentration, pH, soil composition, organic matter and the electronegativity of the virus.

Waste water is applied to land through (a) spray irrigation, (b) overland flow, and (c) infiltration percolation (Jones, 1982)

Spray irrigation: Waste water is sprayed on to land at some pre-arranged rates (cms/week). The material then infiltrates and percolates within the boundaries of the disposal site. In USA, spraying is used on flat lands, or on slopes upto 20%, or in wooded sites (as in California).

Overland runoff: In this arrangement, waste water moves along a flow path as a sheet flow down the slope. Slopes are usually less than 10%. The waste water infiltrates the top two feet of the soil. Vegetation usually picks up P and N from waste water. The water flows through the furrows, and plants are grown on the ridges between the furrows. The application rate may be 20–30 times of spray irrigation.

Rapid infiltration: The principal purpose of this kind of application is groundwater recharge. Normally, the liquid applied in this method is the disinfected effluent which has undergone secondary treatment.

As fresh water becomes scarcer, the need and justification for the reuse of wastewater is getting strengthened. The place of wastewater treatment and reuse in meeting the urban demands for water, could be appreciated from the following scheme.

The sources of fresh water are:

- stream flow,
- groundwater,
- agricultural tradeoff, and
- interbasin transfer.

These waters are subjected to raw water treatment before being supplied to the urban areas. Wastewater gets produced in the cities when the fresh water is used for domestic, municipal and industrial purposes.

The wastewater has to be treated in the following ways before being reused:

Primary,
↓
Secondary → wastes from the secondary treatment end up as urban effluent,
↓
Tertiary → wastes from the tertiary treatment end up as urban effluent.

In most cases, water would be ready for reuse after the tertiary treatment. In a few cases, desalting may have to be undertaken after the tertiary treatment before the water is reused.

Allowable limits for wastewater for specific uses are shown in Table 2.6.1 (source: Kramer, 1985, p. 1669). Climate is a critically important factor in deciding about the method of application.

Table 2.6.1 Allowable limits for wastewater for specific uses

Description *pB*	*1* *4.5-9*	*2* *4.5-9*	*3* *4.5-9*	*4* *4.5-9*	*5* *4.5-9*	*6* *4.5-9*	*7* *4.5-9*	*8* *4.5-9*	*9* *4.5-9*	*10* *6.5-9*
CFU/100 ml										
Geometric mean*	1000	1000	1000	1000	1000	200	2.5	2.2	1000	200
Single sample not to exceed	4000	4000	4000	4000	2500	1000	75	25	4000	800
Turbidity (NTU)	–	–	–	–	–	–	5	1	5	1
Enteric virus (PFU)	–	–	–	–	–	–	125/401	1/401	125/401	1/401
Entamoeba histolytioa	–	–	–	–	–	–	–	N.D.	–	N.D.
Giardia lamblia	–	–	–	–	–	–	–	N.D.	–	N.D.
Ascaris lumbricoides	–	–	–	–	–	–	N.D.	N.D.	N.D.	N.D.
Common large tapeworm	–	–	N.D.	N.D.		–	–	–	–	–
Trace substances	R 9-21-209	R 9-21-209	R 9-21-209	R 9-21-209	R 9-21-209	R 9-21-209	R 9-21-209	R 9-21-209	R 9-21-209	R 9-21-209
Organic chemicals	–	–	–	–	–	–	–	–	–	–
Radiochemicals	–	–	–	–	–	–	–	–	–	–
Treatment likely to achieve standards	A	A	A	A	A	A	B	C	A	A

CFU – Colony Forming Units; NTU – Nephelornetric Turbidity Units; PFU – Plaque Forming Units; N.D. – None Detectable (i.e. should not be present); A – Secondary disinfection, B – Secondary disinfection, infiltration; C – Soil aquifer treatment of secondary disinfection, coupled with chemical coagulation.

*Geometric mean of minimum of five samples. 1 – Orchards; 2 – Fiber, seed and forage; 3 – Pastures; 4 – Livestock watering; 5 – Processed foods; 6 – Landscape area, restricted access; 7 – Landscape area – open access; 8 – Food to be consumed raw; 9 – Partial body; 10 – Whole body.

Source: Kramer 1985, p.1669.

In Phoenix, Arizona, USA, large quantities (225 million gallons per year = $910 \times 10^3\ m^3\ yr^{-1}$) of wastewater is applied.

2.6.5 GEOPURIFICATION

Geopurification or Soil – Aquifer – Treatment (SAT) is a low-technology treatment system, whereby the sewage effluent is subjected to groundwater recharge with infiltration basins. After SAT, the water will have very low suspended solids content, essentially zero BOD, much reduced concentrations of N, P, organic compounds and heavy metals, and almost zero levels of pathogens.

Four types of Soil-Aquifer Treatments are possible:

1. Natural drainage of renovated water into stream, lake or low-lying area,
2. Collection of renovated water by subsurface drain,
3. Infiltration areas in two parallel rows and lines midway between, and
4. Infiltration areas in centre surrounded by circle of wells (Bouwer, 1994).

Bouwer (1994) shows how there is improvement in quality when mildly chlorinated secondary effluent (activated sludge) is let into infiltration basins, and the treated water then picked up through the recovery wells. Such water can be used for unrestricted irrigation.

Studies in the Dan region sewage reclamation project of Israel show that the long-term (20 yr) use of SAT reduces the trace element sorption capacity of the soils (Roehl & Banin, Abstract, ICOBTE V, Vienna, July 1999). The soils undergo leaching of carbonates and manganese oxides and the accumulation of organic matter. The recharged soils have been found to have drastically reduced sorption capacity for (say) copper. This has been attributed to the acidolytic dissolution of the carbonates and manganese oxides, both processes being driven by the decomposition of the additional organic matter arising from recharge.

2.6.5.1 Regulations regarding wastewater reuse

Regulations regarding wastewater reuse may either specify water quality requirements or prescribe the treatments required for specific reuse. The second approach is generally preferred, as it is easy to enforce.

Secondary treatment is required for:

- Irrigation for fibrous or forage crops not intended for human consumption,
- Irrigation of orchard crops by methods which do not result in direct application of water on fruit or foliage,
- Watering of farm animals other than producing dairy animals.

Secondary treatment and disinfection are required:

- Irrigation for any food crops where the product is subjected to physical or chemical processing sufficient to destroy pathogenic organisms,
- Irrigation of orchard crops by methods which involve direct application of water to fruit or foliage,

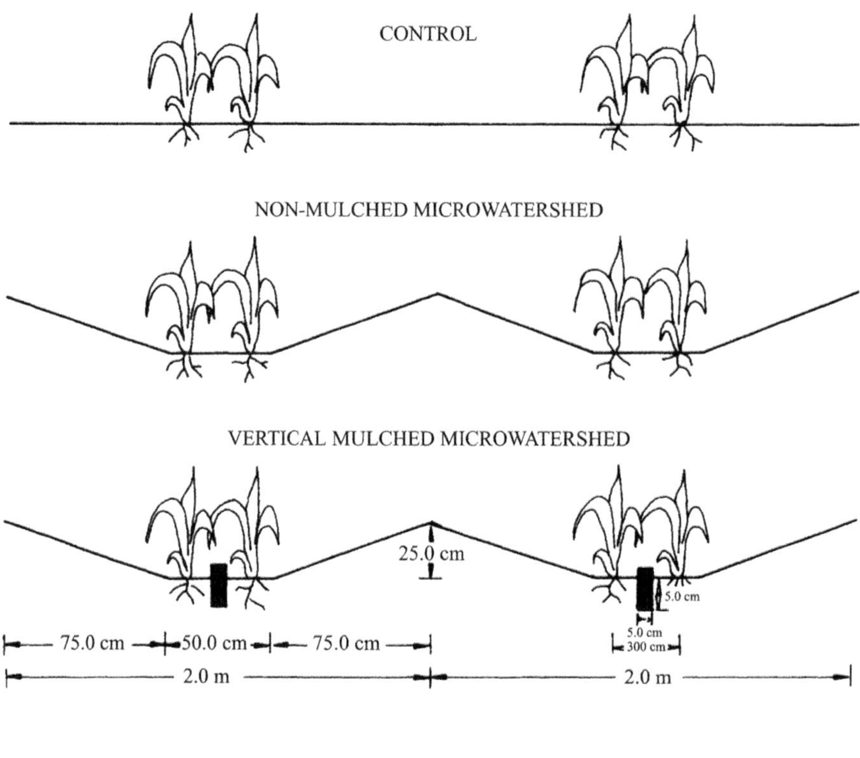

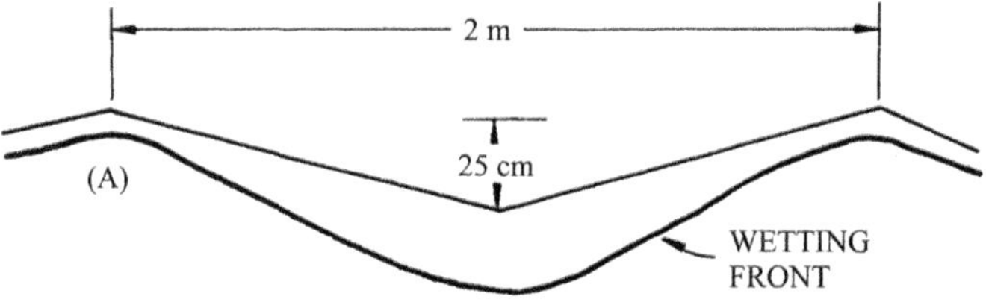

(B) MICROWATERSHED WITH MULCH

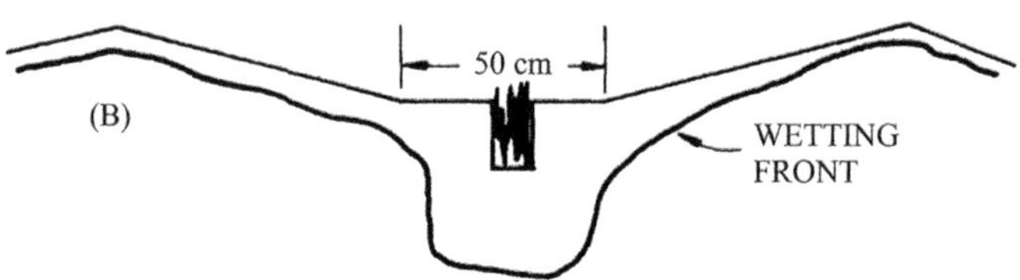

Figure 2.6.3 Different types of wastewater applications
Source: Jones, 1982

- Irrigation of golf courses, cemeteries, etc.,
- Watering of producing dairy animals,
- Water stored in impoundments intended for partial body contact recreation.

Tertiary treatment and disinfection required for:

- Waters stored in impoundments intended for full body recreation,
- Irrigation of school grounds, playgrounds, lawns, etc. where children congregate,
- Irrigation of food crops which are consumed in their raw or natural state.

Several countries have prescribed regulations about the use of waste water in irrigation (Sepp, 1976). Waste water is sought to be used in a manner which while providing irrigation to plants, will not have adverse health consequences to the consumers of plant-derived foods. Two cases are cited:

Israel: Settled sewage (primary effluent) may be used on industrial and fodder crops, pastures and hay, vegetables eaten in a cooked state, fruit trees, ornamental plants or seed plants. Irrigation should be stopped one month before the harvest in the case of apples, pears and plums, broccoli and cauliflower when furrow irrigated, and tomatoes used for canning if furrow irrigated.

California (USA): It is forbidden to use raw sewage to irrigate crops. Settled but undisinfected sewage may be used only on industrial, grain and fodder crops, and on vegetables grown for seed purposes. Such sewage may not be used on water growing vegetables, berries, low-growing fruits and vineyards and orchards during fruit growth. Completely treated, well oxidized, and reliably disinfected effluent conforming to the bacterial requirements of US Public Health Service (PHS) for drinking water standards, is allowed to be used on all crops.

2.6.6 ECONOMICS OF WASTEWATER REUSE

Technology is available for the primary and secondary and tertiary (coagulation, sand filtration and chlorination) treatment of wastewater to bring down the faecal coliform concentrations to essentially zero, but that is expensive (about USD 200–500/1,000 m^3; Richard et al., 1992). Desalination of water is equally expensive – about one USD per 1,000 m^3 for every 10 $mg\,l^{-1}$ salt removed with reverse osmosis, or about 2,000–3,000 per 1,000 m^3 for seawater (Bouwer, 1994).

Al-A'Ame & Nakhle (1995) gave a case study of the relative economics of wastewater reuse versus desalination. In Jubail, Saudi Arabia, there is a municipal wastewater plant of 19 MGD (million gallons per day) (73,625 $m^3\,d^{-1}$) capacity, and an industrial wastewater plant of 11 MGD capacity (42,625 $m^3\,d^{-1}$).

The wastewaters are subjected to biological treatment followed by pressure filtration. The effluents have the following characteristics:

	TDS	*TSS*	*BOD*	*SAR*
Municipal	936	4.4	2.4	7.4
Industrial	762	2.1	2.4	10.5

Treatment of these wastewaters costs less (USD 2,000 per 1,000 m^3) than desalination (about USD 2,670 per 1,000 m^3 in Jubail). The treated effluent is largely used for landscape irrigation. Thus, wastewater reuse in Jubail is not only cost-effective but is environmentally beneficial. Desalination is a viable option only for the energy-rich countries of the Middle East.

Developing countries have neither the capital nor human resources needed to build, maintain and operate plants for either Advanced Wastewater Treatment or desalination. For those countries, the WHO (1989) guidelines could be used, namely, maximum faecal coliform concentrations of 1,000 per 100 ml and upto one helminthic egg per liter. Low technology treatments like lagooning and bioponds with detention times of more than a month are effective in the removal of most of the pathogens in the wastewater, after which the effluent can be recharged to groundwater.

Primary and secondary treatments typically cost USD 100/1,000 m^3. Costs of post-treatment technologies (carbon filtration, reverse osmosis on half the flow and disinfection) for a 0.4 million m^3/d plant, work out to about USD 230/1,000 m^3. The costs would increase to USD 300/1,000 m^3 if the carbon absorption were deleted, and the entire flow went through reverse osmosis followed by disinfection. As membrane technologies are improving and becoming cheaper, the latter may turn out to be the most cost-effective option in future. The total cost of converting raw sewage into potable water is about USD 400/1,000 m^3 (Bouwer, 1993).

Soil – Aquifer Treatment (SAT) by itself is inexpensive, since only pumping costs from recovery wells are involved – about USD 5/1,000 m^3 if the groundwater is shallow, and USD 50/1,000 m^3 1,000 if it is deep (50 m) (Bouwer, 1993). The preferred option is a sequence which avoids the most expensive component, namely, the tertiary treatment component (AWT), and substitutes it by SAT.

2.6.7 HEALTH HAZARDS IN WASTEWATER REUSE

It is known that waste waters from urban areas contain a variety of faecally-excreted human pathogens including helminths, protozoans, bacteria and viruses. Waste water may contain high concentration of pathogens (10^7–10^4organisms per gm of faeces) with long persistence times (weeks to months) (Gunnerson et al., 1985). Thus, when waste water is used for irrigation, the concentrations of the pathogens in the irrigated soil and irrigated crops is sufficiently high and their persistence sufficiently long as to create a health hazard to the exposed population groups (WHO, 1989).

Gunnerson et al. (1985) gave the following summary of health hazards arising from the use of raw or only partially treated waste water in the Developing countries:

- To the general public consuming salad or vegetable crops irrigated with raw waste water: ascariasis, trichuriasis, cholera, and possibly tapeworm (from eating meat of the cattle grazing on waste water irrigated pasture) infections, leading to cholera, typhoid fever, giardiasis, etc.
- To workers engaged in waste water irrigation: ancyclostomiasis (hookworm), ascariasis, possibly cholera, and to a much lesser extent, infection caused by some other bacteria or viruses.

- To the general public living in the proximity of waste water irrigation projects, particularly sprinkler irrigation projects using raw or poorly treated waste water: Minor transmission of diseases, particularly to children, caused by enteric viruses; also possibly limited transmission of shigellosis and other bacterial diseases due to contact.

Gunnerson et al. (1985) ranked the pathogenic agents (connected with waste water irrigation) in the following order:

- High risk (high excess incidence of infection): Helminths (*Ancylostoma, Ascaris, Trichuria* and *Taenia*),
- Medium risk (medium excess incidence of infection): Enteric bacteria (cholera, typhoid, *shigella*, etc.),
- Low risk (low incidence of excess infection): Enteric viruses.

2.6.8 USE OF SEWAGE SLUDGE AS FERTILIZER

Sewage sludge is a byproduct of wastewater treatment. Agricultural utilization of the sewage sludge not only constitutes a sustainable outlet for the disposal of the sludge, but it also enhances crop productivity as the sludge contains valuable micro-nutrients, trace elements and organic matter. Sometimes the sludge may contain toxic elements and pathogens which may enter the food chain and cause diseases in the humans and animals.

Sewage sludge is used in agriculture as a fertilizer and soil conditioner. It has been observed that sludge application reduces the surface runoff, and gives some protection against soil erosion.

The composition of sewage sludge varies greatly, depending on the source of sewage (domestic, industrial, etc.). The mean concentration level and range (percent content of dry matter) of the fertilizer content are as follows: N – 2.2 (1.5–3.5), P – 1.7 (1.0–2.5), K – 0.15 (0.05–0.3). The undesirable consequences of sludge application arise from the heavy metal content of the sludge, which may be taken up by the plants.

The metals present in sewage sludge may be divided into three categories on the basis of their availability to plants:

- Unavailable forms, such as insoluble compounds (oxides or sulfides);
- Potentially available forms, such as insoluble complexes, metals linked to ligands, or forms attached to clays and organic matter; and
- Mobile and available forms, such as hydrated ions or soluble complexes. A soil which has high pH (i.e. alkaline) and high cation exchange capacity, will be able to immobilize the metals added to it via the sewage sludge. A soil which is presently alkaline may not always remain that way. Thus, if at a later stage the pH drops (i.e. becomes acidic), the metals may get released.

Elements such as iron, zinc and copper are essential elements needed by man. The same elements may become toxic at high concentrations, however. Cadmium is a highly toxic element. Because of its geochemical affinity with zinc, it enters the plants along with zinc.

In a study made in Norway, sludge application at rates of 60–120 t (dry matter) per ha increased the Ni and Zn contents of plants, but had no significant effect on the concentration levels of Cd and Pb. Excessive application of sludge could increase the content of heavy metals in the crops to levels which render them unfit for human consumption.

Liming (1.5–6 t of $CaCO_3$ ha^{-1}) reduces the heavy metal uptake by plants *in the short run.* The heavy metals will continue to persist in the top soil, however. It has been reported that sludge-amended soils almost invariably contain higher levels of Cd than garden soils. Potential increase in the cadmium content of plants due to the application of sewage sludge constitutes the most important health hazard, which can be mitigated by increasing the available content of zinc in the soil. Sludge should not be applied to soils which are used for growing vegetables.

2.6.8.1 Case study of sewage sludge in Cairo, Egypt

As the Greater Cairo Wastewater project becomes fully operational, substantial quantities of sewage sludge are being produced. The sludge was treated in different ways to understand the chemical changes arising there from. Experiments were conducted to determine to what extent the application of treated sludges affect the heavy metal concentrations in plants. The study confirmed the earlier observation (Hernandez, 1991) that the application of aerobic sludge resulted in the reduction of heavy metal content in the growing plant, and is therefore to be preferred. Incinerator sludge brought about the largest decrease in the heavy metal content of the plants, possibly because of the formation of oxides or because of the alkalinity of the soil treated with incinerator sludge.

REFERENCES

Al-A'Ame, M.S. & G.F. Nakhle, 1995. Wastewater reuse in Jubail, Saudi Arabia. *Water Res.* 29(6): 1579–1584.

American Waterworks Association (AWWA) 1985. *Future of Water Reuse*, Proc. Symp. on Water reuse, Aug. 26–31, 1984, San Diego, Calif., USA, v. III.

Ayers, R.S. 1975. Quality of water for irrigation. *Proc. Irrig. and Drain. Div., Speciality Conf., Amer. Soc. Civil Engineers*, Logan, Utah, 13–15 Aug., 24–56.

Bouwer, H. 1993. From sewage farm to zero discharge. *Europ. Water Pollu. Control.* 3(1): 9–16. Amsterdam: Elsevier

Bouwer, H. 1994. Role of geopurification in future water management. In *Soil and Water Science: Key to Our Understanding Our Global Environment.* Soil. Soc. Amer. Sp. Publ., 41: 73–81.

Dean, R.B. & E. Lund, 1981. *Water Reuse – Problems and Solutions.* London: Acad. Press.

Gunnerson, C.G., H.I. Shuval & S.Arlosoroff, 1985. *Future of Water Reuse*, Proc. Symp. on Water Reuse, Aug. 26–31, 1984, San Diego, Calif., USA, III, 1576–1605.

Hang Xu et al. 1991. Experimental studies on the purification of wastewater in water peanut ponds. In B.Z. Wang et al., (ed.) Low-cost and energy-saving wastewater treatment technologies. *Water Sci. & Tech.*, 24 (5), 97–109.

Hernandez, T., J.I. Moreno & F. Costa, 1991. Influence of sewage sludge application on crop yields and heavy metal availability. *Soil Sci. Plant Nutr.*, 37(2): 201–210.

Jones, P.H. 1982. Wastewater treatment technology. In Laconte, P. & Y.Y. Haimes (eds.), *Water Resources and Landuse Planning: A Systems Approach*, pp. 93–132. The Hague: Martinus Nijhoff Publishers.

Kramer, R.E. 1985. Regulations for the reuse of wastewater in Arizona.*Future of Water Reuse*, Proc. Symp. on Water Reuse, Aug. 26–31, 1984, San Diego, Calif., USA, III, 1666–1672.

Laconte, P. & Y.Y. Haimes (eds.) 1982. *Water Resources and Landuse Planning: A Systems Approach*. The Hague: Martinus Nijhoff Publishers.

Matters, M.F. 1981. Arizona rules for irrigating with sewage effluent. In *Proc. Sewage Irrigation Symp.*, Phoenix, Arizona. US Water Conservancy Lab., 6–12.

National Academy of Sciences and National Academy of Engineering. 1972. Report no. 5501-00520. Supt. of Documents.

Richards, D., T. Asano & G. Tchobanoglous, 1992. *The cost of wastewater reclamation in California*. Rep. from Dep. of Civil and Environ. Engg., Univ. of California, Davis.

Sepp, E. 1976. *Use of sewage for irrigation – A literature review*. Bureau of Sanitary Engineering, California, USA

Shi, S., & X. Wang, 1991, The purifying efficiency and mechanism of the aquatic plants in ponds. In B.Z. Wang et al., (ed.) Low-cost and energy-saving wastewater treatment technologies. *Water Sci. & Tech.*, 24(5), 63–73.

Shiklomanov, I.A. 1998. World Water Resources: A new appraisal and assessment for the twentyfirst century. Paris: UNESCO.

Wang, B. 1991. Generating and resources recoverable technology for water pollution control in China. In B.Z. Wang et al., (ed.), *Low-cost and energy-saving wastewater treatment technologies*, *Water Sci. & Tech*. 24(5): 9–19.

World Health Organization (WHO). 1989. *Health guidelines for the use of wastewater in agriculture and aquaculture*. Tech. Bull. Ser. 77, Geneva: WHO.

Chapter 2.7

Etiology of diseases arising from toxic elements in drinking water*

U. Aswathanarayana (India)

A caveat should be entered straightaway about the scope of the chapter – *this chapter covers only those diseases (such as, arseniasis and fluorosis) where drinking water is the principal route of intoxication*. The chapter does not discuss numerous water-based, parasitic, bacterial, viral, etc. diseases, such as, gastrointestinal disorders, cholera, typhoid, malaria, schistosomiasis, etc.

2.7.1 ROUTES AND CONSEQUENCES OF INGESTION OF TOXIC ELEMENTS

The potential exposure routes through which toxic substances reach humans are given in Table 2.7.1.

Drinking water may be contaminated with radionuclides near nuclear reactors or nuclear facilities or as a consequence of nuclear accidents (such as, Chernobyl and Fukushima, Japan). The Maximum Contaminant Levels allowable in drinking water are: Combined Radium-226 and Radium 228 – 5 pCi/L; Gross Alpha Particle Activity (excluding radon and uranium) – 15 pCi/L. The Maximum Permissible Concentrations (MPC) in terms of ($Ci\,ml^{-1}$) of selected radionuclides in drinking water is given in Table 2.7.2.

*Basic data and concepts in this chapter have largely been drawn from the author's own work, "*Water Resources Management and the Environment*", 2001, pp. 311–338.

Table 2.7.1 Potential exposure routes of contaminants from various sites

Exposure route	*Domestic*	*Commercial/industrial*	*Recreational*
Food			
Ingestion	L	–	L
Groundwater			
Ingestion	L	A	–
Surface water			
Ingestion	L	A	L, C
Dermal contact	L	A	L, C
Sediment			
Dermal contact	C	A	L, C
Air			
Inhalation of vapor phase chemicals			
Indoors	L	A	–
Outdoors	L	A	L
Inhalation of particulates			
Indoors	L	A	–
Outdoors	L	A	L
Soil/dust			
Incidental ingestion	L, C	A	L, C
Dermal contact	L, C	A	L, C

L: Lifetime exposure, C: Exposure in children may be significantly greater than in adults, A: Exposure to adults (highest exposure is likely to occur during occupational activities), –: Exposure of this population via this route is unlikely to occur.
Source: US EPA 1989

Table 2.7.2 Nuclear and health physics data for selected radionuclides

Radionuclide	*Half-life (years)*	*Major radiation*	*Critical organ*	*Biological half-life*	*MPC* ($\mu Ci\, ml^{-1}$)*
^{3}H	12.26	α	Total body	12 days	3×10^{-3}
^{90}Sr	28.1	β	Bone	50 years	3×10^{-6}
^{129}I	1.7×10^{7}	β, γ	Thyroid	138 days	6×10^{-8}
^{137}Cs	30.2	β, γ	Total body	70 days	2×10^{-3}
^{226}Ra	1,600	α, γ	Bone	45 years	3×10^{-8}
^{239}Ra	24,400	α	Bone	200 years	5×10^{-6}

*MPC = Maximum Permissible Concentration for water consumed by the general public without readily apparent ill effects (these figures are revised as the toxicity of the radionuclides is better understood).
Source: Paper 1.1 in Laconte and Haimes 1982

2.7.2 ARSENIASIS

2.7.2.1 Introduction

Arseniasis (also called arsenicosis or arsenicism) has emerged as an environmental issue of global concern. Arseniasis (manifested in the form of skin lesions, vascular damage, cancers of bladder, lung, liver and kidney, etc.) arises from the ingestion of

Table 2.7.3 Incidence of arseniasis in Asia

Area	*Source*	*Population at risk*	*Non-cancer manifestations**	*Cancer manifestations***
Bangladesh	Well water	50,000,000	M/K, D, G, B	S
China				
Guizhou	Burning of high-As coal	200,000	M/K, G, P	S, Li
Inner Mongolia	Well water	600,000	M/K, G, P	A, S, Lu, Li, U, K
Shaanxi	Well water	1,000,000		
Xinjiang	Well water	100,000		
Yunnan	Metal smelting	100,000	M/K, G, P	A, S, Lu
India				
West Bengal	Well water	1,000,000	M/K, D, G, B,P	S
Japan				
Toroku/Matsuo	Metal smelting	217 patients	M/K, D, G, B,P	A,S, Lu, U, K
Nakajo	Contaminated water	44 patients	M/K, G, B, P	A, Lu, Li, U, K
Taiwan				
Southwest coast	Well water	100,000	M/K	A,S,N, Lu, Li, U, K, P
Northeast coast	Well water	100,000	M/K	A,S,N, Lu, Li, U, K
Thailand				
Ronpibool	Tin mining	1000 patients	M/K	A
Philippines				
Mindanao	Geothermal drilling	?		

*Non – cancer: M/K – melanosis/keratosis; D – dermatitis; G – gastroenteritis; B – bronchitis; P – polyneuropathy.
**Cancer: A – all sites; S – skin; Nasal cavity – N; Lung – Lu; Liver – Li; Urinary bladder – U; Kidney – K; Prostate – P.
Besides these, the following cancers have been reported from the regions noted against them: esophagus – Inner Mongolia, cervix uteri – Nakajo, and stomach-northeast coast of Taiwan.
Source: Chen et al., 1999

excessive quantities of arsenic, through drinking water, and inhalation. Chen et al. (1999) summarized the present situation about the incidence of arseniasis in Asia (Table 2.7.3).

The natural sources of arsenic are: volcanic exhalations and products, forest fires, and weathering of As-bearing minerals, while anthropogenic sources are: fertilizers and pesticides, slag piles and mining wastes, combustion of fossil fuels and smelting of non-ferrous metals (Fig. 2.7.1). In terms of geologic setting, the natural sources of arsenic causing arseniasis are: estuarine sediments in West Bengal and Bangladesh, Quaternary volcanics in Chile, black shale in Taiwan, etc. Anthropogenic sources causing arseniasis are: coal burning as in China, and mining and smelting as in NWT, Canada, etc.

2.7.2.2 Pathways of arsenic to humans

The following are the pathways of arsenic to man, through:

- Drinking water – as in Bangladesh, West Bengal (India), Inner Mongolia, Shaanxi, Xinjiang (China), southwest and northeast coasts of Taiwan, western USA, etc.
- Inhalation of arsenic-containing aerosols – coal burning, as in Guizhou (China), copper smelting and arsenic mining, as in Yunnan (China), NWT (Canada), etc.

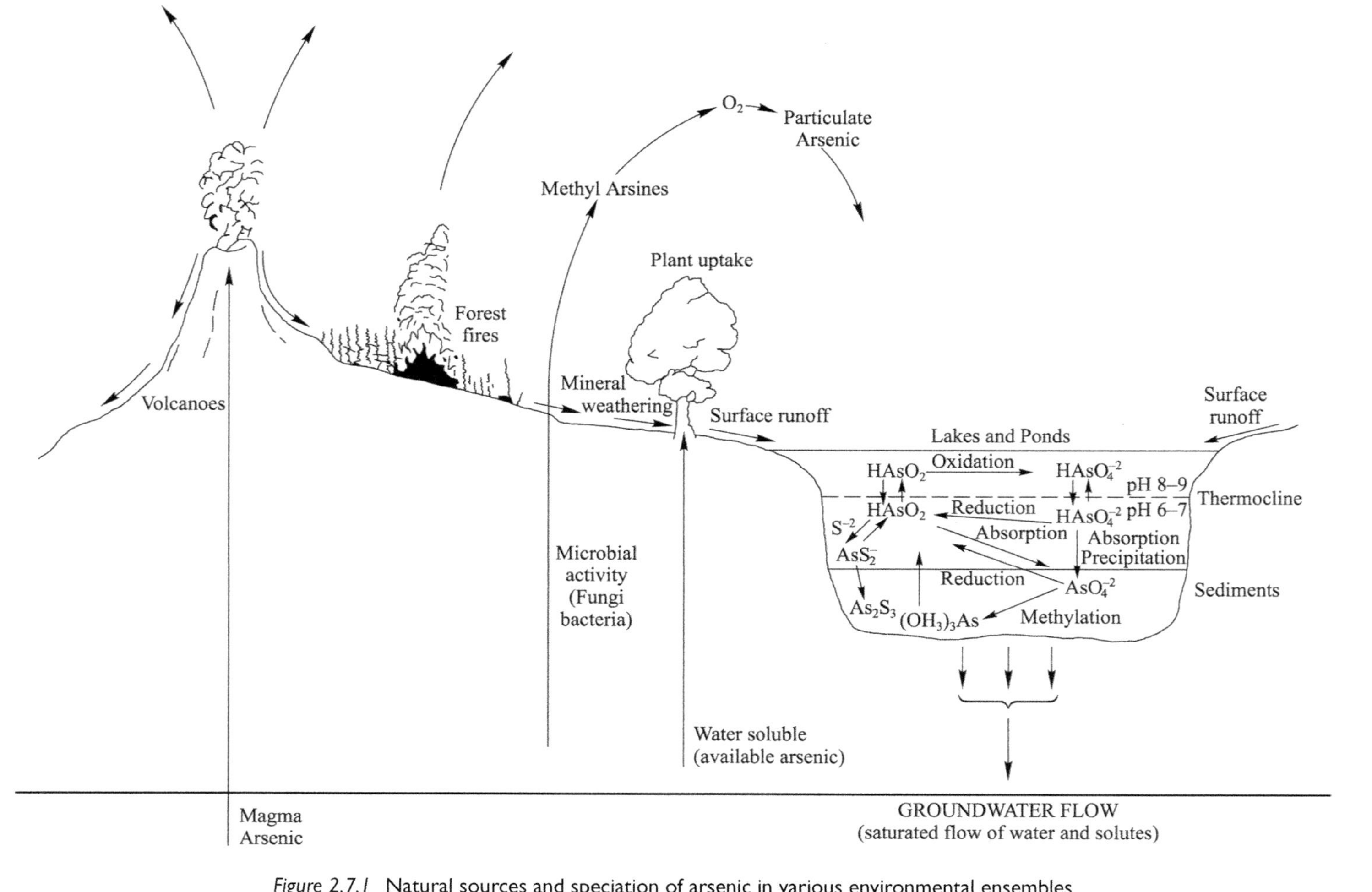

Figure 2.7.1 Natural sources and speciation of arsenic in various environmental ensembles
Source: Piver, 1983

- Dermal uptake: Although the human skin does act as a barrier, it cannot completely block the entry of environmentally harmful substances from entering the body. For instance, arsenic-induced skin cancers resulting from the use of arsenical pesticides have been reported among the wine growers of Beaujolais. Similarly, rice farmers, fishermen and salt workers whose feet and legs are exposed to arsenical waters for long periods, have been found to develop arsenical skin lesions. The dermal pathway explains the prevalence of "Blackfoot disease" in Taiwan where rice farming is practiced, but not in arid Chile, though in both the cases the water has high arsenic content.
- Diet: The daily dietary intake of inorganic arsenic ranges from 8–14 $\mu g\, d^{-1}$ in USA, and 5–13 $\mu g\, d^{-1}$ in Canada. It is probably 50 $\mu g\, d^{-1}$ in the case of Taiwan. Though the content of arsenic in fish (0.1–64 mg As kg^{-1}) and shellfish (0.2–126 mg As kg^{-1}) is high, it is present in a relatively non-toxic form (e.g. arsenobetaine: $(CH)_3$ As^+ CH_2COOH), which is readily excreted. Herbal medicines of Indian and Chinese origin which are becoming increasingly popular in the western countries, have high content (25–107,000 mg As kg^{-1}) of arsenic, all of which is in inorganic form. No significant incidence of arseniasis has been reported any where in the world based on diet alone.

Chandrasekharam (2008) traced the pathways of arsenic from water to food in West Bengal, India. Apart from drinking water, food is now known to be a source of arseniasis in West Bengal. Bioaccumulation of arsenic in food crops is strongly influenced by the arsenic content of the irrigated water, and chemical, physical and microbial characteristics of the soil. The arsenic content of vegetables and cereals grown using arsenious groundwater (85 to 108 μg/L)in West Bengal is 300% greater compared to the mean concentration generally reported in vegetables and cereals elsewhere in the world. For instance, people consuming 100 gm of "arum", a leafy vegetable, that could contain 022 mg/kg of arsenic, will reach the maximum daily allowable limit of arsenic by eating this leafy vegetable alone. Even food cooked with arsenic-contaminated groundwater showed high values (0.12–1.45 mg/kg) that fell well above the limit prescribed by WHO. Arsenic accumulation in rice roots is high where tube wells are used for irrigation. Where the roots of the rice plants are ploughed back into the soil after harvesting, the arsenic in the rice roots gets mobilized and infiltrates into the shallow aquifers. Thus arsenic from the groundwater, after entering the food chain, gets concentrated in the roots and enters the shallow aquifer thus establishing an "*Arsenic flow cycle*" in the rice fields. This is a catastrophe waiting to happen. This can only be prevented by using canal water for irrigation, and using rooftop rainwater harvesting for drinking water.

2.7.2.3 Where do we go from here?

The important issues in the case of arseniasis are posed in the form of four questions.

1. *Is arsenic an essential element?*
   Animal experiments conclusively show that arsenic has a physiological role affecting methionine metabolism. The essentiality of arsenic for human beings is,

however, not established. The central question is whether arsenic is like selenium (i.e. an essential element, with a Recommended Daily Intake, though toxic in high doses) or whether arsenic is like cadmium which is toxic at any level (gun kills – there is no such thing as a benevolent gun).

2. *Is there a threshold for arsenic exposure for non-cancer and cancer end-points?* There are sound reasons to believe that the dose – response relationship in the case of arsenic is non-linear. For instance, Guo et al. (1994) found the incidence of bladder cancer in Taiwan increases sharply above 500 $\mu g\,l^{-1}$. On the basis of the statistical analysis of data from Taiwan and elsewhere, Stöhrer (1991) concluded that there is a common threshold (400 $\mu g\,l^{-1}$ (?)) for skin cancers, internal cancers and non-cancer endpoints. While the incidence of arseniasis above 200 $\mu g\,l^{-1}$ is well established, there is much controversy about the risks of intake of less than 100 $\mu g\,l^{-1}$ (Hindmarsh, 2000).
3. *How do nutrition/socio-economic status, genetics, etc. modify the risk?* There is a clear correlation between the susceptibility for arseniasis (blackfoot disease in Taiwan and arsenical melanosis in West Bengal) and poverty and inadequate nutrition. That the victims having diffuse melanosis (preliminary stage of skin lesions) recovered when they discontinued the use of contaminated water and got nourishing food, demonstrates the essential correctness of the above surmise (SOES, 1996).

   Chile is a spectacular example of the role that genetics could play in modifying the risk. Arroyo of Chile (quoted by Brown, 1999) raised two pertinent questions:

   - Why is there no arseniasis among the people of Atacameño who have been drinking water with high levels of arsenic (600 $\mu g\,l^{-1}$) for many decades, and
   - Why do some people in Antafagosta develop arsenic-related diseases and others do not, though they are exposed to the same levels of arsenic?

4. *What is the best treatment for chronic arsenic intoxication?* Therapy with d-penicillamine did not prove effective. On the other hand, oral treatment with retinoids and the use of selenium as an antioxidant nutrient, hold promise to mitigate the cutaneous arsenicism and other forms of arsenic toxicity. Use of antidepressants for pain relief in regard to peripheral neuropathy, and the application of topical keratolytics for palmar-plantar hyperkeratoses, may offer short-term symptomatic relief (Kossnet, 1999). Field experiments in Chile, Inner Mongolia (China) and Romania showed that treatment with DMPS (2,3-dimercatopropane-1-sulfonate) increases the urinary excretion of arsenic in humans, without any adverse effects (Aposhian, 2000 – p. 54, Abstract in the Fourth Arsenic Conference, San Diego, June, 2000).

2.7.3 FLUOROSIS

Fluoride disorders and fluorosis are caused by excessive ingestion of fluoride, mainly through drinking water route.

Fluorine (at. wt.: 9; electron configuration: 2–7) has only one oxidation date (F^{-1}). Fluorite ($Ca F_2$) is the principal mineral of fluorine. The dissolution of fluorite takes place in the following manner:

$$Ca F_2 \leftrightarrow Ca^{2+} + 2F \tag{2.7.1}$$

Groundwaters with higher natural concentrations of Ca^{2+} are characterized by lower concentrations of F^-, and vice versa. Handa (1975) verified this observation in the field in the case of fluorous waters associated with Sirohi granites, Rajasthan province, western India. Similarly, in northern Tanzania, groundwaters in low-ca, alkali basalt aquifers are invariably more fluorous than groundwaters in high-ca, tholeiitic basalt aquifers. On the basis of this consideration, it is possible to identify broadly the areas where the groundwater is likely to be relatively safer (i.e. low fluoride concentration).

2.7.3.1 Geochemical distribution of fluorine

Fluorine in surface and ground waters is derived from the following natural sources:

- Leaching of rocks in the area: fluorine is concentrated in granites (750 ppm), and phosphatic fertilizers (3–3.5%),
- Dissolution of fluorides from volcanic gases by precipitation,
- Fresh or mineral springs,
- Marine aerosols and continental dusts, etc.

Anthropogenic sources of fluoride are:

- Industrial emissions such as freons, organo-fluorine compounds produced by the burning of fossil fuels, and from dust in the cryolite factories,
- Industrial effluents, and
- Runoff from farms using phosphatic fertilizers extensively.

In areas like northern Tanzania, the high fluoride content of the waters arises entirely from natural sources, and hence it is a case of "natural pollution". On the other hand, fluorine-bearing cryolite dust (as in Denmark in the past) is wholly anthropogenic. The high fluoride contents in parts of Greenland is attributed not only to the leaching of alkalic rocks by the stream waters, but also to the effluents from an industrial unit processing the fluorine-bearing ores. This is a good example of a situation where the 'natural' and anthropogenic "pollutions" cannot be separated.

Surface waters generally contain about 0.2 mg l^{-1} of fluoride, whereas the fluoride content of the groundwaters depend upon the nature of the aquifer. It may vary from 0.4 mg l^{-1} in aquifers composed of shales, to 8.7 mg l^{-1} in areas of alkalic rocks. Effluents from phosphate mining may contain upto 100 mg l^{-1} of fluoride. Plants have high fluoride content in the range of 0.1 to 10.0 ppm (dry weight). Some plants accumulate excessive quantities of fluoride – for instance, the fluoride content of tea could be 100–760 ppm. Among food items, fish concentrates may contain 20–760 ppm of fluoride. A saline encrustation, trona ($Na_2CO_3 \cdot NaHCO_3\ 2H_2O$), called 'magadi' in Swahili language, is used in the cooking of beans, and as salt. 'Magadi' has very high fluoride content, ranging from 200 to 6,000 ppm (Jan. 2000 issue of *National Geographic* of

USA carries a spectacular photograph of huge blocks of trona forming in Lake Natron, a rift lake in northern Tanzania). Poor people use trona in lieu of sea salt because trona improves the cookability of the beans, and is collectible at no cost.

While fluorosis arising from the drinking of fluorous waters continues to persist in northern Tanzania, fluorosis arising from the inhalation of cryolite dust in Denmark has been eliminated.

Groundwaters in several parts of India (provinces of Andhra Pradesh, Madhya Pradesh, Rajasthan, etc.) are fluorous. The problem is got over either through the defluoridation of water using alum-lime treatment or by passing the water through activated charcoal made from coconut shell. In the outlying parts of Hyderabad city in southern India, the tubewell water is used for most purposes (bathing, washing, gardening, etc.), and potable water drawn from Manjira river and supplied through tankers on payment basis, is used for drinking. This is an eminently sensible approach.

2.7.3.2 Pathways of fluorine to man

Fluorine is an essential element with Recommended Daily Intake (RDI) of 1.5 to 4.0 mg d^{-1}. Health problems may arise from deficiency (dental caries) or excess (dental mottling and skeletal fluorosis). Unlike other elements for which food is the principal source to the extent of (say) 80%, water is the principal source of fluoride. The diadochic relationship between F^- and OH^- ions in many silicates and phosphates arises from the close similarity in their ionic charge and radius (F^-: 1.36 Å; and OH^-: 1.40 Å). This property explains the replacement of hydroxylapatite in teeth and bones by fluorapatite in the case of persons who are exposed to higher levels of intake of fluoride.

2.7.3.3 Fluorosis in northern Tanzania – a case study

Aswathanarayana et al. (1986) gave a detailed account of the etiology and epidemiology of fluorosis in northern Tanzania.

The natural waters in northern Tanzania are characterized by very high contents of fluoride, which are among the highest in the world: Maji ya Chai river (12–14 mg l^{-1}) (in Swahili language, maji means water, and chai means tea; the color of the river water looks like tea decoction!), Engare Nanyuki river (21–26 mg l^{-1}), drinking water pond of the Kitefu village (61–65 mg l^{-1}) (as should be expected, every child in the village has mottled teeth!), thermal springs of Jekukumia (63 mg l^{-1}), and the soda lakes of Momella which are the habitat for hippos (upto 690 mg l^{-1}). The high fluoride content of water of several streams is attributed to their being fed by thermal springs.

Episodic, massive influx of fluoride which arose due to the leaching of the highly soluble villiaumite (NaF) present in the volcanic ash, exhalations and sublimates related to Miocene to Recent volcanism (for instance, the ash of the unique carbonatitic volcano of Oldoinyo Lengai in the region which erupted as recently as 1960, has fluoride content of 2.7%; minor eruptions of this volcano have continued, and as the area is remote and uninhabited, the eruptions are known only from the satellite images).

Figure 2.7.3 is a plot of the δD vs $\delta^{18}O$ of the fluoride waters of the northern Tanzania. Its slope of 5.5 is indicative of excessive evaporation, and is sharply different

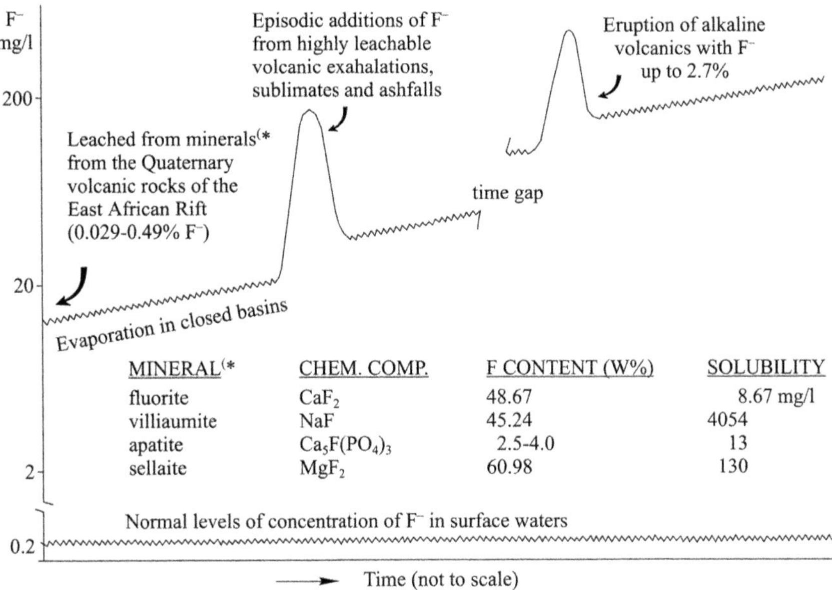

Figure 2.7.2 Schematic diagram of the geochemical model to account for the high fluoride content of the natural waters of northern Tanzania
Source: Aswathanarayana, 1995, p. 244

from the slope of 8 which is characteristic of meteoritic waters. Whereas the cluster related to brackish springs and rivers is characterized by negative δD and negative $\delta^{18}O$ values, the cluster pertaining to the saline lakes (with high F and Na, and low Ca) shows positive δD and positive $\delta^{18}O$ values. One freshwater lake with low F and Na and high Ca does not plot with the saline lakes cluster, but has similar slope and isotopic characteristics as other lake waters.

It has been estimated that the rate of ingestion of fluoride in northern Tanzania is about 30 mg d^{-1}, contributed as set out below.

This is about 15 times more than the estimated intake of 2 mg d^{-1} of fluoride per capita per day in temperate countries (1–2 l d^{-1} of water with 1–1.5 mg l^{-1} of flouride).

Source	*Contribution*
3 l of drinking water, with 8 mg l^{-1}	24 mg d^{-1}
10 g of locally grown tea, with 200 ppm of fluoride	2 mg d^{-1}
5 g of "magadi" with 1,000 ppm of fluoride	5 mg d^{-1}
Miscellaneous (through diet)	2 mg d^{-1}
Total	33 mg d^{-1}

The endemicity of fluorosis in northern Tanzania is a consequence of prolonged ingestion of more than 8 mg $d^{-1}$of fluoride. Fluorosis may be manifested as dental

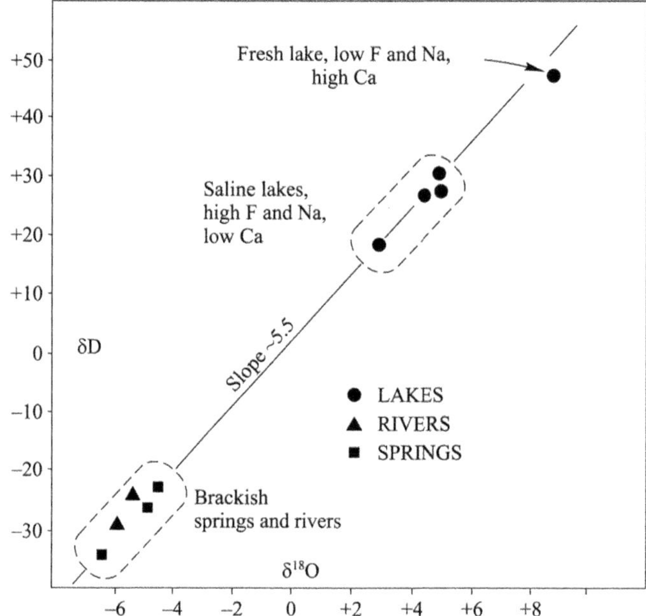

Figure 2.7.3 Plot of δD vs. $\delta^{18}O$ of the fluoride waters of northern Tanzania
Source: Aswathanarayana, 1995, p. 244

mottling, and in more severe cases, as skeletal fluorosis (*genu valgum*, characterized by bow legs, knock-knees, stiffness of trunk, impeded movement of limbs, severe joint pains, etc.). Variability in the severity of fluorosis in the population whose drinking water is drawn from the same source, is attributed to the following factors, acting singly or jointly:

- Fluorosis may be aggravated by the inadequate intake of nutrients, such as, proteins, ascorbic acid, calcium, etc. Thus, low-income groups who are unable to afford nutritious food, are at higher risk than affluent people who can afford nutritious food,
- Children are more affected than adults, probably because their teeth and bones (in which F^- substitutes for OH^-) are still forming,
- Men, particularly those involved in doing manual labor in the sun, are more affected than women, probably because such men tend to drink more water, and therefore ingest more fluoride. There is evidence to suggest that drinking of water with 8 $mg\,l^{-1}$ of flouride protects post-menopausal women from osteoporosis (this is probably because fluorapatite resists resorption better than hydroxylapatite).

To prevent dental caries, some communities fluoridate their drinking water supplies in order to ensure about 1 $mg\,l^{-1}$ of fluoride. Recent studies indicate that the substitution of OH^- by F^- in some metabolites, could render them carcinogenic. On this ground, some authorities have raised doubts about considering fluoride as an essential element.

2.7.3.4 Concurrent endemicity of fluorosis and arseniasis

Waters in Chile have high As content (400–600 $\mu g\,l^{-1}$) but normal fluoride content (0.30–0.50 $mg\,l^{-1}$), and hence while arseniasis is endemic in Chile, fluorosis is not. On the other hand, both arseniasis and fluorosis are endemic in Inner Mongolia (China) and possibly in the Kilimanjaro region of northern Tanzania. In the case of northern Tanzania, drinking water is the only route of ingestion of fluoride and arsenic. In the case of Inner Mongolia, the ingestion of arsenic and fluoride is not only through the drinking water but also through the inhalation of smoke from the burning of arsenious coal.

As and F form complexes in the aqueous environment – $As\,F_3^0$ (aq) and $As\,F_5^{2-}$ (aq) – but the pathways of these complexes to man, and their metabolic consequences are poorly known. The behavior of these complexes in the aquatic environment is of particular relevance in the understanding of the concurrent endemicity of fluorosis and arseniasis in Inner Mongolia, and possibly in northern Tanzania.

Alumina-based, metal oxide adsorption media are commercially available to remove both arsenic (arsenite and arsenate) and fluoride simultaneously.

2.7.4 RISK ASSESSMENT

Risk is estimated on the basis of Poisson statistics.

Chen et al. (1992) studied the relationship between the content of inorganic arsenic in drinking water, and the incidence of liver cancer, lung cancer, bladder cancer and kidney cancer among the population at risk in Taiwan. For each cancer, the number of deaths during a 13-year period were analyzed, in terms of four age categories (25, 40, 60 and 80 yrs.) and four water categories (0.02, 0.2, 0.45 and 1.0 $mg\,As\,l^{-1}$). The data for males and females were modeled separately. Two lag periods (0 and 15 yrs) were considered.

The number of cancer deaths in the *i*-th age and *j*-th exposure category were modeled as a Poisson variable with the expected value

$$Py_{ij} * a_i[1 + \beta * (t_i - \text{Lag}) * e_j] \qquad (2.7.2)$$

where Py_{ij} is the number of person-years of observation in the *ij*-th cell, t_i is the mid-point of the *i* th age interval, e_j is the assumed average inorganic arsenic concentration in the *j*-th exposure category, and β is the parameter that quantifies the carcinogenic potency of inorganic arsenic. The model provided an acceptable fit ($p > 0.05$) for the various cases.

The risk of cancer from the lifetime consumption of inorganic arsenic in drinking water is estimated from the following parameters:

Ed_{10}: Concentration of inorganic arsenic in water ($mg\,l^{-1}$) and in air ($\mu g\,m^{-3}$) from which a person exposed constantly over a lifetime would have a 10% increased risk of dying of a particular cancer.

LED_{10}: 95% statistical lower bound on the Ed_{10}.

Clewell et al. (1999) gave the following consolidated comparison of risk estimates for inorganic arsenic for oral and inhalation routes:

Route	*LED_{10}*	*MCL**	*LED_{10} – Linear unit risk*	*Current USEPA unit risk*
Oral	$10\,\mu g\,l^{-1}$	$50\,\mu g\,l^{-1}$	$0.01\ (\mu g\,l)^{-1}$	$5 \times 10^{-5}\ (\mu g\,l)^{-1}$**
	$144\,\mu g\,l^{-1}$		$7 \times 10^{-4}\ (\mu g\,l)^{-1}$	
Inhalation	$168\,\mu g\,m^{-3}$	N.A.	$6 \times 10^{-4}\ (\mu g\,m^{-3})^{-1}$	$4.3 \times 10^{-3}\ (\mu g\,m^{-3})^{-1}$

*Maximum Contaminant Level in drinking water, prescribed by US EPA, **Based on skin tumour incidence (all others based on internal cancers)

The conclusions of Clewell et al. (1999) are as follows:

1. Though it is generally agreed that the carcinogenicity of inorganic arsenic is non-linear, the exposure level at which non-linearity might occur is not known,
2. There is a significant risk involved in the consumption of water with MCL of US EPA, and
3. Inhalation does not appear to entail significant risks of cancer.

REFERENCES

Aswathanarayana, U., P. Lahermo, E. Malisa & J.T. Nanyaro 1986. High fluoride waters in an endemic area in northern Tanzania. In I. Thornton (ed.), *Environmental Geochemistry*: 243–249. London: Royal Society.

Aswathanarayana, U. 1995. *Geoenvironment: An Introduction*. Rotterdam: A.A. Balkema.

Chandrasekharam, D. 2008. Pathways of arsenic from water to food, West Bengal, India. In "*Food and Water Security*", (Ed. U. Aswathanarayana), pp. 63–70. London: Taylor & Francis Group.

Chen, C.J. et al. 1999. Emerging epidemics of arseniasis in Asia. In Chappell, W.R., C.O. Abernathy & R.L. Calderon (eds.), *Arsenic Exposure and Health Effects*, pp. 113–121. Amsterdam: Elsevier.

Chen, C.J. et al. 1992. Cancer potential in liver, lung, bladder and kidney due to ingested inorganic arsenic in drinking water. *Br. J. Cancer* 66: 888–892.

Clewell, H.J. et al. 1999. Application of risk assessment approaches in the US EPA proposed cancer guidelines to inorganic arsenic. In Chappell, W.R., C.O. Abernathy & R.L. Calderon (eds.), *Arsenic Exposure and Health Effects*, pp. 99–111. Amsterdam: Elsevier.

Guo, H.-R. et al. 1994. Arsenic in drinking water and urinary cancers: A preliminary report. In Chappell, W.R., C.O. Abernathy & C.R. Cothern (eds), *Arsenic Exposure and Health*, pp. 119–128. Northwood, England: Science & Technology Press.

Hindmarsh, J.T. 2000. Arsenic, its clinical and environmental significance. *J. Trace Elem. in Experim. Medic.* 13: 165–172.

Kossnet, M.J. 1999. Clinical approaches to the treatment of chronic arsenic intoxication from chelation to chemoprevention. In Chappell, W.R., C.O. Abernathy & R.L. Calderon (eds.) *Arsenic exposure and health effects*. pp. 349–354. Amsterdam: Elsevier.

Laconte, P. & Y.Y. Haimes (eds.) 1982. *Water Resources and Land-use Planning: A Systems Approach*. The Hague: Martinus Nijhoff Publishers.

Piver, W.T. 1983. Biological and environmental effects of arsenic. *Top. Environ. Health*. New York: Elsevier.

SOES (School of Environmental Studies) 1996. Bangladesh's arsenic calamity may be more serious than West Bengal. A Report from the School of Environmental Studies, Jadavpur University, Calcutta, India.

US EPA (1989). *Exposure Factors Handbook*. Washington, D.C.: US EPA.

Chapter 2.8

Water and agriculture: Usefulness of agrometeorological advisories

L.S. Rathore
India Meteorological Department, New Delhi, India

N. Chattopadhyay & S.V. Chandras
India Meteorological Department, Pune, India

2.8.1 INTRODUCTION

Among the developing countries, India has pioneered in the use of agrometeorological advisories in optimizing water use in agriculture. Agrometeorological advisories have a great potential to meet the challenges of the climate change impacts on water and agriculture. This chapter gives an account of Indian experiences in this regard which are relevant to other countries, particularly those with monsoon climate. Indian agriculture has to provide food for 1.2 billion people and fodder for about a billion animals.

The Green Revolution in India has been made possible through the introduction of high-yielding Mexican Dwarf wheat and IR8 rice varieties. The dwarf Mexican wheat varieties have an average yield of 5–6 $t\,ha^{-1}$, as against 1.5 $t\,ha^{-1}$ of the traditional varieties. Similarly, the IR8 rice has yields of 5 $t\,ha^{-1}$ without use of fertilizer and 10 $t\,ha^{-1}$ under optimal conditions. Consequently, the wheat production which was 10 million tonnes in 1960, rose to about 86 Mt. in 2010–2011. Similar increases have been recorded in the case of rice (95 Mt), coarse cereals (42 Mt) and pulses (18 Mt), totaling 241 Mt. in 2010–2011 (*The Hindu*, July 20, 2011). At this rate, the goal of 280 Mt. of food grains for 2020, as envisioned by the Prime Minister, appears to be realizable.

The Green Revolution has a dark side. As the inputs of irrigation and fertilizer are necessary to realize the benefits of the high yielding varieties, groundwater has been intensively used for irrigation, particularly in Punjab and Haryana. During the last 30 years, the number of tube wells in India went up from 2 million to 23 million, with the consequence that the water table has been going down from 1 to 3 m per year. This led to extensive depletion of groundwater and soil salinization. GRACE

(Gravity Recovery And Climate Experiment) satellite technology has made it possible to monitor the depletion of groundwater. Tiwari, Wahr, and Swenson (2009) found that northern India and its surroundings which is the home of 600 million people and the most heavily irrigated region in the world, has been losing groundwater at the rate of $54 \pm 9\,km^3/yr$. This is probably the largest groundwater loss in any comparable region in the world. Evidently this cannot go on indefinitely. There is urgent need to reverse this trend.

The intensive use of fertilizer led to high nitrate levels in ground water in Haryana –114–1,800 mg/l, as against the permissible level 50 mg/l, thus rendering the groundwater non-potable.

2.8.2 IMPACT OF CLIMATIC VARIABILITY ON AGRICULTURAL WATER CHALLENGES

India is monsoon land par excellence. Monsoons have profound influence on the economy and quality of life in India. India has two crop seasons – *Kharif* (June–Oct.), and *Rabi* (Nov.–Apr.). During the South West Monsoon (June to September), India receives about 75% approx. of the annual rainfall. Thus, variability in onset, withdrawal and quantum of rainfall during the monsoon season has profound impacts on water resources, and hence has significant bearing on power generation, agriculture, economics and ecosystems in the country. Majority of the food grain production in the country still depends on rainfed agriculture. Stagnation/decline in yields is due to the inter-annual and intra-seasonal climatic variability. Agricultural production is frequently affected by extreme weather events such as droughts and cyclones. Climate variability directly affects crop production, primarily by driving supply of soil moisture in rainfed agriculture, surface water runoff and shallow groundwater recharge in irrigated agriculture. As the non-biological response is nonlinear and generally concave over some range of environmental variability, climate variability tends to reduce average yields. Climate induced vulnerability of agriculture has contributed to stagnating farm productivity in the country.

Wide variation of rainfall affects the crops in *kharif* season. Droughts of 1972, 1987, 2002 and 2009 caused heavy crop damage. The year 2002 is a classical example to show how Indian food grain production depends on rainfall in July – it was declared as all-India drought, as the rainfall deficiency was 19% against the long period average of the country and 29% of area was affected due to drought. On the other hand in the year 2009, although the rainfall was deficient by 20% but through efficient management and dissemination of information (including agrometeorological information) the fall in farm production has been averted to a large extent. Figure 2.8.1 shows annual rainfall trend in India during last 100 years.

Extreme rainfall events also increased over the West Coast of India (based on analysis of 100 years of data; 1901–2000). According to Rupa Kumar (2002), the summer monsoon rainfall during 1901–2000 has shown decreasing trends in the sub-division of northeast India, Orissa and East Madhya Pradesh while increasing rainfall trends have been noted in Konkan and Goa, coastal Karnataka along the west coast and in Punjab, Haryana and Delhi.

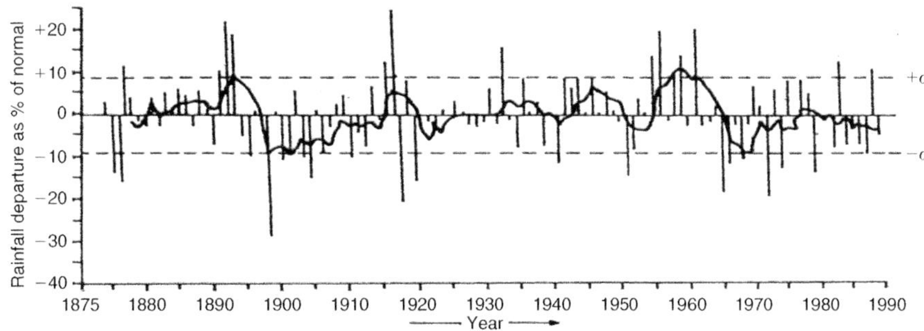

Figure 2.8.1 Annual rainfall of India. Thick line – five year running mean
Source: Thapliyal & Kulshrestha, 1991

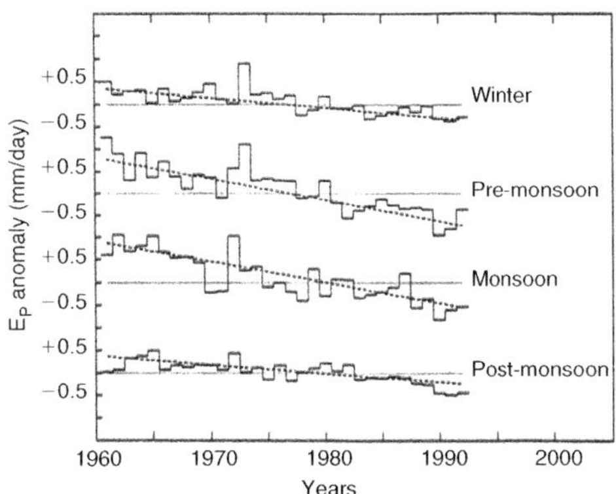

Figure 2.8.2 Regionally averaged annual E_p anomalies (mm day^{-1}) for the period 1961–1992 with respect to the 1976–90 mean for different seasons over India. Number of stations is ten between 1961–75 and 1991–92 and 19 between 1976 and 1990. Dashed lines show best fit linear trend
Source: Chattopadhyay & Hulme, 1997

In spite of general increase in temperature over recent decades, there has been decreased trend in Pan Evaporation (Ep) in almost all the parts of India – it is particularly significant in pre-monsoon and monsoon season (Chattopadhyay & Hulme, 1997). Figure 2.8.2 shows regionally averaged annual Ep anomalies (mm day^{-1}) for the period 1961–1992 with respect to the 1976–90 mean for different seasons over India and Fig. 2.8.3 shows regionally averaged annual PE anomalies (mm day^{-1}) for the period 1976–90 with respect to the 1976–90 mean for different seasons over India. Seasonal and spatial pattern of changes in Potential Evapotranspiration (PE) are similar to those for Ep, but magnitude of changes is less. In monsoon and post monsoon seasons PE has decreased over the whole country, whereas no well-defined trend is recognizable in the winter and pre-monsoon season.

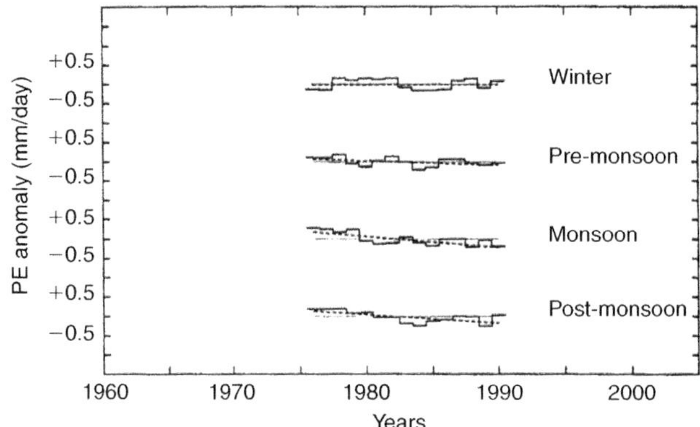

Figure 2.8.3 Regionally averaged annual PE anomalies (mm day^{-1}) for the period 1976–90 with respect to the 1976–90 mean for different seasons over India. Number of stations averaged is ten. Dashed lines show best fit linear trend
Source: Chattopadhyay & Hulme, 1997

Farmers will continue to face considerable climate risk, and extreme events can reverse development gains made over many years. A single drought or flood could set back all the agricultural development progress that result from improved local water management. We term this remaining climate risk, "residual risk," and recommend that a strategy for investing in agricultural water management should include a multi-pronged approach to dealing with the full range of climate variability, including not only moderate years but also the extremes. Improved control of water resources is a fundamental method for mitigating the impacts of climate variability.

2.8.3 USEFULNESS OF AGRO-CLIMATIC INFORMATION IN WATER USE

To develop new genetic strains and evolve the most effective agricultural practices, much climatic information is required. This relates not only to rainfall and atmospheric temperature, humidity etc. but also radiation, evaporation and soil moisture. General meteorological observations commenced in India in the second half of nineteenth century. However, systematic agrometeorological observations were initiated after the setting up of Agriculture Meteorology Division at India Meteorological department, Pune in the year 1932. Sufficient data are now available for presenting the agroclimatic information in the country. In order to derive maximum benefits from the available resources and climate conditions, agro-climatic and meteorological information are necessary in farm planning to ensure sustainable crop production. To provide weather-based Agro Advisories to farmers in India is a complex proposition because of variety of climatic zones leading to a variety of cropping patterns. Area specific agro-met advisory services are, therefore, needed to meet the requirements of the farming community in various crop-climate zones. Such crop-cum-area specific advisories can help the

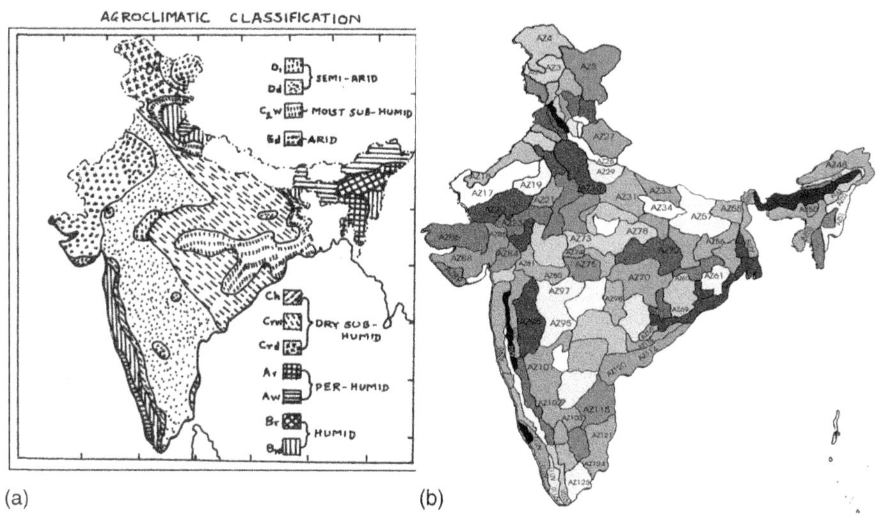

Figure 2.8.4 (a) Agroclimatic Classification in India, (b) Agroclimatic Zones in India under NARP

farmers to carryout farm operations in tune with the prevailing and prognosticated weather.

Considerable work has been done on agro-climatic classification (Fig. 2.8.4(a)) in order to bring out agricultural potential of a region. Agro-climatic zones have been delineated using Thornthwaite Moisture Availability Index (MAI) and other methods. Studies were made in India Meteorological Department to classify agro-climatic zones based on Penman method and water balance technique. Penman method has been used to estimate evapotranspiration of about 230 stations located in India (1971). These estimates are used to compute water balance of these stations in India, wherever available. The type of the soil of each station is found out from the soil maps of India published by National Atlas Organization, Govt. of India (1957). Water balance computations are done according to the latest approach of Thornthwaite. The moisture index is calculated without using the weighting factor for aridity index and the complete climatic classification is done for about 230 stations according to the revised Units (1966). Using these data, maps of moisture and thermal regimes of the country prepared.

In order to plan agricultural activities more accurately, each region (15 Resource Development Regions proposed by Planning Commission) is divided into sub regions based on soil, climate (temperature, rainfall) and other agrometeorological characteristics. Under NARP project a total of 127 agro-climatic zones have been identified in India (Fig. 2.8.4(b)). Physiographic divisions, rainfall pattern, soil type, availability of irrigation water, existing cropping pattern and administrative units have been taken into account in the identification of the agroclimatic zones. The delineation of zonal boundary of the NARP was carried out mostly in terms of districts and in some cases talukas/tahasils or subdivisions.

The most useful categorization for agricultural purposes is the agro-climatic zones which used the water balance concept which assist in determining the length of crop growing season (including growing period onset dates) at certain probability levels (National Meteorology Service Agency-NMSA, 1996). Seasonal forecasts for temperature and rainfall for a specific agroclimatic zone is very useful in providing advisories well in advance. Water resources in each agroclimatic zone should be used most judiciously to ensure sustainable agriculture development and productivity. This, in turn requires knowledge of crop water requirements (CWR) in various agro-climatic zones of the province. Lack of this information often results in farmers over or under irrigating their fields with consequent loss in yields and production.

2.8.4 FARMER-CUSTOMIZED AGROMETEOROLOGICAL ADVISORIES

There is substantial opportunity for risk minimization in crop and livestock production through delivering relevant weather and climate information, agrometeorological products and farm management advisories to assist farming communities where complete water control is not feasible. Farming communities in different areas have their varying needs of weather based agro-advisories in view of the variety of climate and cropping patterns. Weather based farm advisories as support system are organized after characterization of agro-climate, including length of crop growing period, moisture availability period, distribution of rainfall and evaporative demand of the regions, weather requirements of cultivars and weather sensitivity of farm input applications. All this is used as background information. Following are the ingredients of a typical Agromet Advisory Bulletin being practiced in the country to reap benefits of benevolent weather and minimize or mitigate the impacts of adverse weather.

1. District specific weather forecast, in quantitative terms, for next 5 days for rainfall, cloud, max/min temperature, wind speed/direction and relative humidity, including forewarning of hazardous weather event likely to cause stress on standing crop and suggestions to protect the crop from them.
2. Weather forecast based information on soil moisture status and guidance for application of irrigation, fertilizer and herbicides etc.
3. The advisories on dates of sowing/planting and suitability of carrying out intercultural operations covering the entire crop spectrum from pre-sowing to post harvest to guide farmer in his day-to-day cultural operations.
4. Propagation of techniques for manipulation of crop's microclimate e.g. shading, mulching, other surface modification, shelter belt, frost protection etc. to protect crops under stressed conditions.

The exercise involves preparing district specific agrometeorological advisory bulletins which are tailored to meet the farmers' need and can aid in his decision making processes. The suggested advisories generally alter actions in a way that improves outcomes. It contains advice on farm management actions aiming to take advantage of good weather and mitigate the stress on crop/livestock.

The Agro-met Advisory Bulletins are issued at district, state and national levels to cater the needs of local level to national level. The district level bulletins are issued by AMFUs and include crop specific advisories including field crops, horticultural crops and livestock. At present these bulletins are issued for 550 districts of the country.

2.8.5 INTEGRATION OF AGRO-CLIMATIC RESOURCES WITH AGRICULTURAL INPUTS

Though, climate data is recorded systematically across the country, it is under utilized because of the scattered location of the databases and impediments in the free flow of data. Thus, compiling climate data of various regions of India at one location will be of immense use. There is a considerable scope for decreasing the vulnerability of agriculture to increasing weather and climatic variability through weather forecast based agro-advisories. India Meteorological Department (IMD), which is now under the Ministry of Earth Sciences (MoES), is operating an Integrated Agro-Meteorological Advisory Service (IAAS) at district level, in India. This represents a small step towards agriculture management in consonance with weather and climate variability leading to weather proofing for farm production.

IMD has started issuing quantitative district level (612 districts) weather forecast up to 5 days since 1st June, 2008. Figure 2.8.5 shows weather forecasting system setup in IMD and preparation and dissemination of Agromet Advisories. The product output comprise of quantitative forecasts for 7 weather parameters viz., rainfall, maximum temperature, minimum temperatures, wind speed, wind direction, relative humidity and cloudiness. In addition, weekly cumulative rainfall forecast is also provided. IMD, New Delhi generates these products using Multi Model Ensemble technique based on forecast products available from number models of India and other countries. These include: T-254 model of NCMRWF, T-799 model of European Centre for Medium Range Weather Forecasting (ECMWF); United Kingdom Met Office (UKMO), National Centre for Environmental Prediction (NCEP), USA and Japan Meteorological Agency (JMA). The products are disseminated to Regional Meteorological Centres and Meteorological Centres of IMD located in different states. These offices undertake value addition to these products using synoptic interpretation of model output and communicate to 130 AgroMet Field Units (AMFUs), located with State Agriculture Universities (SAUs), institutes of Indian Council of Agriculture Research (ICAR) etc. on every Tuesday and Thursday.

Under AAS, the needs of farming community are delineated through ascertaining information requirement of diverse groups of end-users. It emerged that the prime need of the farmer is location-specific weather forecast in quantitative terms. Hence, the same was developed and made operational in June, 2008. Operational mechanism involving different organizations for implementation of IAAS is shown in Fig. 2.8.6. Thereafter, mechanism was developed to integrate weather forecast and climatic information along with agro-meteorological information to prepare district level agro-advisories outlining the farm management actions to harness favorable weather and mitigate impacts of adverse weather. A system has also been developed to communicate and disseminate the agro-meteorological advisories to strengthen the information outreach. The institutional dissemination channels such as farmer association,

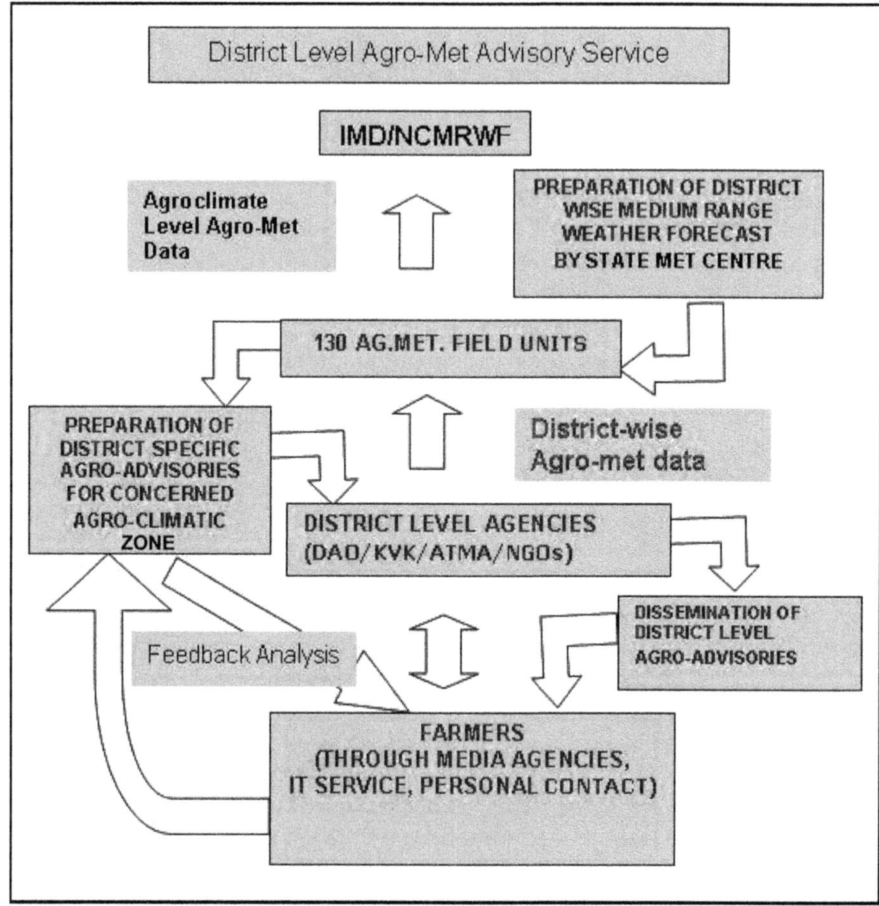

Figure 2.8.5 Weather forecasting system setup in IMD and preparation and dissemination of Agromet Advisories

non-governmental organizations (NGOs), input suppliers, progressive farmers are also employed.

District-specific medium-term forecast information and advisories help maximize output and avert crop damage or loss. It also helps growers anticipate and plan for chemical applications, irrigation scheduling, disease and pest outbreaks and many more weather related agriculture-specific operations. Such operation include cultivar selection, their dates of sowing/planting/transplanting, dates of intercultural operations, dates of harvesting and also performing post harvest operations.

Agrometeorological advisories help in increasing profits by consistently delivering actionable weather information, analysis and decision support for farming situations, such as: to manage pests through forecast of relative humidity, temperature and wind; manage irrigation through rainfall & temperature forecasts; protect crop from thermal stress through forecasting of extreme temperature conditions etc.

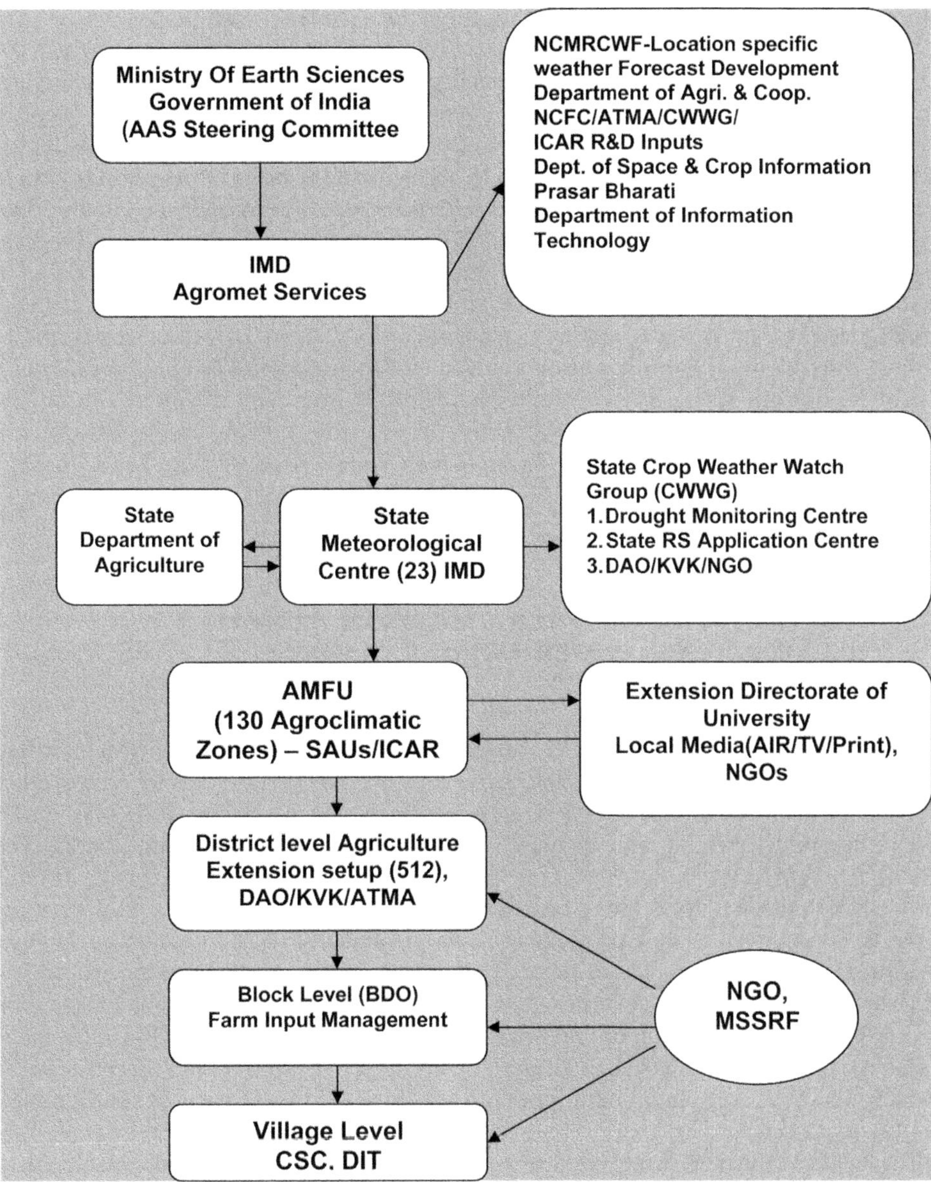

Figure 2.8.6 Operational mechanism involving different organizations for implementation of IAAS

Under the AAS system, more focus has been started to be given to use the crop/soil simulation models to decide crop management strategies, for a given weather condition. Agricultural scientists at Agrometeorological Field Units have started using crop simulation models as a decision support tool for helping with weather forecast based farm management decision making as they are more objective. The user community is advised on how best they can avail advantages of meteorological parameters for their use in crops' growth and development and for the optimum yield.

R. Nemani of NASA and his colleagues (2008) developed the TOPS system which is a data and modeling software system designed to seamlessly integrate data from satellite, aircraft and ground sensors with weather/climate and application models to expeditiously produce nowcast and forecasts of ecological conditions. TOPS can be adapted for various agroclimatic situations to help agricultural administrators and farmers in crop management. For instance, Leaf Area Index (LAI) derived from satellite data can be used to compute water use and irrigation requirements to maintain crops at optimal level of water requirements. By integrating the leaf area index, soil moisture, and soil temperature data, daily weather and weekly weather forecasts, TOPS can estimate spatially varying water requirements within the farm, so that water managers can adjust water delivery from irrigation system. TOPS system has been used successfully to advise individual vineyards in California, USA.

In due course, as more satellite information is available, and the use 3G broad-band communication spreads, customized agrometeorological advisories can be provided to the farmers on their cell phones.

2.8.6 PROJECTION OF WATER STATUS IN INDIAN AGRICULTURE UNDER FUTURE CLIMATE CHANGE SCENARIO

Using a number of climate models, different scenarios have been generated for the future climate change in India. It has been projected that average surface temperature will increase by 2–4°C during 2050s, with marginal changes in monsoon rain in monsoon months (JJAS) and large changes of rainfall during non-monsoon months. The number of rainy days are set to decrease by more than 15 days and the intensity of rains is likely to increase by 1–4 mm/day. Increase in frequency and intensity of cyclonic storms is projected. The hydrological cycle is predicted to become more intense, with higher annual average rainfall as well increased drought (Bhattacharya, 2006). Figure 2.8.7 shows rainfall projections at different seasons. There is a predicted increase in extreme rainfall and rainfall intensity in all three river basins (Ganga, Godavari & Krishna) towards the end of the 21st century. Number of rainy days may decrease in the western parts of the Ganga basin, but with increases over most parts of the Godavari and Krishna basins. Thus, surface water availability may show a general increase over all 3 basins. though future population projections would need to be considered to project per capita water availability. According to Lal et al. (2001) an annual mean area-averaged surface warming over the Indian subcontinent will range between 3.5 and 5.6°C over the region by 2080. These projections show more warming in winter season over summer monsoon. The spatial distribution of surface warming suggests a mean annual rise in surface temperatures in north India by 3°C or more by 2050. The study also suggests that during winter, the surface mean air temperature could rise by 3°C in north and central parts and 2°C in southern parts by 2050. In case of rainfall, a marginal increase of 7 to 10 per cent in annual rainfall is projected over the subcontinent by the year 2080. However, the study suggests a fall in rainfall by 5 to

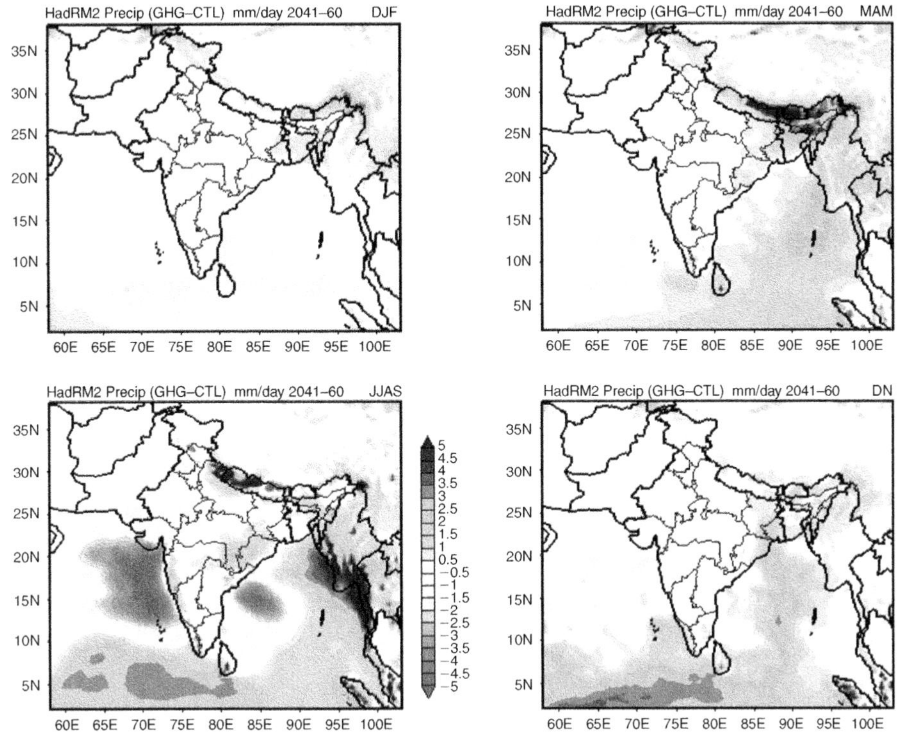

Figure 2.8.7 Rainfall projections for different seasons
Source: Bhattacharya, 2006

25% in winter while it would be 10 to 15% increase in summer monsoon rainfall over the country.

Future changes in PE over India and adjoining countries project an increase in all the global climate models. In the winter seasons, maximum models show increasing trend in PE over southern and central India up to around 25°N. In most of the model experiments maximum winter increase in PE is of the order of 3–4% per degree Celsius of global warming and is seen in peninsular and most central parts of India. In the monsoon season maximum increase in PE is over northwestern India. Inter relationship between PE and rainfall was assessed by mapping the number of GCM experiments which yield increased P/PE ratio for the monsoon season. A number of GCMs agree that P/PE ratio becomes more favorable over northeastern India and changes in this ratio are less favorable in post monsoon season and in the extreme south in the country. (Chattopadhyay & Hulme, 1997). Figures 2.8.8 and 2.8.9 show calculated change (%) in mean seasonal PE for CCC experiment and the GFDL experiment for 1°C of global warming. Figures 2.8.10 and 2.8.11 show mean seasonal change in UKTR experiment and number of GCM experiments which yield an increase in P/PE ratio for each season.

Although the effects of climate change from anthropogenic forcing on the use of water resources in the world remain difficult to project, anticipated climate change

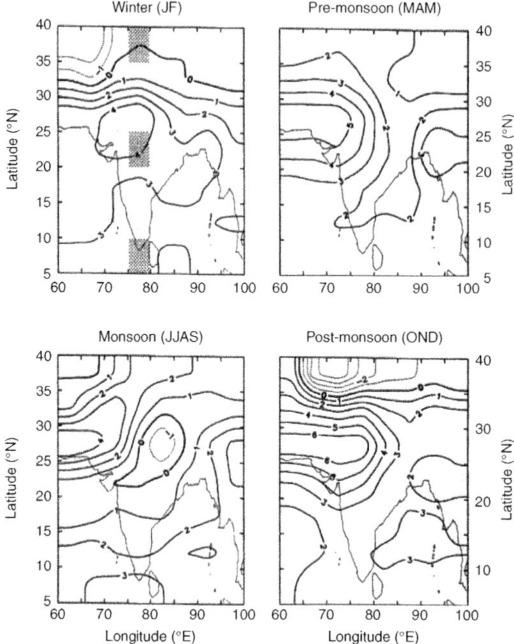

Figure 2.8.8 Calculated change (%) in mean seasonal PE for 1°C of global warming for the CCC experiment
Source: Chattopadhyay & Hulme, 1997

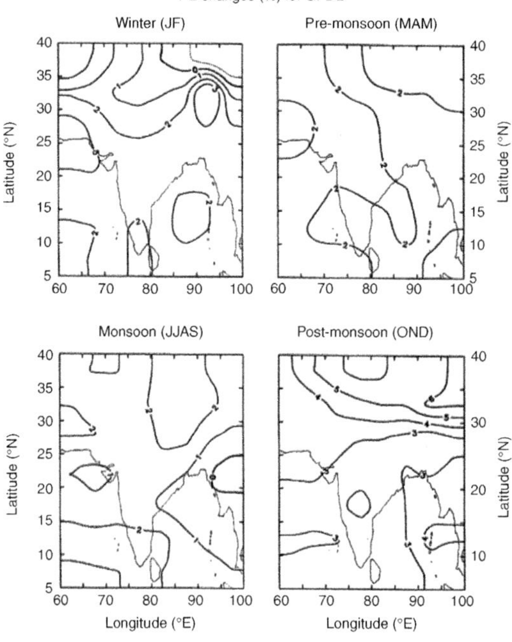

Figure 2.8.9 Calculated change (%) in mean seasonal PE for 1°C of global warming for the GFDL experiment
Source: Chattopadhyay & Hulme, 1997

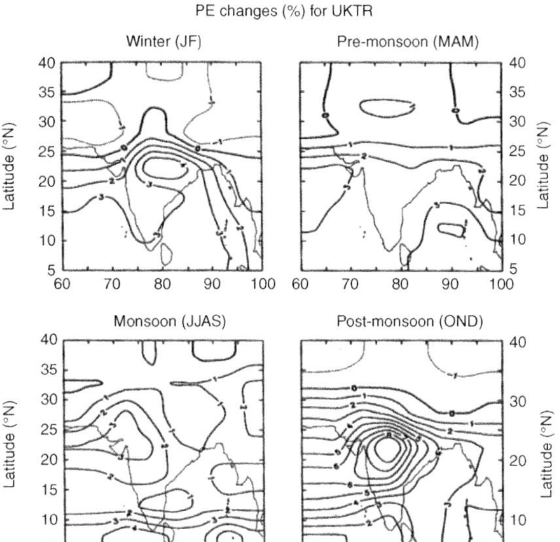

Figure 2.8.10 Calculated change (%) in mean seasonal PE for 1°C of global warming for the UKTR experiment
Source: Chattopadhyay & Hulme, 1997

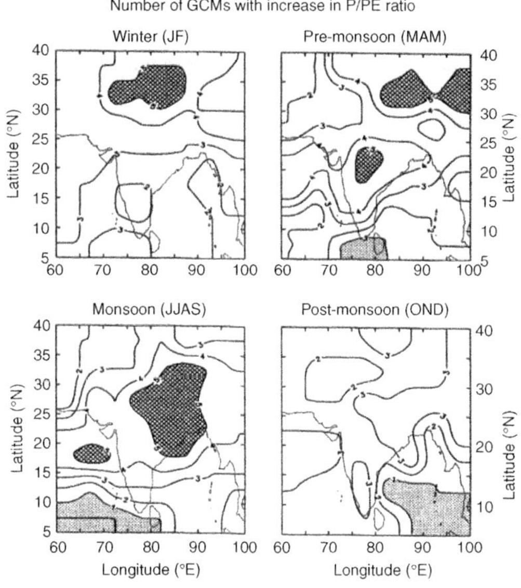

Figure 2.8.11 Number of GCM experiments which yield an increase in P/PE ratio for each season. Maximum number is six. Areas of agreement in the sign of the change between all six GCMs are shaded
Source: Chattopadhyay & Hulme, 1997

combined with other drivers of change is likely to intensify current agricultural water management challenges in India. The effects of population growth and increasing water demand, which are often but not always coupled, are likely to be a more significant source of water stress than climate change when considering changes to mean precipitation and runoff. Increasing temperatures in all regions are expected to increase evaporative demand, which would tend to increase the amount of water required to achieve a given level of plant production if crop phenology and management are to be held constant. However, if cultivars and planting dates were to remain unchanged, accelerated crop development in response to temperature increases would tend to have the opposite effect on water requirements. Increased temperatures are also expected to increase evaporative losses of surface water resources. The magnitude and even direction of projected changes in precipitation are quite uncertain. Increases in rainfall are likely for much for the summer monsoon over much of peninsular and eastern India. There is, however, a consensus that climate change will tend to increase the variability of rainfall and decrease the natural storage provided by snow pack and glaciers, such as in the Himalaya that feed the rice-wheat belt of the Indo-Gangetic Plains. The concern pertaining to climate change impacts should not overshadow the present challenges that climate variability poses to agricultural development. While the future impacts of climate change remain uncertain, climate variability persists as a challenge to development and as an impediment to meeting the Millennium Development Goals. Over the next decades, while climate change trends may begin to have some effect, droughts and floods associated with climate variability will continue to ravage vulnerable communities in developing countries. Fortunately, there is much that can be done now to reduce that vulnerability.

2.8.7 HOW TO PRODUCE MORE FOOD (THROUGH OPTIMIZATION OF SOIL-WATER-PLANT SYSTEM)

In many parts of the world, crop production only reaches 20% of the yields achieved in the developed world. Optimising the yields from existing crop production would enable farmers to grow more food with a comparable water footprint. To achieve this, we need effective dissemination of agronomic knowledge as well as farmer access to inputs such as good quality seed and crop protection to reduce pre- and post-harvest losses due to pests and disease.

Beyond improving crop yields, we need to reduce the water used to produce food. The most effective way to increase water efficiency is through a step-by-step approach, from the source of water to the eventual plant biomass or animal meat produced. Efficiency-improving measures include repairing leakages along canals and pipes; improving or upgrading irrigation systems; and promoting plant canopy growth to reduce water evaporation from soil. Plant science technologies also help. Herbicide use with biotech crops enables conservation tillage, which increases the moisture retention in soil and reduces erosion. Biotechnology holds significant promise for improving the water efficiency of agriculture through drought tolerant or water efficient crop varieties.

From modern plant varieties that produce higher yields, to the adoption of conservation tillage that preserves soil moisture, plant science innovations are already leading progress on global water conservation efforts, and hold tremendous potential for the future. According to the United Nations, a one per cent increase in water productivity in food production alone can make up to an extra 24 litres of water availability per person per day. With the global population projected to reach 9.1 billion people by 2050 from about 6.8 billion in 2010, the Food and Agriculture Organization (FAO) has estimated that we will need a 70 per cent increase in world food production. It's also estimated that by 2030, close to half the world's population will be living under severe water stress, making conservation efforts essential (Source: Crop Life International, Belgium). Agricultural water consumption can be optimised by improving existing irrigation systems, enhancing the water use efficiency of crops, and by properly maintaining the existing systems to avoid malfunctioning. Regarding crop irrigation, optimal water efficiency seeks to minimise water losses due to evaporation, evapotranspiration, runoff or subsurface drainage. Improving the water use efficiency of crops means selecting crops that are adapted to the ambient climate, e.g. crops that are drought tolerant or adapted to dry climates. The optimisation of water use in agriculture can sometimes be achieved by simple means and is also of economic importance (water savings). However, farmers need to be motivated by the right incentives and policies and may require technical assistance.

Improved farming techniques, such as increased drip and sprinkler irrigation, combined with best practices such as no-till farming, improved drainage, better seed varieties and optimized fertilizer use are all part of agriculture's water conservation efforts.

2.8.8 HOW TO DO WITH LESS WATER (IN AGRICULTURE, INDUSTRY AND DOMESTIC PURPOSES)

Along with the increase in population, it is expected that nearly 61% of the population will be living in urban areas by the year 2050 as per high-growth scenario and 48% in low growth scenario. Based on these norms and projection of population, it is estimated that by 2050, water requirements per year for domestic use will be 90 km^3 for low demand scenario and 111 km^3 for high demand scenario. It is expected that about 70% of urban water requirement and 30% of rural water requirement will be met by surface water. Large canal infrastructure network for providing irrigation has been the prime goal of the Government of India, since the first five-year plan, which continued up to seventh five-year plan. Groundwater utilization for irrigation in waterlogged areas can help to lower the groundwater table and reclaim the affected soil. It is desirable that the irrigation needs for fulfilling crop water requirements should be satisfied by judicious utilization of available canal water in conjunction with groundwater so as to keep the water table within the acceptable range. The use of sprinkler irrigation saves about 56% of water for the winter crops of bajra and jowar, while for cotton, the saving is about 30% as compared to the traditional irrigation.

Some parts of the country facing frequent drought are adopting the dry land farming practices to grow crops which require less amount of water. However, there is a need to take up such studies for assessment of available water resource for different agro-climatic regions of India and for various adaptation scenarios under the changing climatic scenarios. Some recommendations to cope with the problems in a systematic and a planned manner are: (i) a nation-wide climate monitoring programme should be developed; (ii) while formulating new projects that influence climate, it should be ensured that no action is taken which causes irreversible harmful impact on the climate; (iii) improved methods for accounting of climate-related uncertainty should be developed and made part of decision making process; (iv) existing systems should be examined to determine how they will perform under the climate situations that are likely to arise; (v) water availability and demands in all regions, particularly in water-scarce regions should be reassessed in the new climate scenario; (vi) a re-examination of the water allocation policies and operating rules should be taken up to see how these need to be updated to handle extremes that are likely to arise; and (vii) there should be proper coordination among concerned organizations so as to freely share the data, technology and experience for capacity building. Another strategy that could be adopted refers to planning of land use especially in new land developments. Areas where water supply priorities are low can be planted with drought-resistant varieties of trees. For this purpose, these variety of trees need to be developed. Another strategy that could be suggested can be in the agricultural sector. Considerable information exists on time distribution of water requirements for various crops and various planting dates. This knowledge is required to be integrated systematically with water supply probabilities to develop planting strategies. The selection of cropping pattern as per availability of water will reduce adverse impacts of drought on potential water consuming crops. The plants suitable for water scare areas can be (i) with shorter growth period, (ii) high-yielding plants requiring no increase in water supply, (iii) plants that can tolerate saline irrigation water, (iv) plants with low transpiration rates, and (v) plants with deep and well-branched roots. (Rakesh Kumar et al., 2005).

2.8.9 CONCLUSION

India's water resources are finite (about 4,000 billion cubic metres). With increasing population, the per capita availability of water has been steadily decreasing. In order to protect our ecology and economy, we must reduce the water footprint of water uses, particularly agriculture. Farmer-specific agrometeorological advisories coupled with water-saving techniques like drip irrigation will help the farmer to achieve higher yields while using less water.

REFERENCES

Bhattacharya, S. (2006) *Climate Change and India*. Proc. International Workshop on Future International Climate Policy, August 9, 2006, University of Sao Paulo, Brazil.

Crop Life International. Avenue Louise 326, box 35 – B-1050 – Brussels – Belgium.

Chattopadhyay, N. and Hulme, M. (1997) "Evaporation and potential evapotranspiration in India under conditions of recent and future climate change". *Agric Forest Meteorol* **87**: 55–73 Impacts, Adaptation and Vulnerability IPCC Working Group II.

Lal, M. (2001), Future climate change: implications for Indian summer monsoon and its variability. *Curr. Sci.*, **81**(9):1205.

Nemani, R. et al. (2008) Remote sensing methodologies for ecosystem management, In *Food and Water Security* (ed. U. Aswathanarayana), pp. 3–20. Leiden: Taylor and Francis/Balkema.

NMSA (National Meteorology Service Agency), 1996, *Climatic and Agroclimatic Resources of Ethiopia*. Vol. 1, No. 1. National Meteorology Service Agency of Ethiopia, Addis Ababa. 137 p.

Rakesh Kumar, R. D. Singh and K. D. Sharma (2005) *Water resources of India*, National Institute of Hydrology, Roorkee, India, pp. 247–667.

Rupa Kumar, K., Krishna Kumar, K., Pant, G.B., Srinivasan, G. (2002) Climate Change-the Indian Scenario. In: Background paper prepared by FICCI, International Conference on Science and Technology Capacity Building for Climate Change, October 20–22, New Delhi, 5–17.

Thapliyal, V. and Kulshrestha, S.M. (1991), Climate change and trends over India, *Mausam* 42: 333–338.

Tiwari, V.M., Wahr, J., and Svenson, S. (2009) Dwindling groundwater resources in northern India from satellite gravity observations. *Geophy. Res. Lett.*, v. 36. L 1840, 17 Sept. 2009.

Chapter 2.9

Remote sensing in water resources management

Venkat Lakshmi

Department of Earth and Ocean Sciences, University of South Carolina, Columbia, SC, USA

2.9.1 BACKGROUND AND SOCIETAL IMPORTANCE

The water availability per capita in the world is constantly changing. A simple explanation would be increasing global population coupled with not a proportionate redistribution in available water decreases the per capita availability of water. With the increase in the global population from 2 billion in 1950 to 6.6 billion current population, for the same water availability, the per capita water availability should have decreased by a factor of 3. However, there are two reasons that this is not true. Firstly, there have been a greater than X3 population increase in certain areas (example: India and China) and secondly, in many regions of the world, the groundwater is a predominant source of water and this is being exploited and is just beginning to run out (example the High Plains of the United States).

The United Nations UNESCO (2006) report on Water outlines all possible concerns over the availability of adequate amount and quality of water for human use. It is impossible to fathom the dire consequences that would befall us if we did not immediately involve ourselves with many conservation practices to safeguard this very important resource. This concept of sustainability is explained eloquently in Varady et al., 2008 which deal with water resources planning and principles of efficient management. At times the challenge of being "green" comes at a severe cost to water resources (Rockstrom et al., 2007) and therefore any future program in sustainability and "green" technology should include water resources in evaluation.

2.9.1.1 Climate change and water

There have been two recent and very important studies concerning the impact of climate change. The first one is the Intergovernmental Panel on Climate Change (IPCC) report. Whereas the entire report is not of substance here, Chapter 3 of Working group 2 on Freshwater resources and their management (Kundzewicz, Z.W. et al., 2007) summarizes the impact of climate change on water resources. This chapter concludes that there is a net negative impact of climate change on freshwater resources worldwide. In particular, there is an increase in winter flow in the ice/snow dominated areas due to earlier melt and a decrease in summer flow. In arid and semi-arid regions, there are lesser freshwater resources due to lower precipitation availability and reduced recharge to the groundwater aquifers. The recharge of groundwater under the climate change scenario is a very uncertain quantity (Holman, 2006; Scibek and Allen, 2006) and cannot be easily measured or modeled using computer simulations. In several regions, increased water temperatures in summer coupled with longer periods of low flows results in nutrients and other pathogen pollution (plus more discharges of non-treated water – low flows are not the only reason, they just make it worse). The second report is the assessment of climate change on United States and includes many impacts related to water. The study finds that precipitation increased at a rate of 6% over the last century (1901–2005) and even though extreme events – floods and droughts decrease in the latter half of the last century, various regions such as the United States Southeast have been suffering from severe droughts 1999 to the present time. Stream flow has increased by 25% in the past 20 years but in the central Rocky Mountain region it has been decreasing by 2% per decade over the last several decades. In particular, the Colorado River has suffered from decreases in runoff (Cayan et al., 2008). This has been coupled with increased winter flows due to earlier snowmelts due to higher air temperatures and lower summer flows. In addition, an often ignored component of fresh water is the snow cover which has been found to have decreased in the spring time (spring snow supplies summer runoff in many rivers in the United States West). Lakes and rivers have been freezing later during the year and many glaciers have been continuously losing mass.

2.9.1.2 Increase/changes in demand for water resources

The IPCC report also found that there has been global increase in water demand due to population growth, growing economic activity, land use changes from their native forest/grasslands to agriculture and/or urban developments. These increase the occurrence of droughts (periods of low flows) and floods (overflow of river over banks) due to excess use of water and increase in built up areas which reduces infiltration and decreases response time of overland flows. The United States Scientific Assessment of climate change has found that there are less reliable sources of water for urban water supplies. Pressures for reallocation of water in regions such as the Southwest from traditional agriculture (California farms) to urban centers (Las Vegas) have become politically charged issues. One of the major issues in the United States Southeast is the intrusion of saline (sea) water into coastal aquifers due to over development and excessive withdrawals from the groundwater source. These and numerous other problems have been outlined in Zekster and Everett, 2004 which is a compendium of all the groundwater resources of the world. It points out various countries that completely depend on groundwater for their water supply (Denmark, Malta and Saudi Arabia); groundwater is a major (greater than 70%) portion of the water resources in numerous countries such as Switzerland, Tunisia, Austria, Belgium and Hungary to name a few. Also, dependence on groundwater as a source of irrigation water in many countries is very high (45% of United States irrigation demands are satisfied by groundwater). Therefore, dependence on groundwater and overuse of this resource which has been slowly accumulated over geological time and reduction of recharge all contribute to the reduction in groundwater. Using computer simulations and outputs from climate change scenarios from the Hadley center model, Alcamo et al. (2006) have determined that by the year 2050, 55–75% of the land area will be under severe stress. This stress will be caused primarily due to increase in domestic water usage due to increased income and population growth was not an important factor and the global amount of irrigated area was assumed to be unchanged.

2.9.2 CURRENT MONITORING METHODOLOGIES

There are various aspects to the monitoring of water availability. One of the most important resources in this monitoring is the use of satellite remote sensing. By virtue of their large spatial coverage and repeated temporal overpass, satellite remote sensing is becoming a much used tool in the area of land surface hydrology (Lakshmi, 2004). There are a number of variables in the land surface hydrological cycle that are observed using satellite sensors. In the scope of data and observations, there are two main items. Firstly, the monitoring of levels of rivers, lakes and flooded regions using satellite data, these mainly deal with the water level and are important during excess flow episodes and reservoir management. Secondly, the routine monitoring of variables such as precipitation, soil moisture, vegetation and surface temperature, these deal with the availability of water in the agriculture or pertain to vegetation health. The second item deals with computer simulations using hydrological models at local scales to investigate the effects of climate change on water resources availability.

2.9.2.1 Inundated areas and flow

The first attempts at mapping inundated areas using the Special Sensor Microwave Imager (SSM/I) 85 Hz vertical polarization minus the 37 GHz horizontal polarization observations (Lakshmi and Schaaf, 2001; Basist et al., 2001) showed promise when used to map the regions affected by the flooding of the Mississippi River in 1993 (Lakshmi and Schaaf, 2001) and several locations in various continents that were cooler than their surrounding areas due to rainfall and/or irrigation and subsequent crop growth (Basist et al., 2001). These microwave techniques (similar to the soil moisture remote sensing described later in this section) are based on the dielectric contrast between wet and dry surfaces. Water has a dielectric constant (real part) of 80 whereas that of dry soil is 5. As the proportion of water, whether it be standing water or water in soil increases, the dielectric constant increases and the brightness temperature (recorded by the microwave sensor at the satellite) decreases (Schmugge et al., 1986). More recent efforts have used radar altimeters (TOPEX POSEIDON) to

Table 2.9.1 Remotely sensed data used in land surface hydrology

Variable	*Sensor*	*Spatial Resolution*	*Temporal Resolution*	*Period of Record*	*Reference*
Leaf Area Index	AVHRR	8 or 16 km	Bi-weekly	1980–Present	Goward, 1994
	MODIS	500 m–1 km	Bi-weekly	1999–Present	Justice, 2002
Soil Moisture	SSM/I	50 km	2/day	1987–Present	Lakshmi, 1997
	AMSR	50 km	2–4/day	2002–Present	Njoku, 1999
Surface Temperature	ASTER	90 m	On-request	2000–Present	Gillespie, 1998
	MODIS	500 m–1 km	2–4/day	2000–Present	Justice, 2002
	AVHRR	1–5 km	1–8 day	1980–Present	Becker, 1990
	TOVS	1°	2/day	1980–Present	Susskind, 1997
	AIRS	50 km	2/day	2002–Present	Susskind, 2003
Surface Air Temperature	TOVS	1°	2/day	1980–Present	Susskind, 1997
	AIRS	50 km	2/day	2002–Present	Susskind, 2003
Precipitation	TRMM	20 km	Daily	1998–Present	Kummerow, 2000
	SSM/I	50 km	Daily	1987–Present	Ferraro, 1997
	AMSR	50 km	Daily	2002–Present	Wilheit, 2003

map the variations of Lake Chad (Sarch and Birkett, 2000) due to human influences and impacts of climate change. The lake variations on an annual basis have a profound impact on the livelihood choices of the communities that live around the lake and depend on the lake for sustenance. The ERS1 a microwave SAR has been used to study the flow due to summer snow melt and river flow in the Ob River in Northern Russia draining into the Arctic Ocean (Papa et al., 2007). The prediction of the time of snowmelt and the increase of runoff of freshwater into the ocean are important variables that characterize the Boreal ecosystem. Incidentally, the Ob' river discharge variations over a 10 year period (1992–2002) have been studied using the TOPEX-POSEIDON sensor (Kouraev et al., 2004) to infer hydrological variability due to snow melt. All of these studies link river discharge to catchment hydrology and hence help in estimation of water resources/river discharge which is very important for the growing season (in the case of agriculture). Another and more recent use of a satellite sensor is the Gravity Recovery and Climate Experiment (GRACE) to estimate using anomalies of gravity the changes in the terrestrial water storage (Rodell and Famiglietti, 1999, 2001). More recent work has attempted to link the satellite estimates of the changes in the land surface and subsurface water components to those estimated independently using continental scale hydrological models using data assimilation techniques (Syed et al., 2008). On the other hand floodplain rivers such as the Amazon are very complex to study by single observations on the ground or by using complex computer models. In this regard, the changes in the water surface elevation has been studied using TOPEX/Poseidon (Zakharova et al., 2006) and a space-borne synthetic aperture radar (JERS-1) (Alsdorf et al., 2007).

A comprehensive review of these techniques is presented in Alsdorf and Lettenmaier (2003).

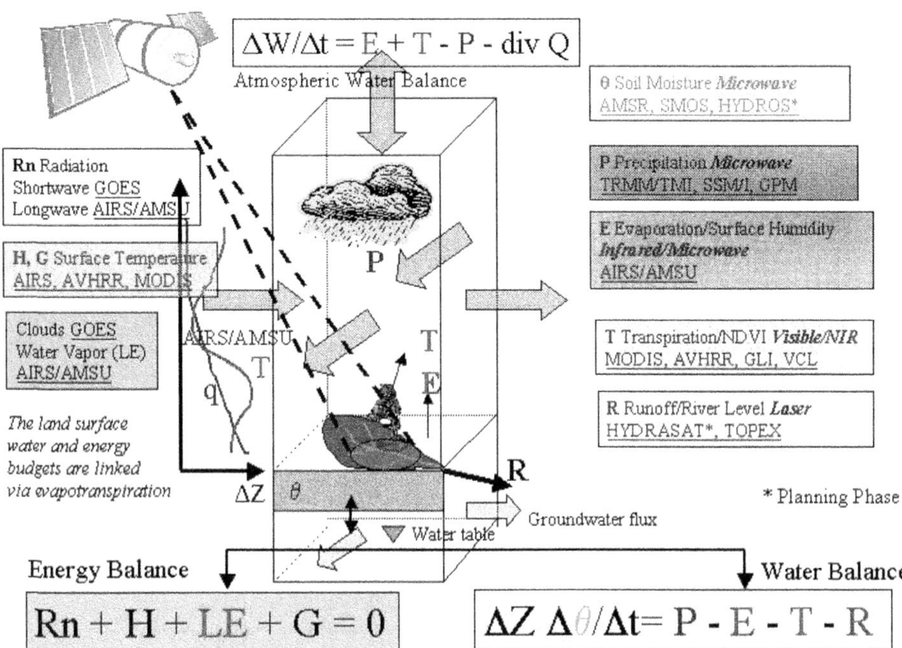

Figure 2.9.1 Satellite data in the land surface hydrological cycle

2.9.2.2 Land surface variables

Land surface variables can be sensed remotely from space. Table 2.9.1 above displays the satellite sensor, spatial resolution and temporal repeat and the period of record for these variables along with an important reference.

We have a wealth of satellite data at various spatial scales, and different temporal resolutions that can be used in putting together a complete picture of the land surface hydrological cycle. Figure 2.9.1 represents both the wealth of data and the dilemma on its usage. It remains as a challenge to the scientific community to reconcile these issues and use this data in the most synergistic methodology possible to help in our endeavor to predict water resources in ungaged basins where no direct observations viz. gages for river flow or water levels, are available.

2.9.2.2.1 *Surface temperature and vegetation index: Moderate resolution imaging spectroradiometer*

Surface temperature (Ts) and Normalized Vegetation Difference Index (NDVI) are derived from the two platforms, viz., AVHRR (Advanced Very High Resolution Radiometer) and Moderate Resolution Imaging Spectroradiometer (MODIS). The predecessor of MODIS, Advanced Very High Resolution Radiometer (AVHRR) has

The land surface hydrological cycle responds to changes in the moisture stores and the fluxes. Specifically, the satellite sensors detect changes in the stored moisture, either in the atmosphere (in terms of atmospheric water vapor profile) or in the soil as in soil moisture or in lakes/ponds or rivers as changes in water level. These changes are modulated by vegetation changes that are observed by numerous satellite sensors.

The different variables of the hydrological cycle can be sensed using single or a combination of sensors on various satellite platforms. It should be the object of the Hydrologic Information Systems to combine information from various sources in a synergistic fashion, spanning across differing spatial resolutions and temporal repeats to piece together the entire, land-atmosphere hydrological cycle.

With the advent of numerous EOS-era sensors, we will have access to multiple sensors for the same variable. It will therefore be a challenge to catalog these data sets in a proper fashion for usage. The model for use of satellite data would be the derived product rather than the raw radiances or brightness temperatures or backscatter values. This would overcome the problem for individual investigators to undertake retrieval algorithms and used standardized algorithms for the hydrologic community.

Satellite data represent a wealth of information. These bridge the gap between point measurements and computer based simulations. Properly utilized, the satellite data would be of great benefit to the hydrological community.

been collecting observations of the earth's biosphere since 1980 (Tucker, 1979, Goward et al., 1994). MODIS is onboard the Terra and Aqua satellite platforms, which were launched on December 1999 and on May 2002, respectively at a sun-synchronous polar orbit, with equatorial crossing times of 10:30 am (Terra) and at 1:30 pm (Aqua). Details of the MODIS land data are available at the MODIS website (http://modis.gsfc.nasa.gov), and the products NDVI, Ts, Leaf Area Index (LAI), and Land Cover Map are available at a 1km spatial resolution from the Land Processes Distributed Active Archive Center (DAAC) website (http://edcdacc.usgs.gov) and they use validation algorithms, NDVI (Tucker, 1979; Myneni et al., 1995); LAI (Myneni et al., 2002); Land cover classification (Defries et al., 1998; Friedl et al., 2002) and surface temperature (Justice et al., 1998; Wan and Li, 1997). The land surface temperature can be derived from MODIS up to 4 times per day (corresponding to the morning and afternoon overpasses of the two satellites); the NDVI and LAI are generally used as bi-weekly composites in order to minimize effects of cloud contamination. Lower spatial resolution data is available from 1980. These data come from the Tiros Operational Vertical Sounder (TOVS) High Resolution InfraRed Sounder (HIRS) sensor (Susskind et al., 1997) at a spatial resolution of 1°. Surface skin and air temperature has been observed using Tiros Operational Vertical Sounder (TOVS) High Resolution InfraRed Sounder (HIRS) sensor (Susskind et al., 1997) at a lower spatial resolution of 1° from the late 1970s and more recently using the AIRS (Advanced Infra Red Sounder) at a 12.5 km spatial resolution twice a day on the AQUA satellite (Susskind et al., 2006).

2.9.2.2.2 *Precipitation*

Precipitation is one of the most difficult variables to measure as it is often spatially variable and may occur over a wide range of time scales. The data come from a number of sources: long-term climatology, direct point measurements, and satellite-based estimates. CPC (Climate Prediction Center) Merged Analysis of Precipitation (CMAP) is a monthly averaged $2.5^\circ \times 2.5^\circ$ global precipitation datasets available from 1979 until today. It is made using satellites estimates from 5 different satellites (GPI, OPI, SSM/I scattering, SSM/I emission and MSU). An enhanced version of the precipitation field is also available which incorporates the precipitation reanalysis from the NCEP/NCAR models. The data can be accessed from http://www.cdc.noaa.gov/cdc/data.cmap.html#detail. Each of these precipitation data sets uses all available data sources to create the best spatial and temporal variation of rainfall data. Monthly precipitation climatology for several locations within the basin is available from the Western Regional Climatology Center (http://wrcc.dri.edu). These data utilize rain gauge measurements of cumulative monthly precipitation for the past 58 years. The calibrated satellite-based estimates of precipitation from the Tropical Rainfall Measuring Mission (TRMM) which are available from the NASA data pool at (http://disc.gsfc.nasa.gov/data/datapool/TRMM). These data are taken from the level-three merged monthly precipitation product which is available at 0.25° resolution and uses supplementary information from rain gauges and infrared estimates to improve data quality.

2.9.2.2.3 *Surface soil moisture*

Surface soil moisture is derived using the Advanced Microwave Scanning Radiometer (AMSR-E) sensor (Njoku et al., 2003; Shibata et al., 2003) on board the AQUA satellite. The AMSR-E observed brightness temperatures at the 6, 10 and 18 GHz are used in conjunction with a radiative transfer model to simultaneously retrieve the surface soil moisture, surface temperature and vegetation water content (Njoku and Li, 1999; Njoku et al., 2003). The AMSR-E/Aqua level 3 global daily 25 km data is stored at the National Snow and Ice Data Center (NSIDC) website (http://www.nsidc.org/data/amsre). AMSR soil moisture estimates have an accuracy of around 10% and cannot be estimated under vegetation biomass conditions of greater than $1.5\,kg\,m^{-2}$.

2.9.2.2.4 *Root zone soil moisture*

LDAS (Land data assimilation system) aims to provide high resolution maps of land surface state and fluxes by using both in situ and remote sensing data (Mitchell et al., 2004). The LDAS system runs multiple offline land surface models (VIC, NOAH, CLM and Mosaic) to generate the final products. The LDAS system is forced using precipitation and radiation from satellite and other merged sources. Soil, vegetation and elevation are parameterized using high resolution datasets (1 km satellite data in the case of vegetation). The input forcings (Cosgrove at al 2003; Luo et al., 2003) and outputs of the LDAS system have been extensively validated (Lohmann et al., 2004; Robock et al., 2003; Schaake et al., 2004). The outputs are available at a $1/8^\circ$ spatial resolution and hourly time step at http://ldas.gsfc.nasa.gov/. The LDAS output product is the total soil moisture of the 1meter soil column.

2.9.2.3 Watershed studies on impact of climate change on hydrology and water resources

There have been numerous studies that investigate the impact of climate change on watershed hydrology. It would be impossible to describe or even to list all such studies but a few are described in this subsection to provide as examples.

It is a common conclusion that one of the most important resources of water, the groundwater aquifers will face decreases in recharge in the face of climate change. Simulations in several catchments in England have shown that when they use the Hadley GCM predictions for future climate change scenarios, the amount of recharge decrease by almost 90% (Herrera-Pantoja and Kiscock, 2008). In addition, there is an increase in the persistence of dry periods. Coastal areas aquifers show a loss of freshwater and saline intrusion (Ranjan et al., 2006). In a study of climate change impacts for watershed in California, USA for the period 1983-present, Vicuna and Dracup (2007) found that the there is an early onset of spring snowmelt streamflow and an increase in streamflow temperatures. A similar conclusion has been reached by Charlton et al. (2006) for catchments in Ireland and the Rhine River Basin (Middelkoop et al., 2001) where they found decreases in runoff, decrease in summertime flow and increased evapotranspiration and increased likelihood of winter floods due to snowmelt as a result of climate change. Other impacts include changes in spatial and temporal availability of freshwater that affect water availability rather than just a shortage of water as in a study on several basins in the Czech Republic (Dvorak et al., 1997). Changes in precipitation due to climate change were found to increase runoff during wet season and decrease the runoff in the dry season in a study for a catchment in Taiwan (Yu et al., 2002).

2.9.3 LAND SURFACE MODELING AND DATA ASSIMILATION

2.9.3.1 Introduction

Land surface modeling has faced limitations in the past due to the lack of spatially distributed data on land surface characteristics as well as variables in water and energy budgets, namely surface temperature and soil moisture.

Soil moisture is a crucial component of both the water and energy budget. The absence of spatially distributed observations of soil moisture makes it very difficult for distributed hydrological model validation especially with respect to the water budget. Comparison of model and the observed streamflow does not ensure distributed water budget validation for the hydrological model. There could be errors in the infiltration and the evaporation, which could balance each other and thereby attain water balance but the individual components (infiltration and evaporation) as well as soil moisture could be incorrect. It is therefore imperative to use other data sets to ensure spatially distributed validity of these models as well as validity of the individual components of the water and energy budgets.

Satellite observed surface temperatures satisfy our requirement of being spatially distributed and having connections to both the water and the energy budgets. Surface temperature influences evapotranspiration (due to the dependence of the saturation vapor pressure on the surface temperature), hence the energy budget. Evapotranspiration is connected to the water budget as it determines the soil moisture content.

Comparison of surface temperature does not ensure that the model simulations of surface soil moisture are correct. There are various reasons for errors in the modeled soil moisture. The primary reason is the errors in the forcing inputs of precipitation and incoming solar radiation. Therefore, we need to compensate for these errors in the input forcings by assimilating the readily available spatially distributed satellite observed surface temperatures. The assimilated surface temperatures will be used to adjust the model computed surface soil moisture. This adjustment will be carried out such that the new assimilated surface temperature balances the energy balance equation.

The subject of assimilation of soil moisture data or assimilation of meteorological data in order to estimate soil moisture more accurately is relatively a new area of study (McLaughlin et al., 1995). Previous studies (Entekhabi et al., 1994; Lakshmi et al., 1997) have demonstrated the use of microwave satellite data in estimating soil moisture. The assimilation of soil moisture from low-level atmospheric variables using a mesoscale model (Bouttier et al., 1993a, b) has shown that the assimilated soil moisture estimates help in the initialization of atmospheric models. Satellite estimates of surface skin temperature are used to adjust for the soil moisture (McNider et al., 1994; Ottle and Vijal-Majdar, 1994) and estimate with greater accuracy the surface fluxes and surface temperature. Other studies (Blyth, 1993) carry out assimilation by nudging the forecast model evaporation fraction using the satellite data and hydrological model computed evaporative fraction. The results are reductions in the predicted 2 m air temperature and vapor pressure after carrying out these assimilations. Another way satellite data has been used is for parameterization of hydrological models. In this regard, microwave satellite data is especially helpful (van den Hurk et al., 1997).

In this section we will compare the model computed surface temperature to the satellite observed surface temperatures and discuss the sources and reasons for these differences. The effect of assimilation on removing the errors caused by incorrect input forcings will be studied. We will carry out spatially distributed comparisons over a large area (roughly $5^\circ \times 10^\circ$) and a time period of a year between the assimilated and the un-assimilated cases. The implications of the technique in the context of using it in land surface models in global climate models will be discussed with regard to the feasibility.

2.9.3.2 Theory

The land surface hydrology can be represented (Mahrt and Pan, 1984; Lakshmi et al., 1997) by a two-layer model as shown in Figure 2.9.2. The water balance for the model can be written as

$$Z_1 \frac{\partial \theta_1}{\partial t} = P - R - E - q_{1,2}$$

$$Z_2 \frac{\partial \theta_2}{\partial t} = q_{1,2} - q_{2,wt} - T - Q_b$$

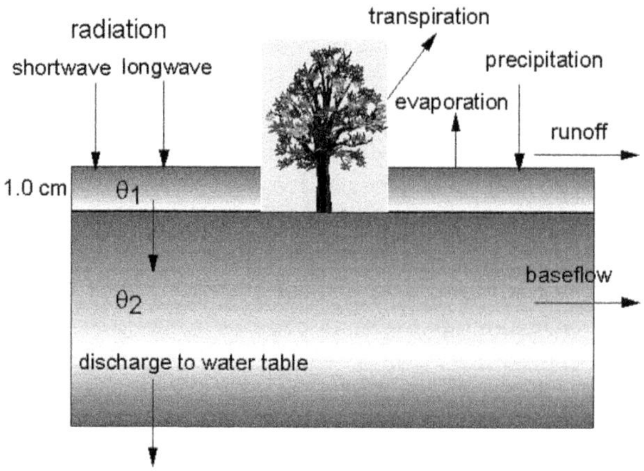

Figure 2.9.2 Representation of water and energy budget in a hydrological model

where q_1 and q_2 are the volumetric soil moistures of the top layer (with thickness Z_1) and the bottom layer (with thickness Z_2), P is the precipitation, E is the bare soil evaporation, R is the surface runoff, T is the transpiration, $q_{1,2}$ is the moisture flow from layer 1 to layer 2 and $q_{2,wt}$ is the moisture flow from layer 2 to the water table. In this model, the transpiration is assumed to occur from the bottom layer only. The moisture flow from layer 1 to layer 2 and the flow from layer 2 to the water table are modeled using the Philips equation accounting for the gravity advection and the moisture gradient. The bare soil evaporation and the vegetation transpiration are estimated using the supply and demand principle, i.e. if there is enough moisture to satisfy the potential value, the evaporation and transpiration occur at the potential rate, else they occur at a rate limited by the amount of available soil moisture.

The energy balance equation for the land surface can be written as

$$R_{sd}(1-\alpha) + R_{ld} - \varepsilon\sigma T_s^4 - \frac{\rho C_p}{\gamma(r_{av}+r_s)}(e_s(T_s) - e_a) - \frac{\rho C_p}{r_{ah}}(T_s - T_a) - \frac{\kappa}{D}(T_s - T_d) = 0$$

where R_{sd}, R_{ld} are the incoming short-wave and long wave radiation respectively. a and e and s are the albedo, emissivity and the Stefan-Boltzmann's constant respectively. ET is the evapotranspiration flux; and T_s, T_a and T_d are the surface temperature, air temperature and the deep soil (50 cm) temperature respectively; $e_s(T_s)$ and e_a are the saturated vapor pressure at surface temperature T_s and actual vapor pressure of the air respectively. r, C_p and g are the density, specific heat and psychrometric constant of air; r_{av} and r_{ah} are the aerodynamic resistances to vapor and heat and r_c is the canopy resistance. k and D are the thermal conductivity and the diurnal damping depth of the soil. The aerodynamic resistances to vapor (r_{av}) and heat (r_{ah}) are taken as equal to each other and are evaluated using established methods (Brutsaert, 1982).

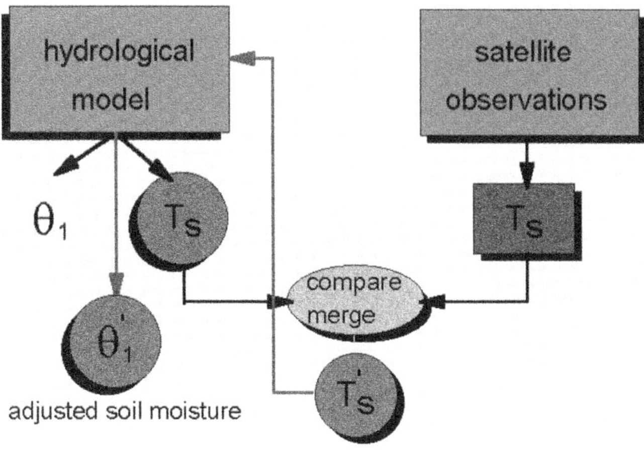

Figure 2.9.3 Assimilation of surface temperature in a hydrological model

Assimilation of surface temperature

The assimilation of surface temperature is carried out using a nudging technique as shown in Figure 2.9.3. The hydrological model produces values of surface temperature and soil moisture. This surface temperature is compared to the corresponding satellite observed value at the time of satellite overpass. (Incidentally, in this study there is at most 2 overpasses over a particular area in one day. However, in the time period, 1989–present, there are upto 4 overpasses in one day). After the comparison of the model derived and the satellite observed surface temperatures, the model surface temperature is "adjusted" to a value midway between the two. (A more realistic way would be to weigh the two values using the error characteristics of the two estimates). This merged surface temperature is used to "recalculate" the soil moisture.

2.9.3.3 Comparison of surface temperature

Description of study area

This study was carried out over an areal extent of 4.75° in latitude and 10.5° in longitude in the southwestern plains of United States. The area includes a small part of eastern New Mexico, most of Oklahoma (except a small part in the north) and northern Texas.

Data sets

The meteorological data of hourly air temperature, dew point temperature, air pressure, wind speed, cloud height (defined as the height of the lowest sky cover layer more than 1/2 opaque), total sky cover and wind speed was obtained from the surface airways stations.

The vegetation data has been obtained from the University of Maryland reprocessed NOAA Global Vegetation Index Data Product (Goward et al., 1994).

The rainfall is obtained from the Manually Digitized Radar (MDR). The Manually Digitized Radar is a program that produces a complete computer generated composite map of the echo characteristics. The MDR data have been generated using information from all the 100 radars around the country (Moore and Smith, 1979; Baeck and Smith, 1994).

Soil type data for the Red River basin was available (Abdulla et al., 1996). Most of the Red River basin is composed of silt loam and loam soil. The soil for the region outside the Red River basin was considered as a uniform silt loam (in the absence of a distributed data set). The Brooks-Corey parameters for a silt loam soil are $q_r = 0.02$, $q_s = 0.50$, $y(q_s) = 0.2$ m, $K_s = 1.89 \times 10^{-6}$ ms^{-1} and $m = 0.2$ (Rawls et al., 1982).

The Tiros Operational Vertical Sounder (TOVS) has flown on NOAA spacecraft since 1978. The radiances have been observed by the High Resolution InfraRed Sounder (HIRS2) and the brightness temperatures of the Microwave Sounding Unit (MSU). The radiances observed by these two sensors have been analyzed to provide daily fields of air temperature and humidity profiles, surface temperature and cloud amounts and altitudes they occur (Susskind et al., 1997). These data sets are available as daily $1° \times 1°$ gridded fields used in this study. The surface skin temperature is computed directly using the channels 8, 18 and 19 (the thermal channels) of the HIRS2 and the Planck equation. The surface air temperature and the specific humidity near the surface are obtained by extrapolating the air temperature profile and the specific humidity profile to the surface pressure level (reference). The data from NOAA 10 satellite is used here. NOAA 10 has a nadir 730 AM/PM local time overpass at the equator. The observations at all other locations, north or south of the equator and all off-nadir observations are at times different from 730 AM/PM. Since air and surface temperatures are very sensitive to time of day, the exact local time at each local was used in conjunction with the surface and air temperatures and the surface specific humidity.

Observed versus simulated surface Temperatures

The satellite retrieved surface temperature corresponding to the NOAA 10 AM and PM overpasses averaged over the entire Red-Arkansas grid box (approximately $5° \times 10°$) is shown as a scatter plot (Figure 2.9.4) with the corresponding simulated surface temperature using a hydrological model. Each point in the plot represents a single day value. However, the value of the surface temperature in each $1° \times 1°$ grid box for the simulation corresponds to the exact time of overpass of the satellite over that grid box. Therefore, time of day effects are eliminated in this manner. The scatter plot shows that the mean difference between the satellite surface temperature and the model surface temperature (bias) is 1.77C for the morning overpass and −3.67C for the evening overpass. The correlation between the satellite and the model simulated surface temperatures is fairly good (0.97 for the morning overpass and 0.98 for the evening overpass). The standard deviation of the difference between the satellite surface temperature and the model simulated surface temperature is 2.5C and 1.79C for the morning and evening overpasses respectively. The slope of the best fit line gives an indication of the range of the two data. Slope values of less than 1 mean that the model

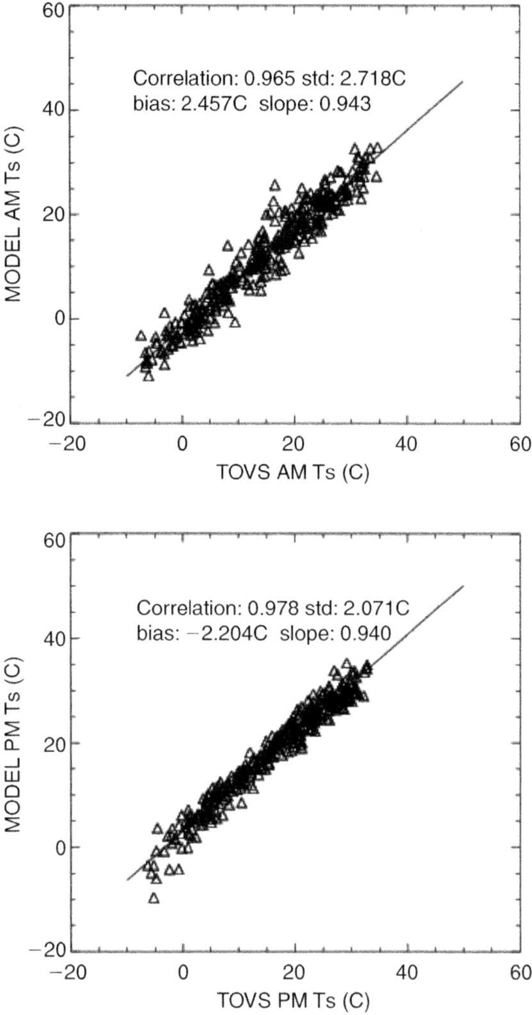

Figure 2.9.4 Comparison of the area averaged model generated and satellite observed surface temperature corresponding to the AM and PM overpass of NOAA11 for the Red River basin grid box over 1 year (August 1987–July 1988)

simulated surface temperature has a lower value of range than the satellite surface temperature. This is the case for both the morning and evening satellite overpasses.

The comparison of the simulated diurnal cycle with the satellite observed surface temperature is shown in Figure 2.9.5 for four different days in the year-long study period for a $1^\circ \times 1^\circ$ pixel centered at 33°30′N, 99°30′W. The simulated surface temperature differs from the satellite observations by a few degrees at the most. The difference between the satellite and the simulated surface temperatures for September 10, 1987 are 2.2C for the morning overpass of the satellite and −3.9C for the evening

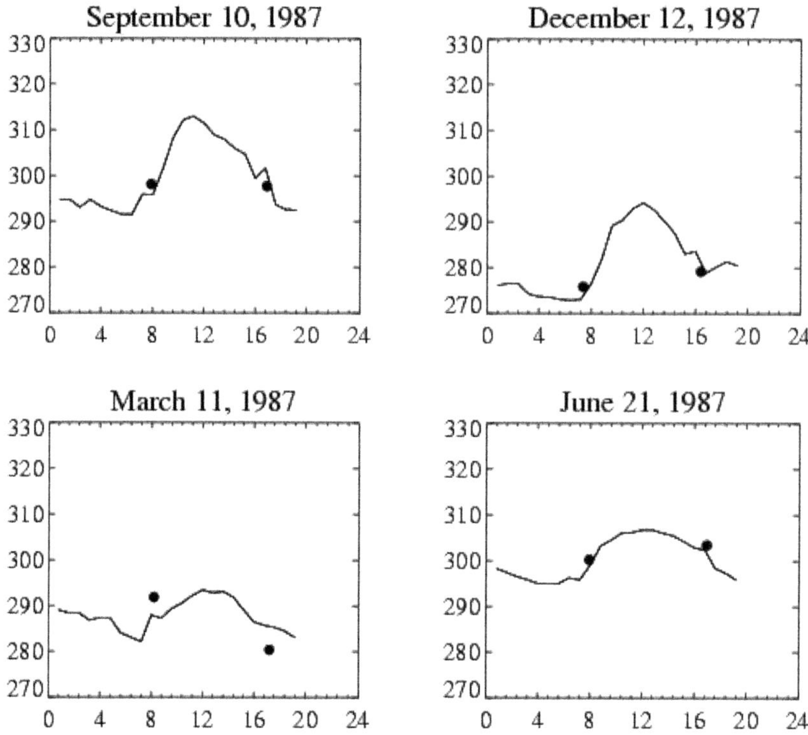

Figure 2.9.5 Comparison of model computed diurnal cycle of surface temperature and the satellite derived surface temperature from NOAA-10 for a 1° × 1° pixel centered at 33°30′N, 99°30′W for four days

overpass. The corresponding numbers for December 12, 1987, March 11, 1988 and June 21, 1988 are 2.8C, 4.0C and 4.4C for the morning satellite overpass and −4.4C, −5.3C and 1.1C for the evening overpass. It is seen that the difference between the satellite and the simulated surface temperatures is positive in the morning overpass (satellite surface temperature is warmer) and negative corresponding to the evening overpass (the model surface temperature is warmer). The only exception is June 21, 1988 evening overpass when the difference is positive (the model is cooler).

2.9.3.4 Simulation and assimilation experiments

Experiment set-up

The assimilation scheme described in Section 2.9.2 is tested using a series of experiments shown in Figure 2.9.6.

The land surface hydrological model is basically run in two modes, viz. without assimilation of satellite surface temperature data and with the assimilation of satellite surface temperature. The data input into the hydrological model described in the previous section is treated as "perfect" inputs. The hydrological model is run along

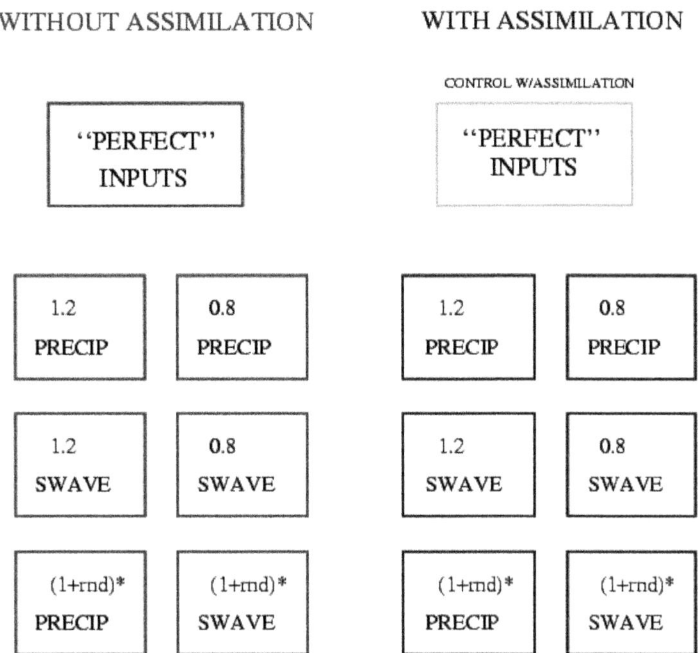

Figure 2.9.6 Schematic representation of the computer runs with and without assimilation

with these sets of inputs with and without assimilation. In addition to this set of runs, the hydrological model is run with the "perfect" inputs perturbed. The precipitation (PRECIP) and the shortwave radiation (SWAVE) are chosen for this study as they have the most impact on the land surface hydrological cycle. The precipitation effects the land surface moisture storage through the infiltration and the runoff and the upper and the lower layer soil moistures. The shortwave radiation is the driving force for the evapotranspiration during the sunlight hours. The precipitation and the shortwave radiation are biased 20% higher (1.2 PRECIP and 1.2 SWAVE), 20% lower (0.8 PRECIP and 0.8 SWAVE) and randomly using a random number (rend) between 0 and 1((1 + rnd) PRECIP and (1 + rnd) SWAVE). In the runs where precipitation is biased, all the other inputs are unchanged (this included the shortwave radiation). The only exception to this is one set of run in which both the precipitation and the shortwave radiation are perturbed randomly simultaneously.

The corresponding runs with and without surface temperature assimilation are compared.

It can be seen that both the bias and the standard deviation of soil moisture in the simulations with +20% precipitation (increased precipitation by 20% termed as 1.2P) and −20% precipitation (decreased precipitation by 20% termed as 0.8P) without assimilation of surface temperature (termed as control w/asml-1.2P weasel or control w/asml – 0.8P weasel) performs much poorly than with assimilation of surface temperature (control w/asml – 1.2P w/alms or control w/asml – 0.8P w/asml) for both

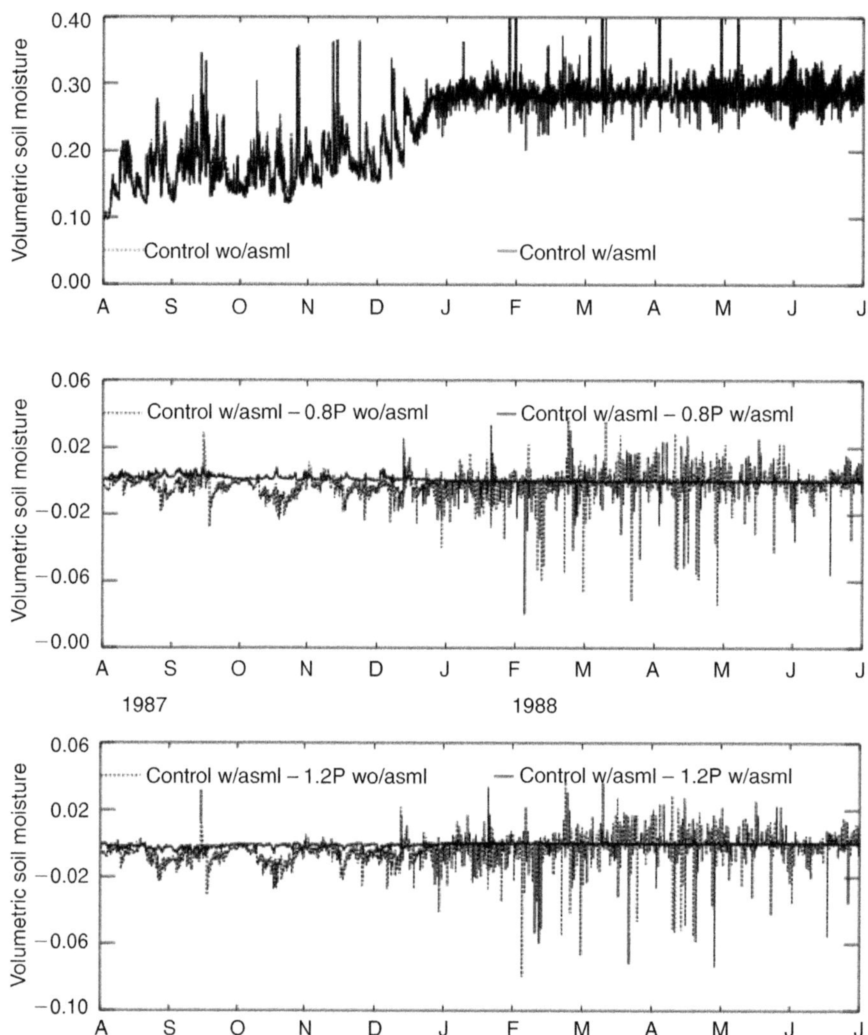

Figure 2.9.7 Temporal evolution of the upper layer volumetric soil moisture averaged over the Red-Arkansas grid box between August 1, 1987 and July 31, 1988 for the control with and without assimilation cases (top panel), difference between the spatial average control with assimilation and the rainfall input biased 20% lower with and without assimilation (middle panel) and difference between the spatial average control with assimilation and the rainfall input biased 20% higher with and without assimilation (bottom panel)

bias (Figure 2.9.7) and standard deviation (Figure 2.9.8). Ideally, both the bias and standard deviation of the differences should be zero. However due to the perturbed input of precipitation, the soil moisture goes off track and has to be reset with the nudging associated with the assimilation function. Similar and more complex assimilation strategies have shown to improved soil moisture and other land atmosphere

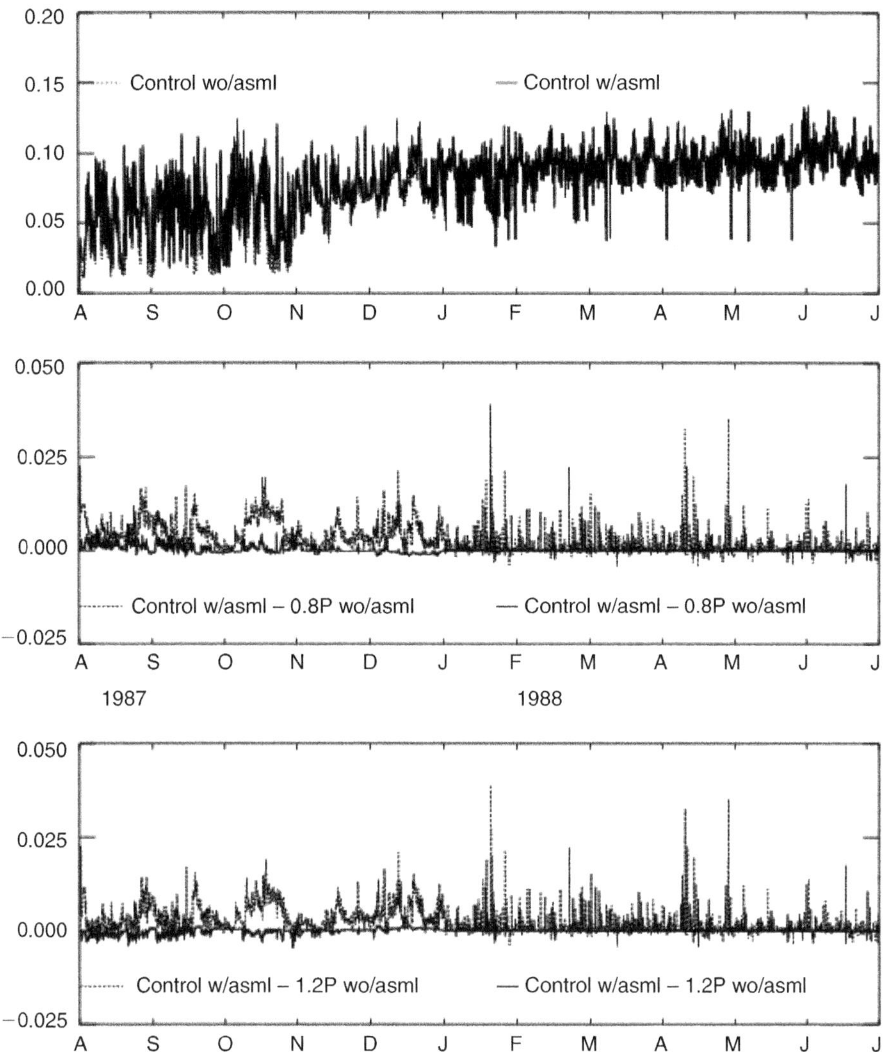

Figure 2.9.8 Temporal evolution of the upper layer volumetric soil moisture spatial standard deviation over the Red-Arkansas grid box between August 1, 1987 and July 31, 1988 for the control with and without assimilation cases (top panel), difference between the spatial standard deviation control with assimilation and the rainfall input biased 20% lower with and without assimilation (middle panel) and difference between the spatial standard deviation control with assimilation and the rainfall input biased 20% higher with and without assimilation (bottom panel)

flux simulations in the presence of erroneous inputs and faulty parameterizations of models or incorrect physics.

Modeling is a powerful tool for estimation of water storage and fluxes and when coupled with data assimilation and satellite remote sensing offers a very strong framework for hydrological analysis.

REFERENCES

Abdulla, F.A., D.P. Lettenmaier, E.F. Wood, and J.A. Smith (1996) Application of a macroscale model to estimate the water balance for the Arkansas-Red River basin, *Journal of Geophysical Research*, 101 (D3), pp. 7449–7459.

Alcamo, J., M. Florke and M. Marker (2007) Future long-term changes in groundwater resources driven by socio-economic and climatic changes, *Hydrological Sciences Journal*, 52(2), pp. 247–275.

Alsdorf, D. E. and D. P. Lettenmaier (2003) Tracking fresh water from space, *Science*, 301, 1491–1494.

Alsdorf, D., P. Bates, J. Meelak, M. Wilson and T. Dunne (2007) Spatial and temporal complexity of Amazon flood measured from space, *Geophysical Research Letters*, Vol. 34, L08402, doi:10.1029/2007GL029447.

Baeck, M.L. and J.A. Smith (1995) Climatological Analysis of Manually Digitized Radar Data for the United States East of the Rocky Mountains, *Water Resources Research*, Vol. 31, No. 12, pp 3033–3049.

Basist, A., Williams, C., N. Grody, T. Ross, S. Shen, A. Chang, R. Ferraro, and M. Menne (2001) Using Special Sensor Microwave Imager to monitor soil wetness, *Journal of Hydrometeorology*, Vol. 2, pp. 297–308.

Becker, F. and Z. Li (1990) Towards a local split window method over land surfaces, *International Journal of Remote Sensing*, 11, pp. 369–393.

Blyth, K. (1993) The use of microwave remote sensing to improve spatial parameterization of hydrological models, *Journal of Hydrology*, 152, pp. 103–129.

Bouttier, F., J.-F. Mahfouf and J. Noilhan (1993a) Sequential assimilation of soil moisture from atmospheric low-level parameters. Part I: Sensitivity and calibration studies, *Journal of Climate*, 32, pp. 1335–1351.

Bouttier, F., J.-F. Mahfouf and J. Noilhan (1993b) Sequential assimilation of soil moisture from atmospheric low-level parameters. Part II: Implementation in a mesoscale model, *Journal of Climate*, 32, pp. 1352–1364.

Brutsaert, W. (1982) *Evaporation into the Atmosphere – Theory, history and applications*, D. Reidel Publishing Co., 299 p.

Cayan, D.R., Maurer, E.P. and Dettinger, M.D. and others (2008) Climate change scenarios for the California region, *Climate Change*, pp. S21–S42.

Charlton, R., R. Fealy, S. Moore, J. Sweeney and C. Murphy (2006) Assessing the impact of climate change on water supply and flood hazard in Ireland using statistical downscaling and hydrological modeling techniques, *Climate Change*, Vol. 74, pp. 475–491, 2006

Cosgrove (and 15 others) (2003) Real time and retrospective forcing in the North American Land Data Assimilation System (NLDAS) project, *Journal of Geophysical Research*, Vol. 108, D22, 8842, doi:10.1029/2002.

Defries, R.S., Hansen, M., Townshend, J.R.G. and R. Sohlberg (1998) Global land cover classification at 8 km spatial resolution: the use of training data derived from Landsat imagery in decision tree classifiers, *International Journal Remote Sensing*, 19(16), 3141–3168.

Dvorak, V., J. Hiladny and L. Kasparek (1997) Climate change hydrology and water resources impact and adaption for selected river basins in Czech Republic, *Climate Change*, Vol. 36, pp.93–106, 1997

Entekhabi, D., H. Nakamura and E.G. Njoku (1994) Solving the Inverse problem for soil moisture and temperature profiles by sequential assimilation of multifrequency remotely sensed observations, *IEEE Transactions on Geoscience and Remote Sensing*, Vol. 32, No. 2, pp. 438–448.

Ferraro, R. (1997) SSM/I derived global rainfall estimates for climatological applications, *Journal of Geophysical Research*, Vol. 102, pp. 16715–16735.

Friedl, M.A., et al. (2002) Global land cover mapping from MODIS: algorithms and early results, *Remote Sensing of Environment*, 83, 287–302.

Herrera-Pantoja, M. and K. Hiscock (2008) The effects of climate change on potential groundwater recharge in Great Britain, *Hydrological Processes*, 22, pp. 73–86, 2008.

Holman, I (2006) Climate change impacts on groundwater recharge-uncertainties shortcomings, and the way forward, *Hydrogeology Journal*, 14, pp. 637–647.

Gillespie, A., J.S. Cothern, S. Rokugawa, T. Matsunaga, S.J. Hook and A.B. Kahle (1998) A temperature and emissivity separation algorithm for Advanced Spaceborne Thermal Emission and Reflection Radiometer (ASTER) images. *IEEE Transactions on Geoscience and Remote Sensing*, 36(4): 1113–1126.

Goward, S. N., S. Turner, D. Dye, and S. Liang (1994) The University of Maryland improved Global Vegetation Index product, *International Journal of Remote Sensing*, 15(7), pp. 3365–3395.

Justice, C.O., et al. (1998) The Moderate Resolution Imaging Spectroradiometer (MODIS): Land remote sensing for global change research, IEEE Transactions on Geoscience and Remote Sensing, 36(4), 1228–1249.

Justice, C.O., Townshend, J.R.G., Vermote, E.F., et al. (2002) An overview of MODIS Land data processing and product status; *Remote Sensing of the Environment*, 83(1–2): 3–15.

Kouraev, A., E. Zakharova, O. Samain, N. Mognard and A. Xazenave, Ob' River discharge from TOPX/Poseidon altimetry (1992–2002), *Remote Sensing of the Environment*, 93 pp. 238–245, 2004.

Lakshmi, V., The role of remote sensing in prediction of ungaged basins, *Hydrological Processes*, Volume 18, Issue 5, Pages 1029–1034, Invited Commentary, 2004

Kummerow, C. and others, The status of the Tropical Rainfall Measuring Mission (TRMM) after two years in orbit, *Journal of Applied Meteorology*, Vol. 39, pp. 1965–1982, 2000

Kundzewicz, Z.W., L.J. Mata, N.W. Arnell, P. Döll, P. Kabat, B. Jiménez, K.A. Miller, T. Oki, Z. Sen and I.A. Shiklomanov, (2007) Freshwater resources and their management. *Climate Change 2007: Impacts, Adaptation and Vulnerability. Contribution of Working Group II to the Fourth Assessment Report of the Intergovernmental Panel on Climate Change,* M.L. Parry, O.F. Canziani, J.P. Palutikof, P.J. van der Linden and C.E. Hanson, Eds., Cambridge University Press, Cambridge, UK, pp. 173–210.

Lakshmi, V., E.F. Wood and B.J. Choudhury (1997) A soil-canopy-atmosphere model for use in satellite microwave remote sensing, *Journal of Geophysical Research*, 102, D6, 6911–6927.

Lakshmi, V. and Schaaf, K. (2001) Analysis of the 1993 Midwestern floods using satellite and ground data, *Transactions on Geoscience and Remote Sensing*, 39(8), pp. 1736–1743.

Lohmann (and 13 others) (2004) Streamflow and water balance intercomparisons of four land surface models in the North American Land Data Assimilation Project, *Journal of Geophysical Research*, 109(D07S91), doi:10.1029/2003JD003517.

Luo, L. (and 14 others) (2003) Validation of the North American Land Data Assimilation System (NLDAS) retrospective forcing over the southern Great Plains, *Journal of Geophysical Research*, 108(D22) 8843, doi:10.1029/2002.

Mahrt, L. and H. Pan (1984) A two-layer model of soil hydrology, *Boundary-Layer Meteorology*, 29, pp. 1–20.

McLaughlin, D. (1995) Recent developments in hydrologic data assimilation, *Reviews of Geophysics, supplement*, pp. 977–984, U.S. National Report to International Union of Geodesy and Geophysics 1991–1994.

McNider, R.T., A.J. Song, D.M. Casey, P.J. Wetzel, W.L. Crosson and R.M. Rabin (1994) Towards a dynamic-thermodynamic assimilation of satellite surface temperature in numerical atmospheric models, *Monthly Weather Review*, 122, pp. 2784–2803.

Middelkoop H., K. Daamen, D. Gellens, W. Grabs, J. Kwadijk, H. Lang, B. Parmet, B. Schadler, J. Schulla and K. Wilke (2001) Impact of climate change on hydrological regimes and water resources management in the Rhine Basin, *Climate Change*, Vol. 49, pp. 105–128.

Mitchell (and 22 others) (2004) The multi-institution North American Land Data Assimilation System (NLDAS) Utilizing multiple GCIP products and partners in a continental distributed hydrological modeling system, *Journal of Geophysical Research*, 109(D07S90), doi:10.1029/2003.

Moore, P.L. and D.L. Smith (1979) Manually Digitized Radar data – Interpretation and application, NOAA Technical Memorandum NWS SR-99.

Myneni, R.B., Hall, F.G., Sellers, P.J., and A.L. Marshak (1995) The interpretation of spectral vegetation indexes, *IEEE Transactions on Geoscience and Remote Sensing*, 33(2), 481–486.

Myneni, R.B., et al. (2002) Global products of vegetation leaf area and fraction absorbed PAR from year one of MODIS data, *Remote Sensing of Environment*, 83, 214–231.

Njoku, E.G., and L. Li (1999) Retrieval of land surface parameters using passive microwave measurements at 6–18 GHz, *IEEE Transactions on Geoscience and Remote Sensing*, 37 (1), 79–93.

Njoku, E.G., Jackson, T., Lakshmi, V., Chan, T., and S. Nghiem, (2003) Soil moisture retrieval from AMSR-E, *IEEE Transactions on Geoscience and Remote Sensing*, 41, 215–229.

Ottle, C. and D. Vijal-Madjar (1994) Assimilation of soil moisture inferred from infrared remote sensing in a hydrological model over the HAPEX-MOBILHY region, *Journal of Hydrology*, 158, pp. 241–264.

Papa, E., C. Prigent and W. Rossow, (2007) Ob' river flood inundations from satellite observations: A relationship with winter snow parameters and river runoff, *Journal of Geophysical Research* Vol. 112, D18103, doi:10.1029/2007JD008541.

Ranjan, P., S. Kazama and M. Sawamoto (2006) Effects of climate change on coastal fresh groundwater recharge, *Global Environmental Change*, Vol. 16, pp. 388–399.

Rawls, W.J., D.L. Brakensiek and K.E. Saxton (1982) Estimation of Soil Water Properties, *Transactions of the American Society of Agricultural Engineering*, 25, pp. 1316–1320.

Robock (and 16 others) (2003) Evaluation of the North American Land Data Assimilation System over the Southern Great Plains during the warm season, *Journal of Geophysical Research*, 108(D22), 8846, doi:10.1029/2002, JD003245.

Rockstrom, J., M. Lannerstad, and M. Falkenmark (2007) Assessing the water challenge of a new green revolution in developing countries, *Proceedings of the National Academy of Sciences*, Vol. 104, No. 15, pp. 6253–6260.

Rodell, M., and J. S. Famiglietti (1999) Detectability of variations in continental water storage from satellite observations of the time dependent gravity field, *Water Resources Research*, 35(9), 2705–2724, doi:10.1029/1999WR900141.

Rodell, M., and J. S. Famiglietti (2001) An analysis of terrestrial water storage variations in Illinois with implications for the Gravity Recovery and Climate Experiment (GRACE), *Water Resources Research*, 37(5), 1327–1340, doi:10.1029/2000WR900306, 2001

Sarch M., and C. Birkett (2000) Fishing and farming at Lake Chad: Response to Lake level fluctuations, *The Geographical Journal*, Vol. 166, No. 2, pp 156–172.

Scibek, J., and D. Allen (2006) Modeled impacts of predicted climate change on recharge and groundwater levels, *Water Resources Research*, Vol. 42, W11405, doi:10.1029/2005WR004742.

Scientific Assessment of the Effects of Global Change on the United States (2008) *A Report of the Committee on Environment and Natural Resources National Science and Technology Council*, May 271pp.

Schaake, J. (and 14 others) (2004) Intercomparison of soil moisture fields in the North American Land Data Assimilation System (NLDAS), *Journal of Geophysical Research*, 109 (D01S90), doi:10.1029/2002JD003309.

Schmugge, T., P. O'Neill and J. Wang (1986) Passive microwave soil moisture research, *IEEE Transactions on Geoscience and Remote Sensing*, 24(1) pp. 12–22.

Shibata, A., Imaoka, K., and T. Koike (2003) AMSR/AMSR-E level 2 and 3 algorithm developments and data validation plans of NASDA, *IEEE Transactions on Geoscience and Remote Sensing*, 41, 195–203.

Susskind, J., Piraino, P., Rokke, L., Iredell, L., and A. Mehta (1997) Characteristics of the TOVS Pathfinder Path A data set, *Bulletin of the American Meteorological Society*, 78(7), 1449–1472.

Susskind, J., Barnet, C., and Blaisdell, J. (2003) Retrieval of atmospheric and near surface parameters from AIRS/AMSU/HSB data in the presence of clouds, *IEEE Transactions on Geoscience and Remote Sensing*, 41(2), pp. 390–409.

Susskind J., C. Barnet, J. Blaisdell, and others (2006) Accuracy of geophysical parameters derived from Atmospheric Infrared Sounder/Advanced Microwave Sounding Unit as a function of fractional cloud cover, *Journal of Geophysical Research – Atmospheres*, 111(D90), D09S17, 2006.

Syed, T., J. Famiglietti, M. Rodell, J. Chen and C. Wilson (2008) Analysis of terrestrial water storage from GRACE and GLDAS, *Water Resources Research*, Vol. 44, W02433, doi:10.1029/2006WR005779.

Tucker, C.J. (1979) Red and photographic infrared linear combination for monitoring vegetation, *Remote Sensing of Environment*, 8, 127–150.

van den Hurk, B. J., W. Bastiaanssen, H. Pelgrum and E. Meijgaard (1997) A new methodology for assimilation of initial soil moisture fields in weather prediction models using Meteosat and NOAA data, *Journal of Applied Meteorology*, Vol. 36, pp. 1271–1283, 1997.

Varady, R. G., K. Meehan, J. Rodda, E. McGovern, M. Iles-Shih (2008) Strengthening global water initiatives, *Environment*, Vol. 50, No. 2, pp. 20–31.

Vicuna, S, and J. Dracup (2007) The evolution of climate change impact studies on hydrology and water resources in California, *Climate Change*, Vol. 82, pp. 327–350.

Yu, P., T. Yang and C. Wu (2002) Impact of climate change on water resources in southern Taiwan, *Journal of Hydrology*, Vol. 260, pp. 161–175.

Wan, Z. and Z.L. Li (1997) A physics-based algorithm for retrieving land-surface emissivity and temperature from EOS/MODIS data, *IEEE Transactions on Geoscience and Remote Sensing*, 35, 980–996.

UNESCO (2006) *Water: A shared responsibility*, United Nations World Water Development Report 2, 601pp.

Wilheit, T., Kummerow, C. and R. Ferraro (2003) Rainfall algorithms for AMSR-E, *IEEE Transactions on Geoscience and Remote Sensing*, 41(2), pp. 204–214.

Zakharova E., A. Kouraev, A. Cazenave and F. Seyler (2006) Amazon River discharge using TOPEX/Poseidon altimetry, *CR Geosciences*, 338, pp. 188–196, 2006.

Zekster, I, L. G. Everett (Editors) (2004) Groundwater resources of the world and their use, IHP Series on Groundwater No. 6, UNESCO, 342p.

Chapter 2.10

Case history and exercises Hydrodynamics of an urban river – Case study of Musi river, Hyderabad, India

B. Venkateswara Rao & V. Varalakshmi
Jawaharlal Nehru Technological University, Hyderabad, India

2.10.1 INTRODUCTION

The chapter seeks to delineate the dynamics of interplay between rainfall, water storage reservoirs, groundwater recharge, and water use patterns in respect of the Musi river, Hyderabad, India, as an aid in developing future strategies of planning of water use.

The Musi basin is a sub-basin of the river Krishna (the basin is nearly closing?) in Andhra Pradesh, India. The present study focuses on the effect of large-scale groundwater withdrawal in the upper Musi basin (Fig. 2.10.1) and its consequences, as manifested by the reduction of inflows into the water supplying reservoirs, namely Himayatsagar and Osmansagar, in the downstream. The adverse effects of the groundwater withdrawals in the upper Musi catchment are clearly felt in the fast developing Hyderabad city which is located downstream of the reservoirs. Changes in catchment hydrology such as decreased runoff and decreasing groundwater levels are more frequently observed over larger and smaller basins in semi arid hard rock regions of India. To what extent these changes are related to climate change or developmental activities in the catchments need to be understood as they are crucial for the formulation of future water management strategies. In fact, though it is based on short term data, the temperature data for the 20 years (1985–2004) (Fig. 2.10.2) in the study area shows that there has been no increase of temperature; interestingly the record shows a slight decrease of temperature.

2.10.2 DESCRIPTION OF THE STUDY AREA

The study area is upper part of the Musi basin (Fig. 2.10.1) a tributary of river Krishna in the Southern India. The area falls in the survey of India topo sheet numbers 56k/7 to

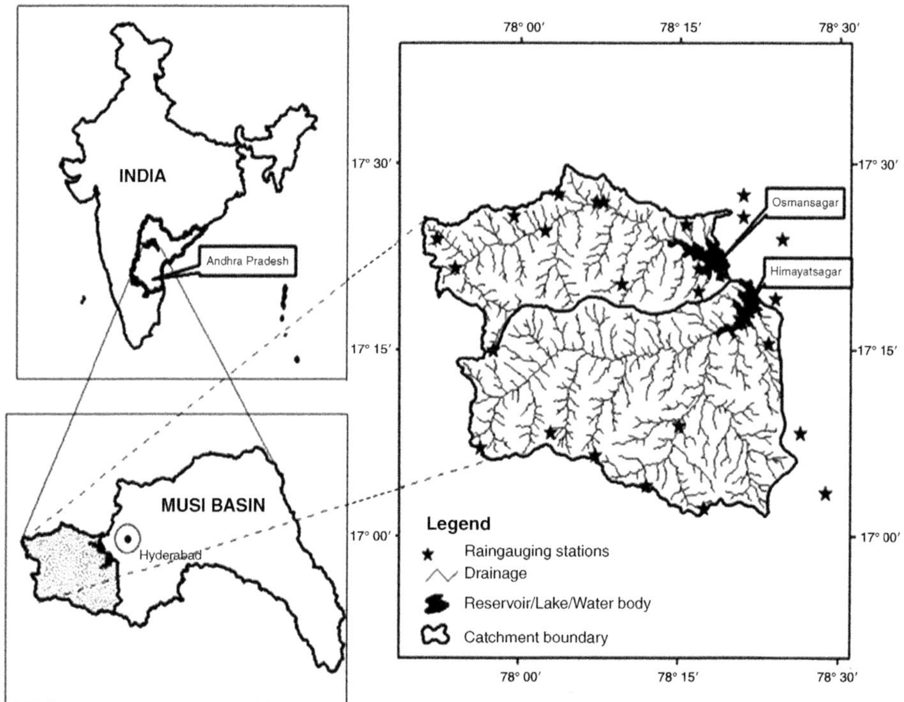

Figure 2.10.1 Catchment area of Osmansagar and Himayatsagar reservoirs

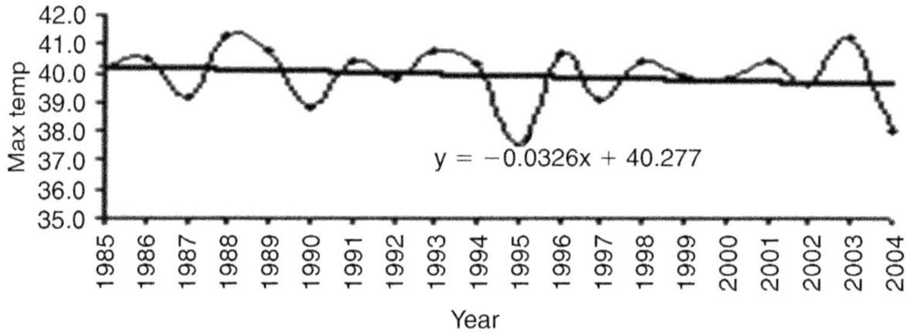

Figure 2.10.2 Annual highest mean maximum temperature variation for the Hyderabad Meteorological station
Source: Praveena, 2005

56k/11 and it is bounded by longitudes 77°49′15″ to 78°56′09″ and latitudes 16°58′12″ to 17°42′70″. The Osmansagar reservoir was constructed in the year 1922 on Musi river, and is located 9.6 km from Hyderabad in western direction. The Himayathsagar reservoir was constructed in the year 1925 on Musa River – a tributary of Musi River and is located 9.6 km in southwest direction from the city.

The drainage pattern is dendritic to sub dendritic and trellis forming up to 4th order streams in Osmansagar catchment with 738 km^2 and 5th order streams in Himayathsagar catchment with 1,308 km^2. Study area has an average altitude of 670 m

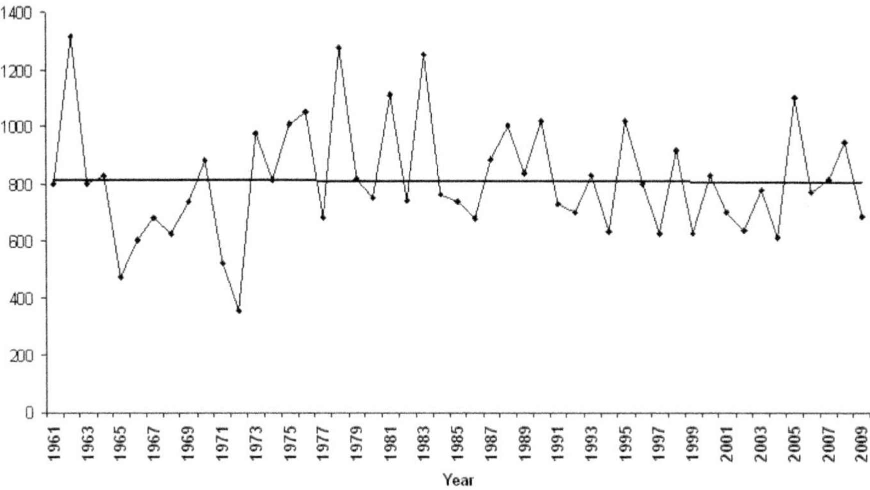

Figure 2.10.3 Variation of rainfall in the catchment area of Osmansagar

in western part, which gradually decreases towards the east till 450 m. The area mainly exposes peninsular gneissic complex that includes a variety of granites, magmatites of various phases and enclaves of older metamorphic rocks belonging to the Achaean age. These are intruded by various acidic (pegmatites, aptite, quartz veins/reefs), and basic intrusives of dolerite and gabbros. Groundwater in the area occurs under unconfined conditions in the open wells to semi-confined conditions in the bore wells if they tap deeper fracture zones.

2.10.3 RAINFALL ANALYSIS OF THE CATCHMENT AREA

The rainfall recorded by various rainguage stations located (Fig. 2.10.1) within and surrounding the catchment areas of Osmansagar and Himayathsagar reservoirs are collected. The rainfall is recorded at 8:30 am daily which is the total amount of rainfall that occurred in the past 24 hours. The data is available for 49 years i.e. from 1961 to 2009 and it is maintained by Bureau of Economics and Statistics and Hyderabad Metropolitan Water Supply and Sewerage Board (HMWS&SB). The depth of rainfall is estimated by using isohyetal method. The time series of average annual rainfall depth so calculated is presented in Fig. 2.10.3 for Osmansagar Reservoir and Fig. 2.10.4 for Himayatsagar Reservoir. From both these figures it can be observed that the rain fall has been more or less oscillating over a normal of 784 mm in the case of Osmansagar and 740 mm in the case of Himayatsagar. For the study period of 49 years (1961–2009), there has been no change in the quantum of rainfall.

2.10.4 ANALYSIS OF INFLOWS TO THE RESERVOIRS

Inflows represent the volume of water that drained from the catchments and reached the reservoirs. The daily water levels at the reservoir site are recorded and are

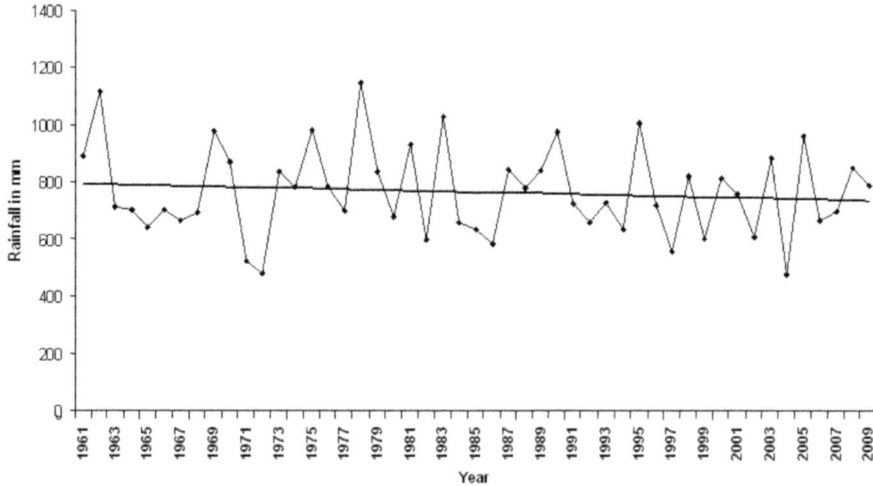

Figure 2.10.4 Variation of rainfall in the catchment area of Himayatsagar

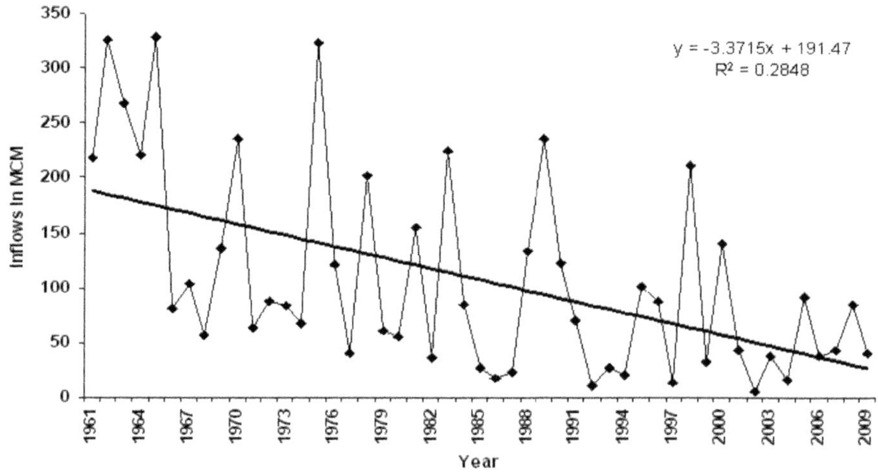

Figure 2.10.5 Inflows received at Osmansagar reservoir

maintained by HMWS & SB. A total of 49 years record commencing from 1961–2009 is taken from the above mentioned office and the inflows are calculated to the reservoir using the methodology proposed by Venkateswara Rao et al. (2010). The resulting inflows are shown as time series graphs (Fig. 2.10.5 and Fig. 2.10.6) for both Osmansagar and Himayatsagar reservoirs

From the above graphs it can be observed that there has been continuous decrease of inflows for the past 49 years in spite of more or less normal rainfall during the same period. The decrease of inflows to the reservoirs may be due to increase in the agricultural production in the catchment through tapping of substantial quantities of

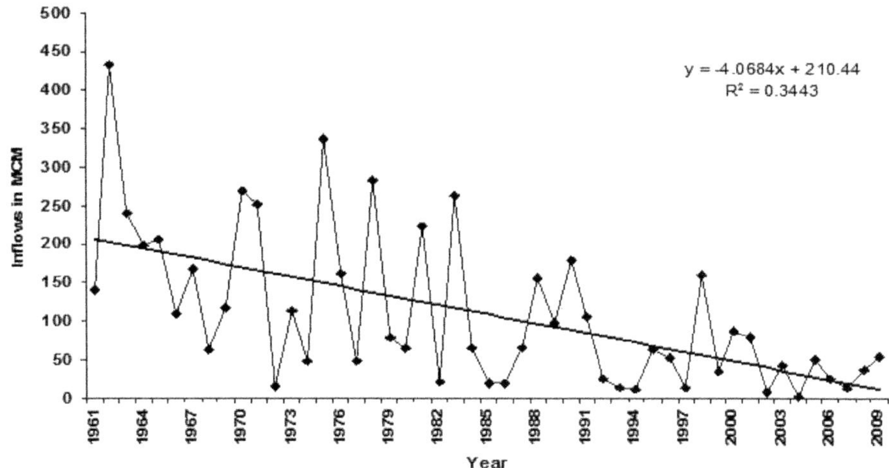

Figure 2.10.6 Inflows received at Himayathsagar reservoir

surface water and groundwater. This increased agricultural production is due to the continuous raising demand of Hyderabad city for vegetables, fruits etc. The recent report of World Bank has also indicated that the increased use of groundwater in the Indian agriculture has led to decreased command area under the tanks (World Bank report, 2005). This is mainly due to the decrease of runoff from the catchment areas of the irrigation tanks due to increased surface and groundwater tapping through various water and soil conservation works in the catchments.

In the present investigation, the above reasons are verified by analyzing the cropping area in the catchment through satellite data and by analyzing groundwater levels in the catchment.

2.10.5 VERIFICATION OF THE CROPPING AREA IN THE CATCHMENTS

In the context of the rapid urbanization of the Hyderabad City, which is close (9.0 km) to the catchments of Osmansagar and Himayathsagar, there has been an ever-increasing demand for agricultural produce (such as vegetables, grains, pulses etc.) that must be necessarily being supplied from these catchments by tapping more surface and groundwater by the farmers in the catchments, resulting in the reduction of inflows to the reservoirs.

For the verification of this observation, further analysis has been carried out by calculating NDVI (Normalized Differential Vegetation Index) using the medium to high spatial resolution multi spectral data provided by remote sensing satellites, such as Landsat TM, IRS-1C/1D LISS-III and IRS-P6 LISS-III for the years 1989, 1994, 1998, 2002 and 2008 for the kharif (June-October: monsoon period) season, and in the years 1975, 1994, 1998, 2003 and 2008 for rabi (November-April: non monsoon period) season. The results are tabulated in Table 2.10.1.

Table 2.10.1 Land Use/Land cover change analysis of satellite data

	Kharif Land use/cover category in hactares			*Rabi* Land use/cover category in hactares	
Year	*Area under crops*	*Area Under Reservoirs/ Lakes/Tanks with water*	*Year*	*Area under crops*	*Area Under Reservoirs/ Lakes/Tanks with water*
1989	75,195.18	4,990.05	1975	7,809.30	2,755.48
1994	76,628.48	2,578.04	1990	12,116.60	1,834.27
1998	92,917.22	4,926.66	1998	17,034.50	624.24
2002	76,100.95	2,695.641	2003	10,268.90	896.17
2008	94,631.20	2,795.10	2008	18,845.96	617.17

From the NDVI analysis it is found that, in the *kharif* season the total cropping area recorded an increase of 25.8% and reservoirs/tanks/lakes occupied with water got reduced by 43% during 1989 to 2008. As against this, in the rabi season during the period 1975–2008, the total cropping area increased by 141.32%, while water-covered areas of reservoirs/tanks/lakes got reduced by 67.47%. This means that during the *rabi* season groundwater is being increasingly utilized in the catchments for irrigation due to non-availability of surface water, thereby depleting the groundwater seriously. Consequently the inflows were reduced to the reservoirs in the recent years due to the fact that whatever the rainfall that occurs in the catchment, it is first utilised for meeting the depleted soil moisture and groundwater, thereby reducing the surface runoff. To verify this fact analysis of groundwater levels were carried out (Varalakshmi, 2010).

2.10.6 WATER TABLE CONTOUR MAPS AND ANALYSIS

Groundwater levels in the Musi basin were observed during premonsoon and the post monsoon seasons for the year 2005. Similarly 20 years data of the groundwater levels of the observation wells and the data recorded by the Digital Water Level Recorders (DWLR) from Andhra Pradesh State Ground Water Department in the upper Musi catchment have been collected. Analysis of change in groundwater levels between premonsoon and post monsoon for the 20 years (1985–2004) has been made to check as to whether the recharge to the groundwater is mostly influenced by rainfall or premonsoon groundwater levels. Further investigations have been made by analyzing the data of the DWLRs installed in the basin. The DWLR records groundwater levels for every 6 hours and the data set constitutes from the year 1998 to 2005. The rainfall is also recorded near the DWLRs. The combined plot of rainfall and water levels below ground level (b.g.l), called Composite Hydrograph, is used for analysis.

Contour maps of water table can be prepared by observing the water levels in open wells or bore wells. These maps are obtained by joining the points, which are having equal groundwater levels. The movement of the groundwater is perpendicular to these lines. The monthly water level data recorded by State Ground water Department of Andhra Pradesh in 23 monitoring observation wells (for the 20 years period between 1985–2004) and 24 piezometer wells (in which some wells were established in 1998,

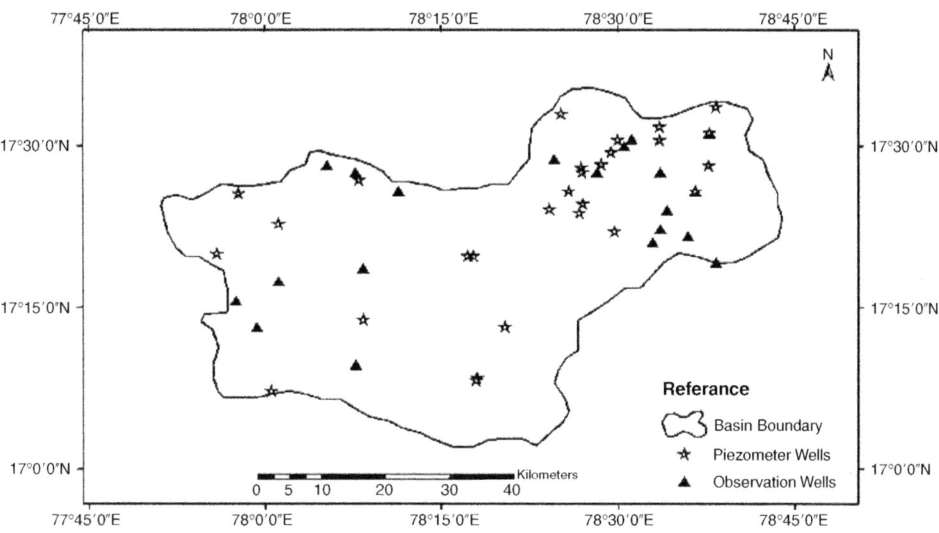

Figure 2.10.7 Location map of observation wells and Peizometer wells in the upper musi basim

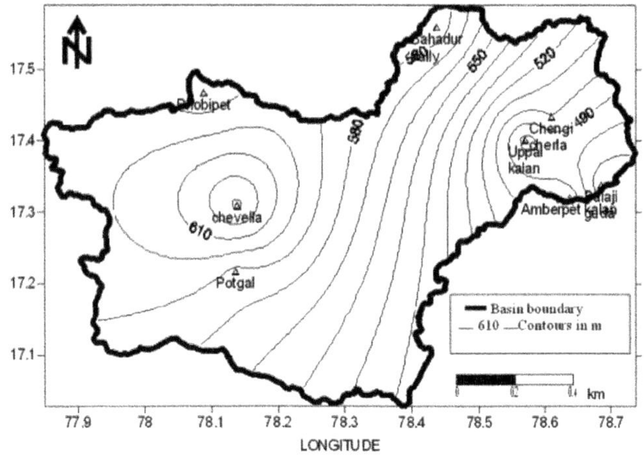

Figure 2.10.8 Water table contour map for the year 1985 post-monsoon

some in 1999 and some wells in 2001) in the study area have been considered for the analysis. The location of piezometer wells and observation wells are shown in Fig. 2.10.7.

The observed water levels give depth in meters below ground level. This observed depth is converted into height of the water table above mean sea level (a.m.s.l.) with the help of topographic contour map of the Survey of India topo sheets. Water table contour maps for pre monsoon and post monsoon for 1985 and 2004 are shown in Fig. 2.10.8 and Fig. 2.10.9 (Venkateswara Rao et al., 2009) respectively.

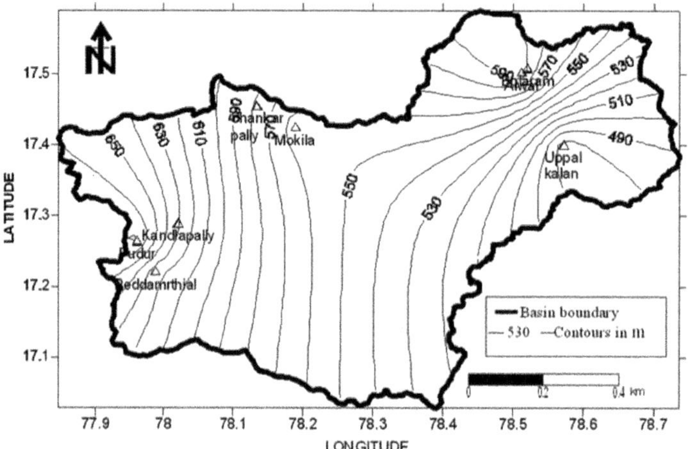

Figure 2.10.9 Water table contour map for the year 2004 post-monsoon

These maps are prepared to identify groundwater potential zones by utilizing the principle that areas with wider contour spacing may have greater groundwater potential (Todd, 1980). From both these maps it can be inferred that basically groundwater levels are following the topography. Groundwater direction is in the mainstream direction. All along the boundary of the upper part of the basin, the contours are widely spaced and these areas are having more potential for groundwater development. Incidentally they are the areas of heavier withdrawal of groundwater. Consequently the groundwater levels are deep, resulting in greater recharge in this region (Fig. 2.10.16). Fig. 2.10.16 is the total Musi basin area while Figs. 2.10.7 to 2.10.9 depicts only upper part of the Musi basin.

From the contour maps of 1985 and 2004 it can also be observed that there has been a progressive increase of area under the successive deeper groundwater contours. For example the 620 contour near Chevella in the contour map of 1985 is occupied by 590 contour in the 2004 contour map indicating depletion of water levels in the upper Musi by as much as 30 m. There is a good recharge zone near Uppal area where the 490 contour is being maintained throughout the 20 years period (Venkateswara Rao et al., 2011).

2.10.7 DISCUSSION ON HYDROGRAPHS OF OBSERVATION WELLS

The hydrographs show the rise and fall in water levels due to monsoon as well as artificial recharge structures such as percolation tanks, and check dams and other rainwater harvesting structures such as desilted tanks, and sunken ponds. The hydrographs of representative observation wells and also Composite Hydrographs of representative piezometer wells fixed with DWLR are described below. These hydrographs are

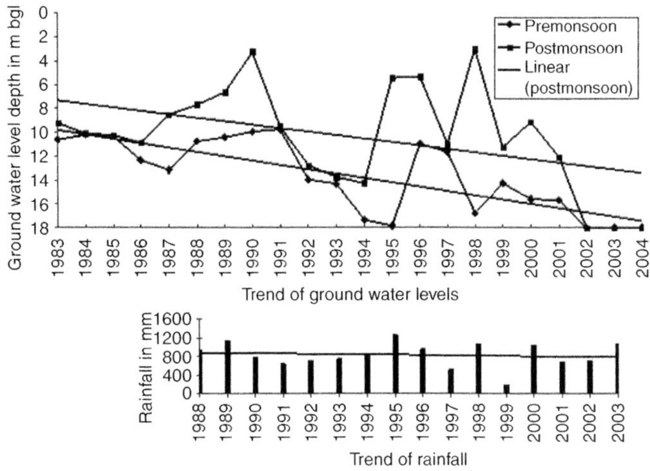

Figure 2.10.10 Groundwater levels and rainfall trend in Shankarapally well

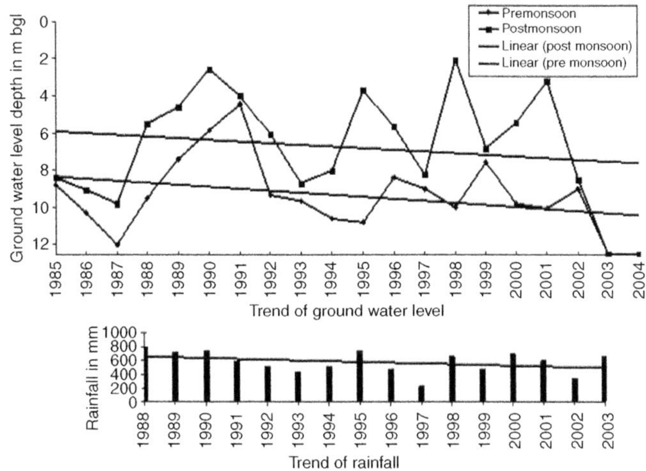

Figure 2.10.11 Groundwater levels and rainfall trend in Shabad well

prepared for 8 observation wells for pre monsoon and post monsoon periods. Two of these are as shown in Fig. 2.10.10 and Fig. 2.10.11. These hydrographs show that the water levels are declining, in spite of normal rainfall for the past 20 years. This is due to (a) high groundwater usage (b) greater well density, (c) less than normal rainfall in some years in some areas (d) specially very low rainfall received during 1999, 2002 and 2004 years. Overall the groundwater levels in the basin have been declining during the study period of 20 years. For example in Shankarpally well, the groundwater level has declined by as much as 8 m (Fig. 2.10.10). This made the open well go dry by the year 2003, forcing the farmer to drill deeper bore well.

2.10.8 COMPOSITE HYDROGRAPHS OF PIEZOMETER WELLS

Composite well hydrographs with rainfall data plotted are shown in Figs. 2.10.12 and 2.10.13. Out of 19 hydrographs, 13 hydrographs are showing a declining trend of water levels. In general the water levels in most of the cases return to their original position after a good rainfall. This phenomenon may be due to rapid recharge taking place due to heavy rainfall and also irrigation returns. If we compare the different years of hydrographs, it is seen that in 1999 when the precipitation was less, the rise in groundwater level was either less or nil. In the years 1998 and 2000 the groundwater levels reacted similarly to the monsoon rainfall.

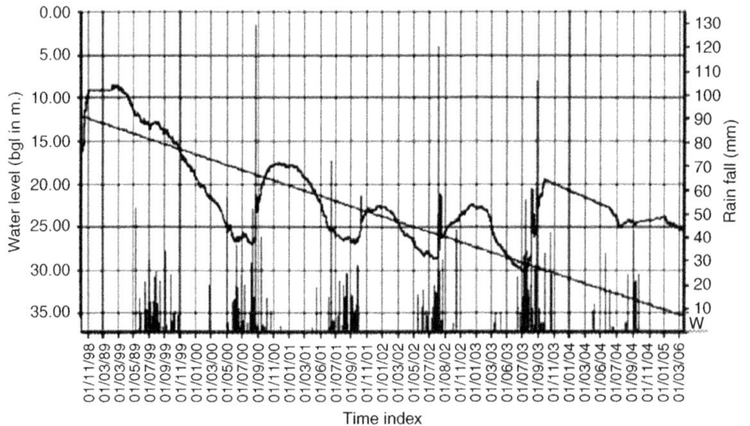

Figure 2.10.12 Composite hydrograph of Moinabad well

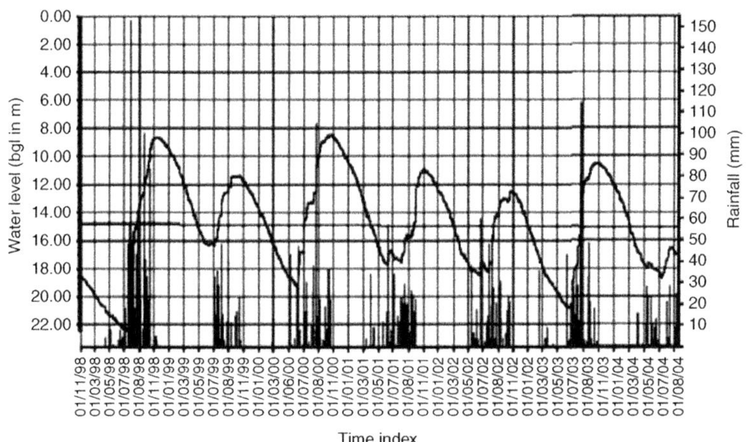

Figure 2.10.13 Composite hydrograph of Vikarabad well

2.10.9 RAINFALL AND WATER LEVEL RISE RELATIONSHIP

It is well known that the position of water table is influenced by the intensity and duration of the rainfall, and total amount of rainfall received in the area. It is also known that a certain amount of rainfall is needed before percolation can occur. Five composite hydrographs from different wells are used to arrive at the threshold values of rainfall required to induce percolation and thereby to raise the groundwater levels of respective places. This exercise was done using the data during July and August months, which receive a good amount of rainfall from south-west monsoon. The relation is obtained by plotting the water level raise as observed in the precision piezometer (fixed with automatic water level recorder) against the rainfall recorded in the nearest rain gauge. In all these cases rain gauge and piezometer are located within few tens of meters of distance between each other. The best-fit lines which are obtained by least square method, show a linear correlation between rainfall and water level raise in wells. One such relation is shown in Fig. 2.10.14.

The regression line for each well has an intercept on the rainfall axis, suggesting that certain minimum amount of rainfall is required for initiating deep percolation which in turn recharges the phreatic aquifer. It can be considered as the minimum amount of rainfall required for recouping the soil moisture deficit in the vadose zone. Rainfall and water level raise relation plots shows that the minimum rainfall necessary for initiating percolation ranges from 8mm to 58mm. It has also been observed that high quantum of rainfall needed for inducing percolation coincide with topographic highs, whereas low values are noticed at topographic lows. R^2 values of these plots suggest that some stations are having higher correlation coefficient and some stations are having lower

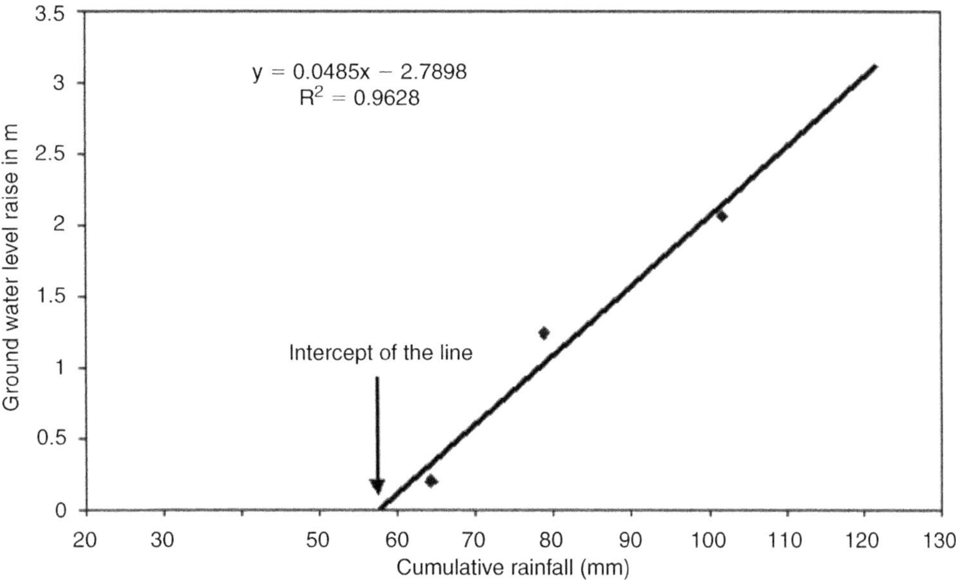

Figure 2.10.14 Rainfall and water level rise relation in Shaikpet well

correlation coefficient, indicating the degree of variability in recharge. This means that sites having higher R^2 value will have greater scope for taking up artificial recharge of groundwater.

2.10.10 INFLUENCE OF PREMONSOON GROUNDWATER LEVELS OVER THE RECHARGE OF RAINFALL WATER TO THE GROUND

Premonsoon and post monsoon groundwater levels are observed in the year 2005 at 90 locations covering entire Musi basin. A graph (Fig. 2.10.15) is drawn between premonsoon water levels and the difference of premonsoon and post monsoon water levels. The graph clearly shows that a positive correlation exists (Fig. 2.10. 15) between pre monsoon groundwater levels and the rise in groundwater levels (the difference between the premonsoon and post monsoon levels) due to rainfall.

That is, the deeper the premonsoon groundwater levels, the more the recharge to the groundwater from the rainfall. In other words premonsoon groundwater level determines the rate of recharge during the monsoon – perhaps it is a new concept in hydrology (Venkateswara Rao et al., 2011). This phenomenon is depicted in the Fig. 2.10.16 covering entire Musi basin.

In these maps black to white shades indicate deep to shallow water levels in pre-monsoon and post monsoon maps while it is high raise of groundwater level to low raise of groundwater level in the difference map (Map depicting the difference between premonsoon and post monsoon groundwater levels). It is evident from these maps that the deeper water level areas in pre monsoon have larger water level raise after post

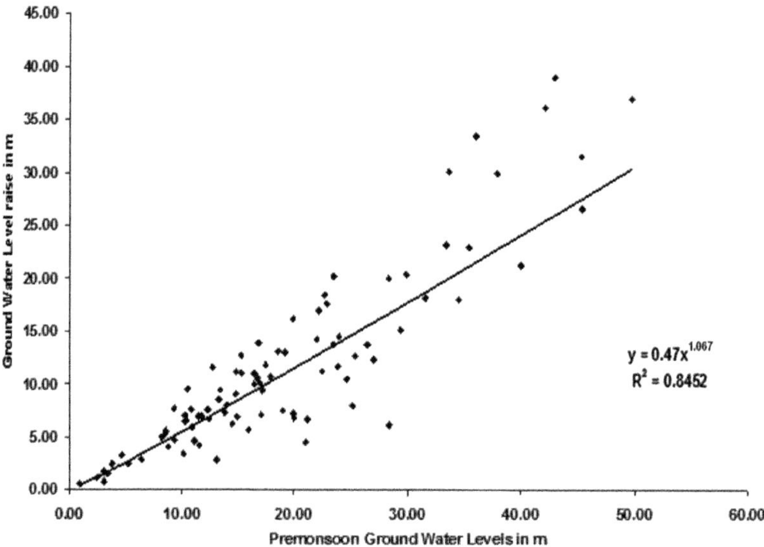

Figure 2.10.15 Relation between pre-monsoon groundwater levels and the difference of pre-monsoon and post-monsoon groundwater levels (Δh)

monsoon. There is another data set comprising the amount of rainfall, premonsoon water level and the rise in groundwater level (Table 2.10.2).

From this data the graphs have been drawn between premonsoon groundwater level versus rise in groundwater level (Fig. 2.10.17) and rainfall versus rise in groundwater level (Fig. 2.10.18) (Venkateswara Rao et al., 2009).

From these graphs it is observed that positive correlation is also found between the rainfall and rise in groundwater levels and also between premonsoon groundwater level and rise in groundwater levels. But it is the premonsoon groundwater level which has greater influence on the recharge, rather than the total amount of rainfall itself, since the slope of this graph is greater than that for rainfall graph. It appears that whatever the rainfall that occurs in the upper catchment is simply percolating down

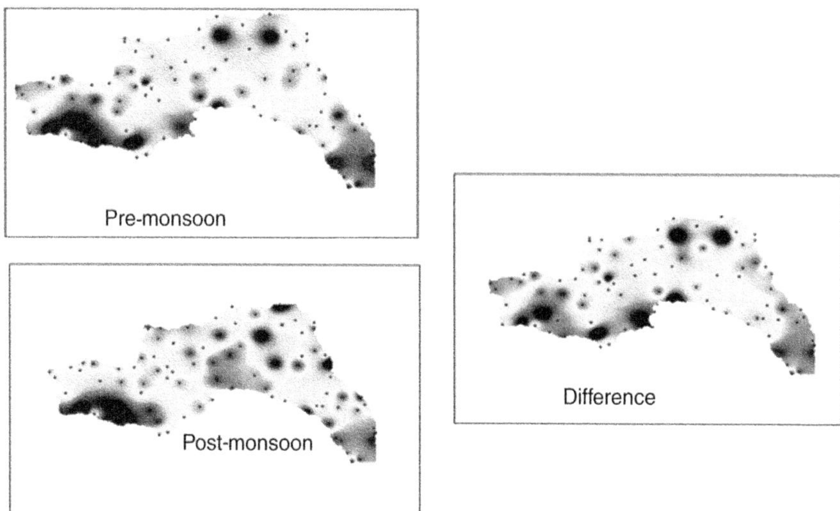

Figure 2.10.16 Musi Basin depicting pre-monsoon, post-monsoon groundwater levels and rise in groundwater levels for the year 2005

Table 2.10.2 Changes of groundwater levels in the upper Musi Basin

Year	*Rainfall (mm)*	*Premonsoon groundwater levels in m bgl*	*Increase in groundwater levels in m*
1998–99	1128	10.86	2.71
1999–00	764	9.27	0.6
2000–01	919	8.69	3.58
2001–02	877	9.5	2.23
2002–03	614	10.25	0.25
2003–04	936	11.13	2.47
2004–05	611	11.49	1.16
2005–06*	685	14.45	4.16

Note: *In this year rainfall is upto September only

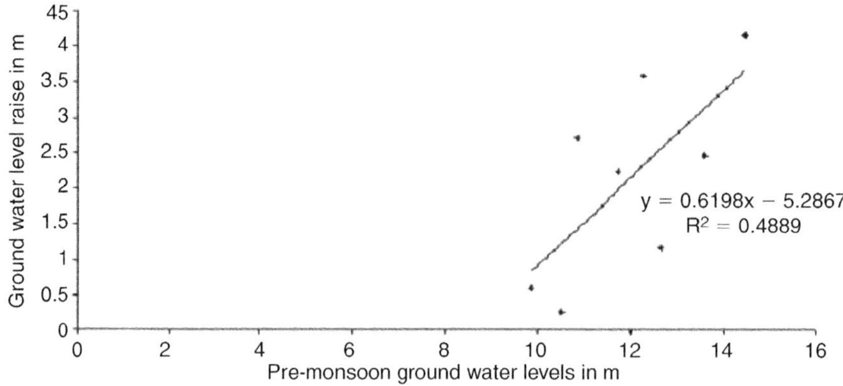

Figure 2.10.17 Relation between pre-monsoon groundwater level and increase in groundwater level

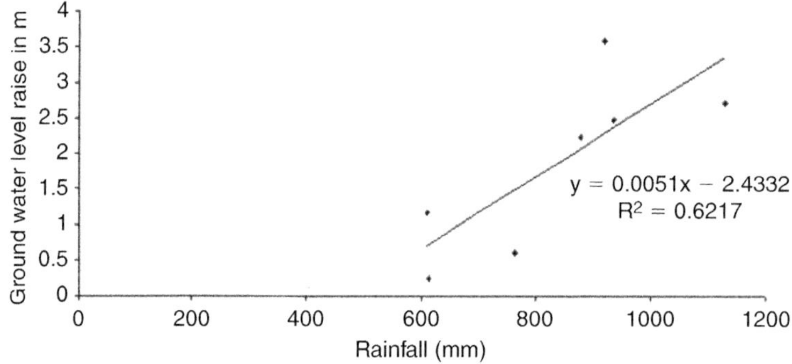

Figure 2.10.18 Relation between rainfall and Increase in groundwater level

to recharge the depleted groundwater table, to meet the soil moisture deficiency and to meet the storage of water in smaller or bigger water conservation structures such as tanks, percolation ponds, check dams, etc., thereby leaving little or no runoff that could reach the downstream. Unless there are extreme rainfall events, it is becoming increasingly less likely for the reservoirs to get filled. Ultimately there will be no water left downstream, thereby closing the entire Basin.

2.10.11 IMPLICATIONS OF THE STUDY AND CONCLUSIONS

The above situation holds well not only for smaller reservoir catchments but also to bigger river basins like Krishna (Fig. 2.10.19) basin in south India and Ganga (Fig. 2.10.20) Basin in north India wherein inflows has been considerably declined after upstream storages have increased.

This means that the well established down stream deltaic irrigation systems of Godavari, Krishna and Cauvery in Southern India will be receiving lesser and lesser surface water. If this scenario continues in future there will be decrease in the discharge

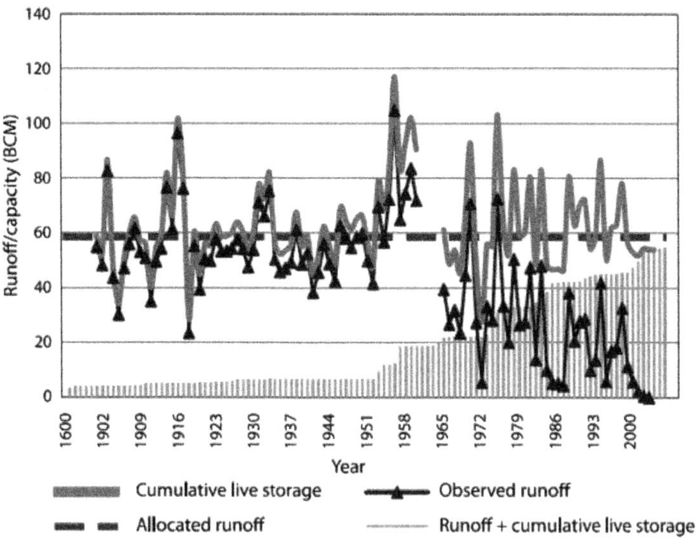

Figure 2.10.19 Discharges of the Krishna basin
Source: Biggs et al., 2007

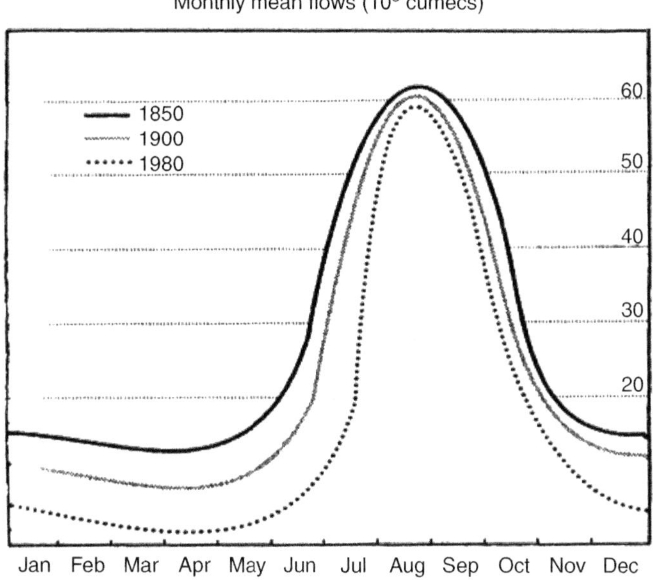

Figure 2.10.20 Mean monthly graph of the Ganges River
Source: Schwarz, et al., 1990, Hydrograph for 1980 and Reconstructed Hydrographs for the years 1850 and 1900

of fresh water from these rivers to the sea and increased danger of sea water intrusion into the deltas threatening the very food baskets of India. This scenario is particularly worrisome in the context of the expected sea level rise due to climate change. Therefore science-illuminated management of water upstream needs to leave some environmental flows down stream to avert the closure of the basin.

On the basis of GRACE (Gravity Recovery And Climate Experiment) satellite data, Tiwari, Wahr, and Swenson (2009) found that northern India and its surroundings which is the most heavily irrigated region in the world, has been losing groundwater at the rate of $54 \pm 9\,km^3/yr$. This is probably the largest groundwater loss in any comparable region in the world. This is not a sustainable situation. Drastic steps need to be taken to reduce the quantum of withdrawal of groundwater.

Since agriculture consumes nearly 80% of fresh water, we have to drastically reduce water use in this sector in order to provide fresh water for drinking and industrial purposes. Therefore it is necessary to adopt water conserving irrigation practices, such as "System of Rice Intensification" (SRI) method, use of drought-resistant rice varieties, aerobic rice which does not need transplantation, and cultivation of low-water need crops. Recycled waste water can be used to grow vegetables and fruit trees. Climate change is expected to lead to droughts and floods. Wisdom lies in storing fresh water at the time of floods, and using it during periods of drought

REFERENCES

Biggs, T.W., Anju Gaur, Christopher A. Scott, Prasad Thenkabail, Parthasaradhi Gangadhara Rao, Murali Krishna Gumma, Sreedhar Acharya and Hugh Turral (2007) "Closing of the Krishna Basin: Irrigation stream flow depletion and macro scale hydrology", *IWMI Report* number 111, 38p.

Praveena, D. (2005) "Variation of Hydrometeorological Parameters of I.M.D Stations In and around Krishna Basin in A.P." Thesis submitted to JNT University. Unpublished.

Schwarz, H.E., J. Emel, W.J. Dickens, P. Rogers and J. Thomson., (1990) "Water quality and Flows". In: B.L. Turner II, W.C. Clark, R.W. Kates, J.E. Richards and W.B. Myers (eds.) *The Earth as translated by Human Action*", Cambridge: Cambridge University Press.

Tiwari, V.M., Wahr, J., and Swenson, S. (2009), "Dwindling groundwater resources in northern India from satellite gravity observations". *Geophy. Res. Lett.*, v. 36. L 1840, 17 Sept. 2009

Todd, D.K. (1980) *Ground Water Hydrology*. John Wiley and Sons, 2nd Ed, Inc, 535p.

Varalakshmi. V. (2010) Catchment Hydrology and Ground water modelling of Osmansagar and Himayatsagar Reservoirs. *Ph.D Thesis submitted to JNT University* Hyderabad 300p.

Venkateswara Rao, B., Varalakshmi.V and Srinivasa Rao, N. (2010) "Effect of Urbanization on inflows to the reservoirs of Osmansagar and Himayatsagar, Hyderabad, India", *International Journal of Water resources and Environment Management*, Vol. 1, No. 1, pp. 67–80.

Venkateswara Rao, B., Varalakshmi,V., and Vijayasarada, S.T. (2009) "Ground water Recharge in hard rock areas of Musi Basin and its impact on down stream reservoirs", *Journal of Applied Hydrology*, Vol. XXII, No. 1, pp. 102–117.

Venkateswara Rao, B., Rajesh Nune, Rajesh M.V.S.S., and Vijayasarada, S.T. (2011) "Large Scale Ground Water Withdrawal and its Consequences on the Closing of the Upper Musi Basin – India", *Int. J. Water*, Vol. 6, Nos. 1/2, pp. 15–28.

World Bank Report (2005) No. 34750-IN, "India's Water Economy Bracing for Turbulent Future", 102p.

EXERCISES

1. In a reservoir the initial water level was 544.159 m. After 24 hrs it decreased to 544.144 m and the same level was maintained for the next 24 hrs. Calculate the inflows that reached the reservoir for these two days. The storage capacity of the reservoir at 543.860 m and 544.360 m is 76.933 MCM, 84.927 MCM respectively. Consider the withdrawal for water supply as 0.012 m.

   Solution:

   In first day water level decreases by 0.015 m and there is no increment in water level. i.e, no inflows in first 24 hrs

   For second day Increment in water level

   $$= (\text{current day water level} - \text{previous day water level})$$
   $$+ \text{withdrawal for water supply in previous 24 hours}$$
   $$= (544.144 - 544.144) + 0.012$$
   $$= 0.012 \text{ m}$$

   Volume of water arrived

   $$= \frac{\text{increment in water level} \times \text{difference in storage capacity of the reservoir}}{\text{Difference of the water level}}$$

   $$= \frac{0.012 \times (84.927 - 76.933)}{(543.860 - 544.360)}$$

   $$= \frac{0.012 \times 7.994}{0.5}$$

   $$= 0.191\, MCM$$

2. Determine the Net draft in MCM for a given area by using the following data:

   Paddy irrigated area in Kharif season = 736 ha
   Paddy irrigated area in Rabi season = 256 ha
   Area under other crops in Kharif = 4,801 ha
   Area under other crops in Rabi = 2,216 ha

   Crop water requirement for paddy in kharif and rabi seasons is 0.95 m and 1.2 m respectively and for irrigated dry crops it is 0.25 m and 0.45 m respectively.

   Solution:

   Draft for Paddy area in kharif season

   $$= \text{Paddy irrigated area} \times \text{Crop water requirement for paddy}$$
   $$= 736 \times 0.95 = 699.20 \text{ ha m}$$

Draft for Paddy area in Rabi season

$= \text{Paddy irrigated area} \times \text{Crop water requirement for paddy}$

$= 256 \times 1.2 = 307.20$ ha m

Draft for irrigated dry crops in kharif season

$= \text{Irrigated dry crop} \times \text{crop water requirement for irrigated dry crop}$

$= 4801 \times 0.25$

$= 1200.25$ ha m

Draft for irrigated dry crops in kharif season

$= \text{Irrigated dry crop area} \times \text{crop water requirement for irrigated dry crop}$

$= 2216 \times 0.45$

$= 997.2$ ha m

$$\text{Total Draft Kharif and Rabi together} = 699.2 + 307.2 + 1200.25 + 997.2$$

$$= 3203.85 \text{ ha m}$$

35% of the applied water goes as the return flow and joins the groundwater $= \dfrac{3203.85 \times 35}{100}$

$$= 1121.34$$

$$\text{Net draft} = 3203.85 - 1121.34 = 2082.51 \text{ ha m}$$

$$= 20.8251 \text{ MCM}$$

3. A county has 331 tube wells and 944 dug wells. Calculate the net draft of that county. Assume unit draft for dug well is 0.65 ha m and unit draft for tube well is 2.05 ha m

Solution:

$$\text{Draft from tube wells} = 331 \times 2.05 = 678.55 \text{ ha m}$$

$$\text{Draft from dug wells} = 944 \times 0.65 = 613.6 \text{ ha m}$$

$$\text{Total draft} = 678.55 + 613.6 = 1292.15$$

35% of the applied water goes as the return flow and joins the groundwater $= \dfrac{1292.15 \times 35}{100}$

$$= 452.25$$

$$\text{Net draft} = 1292.15 - 452.25$$

$$= 839.9 \text{ ha m}$$

$$= 8.39 \text{ MCM}$$

4. The mean monthly temperature (t_m) of May is 32.6°C at a place of latitude 17°N. Using Thoranthwait formula, compute the evapotranspiration at this place for the

month of May 2005, given the monthly percentage of day time hours $(\rho) = 9.14$ and yearly heat index $(I) = 151.27$

Solution:

According to Thornthwaite (1948) formula,
Potential Evapotranspiration $= 1.6b(10\,t_m/I)^a$
where "a" and "b" are the factors to be evaluated with the following formulae

$$\begin{aligned} a &= (6.75\times10^{-7}\times \mathrm{I}^3) - (7.71\times10^{-5}\times \mathrm{I}^2) + (1.792\times10^{-2}\times \mathrm{I}) + 0.49239 \\ &= (6.75\times10^{-7}\times 151.27^3) - (7.71\times10^{-5}\times 151.27^2) \\ &\quad + (1.792\times10^{-2}\times 151.27) + 0.49239 \\ &= 3.7753 \end{aligned}$$

$$b = \text{Possible sunshine hours for the particular month/30 (or 31)} \times 12$$

Possible sunshine hours in a month

$$= \text{monthly percentage of day time hours} \times 365 \times 12/100$$

$$\text{Possible sunshine hours in May} = \frac{9.14\times365\times12}{100} = 400.332$$

$$\text{Therefore "}b\text{" for the month of May} = \frac{400.332}{12\times31} = 1.076$$

$$\begin{aligned} \text{Potential Evapotranspiration} &= 1.6b(10t_m/I)^a \\ &= 1.6\times1.076\times(10\times32.6/151.27)^{3.77} \\ &= 31.12\,\text{cm} \\ &= 311.2\,\text{mm} \end{aligned}$$

5. The confined aquifer at a place extends over an area of 1200 km^2. The piezometric surface fluctuates annually from 25 m to 14 m above the top of aquifer. Assuming the storage coefficient as 0.0003, what ground water storage can be expected annually? Assuming an average well yield of 32 m^3/hr and about 216 days of pumping in a year, how many wells can be drilled in the area.

Solution:

Change in Ground Water Storage

$$\begin{aligned} &= \text{Area of the Aquifer} \times \text{Change in piezometric surface} \times \text{storage coefficient} \\ &= (1200\times10^6)(25-14)0.0003 \\ &= 3.96\times10^6\ \text{m}^3 or 3.96\ \text{million m}^3 \end{aligned}$$

Annual draft $= 32\times24\times216 = 165888$ or 0.165 million m^3

No. of wells that can be drilled in the area $= 3.96/0.165 = 24$

6. Toxic effluents from a tannery (containing chlorides, sulfides and chromium) have contaminated an aquifer with the following characteristics: thickness (D): 40 m; precipitation surplus entering the aquifer at the upper reaches (P): 0.3 m yr^{-1};

porosity fraction (ε): 0.3. Polluted water infiltrates the aquifer at 400–600 m from the divide. A drinking water well located at 500 m from the divide, may be affected. Calculate:

- Thickness of the plume of the effluent,
- Its mean depth, and
- Expected arrival time of the pollution at the well.

(Source: U. Aswathanarayana, 2001, "*Water Resources Management and the Environment*", p. 275)
Infiltration reach $(R) = 600 - 400 = 200$ m

Thickness of the effluent plume

$$= \frac{\text{Infiltration reach (200)}}{\begin{array}{c}\text{Distance of the well}\\ \text{from the divide (500)}\end{array}} \times \text{Thickness of aquifer (40 m)} = 16\text{ m}$$

Mean depth of the effluent plume

$$= \frac{\text{Mean distance from the divide (500 m)}}{\begin{array}{c}\text{Distance of the well from the}\\ \text{divide (500 m)}\end{array}} \times \text{Thickness of aquifer (40 m)}$$

$$= 40\text{ m}$$

Expected time of arrival of the pollution at the well of water infiltrated at 400 m

$$= \ln\left(\frac{x}{x_o}\right) = \frac{Pt}{D\varepsilon} = \sim 9\text{ years}$$

$$= \ln \frac{\text{Water infiltrated at (400 m)}}{\text{Distance of the well from the divide (500 m)}} = \frac{0.3 \times t}{40\text{ m} \times 0.3}$$

$$= \ln\left(\frac{400}{500}\right) = \frac{(0.3 \times t)}{(40 \times 0.3)};\ 0.223 = \left(\frac{t}{40}\right);\ t = \sim 9\text{ years}$$

7. Excessive use of nitrogenous fertilizers in a locality, have contaminated an aquifer ($D = 20$ m, $\varepsilon = 0.4$, $P = 0.5\text{ m yr}^{-1}$) which consequently contains 200 mg l^{-1} of nitrate. Assuming that the use of fertilizer will be stopped with immediate effect, estimate the time (t) required to reduce the nitrate concentration to 10 mg l^{-1} (US EPA prescribed Maximum Contaminant level in drinking water).
(Source: U. Aswathanarayana, 2001, "*Water Resources Management and the Environment*", p. 277)
Residence time of the nitrate in the aquifer $(t_R) = D\varepsilon/P = (20 \times 0.4)/0.5 = 16$ yr

$$10\text{ mg l}^{-1}/200\text{ mg l}^{-1} = 0.05 = \varepsilon^{-t}/t_R$$
$$t = -t_R \ln 0.05 = 16 \times 3 = 48\text{ yrs}$$

8. A landfill is sealed with a clay barrier. Estimate the potential of the benzene in the leachate from the landfill to contaminate the drinking water resources, on the basis of the steady state of flux of benzene in waste site, given the following: size of the waste site $50\,m \times 40\,m$; thickness of the clay liner: 1 m; porosity of clay (ε): 0.5; concentration of benzene in the waste site: $1\,g\,l^{-1}$; concentration of benzene in the groundwater flow below the liner: $0.01\,g\,l^{-1}$; free water diffusion coefficient of benzene (D_f): $7 \cdot 10^{-6}\,cm^2\,s^{-1}$.

   (source: U. Aswathanarayana, 2001, "*Water Resources Management and the Environment*", p. 281)

$$\text{Effective diffusion coefficient } (D_e) \text{ of benzene in water}$$
$$= D_f \cdot \varepsilon = \text{Fick's Diffusion constant} \times \text{porosity}$$
$$= 7 \cdot 10^{-6}\,cm^2\,s^{-1} \times 0.5 = 3.5 \cdot 10^{-6}\,cm^2\,s^{-1}$$

$$\text{Flux } (F) = -D_e \cdot \varepsilon.(\partial C/\partial x) = -3.5 \cdot 10^{-6} \times 0.5 \times (1 - 0.01)/100$$
$$= \sim 1.8 \times 10^{-8}\,mg\,cm^2\,s^{-1}$$

For one year (3.156×10^7 s) and areal extent of the landfill ($2000\,m^2 = 2 \times 10^7\,cm^2$), the flux would be $(1.8 \times 10^{-8}) \times (3.156 \times 10^7) \times (2 \times 10^7) = 11.4 \times 10^6\,mg = 11.4\,kg$.

According to US EPA Drinking water standards, the Maximum Contaminant Level (MCL) allowable for benzene is $0.005\,mg\,l^{-1}$. Thus 11.4 kg of benzene has the potential to contaminate about $2.3\,M\,m^3$ of water resources.

REFERENCE

Thornthwaite, C.W. (1948). An approach toward a rational classification of climate. *The Geographical Review*, 38, No. 1, pp. 55–94.

Chapter 2.11

Basic research and R&D

Balaji Rajagopalan
Department of Civil, Environmental and Architectural Engineering & Cooperative Institute for Research in Environmental Sciences, University of Colorado, Boulder, CO, USA
Casey Brown
Department of Civil and Environmental Engineering, University of Massachusetts, Amherst, MA, USA

2.11.1 BACKGROUND – TRADITIONAL WATER RESOURCES MANAGEMENT

Building reservoirs and dams have been the traditional approach to solving water problems and 20th century saw the most in terms of dam building. Dams are among the most pervasive and beneficial structures on Earth. They provide energy, flood protection, and water for domestic, industrial, and agricultural use as well as recreation. Seeking these benefits, the US has built thousands of dams over the past 100 years, with a peak in the 1960's. This made good sense as a means to provide reliable water supply for irrigation and industry besides power generation in many instances. To their credit these dams have delivered significant socio-economic progress around the world. After the building frenzy subsided, mainly due to the fact that all the easy sites were built out, the drawbacks of the large dams have started to come into sharp focus (WCD, 2000). Dams also have costs, however, some of which have been fully recognized only over the last few decades, including physical, chemical, and biological alteration of downstream ecosystems as well as risks associated not only with design features but also national security, aging, and seismic activity. As a result, significant opposition to large dam projects has gained traction – forcing planners to take a holistic approach to water resources management. We are at a cross roads in terms of water resources and development given our increasing understanding of climate variability, socio-economic factors, environmental impacts and the need for sustainable growth. This calls for a new paradigm for water resources management and development.

2.11.2 NEW PARADIGM FOR WATER RESOURCES MANAGEMENT

It is increasingly evident that the era of building our way out of the water crisis is over. This means we have to manage this precious resource smartly. To this end focus needs to be placed on the two components – (i) Understanding uncertainty and risks in water resources supply and demand, (ii) Managing the uncertainty and risks in a holistic manner for a sustainable growth.

2.11.2.1 Uncertainty and risks in water resources supply and demand

One of the key challenges to water resources management is the uncertainty and variability of water supply, demand and hydrologic hazards – especially in light of the realization that their future variability may not reflect that which we've already experienced. In addition, there is uncertainty related to changes in policy, societal values and the interactions among these factors. While efforts to reduce these uncertainties are needed, a significant and irreducible uncertainty will always remain. Therefore, a strategy for managing uncertainty is paramount for effective water resources management.

Typically, the planning and management of water resources has focused on managing risks, such as the risk of droughts, delivery shortfalls and floods, among others. In this framework, risk is the product of probability and impact. The implications of hydrologic change imply that our ability to estimate hydrologically-related probabilities and consequently, future risks may be greatly compromised.

Key sources of uncertainty related to water resources are the uncertainty associated with supply, demand and the policy context. The uncertainty of water supply results originally from the natural variability of the earth's climate system and the hydrologic cycle as mentioned in the background. This natural variability on broad spatial and temporal scales is driven by large scale climate forcings (e.g. El Nino Southern Oscillation (ENSO), Pacific Decadal Oscillation (PDO), North Atlantic Oscillation (NAO), Atlantic Multidecadal Oscillation (AMO) etc.). Anthropogenic influences are also introducing new sources of uncertainty such as those from emission of greenhouse gases and aerosols. Together the natural and anthropogenic influences affect rates of runoff and evapotranspiration and create a nonstationary climate and hydrologic context within which water resources must be managed (e.g. Milly et al., 2005). This is particularly notable because the current principles for water resources management are based largely on an assumption of hydrologic stationarity, the assumption that the future hydrologic record can be adequately simulated with the statistics of the historical record. There is a growing body of literature providing understanding of hydrologic variability around the world driven by the large scale climate drivers mentioned above.

The uncertainty associated with water demand is largely related to both the difficulty of anticipating demographic change and our limited understanding of how water demand responds to changing climate and policy conditions. In the US, projections have often overestimated demand, as rising water rates and improved efficiency in household fixtures were not adequately accounted for. Better understanding of the dynamic relationship between water demand and supply and more sophisticated means of modeling changes in demand dynamically are needed. To do so the underlying drivers

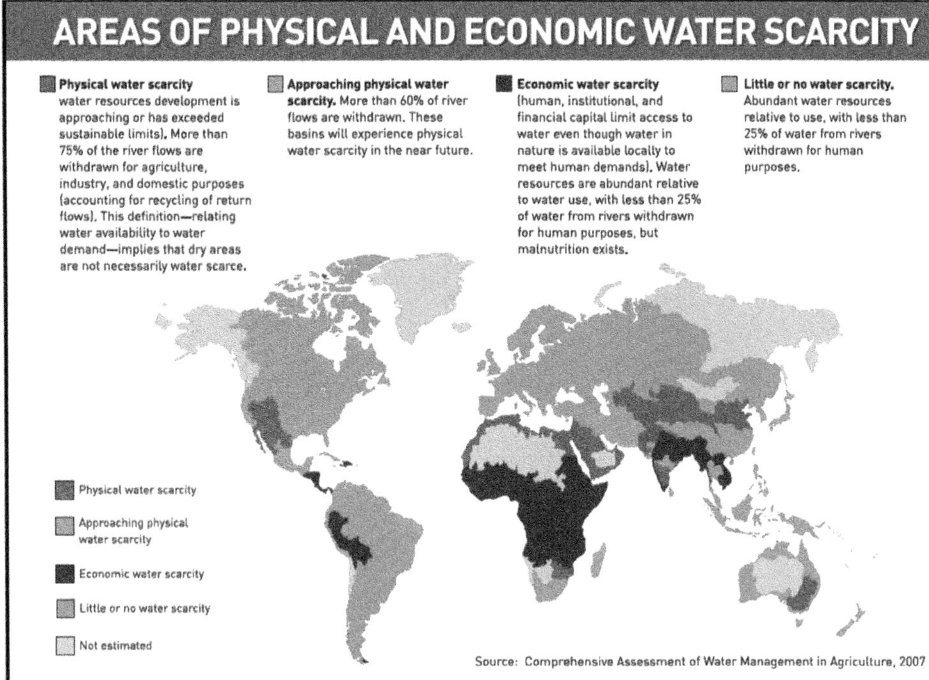

Figure 2.11.1 Global physical and economic water scarcity

of demand growth (or decrease) must be understood – population growth and socioeconomic growth. As economies grow water use increases significantly. This typically leads to water demand increases. The developing world is projected to have much higher growth in total water demand than the developed world. Increasing demand is projected to occur in irrigation, domestic and industrial sectors – all linked to economic growth. Figure 2.11.1 shows the global physical and economic water scarcity – which is a combination of socio-economic growth and water availability. It can be seen that both physical and economic water scarcity is almost entirely in regions with high population growth and situated in the tropics. Economic water scarcity is largely a result of insufficient infrastructure. Economic growth can also lead to more economically efficient management of water resources, thereby reducing the amount of water used while generating higher value – this is illustrated in Figure 2.11.2 for several countries.

There is considerable uncertainty in these projections especially in the projection of socio-economic factors such as population, economic growth etc. Without accompanying increases in the efficiency of water use, growth in water demand may be unsustainable and ultimately be an impediment to continued economic growth.

Climate variations have a significant impact on water supply and are a primary source of uncertainty. Climate models project decreases in precipitation and consequently in streamflow in several places around the World. Figure 2.11.3 shows the water availability (which combines climate variability, precipitation, temperature,

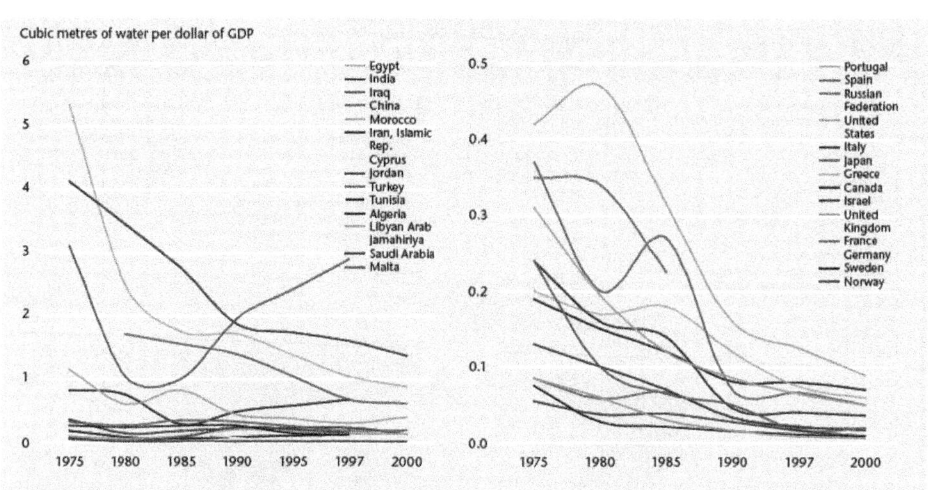

Figure 2.11.2 Water usage per dollar of GDP

topography and socio-economic factors) during recent period and 2050s. These projections of water availability show a reduction in the future due to climate change in tropical regions. While this is a projection of surface water availability which is a major source of water for almost all of human activities, the groundwater scenario is likely to be similar or worse – as reduced precipitation leads to reduced recharge and with exploitation from socio-economic growth it will be stressed more.

Projections of water supply due to climate variability come with large uncertainty. A major question that emerges is should the uncertain climate change projections be used for planning and adaptation decisions. While the information used for water planning has always been uncertain, the current rate of potential change and degree of uncertainty associated with it is probably unprecedented. Managing this uncertainty is one of the great water challenges of the century.

2.11.2.2 Managing uncertainty and risks in water resources

A general framework for risk management can be described in three steps: hazard characterization, risk assessment and risk management. Here we define risk as the product of the probability of an event and the consequences of the event. The first step is to characterize a given event, often considered a hazard, in terms of the impacts or consequences it may have. The next step, called risk assessment, involves the actual calculation of the risk based on the estimation of the probability of occurrence. The final step of risk management is the development of a strategy for addressing the risks identified and quantified in the previous steps. Cost benefit analysis has traditionally been used in the development of such a strategy and the process of risk quantification accommodates a quantitative cost benefit analysis. However, if there are costs or benefits that are not easily quantified then other multi-objective decision making methods may need to be employed.

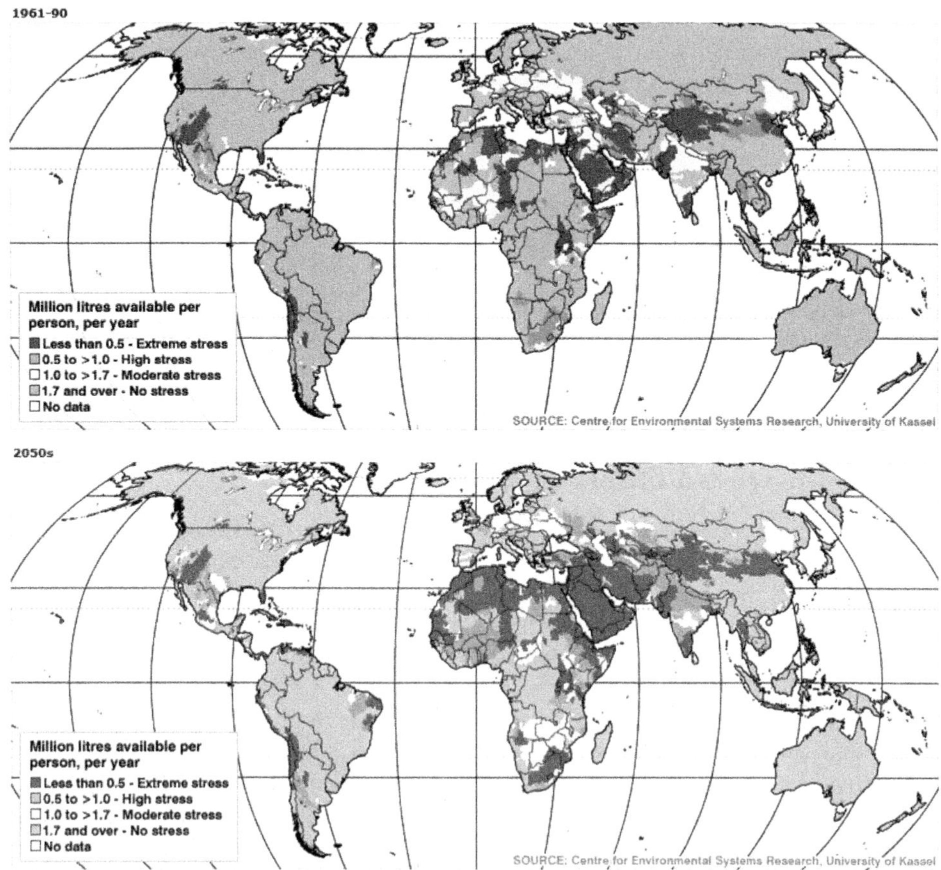

Figure 2.11.3 Water availability per person during 1961–2000 and in the 2050s from climate change projections

Understanding climate variability can help quantify the risks associated with climate forcings at interannual and decadal time scales. Robust decision process is an emerging idea for planning under uncertainty. Robustness here is defined as a plan that works well under a wide range of possible futures. Instead of choosing the plan that performs best for a best guess estimate of the future, a robust plan would perform well enough in many possible realizations of the future. The utility of robustness as a planning criterion reflects the growing awareness of the inability to predict the future (Brown, 2010).

Institutional complexities inhibit flexibility in planning. Especially large and old institutions tend to have entrenched rules and stake holders that resist change and science/knowledge based decision making. In this regard developing countries may have a comparative advantage in that, as they are developing institutions they can instill flexibility and learn from the drawbacks of the matured systems. The Colorado River water resources management system is a good example. Methods such as Bayesian networks in which all the participants and their probabilistic interactions can be represented and, the system performance assessed (Bromley, 2005) provide an excellent framework to lead to a participatory solution.

The report (van der Keur et al., 2006) from NeWater project (www.newater.info) provides an excellent summary on identification of major sources of uncertainty in Integrated Water Resources Management practice.

2.11.3 R&D FOR MANAGING WATER RESOURCES UNDER UNCERTAINTY

There is a growing list of opportunities and methodologies available for addressing the new challenges to water resources management. Listed below are items that are emerging in the literature that needs to be researched and applied to water resources management. Especially for developing countries these innovations are critical.

2.11.3.1 Improved observed and modeled data

Monitoring all the components of the hydrologic cycle is extremely important. Without a robust data collection, monitoring and sharing system there cannot be an efficient water resources management. Developing countries are especially plagued by lack of good data network. Remote sensing, ground based monitoring and modeling that allow real time observations and forecasts at multiple time scales are new and potentially important resources to improve water resources management. However, gaps between the production of this information and its use in practice are wide.

Seasonal climate forecasts are increasingly skillful and there are several agencies that provide probabilistic seasonal forecasts for the entire globe (e.g. http://www.iri.columbia.edu). Innovative tools can translate these to hydrologic and hydroclimate forecasts. For example, researchers have established connections in river basins around the globe between streamflows and large-scale climate forcings, particularly El-Nino Southern Oscillation (ENSO), Pacific Decadal Oscillation (PDO), and Pacific North American (PNA) pattern (Grantz et al., 2005; Regonda et al., 2006 and references therein.) Using these connections, novel forecasts methods have been developed to produce skillful streamflow forecasts at a few months lead time (e.g. Hamlet and Lettenmaier, 1999; Clark et al., 2001; Grantz et al., 2005; Regonda et al., 2006). Additionally, new stochastic simulation techniques (Rajagopalan et al., 2010) can be used to generate ensembles of hydrologic scenarios in a water resources system conditional on large scale seasonal climate forecast. The ensembles can be used in a decision tool to provide estimates of system risks at the seasonal to interannual time scales and thus enable efficient decisions. A good demonstration of these can be found in Grantz et al. (2007) and Regonda et al. (2011).

2.11.3.2 Paleo data

Observational data are limited in length and thus, cannot provide reliable estimates of various risks such as droughts and floods necessary for planning and management. Paleo data provides a valuable and rich resource to quantify historical risk. There is extensive research in this regard to using tree rings records to reconstruct past

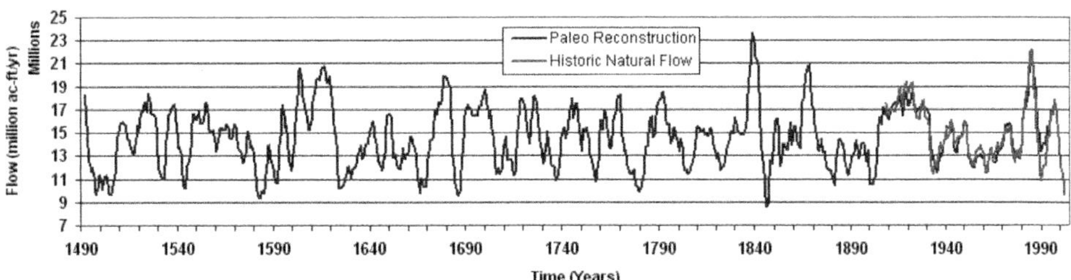

Figure 2.11.4 Paleo reconstruction of Colorado River streamflow

hydrologic variability – an excellent example is the long reconstruction of paleo streamflow on the Colorado River Basin (Woodhouse et al., 2006; see Figure 2.11.4 above).

The paleo data indicates higher risk of extended dry periods while the observation period suggests much wetter state. There are emerging methods to combine observed and paleo information to obtain robust risk estimates (Prairie et al., 2008).

2.11.3.3 Decadal scale patterns of variability

Paleo data coupled with climate data of forcings such as PDO, ENSO, AMO, mentioned earlier, can shed light on multidecadal hydrologic variability. Wavelet and other spectral based analysis provide an excellent framework in this regard (Nowak, 2011). Better understanding of these modes of variability could enhance the preparedness of water resources management. The existence of decadal scale variability is the historical hydrologic record implies that a water resource system could benefit from flexibility to manage periods of relatively greater and less water resources availability (Nowak, 2011).

2.11.3.4 Water resources management by dynamic operations

Traditional water resources management principles are based on the assumption that the probability distribution for hydrologic states is the same every year, i.e. the assumption of climatology, that the long term statistics are dominant in every year. These scientific advances in terms of understanding and making realistic projections of long term streamflow variability as mentioned above, make possible the use of dynamic operational strategies such as reservoirs or water system rule curves that change in response to forecasts or observations, flexible water allocation and pricing approaches, such as in response to scarcity conditions, and the use of economic mechanisms to facilitate water exchanges. Adaptive reservoir operations are gaining in recognition (see e.g. Yao and Georgakakos, 2001, Bayazit et al., 1990), where in rule curves are modified to produce more effective management given improved streamflow forecasts.

2.11.3.5 Adaptive management

Adaptive management is a method for addressing the uncertainties associated with climate change, other hydrologic change and our limited ability to correctly anticipate the future in complex water resources systems. The approach was developed in ecology.

The general principle of the approach is "learning by doing" and includes emphasis on monitoring and the ability to change decisions and update hypotheses based on the observations. It may be appropriate in cases where reversible decisions are possible so that a decision may be changed as climate and socioeconomic conditions evolve and better understanding of the system is achieved. Sankarasubramanian et al. (2009) demonstrate that utility of climate information even of modest skill, when coupled with adaptive operations, can benefit water allocation projects with high demand to storage ratio as well as systems with multiple uses constraining the allocation process. The ability to ingrain flexibility into a system improves the opportunities for making use of adaptive management principles.

2.11.3.6 Robust decision-making methodologies

Possible future conditions may be radically different and more variable than in the past due to climate change, and the risks and uncertainties are difficult to quantify. The near future is less uncertain. Robust Decision Making (Lempert et al., 2006) is a decision making process that addresses "deep uncertainty" by favoring management plans that perform well over a range of possible future conditions (Brekke et al., 2009). Decision-scaling is a method for decision making under climate uncertainty that utilizes subjective probabilities based on climate information and hedges the risks of the prediction uncertainties. Adaptive management can be coupled with robust decision making by continually monitoring the performance of a robust plan. At intervals in the future, if the performance decreases, the plan can be modified to improve the performance. The modification would be made with improved knowledge of future possible conditions. The application of decision scaling to adaptive management is currently being adopted as the approach to the plan for regulation of outflows from Lake Superior (Brown et al., 2011).

2.11.3.7 Information and communication technology

Advances in Information and Communication Technology offer great potential for improving the management of water resources. Their low cost has made them ubiquitous including widespread use in developing countries. Potential applications include forecast transmittal, early warning systems and crowd sourcing for data collection and monitoring.

2.11.3.8 Advanced Decision Support Systems (ADSS)

Application of many scientific advances such as probabilistic forecasts requires a decision support system in which current and future states of the hydrologic system can be modeled, stochastic hydrology can be generated that reflects a wide range of assumptions, various management and infrastructure alternatives can be evaluated. The decision support tools should support quantification of risk and reliability, and the ability to measure performance of a management plan with respect to agreed-upon criteria such as reliability of supply and risk of damage to the human or ecological systems. The ADSS must be available for participatory involvement of a range of scientific, government and management agencies as well as stakeholders of varying

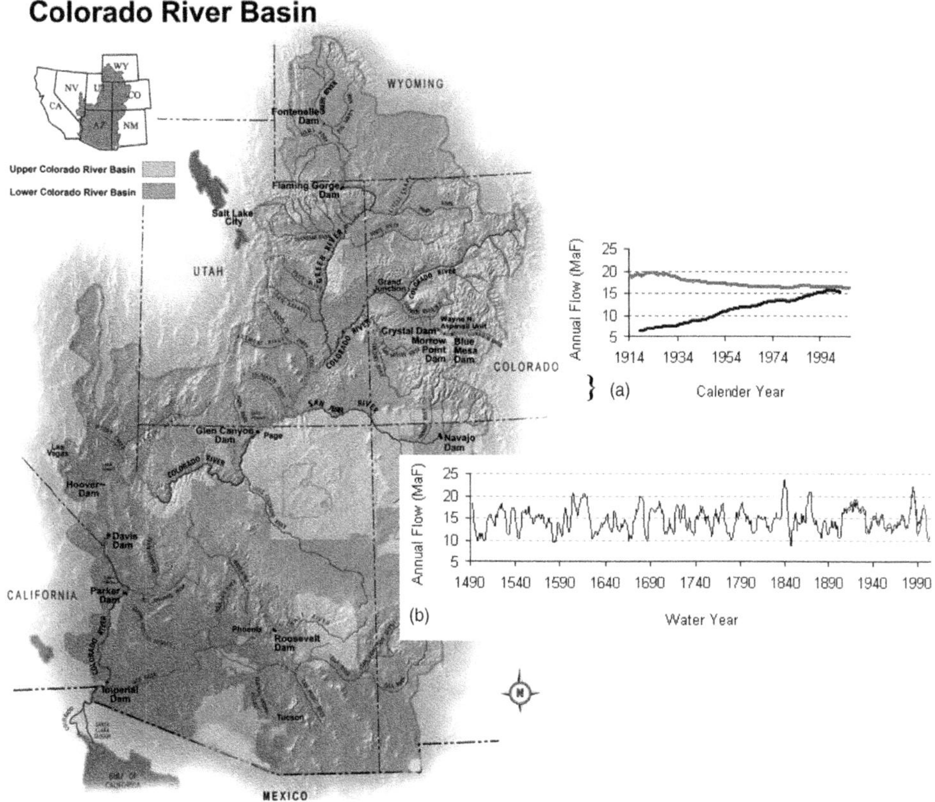

Figure 2.11.5 Colorado River Basin including major dams. (a) Inset shows water demand and losses and supply in the Colorado River Basin. A 9-year running average of the natural flow above Imperial Dam, AZ (upper line). A 9-year moving average of the total consumptive use in the Lower and Upper basin, including reservoir evaporation from Flaming Gorge, Blue Mesa, Morrow Point, Lake Powell, Lake Mead, Lake Mohave, and Lake Havasu, and losses due to native vegetation between Davis and Imperial Dams (lower line). For all years Mexico deliveries are assumed 1.5 million acre-feet. (b) 5-year moving average of paleo reconstructed flows at Lees Ferry, AZ [*Woodhouse*, et al., 2006]. The red line is the natural flow from the observational period. This is figure is from Rajagopalan et al., 2009

interests. Systems have been developed and applied successfully in several basins and are under development in others across the globe. An advanced and robust ADSS is the reservoir operations and management tool, RiverWare (Zagona et al., 2001). This tool is being widely used for intricate management and future planning of network of large reservoirs in the United States.

2.11.4 COLORADO RIVER MANAGEMENT – CASE STUDY

Western United States is mountainous, predominantly semiarid and, gets its water supply from snowmelt. Its population has steadily increased during the latter half of

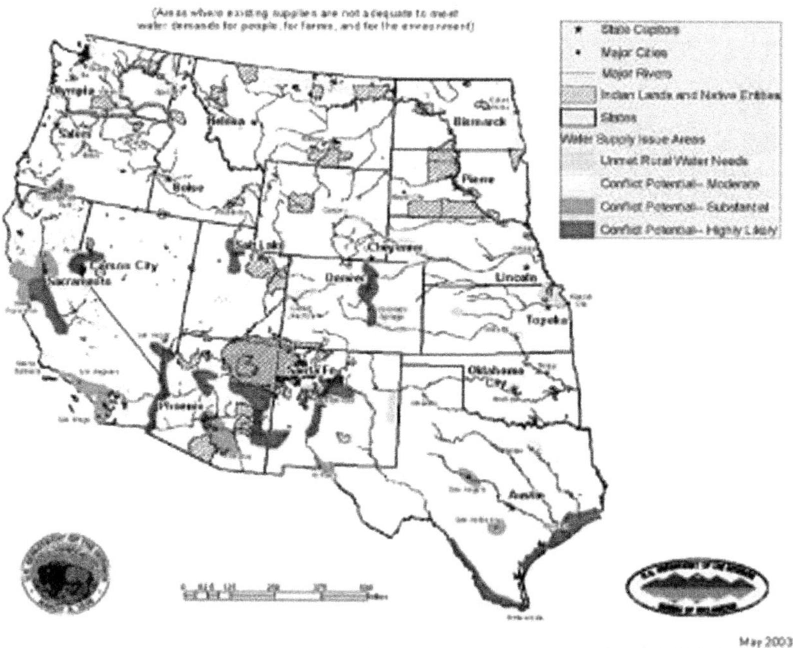

Figure 2.11.6 Potential water supply crises by 2025

20th century. The large population centers are in the desert south west while the water is supplied to them by the Colorado River originating in the higher elevations of the Rocky Mountains far to the north. The water in the river is divided among seven basin states and two countries (US and Mexico) – thus, water management is highly contentious (Figure 2.11.5 from Rajagopalan et al., 2009). The river has extensive storage (almost four times the annual average flow in the river) which is managed by Federal water agency under a complicated set of decrees and rules known as the "law of the river". The law of the river worked well for much of the 20th century when demand was much lower than supply. But a "perfect storm" comprising of economic growth, increasing population and water demands coupled with severe sustained drought in recent decade is stressing the physical system and the agreements. Water managers and stake holders in the basin have adopted much of the R&D suggestions described above in improving the management of water resources system in the basin. The combination of observed and paleo data, streamflow models, open and advanced decision support system and a collaborative attitude from the myriad of stake holders led to the historic drought management agreement (USDOI, 2007) which is regarded as a model for effective and efficient water resources management approach for others around the world to emulate.

Paleo reconstructions of streamflow in the basin (See Figure 2.11.5 and Woodhouse et al., 2006) indicate drier epochs with regular frequency and in fact, they indicate that the 20th century was one of the wettest when the "law of the river" was drafted.

Climate change projections indicate a substantial decrease in annual average streamflow which will only exacerbate the situation. Many studies, recently Rajagopalan et al. (2009), underscore this and also suggest that flexible and innovative management of the system and collective stakeholder participation can instill effective adaptive management practices that can help mitigate the risk. Studies indicate (Figure 2.11.6) that the region is likely to experience severe water supply crises in the not too distant future if management practices are not modified. The R&D measures proposed will be of immense use in developing a robust management and planning strategy in the face of potential stresses in the future.

ACKNOWLEDGEMENTS

The authors also authored Chapter 4 of the World Water Development Report IV from UNESCO, this will be released in early 2012. The material presented in the current chapter benefited from the WWDR chapter. The authors are grateful to UNESCO for the opportunity.

REFERENCES

Bayazit, M. and Unal, N. E. (1990), Effects of hedging on reservoir performance, *Water Resources Research*, 26(4), 713–719.

Brekke, L.D., Kiang, J.E., Olsen, J.R., Pulwarty, R.S., Raff, D.A., Turnipseed, D.P., Webb, R.S., and White, K.D., (2009), Climate change and water resources management – A federal perspective: U.S. Geological Survey Circular 1331, p. 65.

Bromley J(ed) (2005), Guidelines for the use of Bayesian networks as a participatory tool for water resources management. A MERIT report. CEH, UK.

Brown, C. (2010), The end of reliability, *ASCE Journal of Water Resources Planning and Management*, Vol. 136, p. 143.

Brown, C., Werick, W. Leger, W. and Fay, D. 2011. A decision analytic approach to managing climate risks – Application to the Upper Great Lakes, *ASCE Journal of the American Water Resources Association* (in press).

Clark, M. P., Serreze, M. C., and McCabe, G. J. (2001), Historical effects of El Nino and La Nina events on the seasonal evolution of the mountain snowpack in the Columbia and Colorado River Basins, *Water Resources Research*, 37, 741–757.

Grantz, K., B. Rajagopalan, M. Clark, and E. Zagona (2005), A technique for incorporating large-scale climate information in basin-scale ensemble streamflow forecasts, *Water Resour. Res., 41*, W10410, doi:10.1029/2004WR003467.

Grantz, K., B. Rajagopalan, M. Clark and E. Zagona (2007), Water Management Applications of Climate-Based Hydrologic Forecasts: Case Study of the Truckee-Carson River Basin, Nevada, *ASCE, Journal of Water Resources Planning and Management*, 133(4), 339–350.

Hamlet, A. F., and Lettenmaier, D. P. (1999), Columbia River streamflow forecasting based on ENSO and PDO climate signals, *Journal of Water Resources Planning and Management*, 125, 333–341.

Lempert, R. J., Groves, D. G., Popper, S. W. and Bankes, S. C. (2006), A general, analytic method for generating robust strategies and narrative scenarios. *Management Science*, Vol. 52, No. 4, pp. 514–528.

Milly, P. C. D., et al. (2005), Global pattern of trends in streamflow and water availability in a changing climate, *Nature*, 438, 347–350, doi:10.1038/nature04312.

Nowak, K., (2011), Stochastic streamflow simulation at interdecadal time scales nad implications to water resources management in the Colorado River Basin, PhD dissertation, University of Colorado, Boulder, CO, 181p.

Prairie, J., Nowak, K. Rajagopalan, B., Lall, U. and Fulp, T., (2008) "A stochastic nonparametric approach for streamflow generation combining observational and paleo reconstructed data." *Water Resources Research*, 44, W06423, doi:10.1029/2007WR006684.

Rajagopalan, B., et al. (2009), Water Supply Risk on the Colorado river: Can Management Mitigate?, *Water Resources Research*, doi:10.1029/2008WR007652

Rajagopalan, B., J. Salas and U. Lall (2010), Stochastic methods for modeling precipitation and streamflow, In *Advances in Data-based Approaches for Hydrologic Modeling and Forecasting*, Ed by B. Sivakumar and R. Berndtsson, World Scientific, Singapore

Regonda, S., B. Rajagopalan, M. Clark and E. Zagona, Multi-model Ensemble Forecast of Spring Seasonal Flows in the Gunnison River Basin, *Water Resources Research*, 42, 09494, 2006.

Regonda, S., E. Zagona and B. Rajagopalan (2011), Prototype decision support system for operations on the Gunnison basin with improved forecasts, *ASCE Journal of Water Resources Planning and Management*, 137(5), 428–438.

Sankarasubramanian, A., U. Lall, F.D.Souza Filho, A.Sharma (2009), Improved Water Allocation utilizing Probabilistic Climate Forecasts: Short Term Water Contracts in a Risk Management Framework, *Water Resources Research*, 45, W11409.

Van der Keur, P., H.J. Henriksen, J. C. Refsgaard, M. Brugnach, C. Pahl-Wostl, A. Dewulf and H. Buiteveld (2006), Identification of major sources of uncertainty in current IWRM practice and integration into adaptive management, Report of the NeWater project – New approaches to adaptive water management under uncertainty.

Woodhouse, C. A., S. T. Gray, and D. M. Meko (2006), Updated streamflow reconstructions for the Upper Colorado River Basin, *Water Resources Research*, 42, W05415, doi:10.1029/2005WR004455.

USDOI (2007), U.S. Department of the Interior, Final environmental impact statement Colorado River interim guidelines for lower basin shortages and coordinated operations for lakes Powell and Mead, Bur. of Reclam., Boulder City, Nev.

WCD, (2000), Dams and Development, a new framework for decision-making, *World Commission on Dams report, UNEP, Earthscan Publications*, 356p.

Yao, H and A. Georgakakos (2001) Assessment of Folsom Lake response to historical and potential future climate scenarios: 2. Reservoir management. *Journal of Hydrology*, Volume 249, Issues 1–4, pp. 176–196.

Zagona, E., Fulp, T., Shane, R., Magee, T., and Goranflo, H. (2001), RiverWare: A generalized tool for complex reservoir systems modeling, *Journal of American Water Resources Association*, 37, 913–929.

Section 3

Mineral resources management*

---

*This section drew extensively from the author's own works, "*Mineral Resources Management and the Environment*" (2003).

Chapter 3.1

Introduction

Mining, like the proverbial serpent in the Garden of Eden, has never been held in high esteem. Most people consider mining as an unmitigated evil, and some who are more realistic, concede that it is a necessary evil (but evil all the same). This is not a new development. In the olden days, mines were invariably worked by slaves – chained to pillars underground, the slaves used to die in a matter of weeks. In the medieval Europe, being condemned to work in the salt mines was a form of punishment worse than death.

Presently, the horrendous consequences of mining are evident everywhere. The landscape in some countries (e.g. USA, Zambia, PNG) is pockmarked with gigantic pits. As pointed out by Förstner (1999), the mass of the mine tailings produced worldwide (18 billion m^3/year) is of the same order as the quantity of sediment discharge into the oceans. As progressively lower grades are worked, the mass of the mine tailings is expected to double in the next 20–30 years. Vast areas are strewn with rock fragments, and in some areas, Acid Mine Drainage has rendered the soil and water so acidic that not a blade of grass grows there. Whole towns (e.g. eastern India) had to be abandoned due to subsidence caused by underground coal mining. Mine workers are exposed to a number of physical, chemical, biological and mental hazards, and mining is ranked as number one among the industries in the average annual rate of traumatic fatalities.

The iron ore mine of Kiruna, Sweden, is an outstanding example of how technology can be applied to minimize environmental pollution, reduce energy consumption and improve productivity.

The high-tech model used by the Industrialized countries to minimize the number of workers and increase the output, is, however, not applicable to the Developing countries, for the following reasons:

(i) the investments needed are high – for instance, a block cave mine may need an investment any where from USD 100 million to 1000 million, besides requiring very highly skilled personnel (Fig. 3.1.1),

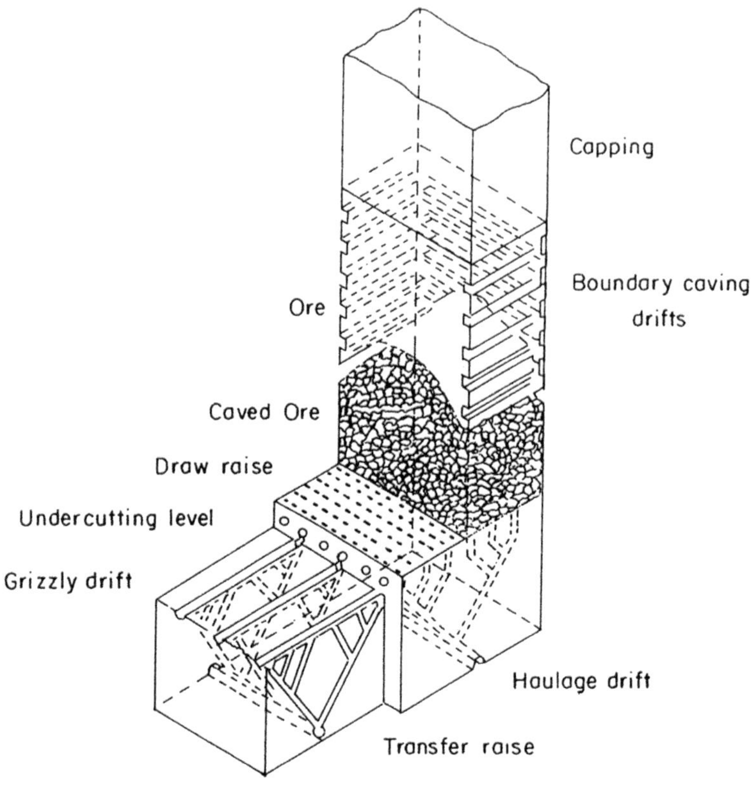

Figure 3.1.1 Mining by block caving
Source: UNEP Tech. Rept./no. 5, 1991, p. 13

(ii) what the developing countries need is *job-led* (and not job-less) economic growth. A sensible strategy for the developing countries is to use the mining industry to promote *job-led* economic growth through the adoption of employment-generating, economically-viable and environmentally-acceptable technologies.

As the most critical issue in regard to the mineral resource management is the severe adverse impact of mining, washing, processing and utilization of ore minerals, this chapter devotes considerable attention to various ways and means of ameliorating the environmental impact.

3.1.1 ENVIRONMENTAL CHALLENGES FACING THE MINING INDUSTRY

An Environmental impact may be defined as a change in the environmental parameters, over a specified period, and in a specified geographical area, resulting from a particular activity compared to the situation which would have existed had the activity not been performed. It is no longer possible for a mine to be started merely because its technoeconomic viability has been demonstrated. The mining project has to be socially acceptable as well.

Sengupta (1993, pp. 22–23) has drawn attention to the "shadow effect" of a mine site. Apart from the degradation of land directly connected to the mine site itself (due to the mine, supporting facilities, waste disposal arrangements, etc.), the shadow effect of the mine site may extend to large areas around the mine site as a consequence of the infrastructure (rail, road, housing, power plants, water storage, etc.) necessary for the performance of the mining operations. Thus, the responsibility of the mining company is not confined to the mine site, but to a large area around it. The mining company has thus to work harmoniously with a variety of land use authorities, concerned with (say) wildlife, forestry, recreation and tourism, fisheries, environmentally-sensitive habitats (e.g. corals, mangroves), parks, reserves, historical sites, native reserves and rights of the indigenous people, urban growth, etc.

Khanna (1999) gave a succinct account of the environmental challenges facing the mining industry. The adverse effects of mining on the geological environment include changes in the landscape, landslides, subsidence, pollution of water and soil, lowering of groundwater, damage caused by explosions, etc. The magnitude of the environmental impact is function of the volume of the material mined, methods of mining, mode of disposal of wastes, environmental protection measures undertaken, etc.

The potential effects of the mining activities on the environment are summarized in Table 3.1.1.

It has been estimated that there are more than 40,000 mines in the world, which process an aggregate volume of 33×10^9 m^3/y of rock (Vartanyan, 1989).

Mining has a negative image – some of the worst industrial disasters happen to be mining related (vide Appendix D in "*Mineral Resources Management and the Environment*" – U. Aswathanarayana, 2003). Mining industry has a characteristic, which is not shared by other industries. For instance, mining has to be undertaken where the ore occurs – direct relocation is not possible. There has been much controversy whether the concept of sustainable development is at all applicable to the mining sector, which is based on the production of non-renewable resources from finite deposits. Mining takes out the ore, but leaves nothing in its place – in other words, mining is inherently unsustainable. On the lines of the definition of the Ecologically-Sustainable Industrial Development (ESID), *Sustainable Mining* may be defined as those patterns of mining that enhance economic and social benefits for the present and future generations *without impairing the basic ecological processes.* This implies that any uses of mineral resources that lead to significant degradation of ecological processes, are deemed to be *ipso facto* unsustainable and hence unacceptable.

Mining industry faces pressure to follow "good environmental practice" from the following kinds of institutions: (1) Environmental pressure groups, such as, Minewatch, Greenpeace, Friends of the Earth and the Mineral Policy Centre, (2) International organizations, such as, the World Bank, UNDP, and the International Council of Metals are using their financial leverage to make the mining companies follow certain guidelines, (3) Most national governments have prescribed regulations for the protection of the environment, amelioration of the mined land, and the responsibilities of the mining company in the event of the mine closure, (4) Mining associations which are developing "codes of practice", and helping the mining companies to implement the "Best Practices" – this is a kind of corporate peer pressure which often has proved

Table 3.1.1 Potential effects of mining activity on the environment

	Surface water pollution	*Underground Water pollution*	*Air pollution*	*Solid waste*	*Excavation*	*Noise and vibration*	*Remarks*
Human health and activity	Soluble contaminants in domestic and/or agricultural use waters. Deposition of solids on agricultural lands, and in the shallow zones of the sea; withdrawal of water for industrial purposes	Soluble contaminants in wells, springs, etc (1)	Dust blown on inhabited or agricultural lands (2)	Hazards related to lack of stability of waste deposits		Effects of noise on human health. Damage to buildings due to blasting vibration	(1) Such impacts on underground waters do not occur generally; it depends on the hydrogeology of the area. (2) Plant, especially on the atmosphere of the underground mine
Fauna	Degradation of aquatic fauna, including the destruction of fish species, accumulation of toxic elements by fish				Loss of habitat	Disturbance of habitat feature (3)	(3) Issues regarding unique habitat features (e.g. migration corridors, watering areas, etc.) for threatened or endangered species, should be specially addressed
Flora	Degradation of aquatic flora		Accumulation in plants of toxic elements carried by dust				Spatial requirements of mining operations are normally quite restricted, but within that area, the disturbance can be quite significant
Land use	Deposition of sand in the river channels and in the shallow zones of the sea			Land disturbance. Land subsidence due to underground mining			

Source: UNEP, 1986

very effective, and (5) The coverage of "mining disasters" in the International media, particularly the Internet, can be so extremely intense that a mining company may be put in a tight corner, and may even have to fold up. In the context of the increasing public consciousness about environmental consequences of any commercial activity, it is no longer possible to take decisions about mining based on commercial rationale alone. A community may wonder whether the economic benefit from a mine is worth the ugly scar that would be left behind when the mine is closed. Previously, mining companies used their public relations exercise to *sell* a project. Now they use community consultation techniques to *develop* the project in harmony with the stakeholders who will be affected by the mine.

Poor communities may accept mining, as it may be the only way for social and economic development. But when once the mine is exhausted, the mine-dependent community is left with a big hole in the ground, plus the environmental problems associated with the contaminated soil and ground. In the past, companies simply closed the mine and walked out. Now-a-days, the communities and the government will not tolerate such a step. The mining companies do indeed have a responsibility for the well being of the community when once the mining ceases. A sensible approach would be for the mining company in cooperation with the government and the community concerned, to plan for a long-term development of the area to enable the sustainable development to continue after the mining ceases. In other words, the financial costs of the environmental and social protection have to be integrated into the business plan right at the start. Companies are finding that this kind of *proactive* approach of a long-term, mutually beneficial relationship with the community is better than a *retroactive* approach which tries to sort out the environmental and social conflicts after they become intractable.

Mining companies are slowly getting reconciled to the fact that there is no way they can avoid issuing reports of their environmental performance, as such reports are demanded by the government regulations, by the public, and by the shareholders. It is good for the image of a company to show that it is environment-conscious. Companies, such Cambior Gold, are taking pride in fulfilling the requirements of Industry standard ISO 14001. This is a good trend.

3.1.2 MINING, ENVIRONMENTAL PROTECTION AND SUSTAINABLE DEVELOPMENT

Miller (1999, pp. 317–332) examines the dilemma facing the developing countries (such as Indonesia) as to how to reconcile environmental protection with sustainable development. The developing countries think that sustainable development as defined by the Brundtland Commission seems to imply a low rate of economic growth that impedes the development of their energy and mineral resources. They regard mining as the "engine of development" to promote technological and economic development of the country. Mining accounts for 80–90% of GNP in some countries in Africa. For instance, Botswana with a population of little over one million, earns almost USD 3 billion from the mining sector, principally diamonds. This works to about USD 3,000 per capita per annum, which happens to be several times the GDP per capita of the neighbouring Mozambique.

The mineral resources that mining exploits are non-renewable, but the resources that are affected by mining, namely, water, land, flora and fauna, are renewable. Sustainable mining has therefore to be understood to mean that the mining has to be carried out in a manner that is ecologically sustainable.

3.1.3 ECONOMICS OF ENVIRONMENTAL PROTECTION IN MINING

Maxwell and Govindarajulu (1999, pp. 7–17) gave a good analysis of the economics of environmental protection in mining, with particular reference to Australia. It has been estimated that mining companies in Australia spend upto 5% of capital and 5% of operating costs for new mining projects to maintain best practice environmental management.

Of late, environment has been attracting considerable interest from the economics. It is generally held that markets do not allocate environmental resources efficiently. This is so because many environmental resources are public goods. There is obviously a need for environmental protection regulation. The point that Maxwell and Govindarajulu (1999) raise is how zealous that such a legislation should be. If it is too demanding, the mining operation would result in less than the optimal level of output. The diagram of Coase (1960, quoted by Maxwell & Govindarajulu, 1999) helps us to understand the economics of the environmental impacts of a mine in terms of curves for marginal damage to the environment (air, water, soil, noise, etc. pollution and aesthetic damage) versus marginal abatement cost (or marginal benefit). The most economically efficient level of environmental damage occurs where the marginal damage and the marginal benefit curves intersect (Fig. 3.1.2).

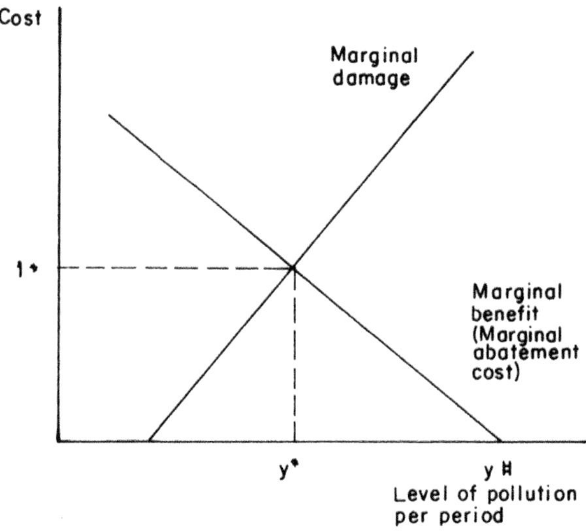

Figure 3.1.2 Economics of environmental protection in mining
Source: Maxwell and Govindarajulu, 1999, p. 13

3.1.4 TECHNOLOGY TRENDS IN THE MINING INDUSTRY

Chadwick (2001) highlighted the implications of technological improvements on the performance of the mining industry. As is happening in other industries, Service and Automation are emerging as two over-riding trends in the mining industry as well. Technology is being increasingly applied to improve efficiency and safety, cut costs, and reduce the adverse impact of mining on the environment. Previously, manufacturers used to supply equipment and spares, and the mining companies were expected to take care of their own maintenance and repair work. Now-a-days, the manufacturers of equipment are undertaking all the related services, such as, managing of spare parts inventories, servicing, maintenance and repair work (even if the equipment concerned is not their own make), and optimizing the utilization of equipment. Mergers and acquisitions in all sectors of manufacture are allowing the manufacturers to achieve the critical mass and provide a truly global service.

3.1.5 AUTOMATION IN THE MINING INDUSTRY

Automation is helping the mining companies to optimize their mining and processing operations, depending upon the prevailing commodity prices. This may take the form of using the equipment in such a manner that render possible the selection and processing of specific ore grades, and extraction of specific metals in a polymetallic deposit, etc. This is similar to the customary practice in oil refining – different crudes are blended and process technologies adjusted, depending upon the product mix that the market needs at any particular point of time.

Automation is revolutionizing exploratory drilling. Drilling can take place autonomously with high hole accuracy, and samples are recovered (and in some systems, analyzed) automatically.

In surface mining, as the capacity of haulers is increasing, there has been concurrent increase in the capacity of loading and ancillary equipment. Blasthole drilling which can be isolated from other surface mining activities, has benefited most from automation. Emulsion explosives have emerged as safe, inexpensive and easy to use alternatives to the old nitroglycerine, water gels, ANFO, etc. These explosives are manufactured in the form of water-in-oil emulsions. As each micro cell of the oxidizer is coated with an oily exterior, the emulsion has excellent water resistance, and could therefore function efficiently under water. Glass microspheres dispersed throughout the basic emulsion serve as bulking agent and help in density control and sensitivity. The consistency of the emulsion can be varied depending upon the blasting applications. Bulk emulsion has a density of 1.25 g/cm^3, VOD of 5,500 m/sec, and energy of 1,030 kcal/cm^2. Studies have shown that the efficiency of the emulsions (93%) is much higher than those of the water cells (70%). *Orica* of USA has developed digital energy control software (*ShotPlus*) for the safe, accurate and efficient control of blasts. *NPV Scheduler* software enables the optimization of open pit mining through the identification of the unique path of extracting minerals in the pit, which will deliver the highest possible Net Present Value.

With the increasing accuracy of GPS equipment, driverless trucks in open pits may indeed become a reality in the not too distant future.

Automation has gone much farther in underground mining. Automated load and haul systems are being increasingly used in Australia, Sweden and Finland. In the *Automine* system operating in Kiruna, Sweden, LHDs with a tramming capacity of 25 t, load themselves from drawpoints under tele-remote control. The automatic tram and dump cycle starts. The LHD takes its load to the overpass, dumps it, and returns to the drawpoint, all autonomously. Under the Sandvik Tamrock's system, the LHD determines its position by dead reckoning. It gets its direction from the onboard gyroscope, and the articulation angle and distance from the drive line (*Mining Magazine*, July, 2000, p. 12). Other navigation systems are based on reflectors suspended from the sides of the drifts, on the basis of which the LHD fixes its position. Tyre life has shown significant improvement, as the LHDs are driven more smoothly under automatic control. Unlike the operator-driven machines, which operate 10–12 hours per day, the automatic machines can work upto 19 hours a day. One operator can control three LHDs. The Kiruna mine expects that its production of 23 Mt/y of ore will be drawn, trammed and dumped by semi-autonomous LHDs (*Mining Magazine*, July, 2000, pp. 12–16).

The following considerations are likely to lead greater emphasis on underground mining: (1) decline in the availability of deposits which are amenable for surface mining, (2) the "greening" of the mining industry which wants to avoid the ugly scars on the surface, (3) reduced extraction of waste by the placement of the tailings back in the underground, particularly as backfill support, and (4) high degree of automation that is possible in underground mining.

Automation in mineral processing allows large plants to be run with minimum staffing. Expert systems not only reduce the personnel costs, but also provide real time information on the processes as they operate, so that the recovery can be fine-tuned depending upon the market situation. Gravity separation is coming into vogue, particularly in situations where the use of cyanide is to be avoided.

3.1.6 TECHNOLOGY-DRIVEN DEVELOPMENTS IN THE MINING INDUSTRY

Technology is being increasingly used to address three major challenges of the mining industry: (1) increase in the global mineral consumption of minerals (for instance, the production of bauxite increased six-fold during the last two decades), (2) the variation in the prices of minerals (for instance, the price of gold which has been oscillating around USD 350–400/oz, is currently around USD 1500/oz), and (3) huge quantities of ever-lower grades are being processed by bulk mining methods (such as opencast mining). Such mining not only leaves big, unseemly pits in the ground and huge waste dumps above the ground, but also leads to increased pollution due to ore processing and increased tonnages of tailings that have to be disposed off.

According to Khanna (1999), the following technological advances have an impact on the issue of Mining and Environment: (1) the total recovery of metals from ore is being improved, with the consequence that the waste material will have lower content of heavy metals, and hence lesser ability to contaminate the environment, (2) process chemicals which are environmentally unacceptable are being replaced by those which are environmentally-friendly and recyclable. Bioleaching has the potential not only to revolutionize the extraction of metals, but also the detoxification of industrial waste products, sewage sludge, and soils contaminated by heavy metals. Several companies have resorted to *in situ* leaching of ores, but the environmental consequences of *in situ* leaching are not yet fully understood.

Sulphur dioxide produced in the course of smelting of sulphidic ores, has been a major pollutant of air. Hydrometallurgical techniques avoid the use of smelting, and thereby eliminate the sulphur dioxide emissions. Copper industry is already using pressure leaching techniques in a big way.

The patented Gold Haber process avoids the use of cyanide in the extraction of gold. The preliminary operation is the same as in conventional cyanide process, namely, crushing and grinding of ore, mixing it with water, and making a slurry. Before pumping the slurry to the leach tanks, Haber's patented reagent suite is added. This involves the use of activated carbon to recover the gold in solution, which is then followed by electrowinning. The acidic tailings are neutralized before disposal. The most serious problem facing the mining industry is the disposal of wastes. The extent of the environmental impact of mine wastes can be illustrated with the example of gold. The world production of gold is about 2500 t. Since the gold content of the mined material is usually of the order of a few gms. per tonne, virtually all the mined material ($\sim$1.5 billion tonnes/yr?) ends up as mine waste which needs to be disposed off.

About 69% of the metallic ores are produced by opencast mining. As more and more low-grade ores are mined, the ratio of waste generated relative to the quantity of mineral produced rises steeply. Waste dumps make the landscape ugly to look at. Potential acid-producing material need to be encapsulated, so that the rainwater and surface runoff leaching the waste dumps do not contain too high concentrations of heavy metals.

When tailings are discharged into impoundments, we not only have to manage the solid wastes but also the water/supernatant. The tailing impoundments need to be dewatered before the rehabilitation of the solid wastes. Failure of tailing impoundments is a kind of disaster that would attract the glare of adverse publicity. Khanna (1999) gives an example of this. A tailings spill at the Marcopper mine in the central Philippines has polluted a 26 km stretch of Makulapnit and Boac rivers. This raised a public furore. The Marcopper Mine was, however, too small a company to be able to afford the amelioration of the problem. Though Placer Dome had only a minority interest in the company, it came forward and cleaned up the rivers. Paste technologies and subaqueous disposal are some of the new technologies that are being developed to reduce the risk of failure of tailings impoundments. The management of waste rock and tailings has become such a major problem that some authorities are proposing that in future mining should be underground only, with the wastes being disposed off wholly underground.

The advanced, low-polluting coal combustion system, called Low NOx Concentric Firing System (LNCFS) reduces the formation of NOx by nearly 40% in older coal burning plants. Power plants equipped with this burner now account for 56,000 MW of electricity in USA. Sales of this system reached over USD one billion (*Mining Magazine*, July, 2001).

Chapter 3.2

Mineral demand in response to emerging technological needs

3.2.1 EMERGING TECHNOLOGICAL NEEDS

A mineral resource is "a concentration of naturally occurring solid, liquid or gaseous material in or on the earth's crust in such form and amount that economic extraction of a commodity from the concentration is currently or potentially feasible" (U.S. Bureau of Mines, 1989). Ore is defined as a "mineral or rock that can be recovered at profit". Gangue is the useless material associated with the ore. Protore is mineralized rock that is too lean to be usable. Thus, a mineral does not remain an ore or non-ore for all time. A mineral can be regarded as ore so long as technology and market demand make it economical to mine it. Alternately, what was yesterday a non-ore may become ore today as technology and market demand make it economically worthwhile to mine it now.

More than two-thirds of the 92 natural elements are metals. Some of them, such as, Au, Ag, Cu, Pb, Sn, Hg, S, etc., have been known and used since ancient times. Improvements in analytical techniques led to the identification of a large number of metals. The specialized and exacting requirements of modern industries led to profound changes in the ways metals are detected, extracted, alloyed and used. New applications for metals are being found all the time, e.g. use of germanium in semiconductors, use of cerium in high temperature superconductors, development of zircalloys in nuclear industry, titanium alloys in aerospace industry, metal glasses, etc. On the other hand, some traditional metals (e.g. Fe and Cu) are being substituted by plastics, fiberglass, ceramics, etc., thus increasing the demand for industrial minerals. The non-metallic minerals are being increasingly used as insulating material, fillers, glasses, and construction material. The ever-increasing need for more fertilizers (due to the need to grow more food for the increasing population of the world) will greatly increase the consumption of fertilizer raw materials, like apatite, potash feldspar, etc. Thus, the demand for a given mineral depends upon technology and markets.

According to Patrice Christmann of BRGM, France, the demand for metals, driven by increasing global population and higher standard of living, is expected to double over the next 15–20 years. Hence between 2010–50, humanity will need to produce more minerals of all kinds than were produced since the origin of humanity till 2010.

The 1972 Club of Rome study, *The Limits to Growth*, projected that some metals would be exhausted by 2000 or 2050. This kind of dire prediction arose from the fact that many people do not appreciate the difference between ore reserves (estimated by mining companies and geological surveys) and geologic resources. The question is not one of availability, but accessibility.

Reserve estimates can fluctuate upward or downward. In 1961, prior to the Club of Rome report, copper reserves were thought to extend to 51 years. This spurred the governments to provide more funding for mineral exploration, with the result that the 1981 estimates for copper reserves stood at 72 years. In 2010, the figure dropped to 39 years.

Apart from minimizing the social, environmental and ethical costs of mining, governments need to develop comprehensive minerals inventory databases, stable regulatory frameworks, new geochemical-geophysical tools, recycling of metals and improving public-private ownerships, in order to meet the emerging challenges in regard to the supply of minerals. An army of geologists have to be trained to look for new or alternative mineral deposits.

The chapter deals with the trends in markets, production and recycling of some selected mineral resources, namely, Rare Earth Elements, Gold, Aluminium, Copper and Lead.

Advances in technology (such as, the need for light-weight wind turbines) led to the development of new materials, such as rare-earth steels, in the recent years. For instance, dysprosium prices went up seven fold since 2003. Gold is one metal for which there has never been a diminishing of demand. The world production of gold was about 1,400 t in 1980's, and about 1,800 t in 1990's. The present world production of gold is about 2,500 t. More countries are producing larger quantities of gold.

On the other hand, metals like copper, lead and aluminium have been in use for a long time. A feature common among the three is secondary production of the metals from scrap. The energy requirement of the secondary production is just a fraction of that of the primary production. It is for this reason that the prices of these metals are reasonably steady.

3.2.2 RARE EARTH ELEMENTS

No group of metals has caused as much political and economic turmoil as the Rare Earth Elements (REE) did in the recent years. That is because they constitute critical ingredients in high-tech industries, and one country, China, holds the monopoly of their world production.

Fifteen lanthanoids and scandium and yttrium are termed as Rare Earth Elements. They are characterized by very high geochemical coherence (and hence the difficulty in separating them), and always occur together. Their importance arises from the high-tech uses they are put to, as should be evident from the following data:

Scandium – Light aluminium-scandium alloy for aerospace components; additive in mercury vapour lamps.
Yttrium – Yttrium-aluminium garnet (YAG) laser, high-temperature superconductors, microwave filters.
Lanthanum – High refractive index glass, flint, hydrogen storage, battery electrodes, camera lenses, fluid catalytic cracking catalyst for oil refineries.
Cerium – Chemical oxidizing agent, polishing powder, yellow colours in glass and ceramics, catalyst for self-cleaning ovens, catalyst in oil refineries.
Praseodymium – Rare-earth magnets, lasers, carbon arc lighting, colorant in glasses and enamels, welding goggles, ferrocerium firesteel (flint) products.
Neodymium – Rare-earth magnets, lasers, violet colours in glass and ceramics, ceramic capacitors.
Promethium – Nuclear batteries.
Samarium – Rare-earth magnets, lasers, neutron capture, masers.
Europium – Red and blue phosphors, lasers, mercury vapour lamps, NMR Relaxation agent.
Gadolinium – Rare-earth magnets, high-refractive index glass, lasers, X-ray tubes, computer memories, neutron capture, MRI contrast agent, NMR Relaxation Agent.
Terbium – Green phosphors, lasers, fluorescent lamps.
Dysprosium – Rare-earth magnets, lasers.
Holmium – Lasers.
Erbium – Lasers, vanadium steel.
Thulium – Portable X-ray machines.
Ytterbium – Infrared lasers, chemical reducing agent.
Lutecium – PET Scan detectors, high refractive index glass.

Till 1948, most of the REE (Rare Earth Elements) production was obtained from placer monazite in India and Brazil. Global production of REE was 80,000 t in 2000. China now produces 97% of the world's REE. Yttrium oxide is derived from ion-adsorption clay in southern China. All of the world's heavy REE, such as, dysprosium, is obtained from polymetallic Baotou deposit in Inner Mongolia.

Steel which is used for the construction of the wind turbines, accounts for 90% of the cost of the turbine. Turbine fabrication costs are being brought down by replacing steel with lighter and more reliable material, and by improving the fatigue resistance of the gear boxes. During the last five years, there has been a phenomenal growth in the use of rare-earth elements in the energy industries. Tiny quantities of dysprosium can make magnets in electrical motors lighter by 90%, thereby allowing larger and more powerful wind turbines to be mounted. Use of terbium can help cut the electricity use of lights by 80%. Dysprosium prices have gone up sevenfold since 2003, with the current market price being USD 116/kg. Terbium prices have quadrupled during the period 2003–2008 to USD 895/kg. The recession brought down the price to USD 451/kg.

On September 1, 2009, China announced plans to reduce export quota to 35,000 t, presumably to conserve resources and protect environment. In 2010, the quota has been further reduced to 14,446 t. There may also be a strong economic motive in China restricting the exports. China prefers to export products higher in the supply chain which bring greater economic benefits – say, rare-earth magnets, and rare-earth-steel rather than the rare earth metal itself.

USA has to intervene with China to ease the situation, because of the critical importance of REE in the Japanese high-tech industries.

As a consequence of increasing demand and tightening restrictions, Australia. Brazil, Canada, South Africa, Greenland and USA are developing REE properties. USA is reopening Mountain Pass Mine in California. Because of the high market price of REE, copper process flow sheets are being modified to produce larger quantities of REE byproducts. Electronic waste may also be a useful source of REE.

Since REE minerals often contain uranium and thorium, extraction and processing of REE minerals leave behind radioactive slurries. In the process of extracting REE from clays in small, rural deposits in south China, many of which are illegal, large quantities of acids are used, which ultimately end up in groundwater. Processing of polymetallic REE ores of Baotou in Inner Mongolia has severely polluted the environment around the deposit.

The financial services company, Goldman Sachs, in their report of 4 May 2011, project that the market should come back into balance and prices should soften somewhat beyond 2013 as new mines come on line.

The University of Tokyo and the Japan Agency for Marine Earth Science and Technology is reported to have found vast deposits of rare earths (~100 Billion tonnes – about a thousand times more than the resources on land) on the seabed of the Pacific Ocean covering an area of 28 million km^2.

3.2.3 GOLD

Gold is not only a jewellery metal, but more importantly, it is a monetary metal. Not only individuals but also countries hoard gold in lieu of money. Gold is presently priced at USD 1500/oz, as against USD 300/oz a few decades back. A surprising development has been that the price of silver is rising at a higher rate than that of gold.

Gold occurs as (i) free gold ores, (ii) gold with iron sulphides, (iii) gold with arsenic and/or antimony minerals (e.g. arsenopyrite), (iv) gold tellurides, (v) gold with copper porphyries, (vi) gold, lead and zinc minerals, (vii) gold with carbonaceous minerals. The crustal abundance of gold is 5 ppb (parts per billion). Its concentration coefficient (factor of concentration of crustal abundance needed to form an economic deposit) is 2000. Mantle-derived mafic-ultramafic rocks have the highest gold contents (upto 10 ppb). In the endogenous environments, gold may be mobilized and transported in the form of thiosulphate complexes $[Au(S_2O_3)]^{3-}$ which result in coarsely crystalline gold alloyed with 50–75% Ag, and chloride complexes ($AuCl_2^{2-}$ and $AuCl_2^-$) which tend to be reprecipitated with concretionary Fe oxides alloyed with <0.5 wt.% of Ag. In the exogenous environments, Au may be transported as Au-humate complexes, and may end up as very fine, lateritic gold. Alluvial placers are of considerable economic importance.

Gold particles may range in size from dispersed (upto 10 μm), small (upto 0.1 mm), medium (upto 1 mm) and large (upto 5 mm). Pure gold is said to have fineness of 1,000 (or 24 Karats). The generally lower fineness of gold in the greenstones (600–900) is attributed to Au–Ag alloy (electrum).

Gold occurs in a large variety of environments (Hutchinson, 1987).

Since pyrite is a precursor for mineralisation of metals, such as gold, it follows that the higher the content of pyrite in the host rock, the greater the possibility of gold mineralisation. A 1:1 correlation has been found between the pyrite content of the host rock and its Au content.

Mössbauer spectroscopy and Raman spectroscopy have emerged as useful tools in looking for gold (Ferrow, 2001). Mössbauer spectroscopy (MES) is useful in determining the valence state, coordination number, spin state, magnetic properties and structure of the minerals. For instance, the MES spectra of Au-poor pyrite samples are characterized by low-spin doublet, while Au-rich samples contain an additional magnetic sextets (produced by the oxidation of pyrite substrate during the simultaneous reduction and sorption of Au). The Raman spectrum of Au-rich quartz is markedly different from the spectrum for Au-poor quartz – the intensity of Raman emission for Si-O-Si and for crystal lattice vibrations is higher for Au-rich quartz than for Au-poor quartz

The most serious problem that hindered the prospecting for gold has been the great difficulty and expense in determining the gold content *in situ* at ppm and sub-ppm levels. The usual practice has been to pan for gold (which requires water) and look for grains of gold visually. Fire assay of gold is accurate, but it cannot easily be done in the field.

Portable XRF devices are now available for the geologist to check the ore grade in the drill core, or this could be done automatically. Niton's new XL-500 Prospector can assay ore samples directly *in situ* (rock face or drill core). It is a single-piece, hand-held analyzer weighing only one kg, including the battery. Typical *in situ* measurements range from 30–60 seconds, and 500–1,000 measurements can be made per day. About 1,000 measurements can be stored in the instrument internally, and can be downloaded as needed for mapping, grade control and other kinds of analyses. Niton also markets a special device for precious metals (called Precious Metals Analyser), for the analysis of Au, Ag, Pt, Rh, Ru, Ir, Pd, Cu, Zn, Ni, Co and Fe in ores, and fire assay can be avoided. Details about Niton instruments can be had from www.niton.com.

McNulty (2001) gave an excellent update on cyanidation.

The cyanidation process was patented in U.K. on October 19, 1887 by J.S. MacArthur and two brothers, W., and R.W. Forrest. Cyanidation changed forever the economics of gold industry. For instance, the application of cyanidation process in the Rand goldfields of South Africa, led to a thousand-fold increase in gold production in a matter of three years – from 300 oz (9.33 kg) in 1890 to 300,000 oz (9,331 kg) in 1893. During the past 20 years, cyanidation accounted for about 92% of the total world production of gold. Cyanidation has the following advantages: (i) it requires only dilute solutions containing typically 300–1,000 ppm (0.3–1.0 g/L) of sodium cyanide, (ii) the pH range used (9.5–11.5) is such that only gold and silver get mobilized, and (iii) it is simple to operate and control.

Heap cyanidation of low grade ores has proved to be efficient and inexpensive, and is extensively used all over the world. The great advantage of cyanide heap leaching is that there would be no discharging of process solutions, and minimum recycling of water. Treatment and discharge of process solutions would not be needed during the operation. In effect, there would be a single, permanent, large heap leach pad. Percolation of pregnant cyanide solutions downwards through hundreds of metres of leached ore can take place, without the solutions undergoing chemical change. On the other hand, other lixiviants (such as, sodium hypochlorite stabilized by sodium chloride, bromine stabilized by sodium bromide, ammonium thiosulphate stabilized by ammonia, and catalyzed by cupric ion, and acidic thiourea) require rigorous control of pH and Eh, and there is always the possibility of side reactions and precipitation of gold.

Cyanidation may be used to extract gold from almost any kind of gold ore. It may be done in different ways: (i) leaching of the run-of-mine or crushed ore, (ii) vat leaching, and (iii) leaching of ground ore, flotation concentrate, etc. in agitated tanks. Activated carbon is being increasingly used to recover gold dissolved by cyanide in ore pulps (carbon-in-pulp process) or in clear pregnant solutions (carbon-in-column process).

The methods of treatment applicable for different kinds of gold associations are summarized by Weiss (*Mineral Processing Handbook*, SME, New York, 1985).

Degussa-Hûls of Germany has developed the following proprietary technologies which has the effect of making the gold recovery more efficient, while at the same time ensuring minimum or nil adverse impact on the environment: PAL – Peroxide-assisted leach to increase gold recovery, CCS – Cyanide control system to optimize cyanide consumption, DETOX – Cyanide detoxification technology to meet the environmental standards. From the environmental point of view, the peroxide-based detoxification technology is most relevant. The detoxification technology can be custom-made for any kind of mining effluent.

Gold was discovered in Boliden, Sweden in 1924. Boliden is currently using INCO SO_2/Air technology (Robbins et al., 2001) under the extremely stringent environmental constraints of discharge of cyanide prescribed by the environmental court: (1) the amount of cyanide and hydrogen cyanide (as CN) should not exceed 5 mg/m^3, (2) the content of total cyanide (as CN) in the processed slurry from the cyanide destruction process may not exceed 2 mg/L over 14 days, and (3) the content of cyanide in the discharge from the clarification pond may as a guiding value for free cyanide not to exceed 0.5 mg/L as a monthly mean (for further details, see Lindstrom et al., 2001, p. 442).

3.2.4 ALUMINIUM

Aluminium is the third most abundant element in the earth's crust. Its principal ore is bauxite, which is composed largely of gibbsite [$Al(OH)_3$] with 65.4% Al_2O_3, boehmite [γ-AlOOH] and diaspore [α-AlOOH)], with 85% Al_2O_3. For purposes of aluminium industry, bauxite ore should preferably contain not less than 45% Al_2O_3, and not more than 20% Fe_2O_3, and 3–5% combined silica. Formation of bauxite is facilitated by a combination of high rainfall (>1,000 mm), high temperature (>20°C), intense leaching, strongly oxidizing environment, subdued topography, long duration of weathering and chemically unstable rock. World bauxite resources are estimated at between 55 and 75 billion tonnes. Because of the large known and potential reserves, there is not much anxiety about depletion. Major uses of aluminium are in the transportation industry, container and packaging industries, building and construction, and electrical and consumer durables. What proportion of aluminium is used for what purpose varies greatly from country to country. Aluminium substitutes for copper in electrical industry, as aluminium is cheaper and lighter than copper. For structural and transportation purposes, aluminium can substitute for steel, and magnesium and titanium alloys. Aluminium can replace wood and steel in construction composites, and glass, plastics and paper in the packaging industry.

Alumina is produced from bauxite, and primary aluminium metal is produced from alumina. World production of bauxite went up to 205 million tonnes in 2008 (up by 2% from 2007 figure). Most of the world's production of bauxite came from Australia (61.4 Mt – million tonnes), China (35 Mt), Brazil (22 Mt), India (21.2 Mt), and Jamaica (14 Mt). The global alumina production was 81.6 Mt in 2008. The price of alumina has been oscillating widely from USD 350–450/t, and dropped steeply to $210–240/t in the later part of 2008 because of recession. The price of primary aluminium in LME was $0.877/lb in Aug. 2009.

In the case of some Developing countries, bauxite and alumina constitute major export items. For instance, exports of bauxite and alumina accounted for more than 56% in 1995 in the case of Guinea. The corresponding figure was 44% in the case of Jamaica.

Both primary aluminium production (from bauxite) and secondary aluminium production (from scrap) have been going up steadily in the previous decades. There has, however, been a decline in production because of the recession that started in 2007. Secondary aluminium production accounts for more than 25% of the primary production. Six countries (United States, Japan, Italy, Germany, United Kingdom, and France) account for approximately 85 per cent of the world's secondary aluminium production. Apart from environmental considerations, the major incentive for recycling is energy saving – production of aluminium from scrap involves 90% less energy than primary production. Also, a recycling plant costs about one-tenth of a primary smelter complex.

Used beverage cans (UBC) constitute the most important source of scrap. The recycling rate of aluminium cans is more than 70% in Japan, about 66 % in USA, 31% in UK, and so on. Some old scrap is also recovered in the form of die casts, particularly in automotive industry.

As bauxite occurs at the surface, it is often recovered by opencast mining. This disturbs the land surface. A good practice is to restore the topsoil after a mine is closed. Still, the restored soil is usually less capable of retaining water than before. Alumina production is associated with the emission of dust and corrosive materials and large amounts of "red mud". Fluoride emissions arise from smelting operations (processing of alumina into aluminium metal). Technologies are available for the minimization of environmental effects.

3.2.5 COPPER

Copper is polymagmatogenic (liquation deposits, post-volcanic massive sulphide deposits, and post-magmatic hydrothermal ore deposits). Supergene enrichment of copper takes place under exogenous conditions. Copper also gets concentrated in the "sabkha" environment.

Copper has two unique attributes – it has been in use by man for more than 10,000 years, and more copper is recovered and put back in use than newly mined ore. Copper is used extensively in the electrical industry (electrical power cables, data cables and electrical equipment), automobile radiators, cooling and refrigeration tubing, heat exchangers, artillery shells, small arms ammunition, utensils, etc. Fibre optics are substituting for copper significantly, while the use of copper in the fax machines, telecommunication equipment continues to increase.

Drinking water with high concentration of copper may induce diarrhea, stomach cramps, and nausea. Long-term exposure of high levels of copper in food and water can cause liver and kidney damage. Chronic toxic effects in the case of animals may manifest themselves in the form of lower fertility, shortened lifespan, and change in appearance in the case of animals, and lower growth rates in the case of plants. For these reasons, one has to be careful in using copper alloys in the water distribution networks.

World reserves of copper are 490 million tonnes, with reserve base of 940 million tonnes. World production of copper in 2007 was 15.6 million tonnes. The principal producers of copper (production expressed in $\times$ 1,000 t in 2007) are: Chile – 5,700; USA – 1,190, Peru – 1,200; Australia – 860; Indonesia – 780; Zambia – 530; Russia – 730; Kazhkstan – 450; and Canada – 585.

In the last decade, the price of copper was the highest ($3.985/lb) in May 2006. As a consequence of recession, the price dropped to $1.30/lb by the end of Dec. 2008. It jumped to $9.409/lb in Dec. 2010.

Beverage cans which are the main source of aluminium scrap, have a lifetime of a few weeks. As against this, products which use copper have a life time of 10 to 25 years. This means that copper scrap comes into the market at a slower rate.

Copper plays a crucial role in the politics and economy of countries like Chile and Zambia. President Allende of Chile was assassinated because of his leftist policies. Zambia has virtually a one-commodity economy. About 90% of the national income is derived from copper and cobalt. When the international prices of copper and cobalt are high, people are happy, and when their prices are down, there is social unrest.

3.2.6 LEAD

Galena (PbS) is the most common ore of lead. In nature, lead and zinc occur together. Pb is transported in the form of complexes, such as ($PbCl^+$) and $Pb(HS)_2$. In the exogenous conditions, Pb and Zn sulphides are oxidized to sulphates. Zn sulphate is soluble and mobile, whereas lead sulphate is insoluble and immobile.

Estimated world reserves of lead are 65 million tonnes, while the reserve base and identified sources are 120 million tonnes, and 1.5 billion tonnes respectively.

World mine production of lead in the form of concentrates is 2,950,000 t in 2008. The principal producers of lead are Australia – 694,000 t, Canada – 150,000 t, China – 660,000 t, Mexico – 140,000 t, Peru – 308,000 t, and USA – 460,000 t.

Recycling accounts for about half of the world's supply of lead. Spent lead-acid storage batteries constitute the principal source of recycled lead. In Canada, 94% of lead-acid storage batteries from motor vehicles are recycled. Energy required for the production of secondary lead is about 25% of that primary lead.

Lead finds its most important use in lead-acid batteries. Such use is 60% worldwide and is as high as 88% in USA. Thus, lead use is closely linked to automobile production. Other uses of lead are blocking x-ray and gamma-radiation in hospitals and shielding in nuclear installations. At one time, lead in the form of lead tetraethyl was used as an anti-knock additive in gasoline. Lead in the automobile exhausts polluted land, soil, water and air. On the basis of isotopic signatures of lead in the Antarctic ice cores, Dr. C.C. Patterson (who was the mentor of the author in California Institute of Technology, Pasadena, USA), convincingly demonstrated that the build-up of lead in the ice cores is traceable to lead emissions in the automobile exhausts. The powerful lead lobby in USA tried to get Patterson removed from his job. It should be said to the great glory of Caltech that they stood firm behind Patterson. Patterson's campaign ultimately led to the banning of the use of lead tetra ethyl as an anti-knock constituent, globally.

The soil near lead smelters, or lead-using factories gets contaminated with lead. When children play in the dust, and some times accidentally eat it, lead gets ingested in the child. The important issue here is the dispersive nature of lead. Lead poisoning damages the nervous system of children. It reduces fertility and causes brain and kidney damage in adults. Lead is also highly toxic to aquatic life.

When the adverse health consequences of lead ingestion got understood, the use of lead for some uses have been drastically reduced. Lead has been replaced by aluminium, plastic and tin in packaging industry. Lead is now confined to uses where it is not dispersed, such as shielding.

Chapter 3.3

Control technologies for minimizing the environmental impact of mining

Though mining has severe environmental impact, there is no way to avoid it. Wisdom lies in designing ways and means of carrying on mining with minimal adverse impact. The following control technologies are available to minimize the environmental impact of mining: Acid Mine Drainage, Tailings Disposal, Dust Control, Low-waste Technologies, Treatment of wastewater, Subsidence, Noise and Vibration, and Mine Closure.

3.3.1 ACID MINE DRAINAGE

Acid Mine Drainage (AMD) gets generated due to the oxidative dissolution of the iron-containing sulphide minerals, such as pyrite. Both purely chemical reactions as well as microbially catalyzed reactions are involved. AMD may arise from the mining of coal, lignite, metallic sulphides, uranium, etc. Under oxidizing conditions, and in the presence of catalytic bacteria, such as *Thiobacillus ferrooxidans*, sulphides are oxidized into sulphuric acid, as per the following equation:

$$\begin{aligned} 4FeS_2 + 15O_2 + 2H_2O &= 2Fe_2(SO_4)_3 + 2H_2SO_4 \\ \text{Pyrite} + \text{Oxygen} + \text{Water} &= \text{Iron sulphate} + \text{Sulphuric acid} \end{aligned} \tag{3.3.1}$$

Surface runoff and groundwater seepages associated with waste piles tend to be highly acidic, and corrosive, and contain high concentrations of iron, aluminum, manganese, copper, lead, nickel and zinc in solution. The discharge of such waters into streams destroys the aquatic life, and the stream water is rendered non-potable.

An understanding of the physical, chemical and biological processes that lead to the production of AMD is necessary for the following purposes: (1) to minimize the production of AMD, (2) to dispose of AMD from the operating mines or for the decommissioning of waste piles, as required by law, and (3) to ameliorate AMD to allow it to be used for beneficial purposes, such as irrigation, industrial and domestic purposes.

Irrespective of whatever technology is used to mitigate the problem of acid mine drainage, it is necessary to study the focuses of oxidation and flow-pattern of waters in the mine, identification of sources of acid mine water, and the pattern of spreading of mine water. Whole-rock analyses and leaching tests can be used to predict the nature and extent of AMD that could develop in a given mine or from a waste pile. Rock samples are leached with water, and the leachate is analyzed for parameters, which indicate the pathways of weathering, namely, pH, specific conductance and sulphate.

The mineralogical and chemical composition of the rock, the pyrite content and the presence or absence of calcareous material are the determining factors. Several countries have prescribed the allowable concentrations in mine effluents: pH: 7, SS (Suspended Solids): 30 mg/l; BOD (Biochemical Oxygen Demand): 30 mg/l; Pb: 0.2 mg/l; Fe: 0.1 mg/l; Cu: 0.1 mg/l. In other words, the mining companies are expected to treat the mine water in such a manner that the discharge stays within the prescribed limits.

In many countries, a company responsible for noncomplying discharges is issued a Notice of Violation by the Environmental Agency of the government concerned. If the company does not take prompt remedial action, it is penalized. The US Bureau of Mines has developed a simple, low-cost, portable and highly efficient system to neutralize the acid mine drainage on site. The only drawback of the system is that it requires at least 130-kPa water pressure, and may not be able to remove manganese if the iron content is low. Apatite can be used to ameliorate AMD. Apatite is soluble only in acid conditions. So it will act only when the AMD becomes sufficiently acid. The phosphate ion can sequester and precipitate Fe^{3+}, Al^{3+}, Mn^{2+}, etc.

3.3.1.1 Principles of mitigation of Acid Mine Drainage (AMD)

Since the supply of both oxygen and water is necessary for the generation of AMD, an obvious way to prevent the formation of AMD is to block the entry of oxygen and water to the mine or waste pile. This is easier said than done, for the simple reason that Fe (III) that may be present in the partly oxidized waste, could serve as an oxidant and still generate AMD. Also, if the pore water in the mine waste is acidic, the mobility of heavy metals gets strongly increased due to their higher solubility and lower tendency for sorption. Thus, if the waste dump contains buffering substances such as calcite, or if lime is added, the development of acid drainage, and the release of heavy metals could be substantially mitigated.

Figure 3.3.1 is a schematic illustration of the causes of, and remedies for, acid mine drainage from sulphidic mine wastes.

MiMi (1998) and Angelos and Niskanen (2001) described several rehabilitation options for the waste dumps. *The common purpose of all of them is to limit the transport of oxygen and air into the waste*:

1. Changing the chemical properties of the waste (such as, separation of pyrite or addition of a buffering substance, such as lime) or physical properties of the waste (such as, compaction to reduce porosity and permeability). This is expensive.

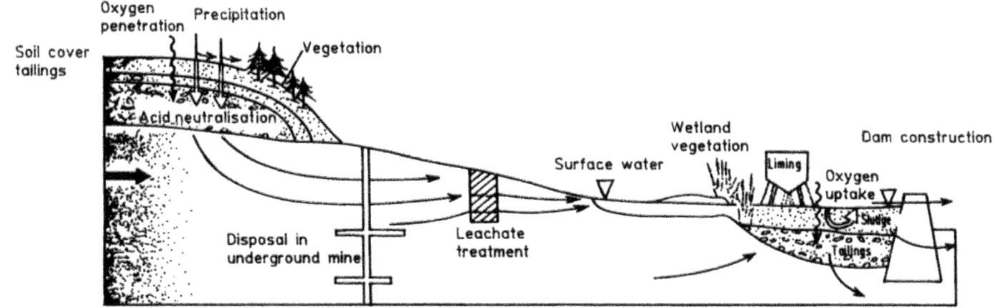

Figure 3.3.1 Schematic outline of the causes of, and remediation for, acid mine drainage
Source: Höglund, 2001, p. 283

2. Flooding of the waste, such that the water table is established above the disposed waste, thereby limiting the transport of oxygen or air into the waste – this is by far the most cost-effective and efficient option, where it possible to implement.
3. Dry covering of the waste.
4. Treatment of the leachate with the objective of reducing the metal concentrations in the water that is discharged from the waste pile.

3.3.1.2 Biologically Supported Water Cover (BSWC)

The water cover has been found to be the most effective in preventing and controlling AMD. This is so because the solubility of oxygen in water is quite low (11 mg/l), and the diffusion rate of oxygen through water is 10,000 times less than through air. The placement of an organic/soil cover between the waste rock and the water cover will not only reduce the oxygen infiltration into the waste rock, but also reduce the metal flux from the waste rock into the water column.

Eriksson et al. (2001, p. 220) evaluated the effectiveness of the water cover at the Stekenjokk tailings pond in northern Sweden using sulphate as conservative tracer for sulphide oxidation mass balance. The water balance for a pond is governed by the equation

$$P + R = O + L + E + \Delta S \qquad (3.3.2)$$

where P is the precipitation on the pond surface (1,187 mm), R is the recharge through surface and subsurface flow (0.9 Mm^3), O is the outlet discharge (1.5 Mm^3), L is the dam leakage ($\sim$0.35 Mm^3/y), E is the potential evaporation from the pond surface (321 mm/y), and ΔS is the net change in the stored volume (which is essentially zero on an annual basis).

After the project was decommissioned, the sulphate concentration in the pond effluent decreased steadily during 1992 to 2000. The pronounced seasonal variations in the sulphate concentrations have been attributed to freezing effect.

Based on the mass balance calculations, it has been found that the resulting oxygen flux through the water cover to the sulphur-rich tailings is less than 1×10^{-10} kg $O_2/m^2/s$. This is an order of magnitude less than the oxygen flux of dry

Table 3.3.1 Types of soil covers and their functions

Cover type	*Primary function*
1. Oxygen diffusion barriers	To limit the transport of oxygen by acting as a barrier against the diffusion of oxygen to the water
2. Oxygen consuming barriers	To limit the transport of oxygen by consuming it before it could reach the waste
3. Low permeability barriers	To limit the transport of oxygen and the formation of leachate by acting as a barrier against the diffusion of oxygen, as well as the infiltration of precipitation
4. Reaction inhibiting barriers	To provide favourable environment to limit reaction rates and metal release

Source: MiMi, 1998

cover which is about 10^{-9} kg $O_2/m^2/s$. The study confirms the effectiveness of the water cover in impeding the formation of ARD. Besides, the water cover cost of USD $2/m^2$ is much cheaper than dry covers which cost USD $12/m^2$.

3.3.1.3 Soil or dry covers

There are four types of soil cover depending upon their primary function (Table 3.3.1).

Experience in Sweden shows that a single layer cover of thickness of 1.0 m results in the reduction of pyrite weathering rate and metal release, in the region of 80–90%. A cover of 2 m of organic waste or lime stabilized sewage sludge can be used as an oxygen-consuming barrier.

Clean (i.e. non-acid generating) wastes from other industries, which are moisture-retaining and oxygen-consuming, can be used as barriers. For instance, in the case of the Luikonlahti mine, magnesite tailings from talc industry, were proposed as soil/dry cover.

Ayres et al. (2002) sought to evaluate three types of dry cover systems, namely, geosynthetic clay liner (GCL), a 0.45 m thick compacted sand-bentonite mixture, and 0.6 m compacted silt/trace clay material, for acid-generating waste rock at Whistle Mine, Ontario, Canada. The waste rock (about 6.4 Mt) is essentially a mafic norite, with an average sulphide content of 3%, and the final contoured surface of the back-filled pit will have a slope of 20%. The parameters affecting the field performance of a sloped cover system are given by MEND (2001). A state-of-art monitoring system has been installed to monitor continuously various climatic parameters, gaseous oxygen/carbon dioxide concentrations, moisture/temperature conditions within the cover and the waste materials, and the quantity of net percolation through each test cover. The observational data obtained from the test plots is made use of to determine the optimum design cover for the waste rock deposit.

3.3.1.4 Passive treatment of Acid Mine Drainage

Passive treatment of AMD makes use of the naturally occurring chemical and biological processes to cleanse the contaminated mine waters, without requiring continuous chemical inputs. The principal passive treatment technologies include constructed wetlands, anoxic limestone drains (ALD), successive alkalinity producing systems

(SAPS), limestone ponds, open limestone channels (OLC), etc. (Angelos & Niskanen, 2001, p. 27). There are numerous permutations and combinations of these techniques.

An open limestone channel followed by a settling pond and filter system has been chosen for the Luikonlahti mine waters. AMD gets ameliorated when acid waters flow through limestone channels, or ditches lined with limestone. The design factors to be taken into consideration are the length of the channel, and the gradient of the channel, which affects the turbulence, and the buildup of coatings. Experience has shown that for channels with slopes of more than 20%, the flow velocities will be sufficient to keep the precipitates in suspension and allow aeration of water.

Filter dams are built with materials having large capacities for the adsorption of heavy metals, such as Zn, Ni and Cu. The optimal absorption takes place at pH levels higher than 6. The purpose of open limestone channels and the settling pond is to remove as much iron as possible, so that the filter dam can take care of heavy metals other than iron.

Gusek (1995) gave a lucid review of the techno-economic aspects of passive treatment of acid rock drainage.

The conventional method of amelioration of acid rock drainage is the liming of the runoff. Liming neutralizes the water and chemically precipitates the metals. However, liming is expensive, leaves behind large quantities of sludge, and has to be continued long after the mine ceased operating. For this reason, much R&D effort has been concentrated in developing low-cost, low-maintenance, passive treatments of AMD. These involve the utilization of vegetation and sediment microbial communities found in wetlands to reduce the acidity and precipitate the metals. The techno-economic viability of the passive treatment is now well established. For instance, the Tennessee Valley Authority (TVA)'s Fabius Mine in Alabama, USA, replaced an earlier lime-treatment plant by a large, passive treatment system. The latter treats 126 l/s (about 2,000 gpm) of coal mine drainage. It has been operating for several years and discharging compliant effluent.

Interestingly, wetlands established for water quality improvement have been found to provide habitat for abundant development of herptofaunal wildlife (Lacki et al., 1992).

It has been known that wetlands are capable of improving the water quality by reducing the contaminants through the precipitation of metal hydroxides, sulphides and carbonates and pH adjustments. Whether these reactions would occur under oxidizing (aerobic) conditions or reducing (anaerobic) conditions would depend on the Eh of the environment, and the chemistries of soil and water. Where natural wetlands are not available, wetlands are constructed. The latter are engineered so as to optimize the biogeochemical processes that take place in the natural wetlands.

Kolbash & Romanovski (1989) gave the design of a constructed wetland. The wetland plants that are most commonly used are *Typha, Schoenoplectus, Phragmites* or *Cyperus.*

The predominant mechanisms by which microorganisms remove soluble metals from solution are as follows: (1) volatilization – whereby microorganisms methylate metals, (2) extracellular precipitation – whereby metals are immobilized by the metabolic products produced by microorganisms. Sulphate-reducing bacteria reduce H_2SO_4 to H_2S, which would readily react with soluble metals to form insoluble metal sulphide minerals, (3) extracellular complexing and subsequent application – whereby chelating agents (known as siderophores) synthesized by microorganisms have a high binding efficiency for some metals, resulting in the generation of metal-binding polymers, (4) binding to bacterial, fungal and algal cell walls, and (5) intra-cellular accumulation (Brierley et al., 1989). Studies made by White and Gadd (1996) showed that the most efficient nutrient regime for bioremediation using sulphate-reducing bacteria required both ethanol as a carbon source and cornsteep as a complex nitrogen source.

Brierley (1990) gave a detailed review of the techniques of bioremediation of metal-contaminated surface and groundwaters. Advances in biotechnology have made it possible to make use of nonliving microorganisms immobilized in polymer matrices to remove low concentrations ($\sim$1 to about 20 mg/l) of heavy metal cations in the presence of high concentrations of alkaline earth metals (Ca^2 and Mg^{2+}) and organic contaminants. The removal process is so effective that the effluent more than satisfies the requirements of U.S. National Drinking Water Standards. Davison (1993) describes a proprietary Lambda Bio-Carb Process which is an *in situ* bioremediation system utilizing site-indigenous, mixatrophic cultures hybridized for maximum effectiveness. Lambda has catalogued about 6,000 microorganisms suitable for the purpose. The system is utilizable in conjunction with wetlands, and is capable of self-adjustment in response to influent changes. It has been successfully used to treat sites contaminated by heavy metals, hydrocarbons, organics, agricultural wastes and other hazardous compounds.

The economics of the passive treatment can be illustrated with a case (Eger & Lapakko, 1989). Drainage from the Dunka mine in the mineralized Duluth complex in northern Minnesota, USA, has increased upto 400 times, the concentration of metals (Ni, Cu, Co and Zn) in the creeks in the proximity. This was naturally unacceptable to Minnesota Pollution Control Agency, and the company concerned had to give an undertaking to achieve the water quality goals. A feasibility study was made of the options for treating 6×10^8 l/y of mine water:

1. A full-scale treatment plant (lime precipitation with reverse osmosis): capital cost: \$8.5 million, and annual operating cost: \$1.2 million.
2. Passive treatment (combining infiltration reduction, alkaline treatment and wetland treatment): capital cost: \$4 million; annual operating cost: \$40,000.

3.3.1.5 In-pit disposal using Sulphate Reducing Bacteria (SRB)

Sulphate Reducing Bacteria (SRB) has been found to be effective in reducing the metal and sulphate concentrations in the mine water. Liquid manure and press-juice from silage were used as nutrients. SRB was obtained from local lake sediments and was enhanced before application. The costs are very low, but the catch in the technology is that would take some years before the treated water is of quality that would permit discharge into local water ways. If a site is an abandoned one, in a remote area, and the costs have to be kept minimal, this technology may turn out to be appropriate (Angelos & Niskanen, 2001, p. 27).

3.3.1.6 Case history of pyritic uranium tailings sites of Elliot Lake, Canada

As a part of the mine waste management and decommissioning studies, Davé and Paktunc (2001) studied the hydrogeochemistry and mineralogy of the inactive and rehabilitated pyritic uranium tailings at Stanrock and Lower Williams Lake sites related to the Elliott Lake uranium mine, Ontario, Canada. The Stanrock sites holds about 8 Mt of pyritic uranium tailings spread over an area of 71 ha. The water table in the area fluctuated between 0.5 to 2 m, rising nearer to the surface in the central section. The water table goes down by about 2 m during the dry summer and winter months. The Lake Williams site is much smaller (about 2 ha) and contains about 20,000 t of tailings. The tailings contained 0.9 to 6.3% pyrite and 0.07 to 5.3% calcite. The pyrite content generally increased with depth. During 1976–77, limestone amendment was applied to the exposed tailings at the surface. The dry tailings were covered with about 1 m. thick layer of glacial sand/gravel and till, which was then vegetated with agronomic species of gases and legumes. The incoming treated water was discharged into the downstream water pond which also serves as a sludge-settling pond. The site was maintained till 1980, but was left on its own since then. Davé and Paktunc (2001) report that the site supports dense, lush vegetation.

The Stanrock tailings essentially consist of quartz, K-feldspar, muscovite, and pyrite, with small quantities of rutile, La-Ce monazite, chalcopyrite, pyrrhotite and galena. The pyrite content of the tailings varied from 0.1 to 12.4%, depending upon the depth (Davé & Paktunc, 2001, p. 133). As the bulk of the unoxidised material which has high acid generation potential is below the water table, its ability to produce AMD is negligible.

An examination of the geochemical characteristics of shallow groundwater along the central longitudinal direction, of the Stanrock tailings show that, except for one central site, the groundwater is characterized by low pH (between 1.8 and 4), high total acidity (1,000–12,000 mg $CaCO_3$/l), and high concentration of dissolved SO_4 (2,000–14,000 mg/l) and Fe (500–6,000 mg $CaCO_3$/l). At the central site, near the surface water streams (between 381,400 and 381,600), the pH of the groundwater is high (6), and the concentrations of total acidity (about 50 mg $CaCO_3$/l), SO_4 (about 2,200 mg/l), and Fe (about 100 mg/l) are low.

The pH, total acidity, Fe and SO_4 contents of the groundwater in the longitudinal direction of the Lower Williams Lake site are given by Davé & Paktunc (2001, p. 134). Compared to the Stanrock site, the pH of the Williams Lake site is much higher (about 6.0–8.0), and concentrations of total acidity, SO_4, and Fe much lower. The saturated conditions that developed within the tailings substrate increased the pH, and the microbial degradation of the organic matter (caused by vegetative cover in the soil layer), besides increasing the total available groundwater alkalinity. Thus, the overall water quality has improved with time.

Davé & Paktunc (2001) conclude that covering the tailings with a vegetated cover layer, and raising the water table can effectively suppress acid generation.

3.3.1.7 Remediation of acid lakes – Case history from former East Germany

The extensive opencast mining of lignite in the former East Germany, has created a large number of acid lakes in the Lusatian mining district, after the mining was abandoned. The pH range in the mining lakes ranged from 2.6 to 3.8. These acidic water bodies were often toxic because of high metal concentrations. It was concluded that increasing the pH by neutralization measures was the most promising way to reduce the metal concentrations. Figure 3.3.2 illustrates the techniques for the abatement of acidification through in-situ technologies.

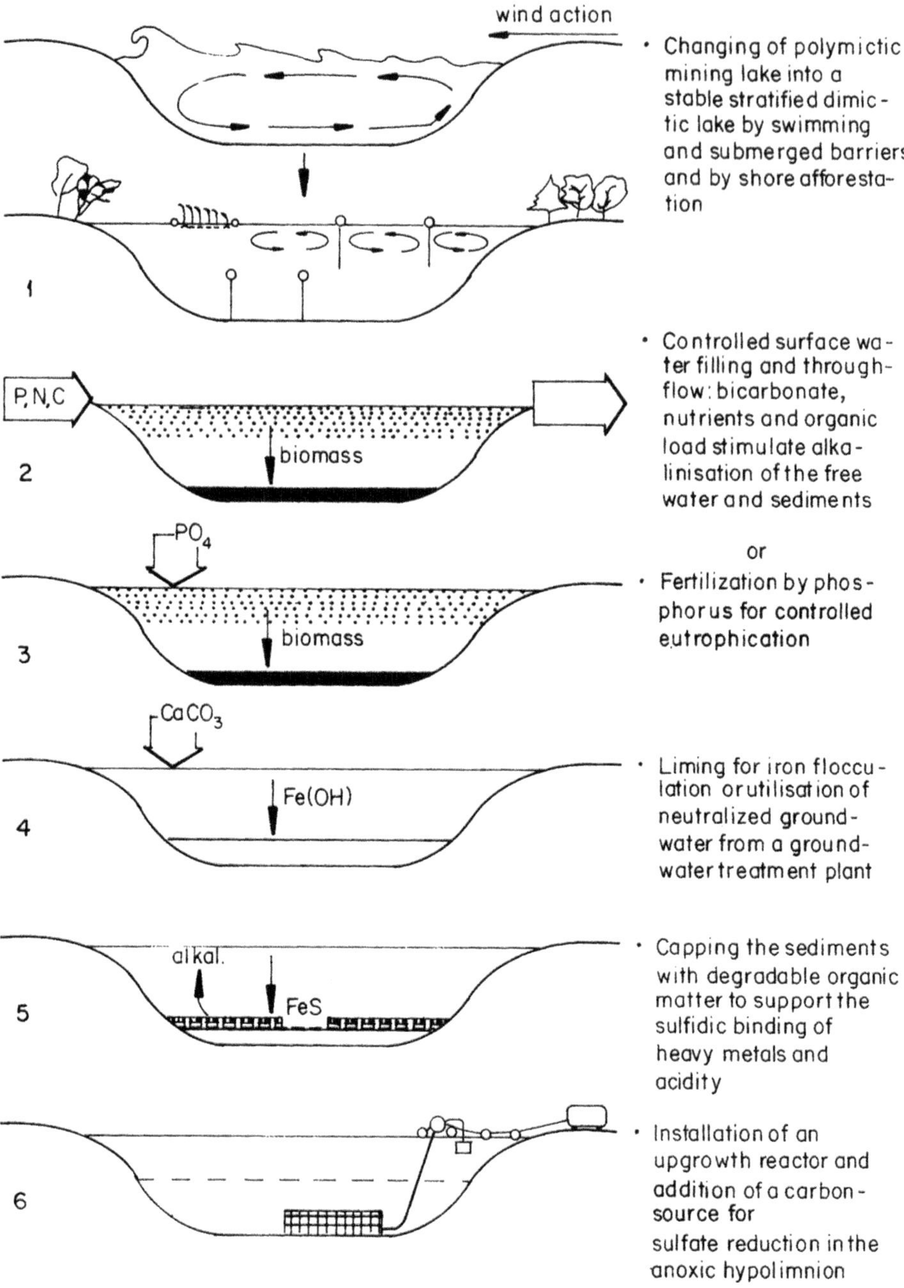

Figure 3.3.2 Abatement of acidification of mining lakes
Source: Klapper and Schultze, 1997, quoted by Stottmeister et al., 1999

3.3.2 TAILINGS DISPOSAL

As stated in the Introduction, the most serious problem facing the mining industry presently is the enormous mass of the mine tailings (about 18 billion m^3/y), which incidentally is the same order as the quantity of sediment discharge into the oceans. As progressively lower grades are worked, the mass of the mine tailings is expected to double in the next 20–30 years. It is not without significance that the failure of the tailings dams figures prominently in the list of major accidents related to mining.

3.3.2.1 Environmental risks from mine tailings

Ellis and Robertson (1999) gave a concise account of the environmental risks from mine tailings:

1. *Chemical contamination*: Tailings may cause acid rock drainage and other undesirable geochemical processes. They may damage the ecosystem and resource use downstream from site. Recovery from a degraded ecosystem is likely to be very slow (of the order of decades or more).
2. *Habitat smothering*: The tailings may smother the living organisms and their habitats. This occurs when the deposition is made at a rate greater than the organisms could cope or grow through the deposits. Recovery of tree growth on land is measured in decades, whereas the recovery underwater may be quicker, i.e. 1–5 years.
3. *Catastrophic system collapse*: Earthquakes or torrential rains may undermine the structural stability of the tailings deposit, and may cause sudden and extensive loss of life and property.
4. *Landform changes*: Tailings may change the landforms and habitats.
5. *Water turbidity and siltation*: The tailings adversely affect the use of rivers and lakes by changing the river channels and flood plains, biological productivity and fisheries resources. Water quality may recover in a matter of days, but the ecosystem consequences may last for years.
6. *Socio-economic changes*: Tailings may cause changes in resource use, and thereby affect the quality of life of the people. The recovery may be complex.

3.3.2.2 Characteristics of tailings

Tailings can be considered as "man-made" soil with properties between those of sand and clay. The grainsize distribution of the coal colliery spoils determine their geotechnical properties, which in there turn influence the design of the tailings deposit.

Vermuelen, Rust and Clayton (2002) summarized the properties of the gold tailings from the literature:

Slurry: "low plasticity, fine, hard and angular rock flour, slurried with process water in a flocculated, slightly alkaline state with soluble salts".

Rheology: (study of deformation and flow of matter): The rheological characteristics of the mine tailings are intermediate between a Bingham plastic and a Newtonian fluid.

Mineralogy: Quartz is by far the most abundant mineral, with small quantities of phyllosilicates, pyrite and other sulphides. Specific gravity ranges between 2.5 and 3.0. Oxidation of pyrite (FeS_2) leads to the production of sulphuric acid, and the acidification of the tailings water. The low pH water is capable of leaching toxic heavy elements from the tailings.

Grading: Generally of silt size range, with small percentages of fine sand and clay-sized particles.

Particle shape and texture: The coarser or sand fraction of the tailings range in shape from very angular to sub-rounded, whereas the fines are invariably angular, with very sharp edges. The surface textures are described as "harsh".

3.3.2.3 Methods of tailings disposal

Tailings from the beneficiation plants in the case of non-ferrous metals are generally in the form of slurry, which is discharged into specially constructed containment structures. The various tailing disposal methods are summarized as follows (source: UNEP Tech. Report no. 5, 1991):

1. *Subaqueous discharge into the tailings ponds*: The great advantage of the method is that transfer of oxygen to the tailings is impeded, thereby inhibiting acid production from the tailings. The disadvantage is that sub-aerial discharge involves lower *in-situ* densities.
2. *Layered methods of tailings disposal*: The tailings slurry is deposited in thin layers of uniform thickness (10–150 mm). The slope of the deposited slurry layers may vary from 0.5 to 1.0%, depending upon the characteristics of the slurry. The fresh tailings are allowed to settle down, and dry – this may take several hours or a few days. The consolidated, gently sloping mass of tailings composed of uniform layers, formed in this manner, will greatly facilitate the de-commissioning of the waste disposal site. Site preparation for this method involves high capital costs.
3. *Thickened tailings disposal*: Tailings may be deposited in the form of cones, with slopes ranging from 2 to 8%, if the tailings slurry is thickened and discharged from spigotting points within the tailings disposal area. For a slope of 6%, the solids content has to be in the range of 55 to 75%. This method allows larger volumes of tailings to be disposed in a small area, but the thickening process involves high operational costs.
4. *Tailings disposal behind a dam*: The dam is usually constructed from the coarse fraction of the tailings. The tailings slurry can be discharged from a single point through a series of spigots. Discharge of tailings through cyclones allows the sands from the tailings to be mechanically separated and used for dam construction. If the beneficiation process involved fine grinding of the ore, the tailings would not contain any coarse materials suitable for dam construction. In such a situation, the dam has to be built with borrowed material. In one sense, this is advantageous in that the quality of the materials and their properties can be controlled, but it carries with it extra operational costs for the excavation and placement of dam material.

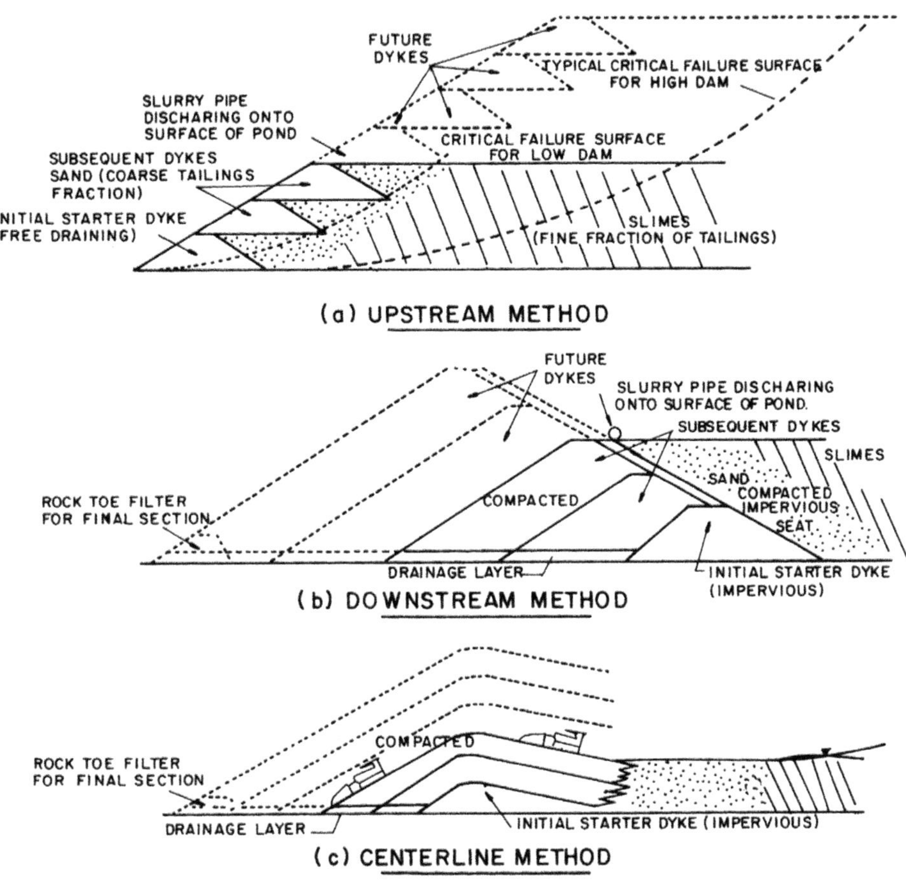

Figure 3.3.3 Different methods of construction of tailings dam
Source: UNEP Tech. Rept., 5, 1991, p. 37

3.3.2.4 Methods of construction of tailings dams

Figure 3.3.3 shows different methods of constructing the tailing dams.

1. *Upstream methods*: Though this method was used extensively in the past, it is not much in vogue now. A typical upstream section incorporates slimes fraction in the dam structure, and the resulting heterogeneous dam is susceptible to failure, particularly under seismic conditions.
2. *Downstream methods*: Under this method, the dam is built of coarse tailings. When cycloned sand is used, the slope will be adequate, the sand will be properly drained, and the dam will retain its stability even under seismic conditions. The drawback of this method is that a large quantity of sand is required.
3. *Centreline method*: As in the case of the downstream method, the centerline is built with coarse fraction of tailings, but with the dam centre line being maintained in the same vertical plane as the dam height is increased. Downstream of the centre

line, the dam will have the same characteristics as the downstream method, and therefore tends to be stable.

Decant towers, siphon systems and barge-mounted pumps are used to release the supernatant water in the tailings disposal facilities.

A tailings dam fails when the peak flow exceeds the hydraulic capacity of the spillways, decants and diversions. The resulting liquefaction and the release of the stored tailings can cause great damage to life, property and the environment. It is therefore essential that the design of the impoundment provides for the spillway and decant structures to take care of the statistical probability of the rainfall/runoff event occurring once in thousand years. If the risks are very high, it is better not to install the tailings disposal facilities where such risks could occur.

3.3.2.5 Disposal of coal mine tailings

Coal mining industry produces enormous quantities of wastes, considering that the world production of coal is about 7,000 million tonnes in 2008. Apart from the solid wastes (shale rock, dolerite, burnt coal, etc.) produced in the course of coal mining, coal preparation plants produce large quantities of coarse and fine particles, and contaminated water.

The usual practice is to impound the tailings and slurry in the lagoons. As breaching of the lagoons constitutes some of the most serious environmental disasters associated with mining, great care should be taken in the design of the lagoons. The stability of, and control of seepage from, the lagoons can be ensured by keeping in mind the following design parameters: (1) Construction of the lagoon at or below the ground level, (2) the banks should have a slope of 34° on the lagoon side, and 26.5° on the outer side, (3) there should be a toe drain to allow the water table drawdown below the outer slope, (4) there should be a free board (water surface to the crest of the lagoon bank) of at least one metre, (5) the inner surface of the bank should be capable of withstanding erosion due to wave action, (6) there should be provision for drawing off the supernatant, as also rainwater due to abnormally heavy rainfall, etc. (Chadwick et al., 1987, pp. 126–127). The supernatant water from the lagoons may be either recycled, or discharged into the natural waterways after treatment.

In some countries, such as Germany and Poland, colliery spoils are tipped with domestic wastes (Skarzynska & Michalski, 1999, p. 189).

Untreated coal colliery spoils are used as bulk material for various types of earth works, such as embankments of roadways, railways, rivers and dams. The water contained in the dumps and infiltrating through the dumps, is likely to be contaminated. In order to prevent such contaminated water from polluting bodies of freshwater, polyethylene sheeting or clay screens should be incorporated with drainage to remove the infiltrating water.

3.3.2.6 Disposal of gold mine tailings

The mode of disposal of tailings is illustrated with the case history of gold tailings in South Africa (Vermuelen, Rust & Clayton, 2002).

The Witwatersrand Goldfields occurring in the Johannesburg area in South Africa are the largest deposits of gold in the world. Despite the reduction in production during the last decade, South Africa has been and continues to be the largest producer of gold in the world. The mineral composition of the typical Witwatersrand gold reef is as follows: Quartz: 70–90%, Phyllosilicates (clays): 10–30%, Pyrites: 3–4%, Other sulphides: 1–2%, Grains of primary minerals: 1–2%, Gold: about 45 ppm.

The mean and the range of the mineral abundances in the Witwatersrand gold tailings (as determined by EDS and XRD analyses) closely follow those of the gold reef (Vermuelen, Rust & Clayton, 2002): Quartz: 75% (59–83%); Muscovite: 8% (7–19%); Pyrophyllite: 5% (1–17%); Illite: 5% (3–11%), with small percentages of clinochlore, kaolinite and pyrite. The specific gravity of the tailings is 2.74 Mg/m^3. The coarse particles (sands) are almost wholly composed of quartz. The slimes also have a preponderence of quartz, with significant amounts of pyrophyllite, muscovite, illite, kaolinite and pyrite.

Grading study of the tailings shows that about 2% are coarser than 200 μm (limit of fine sand), 10% finer than 2 μm (clay-sized), and at least 50% slimes. The median particle size (D50) ranged between 6 and 60 μm. The behaviour of the tailings is largely dependent upon the fines fraction.

The shape of the particles (such as angularity) is as important as the size in determining the engineering behaviour of the tailings. Under load, the angular corners break and crush and angular particles tend to resist displacement, whereas more rounded particles are less resistant to displacement, but may be less likely to get crushed depending upon the surface texture. The coarser tailings sands are characterized by highly angular to subrounded, bulky but flattened particles. In contrast, the slimes, which are composed of clay minerals, consist of thin and plate-like particles. The engineering behaviour of slimes is akin to that of clay of intermediate plasticity. Also, the slimes can be flocculated, indicating the effects of the surface forces.

Electron micrographs of coarser or sand tailings show either smooth surfaces or rough or irregular surfaces. Some particles show the typical conchoidal fracture of quartz. Sand particles with irregular surfaces may have developed as a consequence of fines attaching themselves to these surfaces. Slime particles have invariably very smooth and flat surfaces.

In South Africa, a typical gold tailings impoundment has two sections: the embankment or daywall and the interior or nightpan. The daywall is meant to provide sufficient freeboard to retain the accumulated water from the deposited tailings, besides taking care of storm water when it rains heavily. The daywall has a number of sections or paddocks. A delivery station fills each paddock starting with the midpoint. When the pulp is delivered into a daywall paddock during the dayshift, it gets distributed by gravity, with the excess or supernatant water being decanted into the nightpan. Since the pulp depth has to be closely controlled, the filling of the paddocks is invariably done during the daytime – hence the name *daywall*. During the night, the tailings are discharged into the nightpan, but this is done from delivery stations located inside the daywall. The next day, the clear supernatant water is pumped out or drawn off by penstock decant. A natural beach develops between the delivery point and the pond from which supernatant water is decanted. The paddocks are filled according to a cyclic system, to allow sufficient time for the desiccation, consolidation and densification of the embankment material (Vermuelen, Rust & Clayton, 2002, p. 47).

As should be expected, the impoundment facility at Mizpah, which takes care of the gold tailings of the Vaal River Operations west of Johannesburg, is very large. It was commissioned in 1993 and receives about 5,000 t of tailings per day. The dam was designed for a final height of 60 m, with a total surface area of 165 ha. The average rate of rise is 2.4 m per year, and one depositional cycle takes about ten days. The dam is of the upstream daywall – nightpan system.

The soil-forming processes on the gold tailings lead to a pronounced vertical layering, with coarse layers ("sands") alternating with fine layers ("slimes"). Besides, there may also be horizontal variability depending upon the properties of the slurry and the depositional programme. In South Africa, the gold tailings impoundments are usually constructed using the daywall-nightpan paddock system. Generally, the more competent coarser material tends to get deposited near the embankment, with finer material in the central part of the impoundment. However, it has been observed in practice that a significant amount of fines are trapped in the daywall and settle on the beach.

3.3.2.7 Use of paste technologies in tailings disposal

Environmental and economic considerations demand a reduction in the volumes and sizes of the tailings dams. The dayfall – nightpan system is a case of manipulating the environment to accommodate the tailings. As against this, thickened tailings and paste technologies are being increasingly used to design the tailings disposal sites to suit the surrounding environment. Paste production process is suitable for solids volume of 40 to 55%. Equipment is commercially available (e.g. GL&V) to produce paste consistency material from mill tailings, for purposes of backfill. Tailings are introduced with a flocculent into a feedwell. A mechanical rake and helix concentrates the solid particles by inductive circulation in a compression zone while preventing ratholing. The paste-like material can be withdrawn from the bottom, and a clear liquid overflows into the launder at the top of the tank.

The paste consistency material can be stored indefinitely.

NALCO has developed a new line of polymers for water clarification. The strong points of OPTIMER® mineral processing flocculent are high settling rates of suspended solids, superior overflow clarification, and maximum underflow compaction and pumpability. The Nalco 98DF063 is a liquid polymer system custom-made for the flocculation of red mud in the bauxite industry. The Nalco patented TRASAR technology has four components of tracer chemicals, control equipment, diagnostic capabilities and on-site services. The system provides not only protection against scale formation but the inert tracer allows continuous diagnostic monitoring of the system volume, mixing studies, system flow, residence time/water travel time and environmental compliance. Such a system not only helps in the efficient operation of the process, but also allows remedial action to be taken before a problem becomes serious.

Sofra & Boger (2002, p. 132) show the relationship between shear rate and shear stress for Newtonian and non-Newtonian fluids. The inelastic Newtonian fluids exhibit a linear relationship between the applied shear stress and shear rate. The flow in the case of the Newtonian fluids gets initiated as soon as the shear stress is applied. On the other hand, concentrated mineral tailings often exhibit a non-Newtonian behaviour, in that they are characterized by an yield stress (τ_y). Thus, flow will occur in non-Newtonian fluids only after the critical stress is exceeded. Sofra & Boger (2002, p. 133) found that (1) the yield stress is a function of concentration for a number of industrial slurries, (2) though there is variation in yield stress for different mineral tailings, all materials exhibit an exponential rise in yield stress with concentration.

It is hence necessary to have a thorough understanding of the rheological characteristics of the tailings for the planning, design, operation and optimization of the dry disposal systems including dry stacking, thickened tailings disposal and paste backfill (Sofra & Boger, 2002).

A number of parameters have to be manipulated in order that the tailings to be deposited have the desired rheological characteristics: (1) material parameters, such as solids concentration, viscosity and yield stress, and (2) operational parameters, such as the flow rate (which is determined by the pipe diameter and throughput), and the shear to which the tailings are subjected (which depends upon the pump type, flow regime, etc.). A tailings management system which is safe, environmentally-responsible and cost-effective, can be designed on the basis of the study of (1) the concentrations required to achieve the optimum spreading and drying of the deposited tailings, (2) the optimum conditions for pipeline transport, and (3) the optimum dewatering of the slurry (Sofra & Boger, 2002).

The higher the angle, the greater the volume that can be filled per unit surface area for a constant dam height. A smaller surface area of a tailings disposal site has a number of benefits. If the tailings are capable of generating AMD, multi-layer capping may be needed, and the expenses for capping and decommissioning could be considerable. Also, a smaller area means less evaporation, which may be an important consideration in areas of water scarcity. Less water in the site improves the dam safety – many cases of dam failure are attributable to the presence of large amounts of water in the disposal area.

Thickened tailings, when discharged as solids concentrations by mass of about 60%, are self-supporting enough to attain a slope of about 5°. Slope angles as high as 10° could be attained when pastes are used (a paste is a tailings mixture with extremely high solids concentration). Pastes have been used for backfilling in underground mines with underground transportation in pipelines and boreholes, and for the surface disposal of tailings of base metals and gold.

A paste should contain more than 50% solids, in order to avoid drainage of water and segregation of particles. A small amount of water means less use of a binder for attaining high strength. The average particle size varies from 20 to 100 μm, with more than 15% of the particles being smaller than 20 μm. In order to attain high solids concentrations, conventional thickeners are used in conjunction with mechanical dewatering techniques. Centrifugal pumps are used for large flow rates and low to moderate working pressures, whereas positive displacement pumps are used for small flow rates and high pressures.

3.3.2.8 Paste technologies in mining backfill

Moellerherm and Martens (2002) gave an account of the use of the tailings as paste backfill in the copper mining industry. Out of the total copper ore production of about 2 billion tonnes in 1998, open pit mining and underground mining accounted for 81% and 19%, respectively. Because of their low costs, block caving and room-and-pillar mining have emerged as the most widely used mining methods.

The share of the backfill techniques in copper mines is as follows: Gravity fill (waste rock): 46%, Hydraulic fill (tailings): 28%, Paste fill (tailings): 16%, and sand fill (sand): 10%. Thus, almost half of the mines (46%) use gravity backfill involving waste rock because of the ready availability of overburden and waste rock due to combined open pit and underground operations.

The parameters of the backfill techniques are summarized in Table 3.3.2.

It may be seen from Table 3.3.2 that the paste backfill has two main advantages: (1) the cavity is filled to the extent of 85%, and (2) it makes use of the tailings from the processing plant, thus reducing the need for surface disposal of the tailings. Thus, the use of backfill techniques has the advantage of minimizing the use of land on the surface, but the disadvantage of higher operating costs because of energy consumption.

3.3.2.9 Underwater placement of mine tailings

The studies made under the Canadian MEND (Mine Environment Neutral Drainage) programme have shown that the placement of tailings underwater can prevent or reduce ARD (Acid Rock Drainage). The tailings disposal under river and lake water is infeasible in most situations as it may adversely affect the productivity of the intensively used ecosystems, and may come into conflict with traditional rights (such as, fishing) of the communities. CCORE (1996) study has shown that after defaunation events, the benthic diversity on the seabed may recover in about one year in the case of fine-grained deposits of muds/silts/clays, and within about five years in the case of coarse-grained deposits such as gravels. Thus deep seabed emerged as a possible site for the placement of mine tailings.

The 1996 Protocol for the 1972 London Dumping Convention allows the dumping in the sea of "*inert, inorganic geological materials*". The submarine placement of tailings is deemed to comply with this convention, because (1) the tailings constitute inorganic, geological material, and (2) they would be inert (i.e. would not be able to generate ARD) under submarine conditions.

Ellis and Robertson (1999) made a detailed study of a number of case histories of underwater placement of tailings.

Table 3.3.2 Parameters of the backfill techniques

Parameter	*Gravity fill*	*Hydraulic fill*	*Paste fill*
Solids content	75–90%	50–75%	85–85%
Water content	7–11%	25–35%	12–25%
Concrete content	3–12%	3–15%	2–5%
Compressive strength	1–20 MPa	0.5–3.0 MPa	1–5 MPa
Density	1.8–2.5 t/m^3	1.4–2.3 t/m^3	2.1–2.35 t/m^3
Fill grade (cavity Utilization)	~80%	~85% (tailings backfill) ~60% sand backfill	~90%
Solid	Overburden, waste rock (coarse grained)	Fine-, and coarse-grained tailings, sand	Fine-, and silt-grained tailings
Examples	Kidd Creek, Norilsk	Neves Corvo, Myra Falls	Louvicourt, Brunswick

Source: Moellerherm & Martens, 2002

Potentially acid generating tailings from the Island Copper Mine of Canada were discharged into a target basin within a fjord at a depth of 100 to 200 m. The bulk of the tailings remained within a basin, and the seabed recovered its biodiversity within 1–2 years. The company used an outfall design, which is now standard for submarine tailings placement. When the tailings pipeline reached the edge of the sea, it discharges into a tank where it is deaerated and mixed with seawater. This has the effect of making the slurry denser and more coherent as it flowed on the seabed. The tailings from the Kitsault Molybdenum mine in Canada were discharged at 50 m depth, with the same outfall design as Island Copper. Here also recovery of moderate successional biodiversity took place in 1 to 2 years, though the species present were not identical to those at nearby reference stations. In the case of the Misima Gold and Silver mine, Papua New Guinea, the tailings were placed in a 1,000 to 1,500 m deep near shore basin in an area of tropical, open coast with coral reefs. The tailings were dispersed to ensure a slow rate of deposition so that the organisms are able to cope with it. Ellis and Robertson (1999) suggested the physical and risk assessments to be made to determine the viability of submarine tailings placement at coastal and island mines.

3.3.2.10 Tailing operations and risk assessment

Tailings and waste disposal which are critical components of mine operation, are associated with risks involving human health, loss of life and property, damage to the ecosystem and biodiversity, etc. It is obviously prudent on the part of a mining company to be aware of risks involved, to what extent they are acceptable, and how to manage the risks at least cost. The risk management process is schematically shown in Figure 3.3.4.

In the case of mine tailings, the hazards are slope failure, contaminated seepage, overtopping due to insufficient dam freeboard, etc. The possibility of a given risk occurring is evaluated qualitatively, ranging from "Very likely" to "Very unlikely", on a scale of, say, 1 to 5.

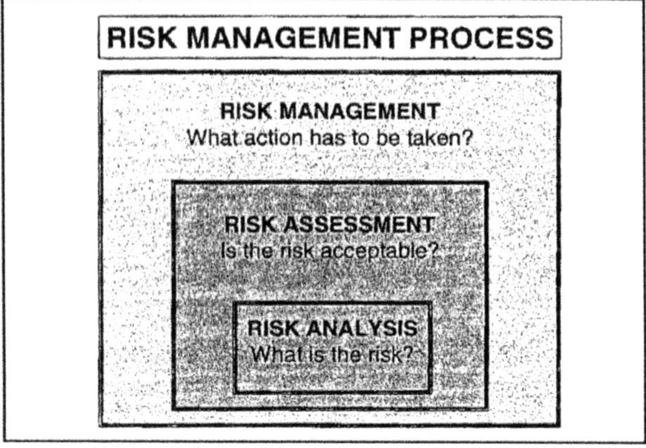

Figure 3.3.4 Risk management process in tailings dam
Source: Alexieva, 2002, p. 296

Risk assessment involves deciding whether the estimated risk is tolerable. Australian National Committee on Large Dams prepared the societal risk criteria curves for dam failure, indicating the limit of tolerability. Risk management has to be ongoing and proactive. Risk management strategy has to be updated when, for instance, the design capacity of the tailings storage facility is increased, or when a new depositional method, or embankment construction method is thought of, and so on.

3.3.3 DUST CONTROL TECHNOLOGIES

Dust is a problem in almost all mineral industries, though the degree of severity of the problem varies from industry to industry. Some generalizations may, however, be made (source: *Mining Mag.*, Sept. 2001, p. 124):

1. Loss of valuable material:Wind erosion from stockpiles may lead to the loss of upto 5% of the stockpiles of (say) coal or mineral concentrates.
2. Environmental problems: Dust can cause air pollution. It can also enter soil and water environments and pollute them.
3. Health hazard: Inhalation of certain kinds of dusts is known to cause diseases, such as, silicosis and pneumoconiosis.
4. Reduced visibility: The haze caused by dust in the air can cause hazardous working conditions for vehicle drivers and plant operators.
5. Explosion/oxidation: Very fine (10–20 μm) combustible particles are liable to explode. Stockpiles of coal can "oxidize" and undergo spontaneous combustion.
6. Machine maintenance: Dust particles can clog machinery parts such as bearings and air filters, and damage them.
7. Capital investment: Greater quantities of dust would require the use of expensive dust control equipment, such as spray bars, pumps and bowsers.

3.3.3.1 Types of dust control techniques

In the case of the iron and steel industry, dust is produced in the process of unloading, storage, recovery and transfer operations involving iron ore, coal, coke, limestone, lime, slag, etc. Dust can be controlled by installing hoods over the conveyor belts which suck in the air and extract dust from it, by smoothing and compacting of coal in the stockyard using a bulldozer, by spraying the stockpiles with water (with the addition of surfactants where available), enclosing the stockyards to prevent dust from being blown away, etc.

Four types of dust control techniques are used in the mineral industries, including the iron and steel industry (UNEP, 1986, pp. 48–49).

1. *Mechanical dust catchers*: These are based on the principle of precipitation of heavy particles by settling (dust catchers) or centrifugal action (cyclones). Mechanical dust produced in the handling of raw materials, particularly in conjunction with blast furnaces.
2. *Electrostatic precipitators*: These consist of electron-emitting electrodes, and electron-collecting electrodes, which are kept at a potential difference (say, 40,000 V). When the dust-laden waste gas circulates at low speed between these

electrodes, the particles of dust are bombarded with electrons. If the particles are sufficiently conducting, they become negatively charged, and get precipitated onto the collecting electrode. From the collecting electrode, the dust particles are either knocked off (dry method) or washed off (wet method).

Electrostatic precipitation is a well-established technique of dust control in the iron and steel industry (main gases in sinter plants, detarring in coking plants, oxygen cutting and scarfing). The efficiency of trapping of fine particles and particles with high resistivity has been improved by (1) adopting high voltages (Nippon Steel uses a voltage of 150,000 V in their sinter strand), (2) redesigning both the electrodes, (3) introducing partitions into precipitators, to minimize the quantity of particles flying off on impact, (4) operating at higher temperatures, and (5) in the case of wet precipitators, using an electrostatic device to improve the spraying action of the liquid and the use wet precipitators which operate at high speed ($\sim$20 m/s).

The improvements in the dust control techniques may be illustrated with the example of iron and steel industry (UNEP, 1986). A dust collector, which can function effectively when a burden of good scrap is loaded into an electric arc furnace, may fail if the charge is oily scrap. But the steel maker may not be in a position to dictate the quality of scrap supplied by a merchant. So the steel maker should be in a position to modify the dust control system as needed.

3. *Filter media*: Bag filters are extensively used in for dust control in industries. For instance, the use of bag filters in the iron and steel industry have made it possible to reduce dust content to less than 10 mg/m^3 N. However, the relative equipment is expensive, requires frequent maintenance and involves energy consumption of 45-kwh/t. Bag filters are used in electric steel plants, for treating the diffuse gases produced by sintering, in blast furnace cast-houses and in steel-making shops. Bag filters made of terylene cloth and felt are no doubt effective, but they can be used only at lower temperatures (say, less than 130°C for terylene). If dust collection has to be done at high temperatures (say 600–1,000°C), fabrics woven from stainless steel fibres or refractory fibres made up of (say) aluminium oxide, have to be used.
4. *High-energy scrubbers*: There are three ways of using water to trap the dust: by collision between water and the dust on the basis of either flow of water or droplets, condensation of water on to the dust (on the analogy of fog), trapping by diffusion (on the principle of Brownian motion involving very fine droplets of water and very fine particles). Dedusting through fine spraying of water into the gas can be achieved either by the gas (high energy scrubbers of, say, the Venturi type, involving a pressure drop of about 250 mb on the waste gas) or by the water (whereby it is injected under high pressure, of the order of 15 bars). Aerodynamic profiling of the Venturis improves the efficiency of the scrubbers with a large pressure drop. A number of new techniques have been developed for bringing the gas or fumes into contact with water. These include a multicellular reactor which contains water gates which the gases have to cross, thus causing a small drop in pressure. In other cases, the classical Venturi device is replaced by a bulb-shaped combining nozzle. The new technique of using columns with perforated plates can be used both for dedusting and desulphurization in the sinter plants.

Aluminium industry discharges huge amounts of fluoride-loaded particulates which can cause dental mottling and skeletal fluorosis in human beings and animals. Aluminium plants produce cryolite mud (at the rate of 0.02 t of cryolite mud per tonne of cryolite used) which contains toxic heavy metals, such as, arsenic, cadmium, nickel, etc.

A. Bernatsky in his book, *Tree Ecology and Preservation,* strongly advocates the use of tree belts around industries to reduce particulate pollution, and noise. One ha of spruce can collect about 32 t of dust from the atmosphere, one ha of pine 36.4 t, and one ha of beech, 63 t.

3.3.3.2 Dust control chemicals

Cognis has developed a number of new surfactants to provide for improved dust control on mine haulage roads while being compatible with solvent extraction and leaching processes. EnviroWet DC-100 is highly biodegradable, and has superior wetting properties relative to the traditional surfactants based on linear alkyl benzene sulphonate and similar compounds.

Two reagents are now available commercially for dust control.

1. ALCOTAC® DS1 is a chemical binder or encrusting agent. When sprayed on fine particles of minerals such as coal, limestone, iron ore, etc., it forms an adherent film, and prevents the creation of airborne dust from the surface of the stockpiles, railcars and road wagons. The film is water resistant, and consequently, there will be no channeling or slumping on the stockpile when there is rain. The chemical also minimizes spontaneous fires in coal stockpiles.
2. ALCOTAC® 1235 is a chemical "wetting agent". When added to water, it will drastically reduce the surface tension of water, and would thereby promote the wetting of fine dust particles. The chemical is so formulated that it has a residual effect after initial application. Consequently, the fine particles are kept wetter longer, thus reducing the frequency of application needed. This reagent has been found to be useful to control the dust at the entrance to crushers, conveyor transfer points or on unmade roads.

3.3.4 LOW-WASTE TECHNOLOGIES

The idea of low-waste technology originated with water – that it is better not to pollute the water during the manufacturing process rather than clean it up afterwards. Low-waste technologies are those that are the least environmentally-degrading, involving pollutants (dust, gas, odour), nuisance (noise, vibration), with least consumption of energy and the use of raw materials.

Waste minimization techniques are schematically shown in Figure 3.3.5.

Low-waste technologies may be categorized into three types: (1) Internal action – this directly concerns the manufacturing process, whereby no waste is produced, and all products are saleable, (2) External action – whereby waste is transformed into saleable products, and (3) Recycling action – whereby waste materials, after intermediate treatment, are reusable as quality raw materials.

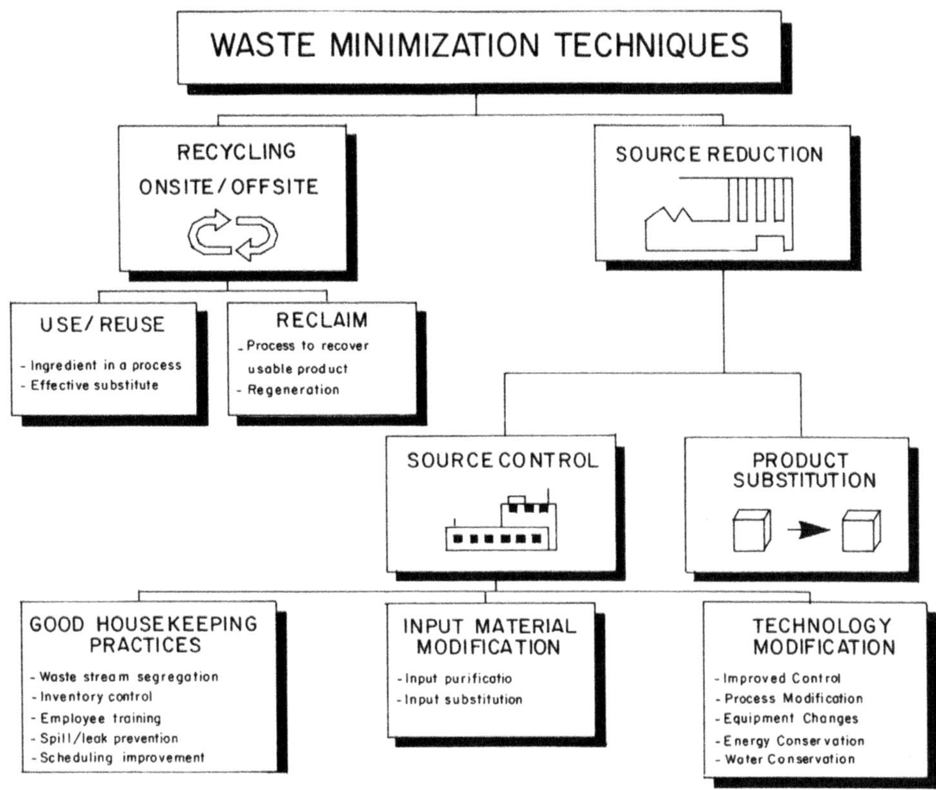

Figure 3.3.5 Techniques of waste minimization
Source: Anonymous

Low-waste technologies in the case of iron and steel industry consist of the following processes (UNEP, 1986):

1. *Pre-reduction or direct reduction of ores*: By this method, coking/sintering/blast furnace stages can be avoided, thereby eliminating the generation of byproducts from the coking plants, blast furnace slag and dust and sludges upstream of the steel shop. Besides, the dusts recovered by gas cleaning can be directly recycled. In Sweden, PLASMARED (SKF) process uses a plasma reactor to reduce the ore.
2. *Scrap preparation*: Scrap can be recycled in the blast furnace and the melting shop without any problem. But the scrap may carry pollutants, such as oils, coatings and alloy elements, which are not environmentally acceptable. The scrap can be cleansed of its pollutants before recycling, by shredding with magnetic separation, cryogenic grinding and preheating to burn off oils and plastic coating (Ceretti process).
3. *Continuous processing*: The blast furnace technology can be considered to be low-waste technology if the slag, dust and sludge could be made use of. Continuous steel making by electrical and other methods, saves energy and is environmentally less polluting. In some iron and steel mills, silica, sulphur and phosphorus are

removed in the pre-treatment processes in the blast furnace launder. In Sweden, liquid pig iron is produced by the pre-reduction in a fluidized bed of fine-grained concentrates, or injection of pre-reduced material by a hollow cathode electrode in an immersed arc furnace. If the furnace is operated by D.C. current, there is reduction in noise.

4. *Low-pollution pickling*: Wastes produced by acid pickling (by HCl or H_2SO_4 for ordinary steels, or HF-HNO_3 for stainless steels) can be minimized in the following ways: (1) Ishiclean process, which is a mechanical-hydroprocess, is virtually pollution-free, (2) Nitric acid pickling could be replaced by fluonitric pickling for stainless steels, as is done in Sweden. If the reheating before pickling is carried out in slightly reducing conditions, it will lead to the formation of scales. Such scales dissolve rapidly in acid and soil the pickling baths. The fouling of the pickling bath by scales can be avoided if the preheating done in an oxidized atmosphere. Common steels pickled in sulphuric acid produce ferrous sulphate ($FeSO_4 \cdot 7H_2O$) which is used in agriculture as a weed killer, and in the treatment of water (flocculation and dephosphorisation). In the hydrochloric acid pickling, ferrous oxide is obtained as a product. This may be recycled in the sinter plant in the steel works.
5. *Blast furnace dust and sludge*: The top gas off the blast furnace is dry dedusted, and then wet scrubbed. This leads to the production of dust and sludge rich in ferrous oxides and carbon, but also containing volatile elements such as zinc and lead. The previous practice has been to recycle the dust and the sludge in the sinter line. This led to operational difficulties due to recirculation of large quantities of zinc. The coarser particles which are generally zinc poor, can be recycled as before. Zinc which tends to be present in fine particles, can be recuperated by cycloning of the top gas before it is wet-scrubbed. It can be either dumped, or sent to the non-ferrous industry.
6. *Remelting of waste materials for special steels*: Valuable trace metals that may be contained in the waste materials (such as, dust, sludge and scale) may be recovered by a combination of the following processes: drying of sludge, blending and mixing with carbon, agglomeration (briquetting or pelletizing), addition to an arc furnace.
7. *Correction of the composition of the slag*: To suit the specifications of the market, it may some times become necessary to adjust the mineralogical and chemical composition of the slag. This may be accomplished during the manufacturing process (e.g. slagging additions to the blast furnace) or by careful tapping or by operating the furnace in a particular thermal regime.

3.3.4.1 Recycling of scrap

The recycling of scrap is explained in terms of the iron and steel industry. Steel production in USA involves the use of 64% of scrap. Each tonne of steel scrap recycled saves 1.1 t of iron ore, 0.6 t of coal and 54 kg of limestone, apart from savings in energy. There are three types of scrap in the steel industry: (1) Scrap arising in the individual steel works, which can be recycled in the same steel mill, without being involved in any commercial transaction, (2) Process scrap produced in the manufacture downstream of steel products, (3) Commercial scrap which helps the steel industry to balance their scrap requirements. Recovery depends upon the useful life of a manufactured product made of steel (for instance, 9–12 years, in the case of automobiles).

Table 3.3.3 Properties of different kinds of scrap

Type of scrap	*Fe (%)*	*C (%)*	*S (%)*	*Density*
Rolling mill off cuts (angles)	99	0.40	0.025	1 to 1.5
Demolition scrap (structure)	99	0.25	0.045	0.6 to 1.4
Shredded scrap (classic process)	95	0.50	0.045	0.9 to 1.1
Cryogenic scrap (shredded at	97	0.17	0.040	0.8
low temperature)	92	1.0	0.050	2.6
Packets of used scrap	80	0.25	0.010	
	82	1.3	0.070	

UNEP, 1986, p.110

The properties of different kinds of scrap, depending upon the source, are summarized in Table 3.3.3.

The waste is made use of in the works itself or is sold. Only a fraction of the tonnage (less than 10%) is dumped.

The ways in which the waste products in the iron and steel industry are recycled are summarized as follows:

1. *Sinter dust*: Dust is produced during the process of sintering, and related handling operations, at the rate of 30 kg/t of sinter. This dust can be recycled in the sinter grate.
2. *Blast furnace slag*: Apart from liquid pig iron, 300 kg of slag per ton of pig iron are produced. As the slag resembles a natural rock in its chemical composition, it is used for building roads, production of cement and to a lesser extent, for thermal and sound insulation.
3. *Oxygen steel-making slag*: Pig iron may be either high phosphoric (P = 1.7%) or low-phosphoric, haematitic (P = 0.2%). Phosphoric slag has good market, as a fertilizer in agriculture. The haematitic slag can be used as limestone adjuster in agriculture, and in road-making, but there is not much market for it. Its low value does not allow it to be transported for long distances. In such a situation, there is no option except to dump the haematitic slag.
4. *Electric arc furnace dust and sludge*: Electric arc furnaces can remelt coated scrap (e.g. galvanized or plastic-coated) and alloy scrap. The dust and sludge recovered from the electric furnace often contain volatile elements, such as Zn and Pb, and are hence useless in the case of steel industry. These elements can, however, be recovered in the non-ferrous industries by various methods, such as reduction in a rotating furnace, soda extraction, injection in a plasma, etc.

3.3.5 TREATMENT OF WASTE WATER

There are a number of ways of treating the large quantities of wastewater produced in the iron and steel industry, namely, recycling, removal of suspended solids, oil, and organic toxic pollutants, etc. (UNEP, 1986, pp. 70–82). These are applicable to other mineral industries as well.

1. *Recycle systems*: Recycling will reduce the pollutant loads at low cost, besides reducing the volume of wastewater that is discharged. However, if the wastewater is recycled too many times, two problems may arise in the recycled water – build-up of dissolved solids and the rise of temperature. High concentration of dissolved solids in the water can cause plugging and corrosion. This can be controlled by treatment of wastewater prior to recycling through the addition of chemicals, which inhibit scaling, or corrosion. If the recycled water is too warm to be used for its intended purpose, it has to be cooled prior to use. This can be achieved by passing the water through mechanical draft cooling towers. Most recycle systems require simple pumping only. They do not need much attention, except routine maintenance. However, if the wastewater concerned has arisen from wet air pollution control devices, the maintenance costs will be high, as the recycled water has to be cleansed of the dissolved constituents, which can cause fouling and scaling.
2. *Removal of suspended solids*: Suspended solids in wastewater may be removed by settling, clarification and filtration.

   When a stream of wastewater is let into a large volume lagoon, the velocity of water is reduced, and the gravitational settling of particles takes place. Settling is a slow process and usually takes days. The process of settling can be speeded up by the addition of settling aids, such as alum and polymeric flocculants. Sedimentation is often preceded by chemical precipitation and coagulation. These enhance the settling process by converting the precipitates into coarser particles, which will settle down faster. The ability of the lagoon to remove the suspended solids (including metal hydroxides) depends on the rate of overflow, density and particle size of the solids, the effective charge of suspended particles and the types of chemicals used for pre-treatment, etc. By allowing sufficient time for retention, by the proper control of pH, and by the regular removal of sludge, it is possible to have an efficient, low-cost system of removal of suspended solids.

   Relative to settling lagoons, clarifiers can remove suspended particles faster and more efficiently. Besides, they occupy less space and provide for centralized sludge collection. They are, however, more expensive to build and maintain. Conventional clarifiers consist of a tank and an arrangement for sludge collection. The tank may be circular or rectangular. The sludge may be collected by a mechanical device, or the sludge may be allowed to accumulate along a sloping, funnel-shaped bottom. In the case of advanced clarifiers, which use inclined plates for sludge collection, it is necessary to prescreen the wastewater to eliminate any materials, which could clog the system. As in the case of settling lagoons, clarifiers use flocculants to speed up settling.

   Filtration is a highly reliable method of wastewater treatment. It is used to remove suspended solids, oil and grease and toxic metals from steel industry wastewaters. It has a number of advantages – low initial and operating costs, small land requirement, no need to add flocculant chemicals which add to the discharge stream, and low solids concentrations in the effluent, etc. Filters may of pressure or gravity type, and may involve one or more media, such as sand, diatomaceous earth, walnut shells and others. Higher flow rates and efficiencies may be achieved by the use of dual or multiple media. In the filtration process, suspended solids and oil accumulate in the filter bed, and impede the movement of wastewater. In order to ensure that the filter bed performs efficiently, it is necessary to backwash the filter. Auxiliary means, such as water jets and air jets, can be employed to "scour" the bed free of solids and oils.

3. *Removal of oil*: This is done through skimming, air flotation and ultraflotation. Pollutants, such as free oil, grease and soaps, float to the surface of the wastewater, and can be removed by skimming. Air flotation and clarification when used in conjunction with skimming, can improve the removal of both settling and floating materials. The removal efficiency of a skimmer depends upon the density of the material to be floated, and the retention time of the wastewater in the tank. Depending upon the wastewater characteristics, retention may take 1 to 15 mins. for phase separation and skimming to be effected. Since skimming is effective in removing naturally floating materials, it constitutes good pre-treatment and improves the performance of the treatments downstream. Some pollutants, such as dispersed or emulsified oil, do not float to the surface by themselves, and skimmimg alone cannot remove them. More sophisticated methods have to be used for the purpose.

   When directed to the filter, oils and greases, either floating or emulsified, are adsorbed on the filter media. If high concentrations of oils are allowed to reach the filter bed, it may get "blinded", and should be promptly backwashed. The purpose of flotation is to cause particles such as metal hydroxides to float to the surface of the wastewater tank where they can be concentrated and removed. The methods of flotation differ from one another in regard to the ways of generating the minute gas bubbles, such as, froth, dispersed air, and dissolved air and vacuum flotation.

   Steel industry wastewaters may contain significant levels of toxic pollutants, such as chromium, copper, lead, nickel, zinc, etc. They can be precipitated by chemical means, and then removed by physical means, such as sedimentation, filtration and centrifugation. Lime or sodium hydroxide can precipitate several toxic metals as metal hydroxides, phosphate and fluoride as calcium phosphate and calcium fluoride respectively. Hydrogen sulphide and sodium sulphide can precipitate many metals as insoluble metal sulphides. The chemicals may be added to a flash mixer or pre-settling tank or they may directly be added to the clarifier. After the solid precipitates are removed, the pH adjustment is made. Chemical precipitation is a simple and effective means of removing many toxic pollutants from wastewater. Complications may, however, arise due to chelating agents, chemical interferences and the problems of storage of hazardous chemicals. When lime is used, it should be in the form of well-mixed slurry.

4. *Removal of organic toxic pollutants*: Activated carbon is made from coal, wood, coconut shells, petroleum base residues, etc. Its ability for adsorption arises from its low pore size (10–100 Å) and consequent high surface area (500–1500 m^2/g). Activated carbon is very effective in removing dissolved organics in the wastewater. The activated carbon can be regenerated and reused through the application of heat and steam or solvent. The wastewater is pre-filtered to remove excess suspended solids, oils and greases before being subjected to carbon adsorption. Suspended solids in the influent should be less than 50 mg/l to minimize backwash requirements. Oil and grease should be less than 15 mg/l. If the influent contains large concentrations of dissolved inorganic material, it may cause scaling, and loss of activity. This can be taken care of by pH control or the use of acid wash on the carbon prior to reactivation.

The advantages of the carbon treatment are its high removal efficiency, and applicability to a variety of organic pollutants. Where the carbon cannot be regenerated because of the high content of adsorbed compounds, it must be disposed off. Microbial treatment involving activated sludge can be used for the removal of pollutants such as ammonia-N, cyanide, phenols (4AAP) and toxic organics present in the wastewaters. The activated sludge system is sensitive to hydraulic and pollutant loadings, temperature and the presence of certain pollutants. Temperature not only affects the metabolic activities of the microorganisms, but also gas transfer rates. Some pollutants are extremely toxic to microbes, and could "kill" them. The activated sludge system significantly reduces the toxic organic pollutants more cheaply relative to the activated carbon. If wastewaters are properly pretreated before being subjected to activated sludge treatment, this process should work well.

5. Advanced technologies for treatment of wastewaters include ion exchange and reverse osmosis, but they may not be economical to treat large quantities of wastewaters.

UNEP (1991, p. 53) gave an account of the method of treatment of metal-containing acid mine water. Metal hydroxide precipitation takes place in the tailings ponds. The capital cost of the system was approx. C$ 800,000 (1985) and the annual operating costs were C$ 550,000 (1985).

3.3.6 SUBSIDENCE

Mining involves the extraction of large quantities of rocks, liquids and gases from the depths of the earth, and therefore causes damage not only on the surface but also to depths of hundreds and thousands of metres. The extent of subsidence varies from a few mm (due to withdrawal of waters from underground aquifers) to more than 6 or 7 m (arising from the extraction of coal from thick seams or due to underground fires). Subsidence may cause direct air circulation due to goaved-out areas, and may cause spontaneous combustion and fires within the goaf areas. Fires starting in one seam in a coal mine may spread to seams above and below it, and to seams in the neighbouring mines (as has happened in the Jharia–Raniganj coalfields in India). The presence of faults and dykes/sills and abandoned old workings may accentuate the problem of underground fires. The subsidence triggered by fires invariably spreads fast. As a consequence of subsidence movements in the underlying seams, the overlying coal seams may be rendered unworkable.

The following impacts of subsidence are common: formation of depressions in the surface, abrupt changes in the road gradients, damage to underground pipelines and cables, damage to surface buildings, plants and pylons, disturbance in the aquifers leading to reduced and contaminated flows, retardation in the growth of vegetation due to reduced availability of water, waterlogging in the central part of subsided area, contamination of surface air due to emissions from the underground fires, flooding of underground mines due to the development of ruptures in the underground waterbodies, etc. (Sengupta, 1993, p. 28).

In the case of surface mining, the extent of geomorphic change is related to the thickness of the overburden covering the deposit, the quantity of barren rock that needs to be excavated per unit of the extracted mineral and the area of the mine. Underground mining may lead to surface subsidence with consequent disturbance to surface runoff, formation of water-filled depressions, and flooding in the coastal areas or near lakes. Mining under water generally involves dredging of loose sediments under water. If the sediments involved are alluvial sediments, then river beds, flood plains and river terraces will be affected. Dredging may leave behind waste dumps and small valleys. The mining of estuaries and intertidal zones (usually for heavy minerals, and diamonds in the case of Namibia) disturbs the balance between land and sea, and may trigger beach erosion.

When the material is removed by underground mining, it triggers ground movement and the consequential deformation of the surface. The nature and extent of deformation depends upon the following parameters: (1) geometry of the mineral deposit – the mining of a massive, flat-bedded deposit will cause more deformation than a vein deposit, (2) the method of mining – longwall mining is more likely to lead to subsidence than room-and-pillar mining, (3) the nature of the mineral deposit, and the nature of the overlying strata – there are less chances of deformation if the mineral deposit and overlying rock are competent, than when they are incompetent.

Subsidence may lead to the following damages:

1. *Fractures*: The fractures may be continuous or discontinuous, and may range in size from millimetres to metres. They can cause severe damage to buildings and installations.
2. *Surface trough*: Continuous deformation may lead to the formation of a surface trough. Uniform displacement does not generally cause much damage. Differential displacement could adversely affect the groundwater flow, and could bring about changes in the gradients of roads, railways, water or gas pipelines, etc. Back filling of underground coal mines by hydraulic stowing of river sand is a common practice in India. Such a stowing reduces the surface subsidence below 10%, protects the aquifers, habitats, farms and fields. The township of Raniganj and the villages around it in the famous Jharia Coalfield in Bihar, India, did not suffer much damage for 75 years so long as the pillars in the underground mines were preserved. The unscientific depillaring of the thick coal seams ("slaughter mining") triggered subsidence, mine fires and environmental pollution in the area. Due to underground mining in the Jharia Coalfield, surface subsidence took place over an area of 32 km^2. The formation of goaf (void space) beneath the surface led to the formation of cracks on the surface, 5 to 10 m long, and about 0.5 m wide. There are also depressions caused by caves-in. At some places, smoke and gases emanate through the cracks. It has been estimated that about 34 Mt of coking coal has already been lost because of underground fires. About 70 fires (out of the initial 110) are still active and blocked out about 50 Mt of coking coal which hence cannot be worked.

Singh, Mathur and Landge (1995) describe how subsidence is controlled in the case of Chapri-Sidheswar mine in the Singhbhum copper belt, Bihar, India: (1) no mining will be carried out at depths of less than 62.5 m – in other words, a 62.5 m cover will be left intact throughout the life of the mine, and (2) the Room-and-Pillar stopes will be supported by 1.5 m long bolts at 1.2 m spacing. The mined out stopes will be promptly backfilled with the sand fraction of the tailings from the concentrator plant. The slurry will have 70% solids by weight.

There may be failure of the pit walls after an open-pit mine is abandoned. For instance, some of the open-pit copper mines in Zambia have steep walls of soft sedimentary rocks hundreds of metres in height, and driving in heavy vehicles near the tip of the mine could easily induce wall collapse. It is therefore necessary to designate a safety zone around the mine.

The safety zones and other measures are designed taking into account the geological, structural, geotechnical and climatic conditions.

Cavities are formed underground when geotechnical methods of mining (such as, leaching, dissolution, fusion) are used. This leads to increase in the porosity, and decrease in the strength of the rocks. The area becomes prone to collapse of roofs and surface subsidence. Instances are known of collapse of rock-salt mines when water entered an abandoned mine and dissolved the salt pillars left there for roof support. Underground gasification of coal in the former Soviet Union (involving a coal seam 5–15 m thick at a depth of 100–130 m, in an area of 1 km^2) gave rise to one of the biggest landslides in the world, with a volume of 0.8 km^3 spread over an area of 8 km^2 (Vartanyan, 1989, p. 42).

Landslides and rock and mud flows are common in the mining areas, especially when the wastes are dumped on the hillsides. For instance, the volume of the mudflow arising from the Yimen copper mine in China, was of the order of 200,000 m^3. Another mudflow of the volume of 100,000 m^3 from a mine in Yunnan, China, destroyed 6.2 km^2 of fertile land on the plains.

Four types of remedial measures are available for mitigating the subsidence in an abandoned mine: point support, local backfilling, areal backfilling and strata consolidation (Sengupta, 1993, p. 439). In the point support method, a large number of grouting holes are drilled, and grouting materials are injected to form the grouting piles and support the roof. Depending upon the engineering method used, the point support method could take the form of gravel columns, grout columns, fly ash grout injection, and fabric formed concrete. The local backfilling does not involve drilling the grouting holes. In this method, small, shallow potholes or surface cracks are filled with gravel, refuse and dirt, by direct dumping. Areal backfilling is meant to protect large urban areas (of the order of hundreds of hectares) from subsidence. This is accomplished by injecting into the underground openings large quantities of grouting materials, such as sand, gravel, coal refuse, mine tailings, fly ash, etc. under pressure.

In the strata consolidation method, the shallow strata beneath the damaged surface structure are grouted or bound into a single rigid unit. If the subsidence continues, the consolidated structure will move as a rigid body without being damaged. There are several ways of bringing about consolidation, such as the use of polyurethane binder, cement grout pad or rock anchor.

3.3.7 NOISE AND VIBRATION

Noise and vibration not only create health hazards, but also cause damage to structures.

The primary purpose of blasting operations in mining is the fragmentation of the rock. Fragmentation takes place when the potential energy contained in the explosive is suddenly released. An unintended and undesirable consequence of the blast is the displacement of the ground in the vicinity of the explosion.

Air blasts refer to air vibrations caused by blasting operations. The severity of the air blast depends not only upon the type and quantity of the explosive used, the degree of confinement and the method of initiation, but also on the climatic conditions, local geology and topography and the distance and condition of the structure that may be affected by the air blast.

Air blast waves may give rise to damage and nuisance. The ground vibrates as a consequence of blasting. The surface of the ground in the vicinity of the blast undergoes displacement. The amplitude of such displacement depends upon the distance of the point from the blast, the energy released in the explosives and the local geological conditions. The extent of damage caused is directly related to the peak particle velocity related to the ground vibration. The lower the frequency of vibration, the greater is the damage for a given peak velocity.

Blasting can generate both dust and gaseous contaminants. The adverse consequences of blasting can be controlled in the following ways: (1) wait for some time before entering the area affected by the blast, (2) wetting down with water before blasting, and (3) ventilation. It is necessary to mention that respirators for particles protect against dust particles only, but not against gaseous emissions, which require gas masks. Planting of dense tree belt has been suggested as a way to reduce noise. It has been reported in the literature (A. Bernetzky) that a tree barrier of 250 m depth can achieve a reduction of 40 dB. In 1980, ILO has issued guidelines about protecting the workers from noise and vibration.

In the case of the steel industry, there are three major categories of vibration, namely, mechanical vibration, vibration by combustion, and aerodynamic vibration.

Technological solutions to mitigate the consequences of vibration depend upon how the vibration is caused (UNEP, 1986, pp. 103–105):

1. *Mechanical vibration*: This kind of vibration arises from rotating and alternating machines, such as, heavy duty fans, blowers and compressors. In the case of rotating machines, the vibration is generally of low frequency. Vibration gets excited when there is a lack of dynamic and hydrodynamic balancing. In the case of reciprocating machines, high frequency vibrations are associated with the movements of various components, such as rods and pistons. The problem cannot be assessed from acoustic studies alone, as noise is caused not only by mechanical vibration as transmitted by structures, but also due to the turbulent flow of fluids. On the basis of theoretical and experimental studies about the dynamic and acoustic behaviour of machines, techniques of reducing vibration through balancing and elastic suspension have been developed.
2. *Vibration due to combustion*: This is particularly relevant to blast preheaters, where vibration may result in unstable combustion. Theoretical studies have shown that the combustion chamber behaves like a tube with one end closed, and another end open (cupola). Vibration gets initiated when the ratio between the length of the oscillation wave and the length of the air and gas ducts of the combustion chamber, reaches a particular critical figure. The instability in combustion gives rise to a pulsing phenomenon with an acoustic wave ($\lambda = 4$–10 Hz) similar to that of the "singing" arc. In most cases, the vibration is not only unpleasant,

but it may be dangerous for the operation of the plant and therefore for the plant staff. The vibration due to combustion could generally be mitigated by using appropriate lengths of the duct. In actuality, the phenomena are more complex. For instance, gas pressure, delay in the combustion, holding temperature in the cupola, fuel injection in the blast furnace, atmospheric conditions, etc. have the effect of modifying the acoustic length, and thereby increasing the possible zones of instability.

3. *Aerodynamic vibration*: Aerodynamic vibration is caused by the transport and distribution of gas and fumes by numerous ducts equipped with regulation valves. There are three specific causes of vibration: "(1) flow turbulence created by decreases in the velocity of the fluids along the inner side wall of the pipes, with fluctuation of pressure at the outer limit, (2) periodic flow phenomena due to pressure modulation by the ventilator or by pulsating combustion of a burner, (3) phenomena of drag and aeroelastic coupling between the flow and the vibration of the obstacle (butterfly valve, for example)" (UNEP, 1986, p. 104). The third phenomenon may cause the structures to emit intense sound on one frequency, if the cavities in which the flow is contained, have similar acoustic modes. Apart from being an acoustic nuisance, resonance may endanger the safety of the staff and cause serious damage to plant.

The principal remedies for suppressing resonance are: (1) avoidance of resonance by appropriate design of piping and valves, (2) mechanical decoupling of the source of vibration from the rest of the ducts, and (3) use of appropriate silencers (such as, diffusion silencers which reduce turbulence, absorption silencers, resonating silencers, which are based on the introduction of uneven, multiple dephasing of the quarter of the length of the wave).

3.3.8 PLANNING FOR MINE CLOSURE

Instances are known of the mine owners just abandoning the mines when the ore runs out. It is critically important that mine closure programme should be incorporated into any mining proposal right at the outset. Proper closure of the mine is absolutely essential, particularly if the mine wastes happen to be acid producing. The leachates from them can play havoc with the waters, soils and biota of the area for many decades, if not centuries.

All access to underground mine workings should be closed properly. Shafts are recommended to be filled with inert material, and sealed with concrete. Adits should be plugged with concrete. If long-term subsidence that could cause damage to buildings is anticipated, appropriate subsidence control measures should be undertaken, if feasible. In the case of mines worked by room-and-pillar method, the vacant spaces inside the mine could be used for high-security storage, warehousing and even for mushroom cultivation.

Open pits for coal and base metals can be very large, and backfilling them with waste overburden may be infeasible or uneconomical. Such pits can be used for purposes of water storage or recreation. An abandoned limestone pit in Vancouver, Canada, has been innovatively developed into a spectacularly beautiful flower garden with waterfalls and aviary.

Now-a-days, governments are under pressure from the public to enforce the mine closure regulations more strictly. In most cases, it is not possible to trace the owners of the abandoned mines, and make them pay for rehabilitation. So the governments concerned have no option except to rehabilitate the mine in public interest with public money.

In some areas, mines constitute the most important economic resource. The closure of mine may have a strong adverse socioeconomic impact. The social dislocation that the mine closure can cause can be mitigated in part through the retraining of the work force to newer employment opportunities, and newer enterprises.

Sengupta (1993, pp. 453–477) gave detailed case histories of decommissioning of gold heap-leaching operations. To plan for closure, it is necessary to model the following aspects: migration routes, through surface water flow through the underdrain, and the groundwater flow through the undersaturated zone, and environmental fate (mixing, dilution/attenuation/precipitation, etc.) of the solutes. The hydrologic event used for risk assessment is the maximum 24-hr rainfall over a 100-year interval.

Metal-complexed, Weak Acid Dissociable (WAD) cyanide, copper, zinc, arsenic, etc. are usually present in the active heaps at levels which could adversely affect the environmental and human receptors. Regulatory agencies invariably prescribe the permissible concentration of WAD cyanide in the heap effluent. Cyanide gets strongly attenuated during unsaturated flow. Experience has shown that when the cyanide levels are brought down to compliant level, heavy metals such as arsenic, will invariably become attenuated and compliant.

Solutes within the immobile or slow flow region would tend to mix with the solutes in the rapid flow or mobile region. The concentration outflow from the heap is determined from the following equation:

$$C_{out} = C_{mobile} + (C_{heap} - C_{mobile})Jt \tag{3.3.3}$$

where

C_{out} = concentration in outflow from the heap,
C_{heap} = average cyanide concentration in the immobile region,
C_{mobile} = average concentration in the mobile flow region,
J = a diffusion term set by the user,
t = elapsed time

The following case history of the Borealis Mine, Hawthorene, Nevada, USA (quoted by Sengupta, 1993, pp. 468–470) is instructive. The general criteria for leach pad closure in Nevada are: (1) WAD cyanide levels of effluent rinse water must be less than 0.2 mg/l, (2) the pH level of the effluent rinse water should be between 6.0 and 9.0, and (3) Contaminants in any effluent from the process water that result from meteoric events must not degrade state waters. The heap was rinsed with fresh water. It was found that the free cyanide levels got reduced from 1.2–3.7 mg/l to ~0.2 mg/l in 2–10 days, and the entire pad was detoxified in 60 days. Among the various detoxification agents tried (such as, ferrous sulphate, alkaline chlorination, etc.), hydrogen peroxide has been found most suitable for the detoxification of heaps.

Chapter 3.4

Health and socio-economic impacts of the mining industry

3.4.1 HEALTH HAZARDS OF THE MINING INDUSTRY

Mining is undoubtedly the most hazardous industrial occupation. For instance, during the period 1980–89, mining ranked as the number one in USA with respect to the average annual rate of traumatic fatalities (with the rate of 31.91 for 100,000 workers), as against 25.61 for the construction industry, 23.30 for the transportation/communications/public utilities industries, and 18.33 for the agriculture/forestry/fishing industries. There are two kinds of health impacts associated with mining: immediate impacts such as accidents, and accumulative and progressive impacts such as stress and pneumoconiosis. Opencast mining is generally less hazardous than underground mining. Industrialised countries tend to use highly automated mining systems, which not only employ lesser number of workers (who have to be highly skilled), but also have the effect of drastically reducing the hazards to which they are exposed. Developing countries cannot afford such high-tech mining systems, so much so that mining accidents are a common occurrence in developing countries such as China and India.

The severity of risk in the case of the mining industry is summarized in Table 3.4.1.

The data regarding occupational illnesses in mining by sector (US data, 1995) is summarized in Table 3.4.2.

An examination of the statistical data (for USA, 1995) in regard to fatalities, nonfatal days lost (NFDL), total accident incident rate, and severity measurements (SM) for underground and surface mines by sector (source: Grayson, 1999, p. 94) leads to the following conclusions: (1) Among all the mining activities for various minerals, the most hazardous is the underground mining of coal, (2) The underground mining of coal, metal and nonmetal has higher severity measure than the corresponding figures

Table 3.4.1 Severity estimates for underground and surface mining by sector (data from USA in 1995)

Sector	*Fatalities*	*Fatal-IR*	*NFDL*	*NFDL-IR*	*Total-IR*	*Overall SM*
Coal underground	25	0.05	5,426	10.57	13.50	772
Coal surface	10	0.03	942	2.75	4.06	288
Metal underground	3	0.04	422	5.79	9.75	498
Metal surface	5	0.03	469	2.52	4.23	390
Nonmetal underground	3	0.08	146	3.83	6.34	712
Nonmetal surface	3	0.05	137	2.41	4.19	433
Stone underground	0	0	75	3.71	5.53	122
Stone surface	11	0.04	1,201	4.15	6.86	345
Sand/gravel surface	4	0.02	881	3.56	5.62	269
Sand/gravel dredge	2	0.04	182	3.77	6.10	3.35

Source: Grayson, 1999, p. 94

Table 3.4.2 Occupational illnesses in mining by sector (US data 1995)

Type/sector	*Coal*	*Metal*	*Non-metal*	*Stone*	*Sand/gravel*
Skin disease	3	4	1	12	3
Dust diseases – lung	207	8	5	9	2
Respiratory – toxic agents	8	1	2	1	0
Poisoning	3	1	0	3	0
Disorders – nontoxic physiological agents	2	12	3	15	5
Disorders – repeated trauma	214	109	32	49	22
All others	21	2	3	5	1
Total	**458**	**137**	**46**	**94**	**33**

Source: Grayson 1999, p. 95

for surface mining for the same minerals, (3) Surface mining of stone has a greater SM than underground mining of stone. On the basis of such analyses, the Mine Safety and Health Administration (MSHA) targets sectors, mines and jobs to enforce the regulations.

By making various improvements in mining technology and mining practices, USA brought down the mining fatalities from 56 in 1990, to 47 in 2000, and to 23 in 2010. China produced 2.8 billion tonnes of coal in 2007, which is about 40% of the world's production. The coal mining fatalities in China which used to be as high as 6,000 per annum, has been brought down to 3,215 in 2008. In the case of India, which is the third largest producer of coal, the coal mining fatalities came down from 200 in 2000 to 35 in 2005.

Table 3.4.2 indicates that coal mining leads mining for other minerals in regard to dust diseases of lungs and trauma disorders. By improving the working conditions in the mines, the number of silicosis cases per year came down from 857.4 during 1968–78 to 284.5 in 1991–92 in USA. Similarly, the number of cases of pneumoconiosis per year, which was 2374.8 in 1968–78, was brought down to 1852.0 per year in 1991–92.

The main environmental consequences of the mining projects are shown in Figure 3.4.1.

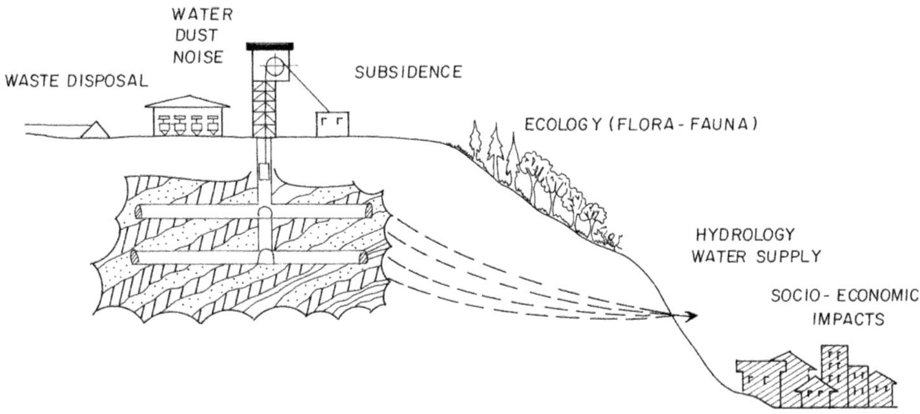

Figure 3.4.1 Main environmental consequences of mining project
Source: UNEP Tech. Rept. No. 5, 1991, p. 84

Health hazards in mining are described with reference to coal mining. There are four types of health hazards (see the excellent account by Chadwick et al., 1987, pp. 203–236, from which the following account has largely been drawn).

1. Physical hazards, e.g. coal dust, silica dust, excessive heat, noise, heavy physical work, contorted body posture,
2. Chemical hazards, e.g. carbon dioxide, carbon monoxide, methane, nitrogen oxide gases,
3. Biological hazards (applicable in some developing countries), e.g. fungus, hookworm,
4. Mental hazards, e.g. shift work, constant danger.

The 3 km deep, Kolar gold mines of South India, constitutes an unusual case where all the above problems are evident at one place, namely, rock bursts, high thermal stress, gas and dust explosions, fires, inundations, hookworm infection, etc. (Pai & Shenoi, 1988).

3.4.2 HEALTH HAZARDS DUE TO DUSTS

Dust is the cause of the many of the cumulative health hazards in the mineral industries, and is hence dealt with in some detail.

The main sources of dust in the mining operations are: Point sources: (1) Ore and waste loading points in trucks, railroad cars, etc. (2) Ore chutes in the haulage systems (bin, conveyors), (3) Screens in outdoor crushing plants, (4) Exhaust from dedusting installations, and (5) Dryer chimneys. Dispersed sources: (1) Waste dumps, (2) Ore stockpiles, (3) Haul roads, (4) Tailings disposal.

The main natural and artificial dusts, associated sources and possible health disorders are summarized in Table 3.4.3.

Table 3.4.4 lists the hazard prevention measures in the case of the coal cycle.

Table 3.4.3 Main natural and artificial dusts, associated sources and possible disorders

Dust type	*Possible source*	*Possible disorders*
Silica (crystalline and amorphous)	Mining, quarrying, sand blasting, abrasives, Glass making, etc.	Silicosis
Coal	Mining, transportation and use, smoke from burning arsenious coal used in domestic cooking	Silico-anthracosis, Coal worker pneumoconiosis, arseniasis
Asbestos	Asbestos cement, insulation, friction materials (brakes, clutches, etc.), floor tiles	Asbestosis, and Pleuropulmonary cancer
Fibrous zeolite	Volcanic tuff	Pleural cancer
Fibrous clays	Quarries, drilling mud, pharmaceutical industry	Fibrosis
Talc	Mining, rubber industry, lubricants, Pharmeceutical industry	Talcosis
Kaolin	Quarrying, ceramic industry	Kaolinosis
Bentonite	Quarrying, drilling	Fibrosis
Aluminium, alumina	Bauxite mines, ceramics, abrasives, paint, metallurgy	Fibrosis
Barytes	Mining, metallurgy, pharmaceutical industry	Barytosis
Beryllium compounds	Metallurgy, aeronautics industry, nuclear industry, solid fuel	Beryllosis (granulomatosis)
Iron oxides	Iron mines, foundries, steel plants	Silico-siderosis, siderosis
Nickel	Mining, polishing	Fibrosis, lung cancer
Chromium	Mining, polishing, electrochemistry	Fibrosis, lung cancer
Cadmium	Mining, polishing, electrochemistry	Fibrosis, urinary tract cancer
Manganese	Mining, polishing, foundries	Fibrosis
Titanium, tantalum, wolfram carbides	Polishing	Fibrosis, lung cancer
All metals	Welding	Fibrosis, lung cancer
Synthetic mineral fibres	Thermal and acoustic insulation, composite materials	Fibrosis
Airborne ash	Coal and oil fired plants, incineration of household and industrial wastes	Fibrosis, cancer
Volcanic ash	Volcanic eruptions	

Source: Archer et al., 1987, p. 171

There are three *E*'s of mitigation of health hazards of the mining industry: Education, Engineering and Enforcement. The goal of the mining industry should be to ensure that the workers could work their entire career without incurring death, disability or serious injury.

3.4.3 MATRIX DIAGRAMS

The purpose of the matrix diagram is to identify which environmental parameters will be affected by the activities of the project and to what extent. All human activities that the project would involve are listed along one axis, and all natural factors that are likely to be affected as a consequence of such activities are listed on another axis. Thus, if there are x number of human activities, and y number of environmental factors, there will be xy number of matrix slots. All slots corresponding to the recognized impacts

Table 3.4.4 Hazard prevention measures

Hazard	*Principle of prevention*	*Preventive measures*
Dust	Suppression at the source, dilution in the air, suppression in the environment	Wetting of the coal face, ventilation. Surfactants on the floor
Noise	Substitution	Other machinery
Heat	Reduction	Ventilation, air-conditioning
Heavy work	Elimination, substitution	Mechanisation/automation, Ergonomic design work
Vibration	Elimination	Remote control
Falls of ground	Elimination	Support to roof and walls
Dangerous machines	Substitution, segregation	Maintenance/replacement, Machined guarding
Blasting	Suppresion at source, dilution of air	Wet methods, ventilation
Gases	Disposal, dilution	Extraction, ventilation

Source: Chadwick et al., 1987, p. 223

are first slashed. Evidently, all impacts would not be of the same magnitude. So the magnitude of impact in the scale of 1 (least impact) to 10 (greatest impact) is indicated in each slashed slot.

Figure 3.4.2 is a matrix diagram which enables the visualization and estimation of how various mining activities, such as, exploration, opencast mining, underground mining, ore processing, tailings, rehabilitation of the mined land, etc. in regard to a given mining project in terms of social environment, physical environment and biological environment. Matrix diagrams are undoubtedly useful, but they suffer from the shortcoming that they cannot bring out the linkages and interactions between various environmental parameters.

The environmental effects of mining, such as the release of the pollutants, degradation of the landscape, disturbance in the habitat, etc. are inter-related. Consequently, change in one particular environmental component (e.g. process technology) will often cause direct and indirect changes in other components (e.g. tailings disposal).

So it is necessary to adopt a holistic approach in the EIA process.

3.4.4 TOTAL PROJECT DEVELOPMENT – A VISIONARY APPROACH

The Total Project Development (TPD) is a new holistic approach to mining (Struthers, 2001, pp. 814–823). Under this approach, a mining project is developed as a part of much wider, multi-activity regional development. All the material extracted by a mining company is put to productive use. Waste rock, mine tailings, excess mine water, etc. are used as raw material for a variety of downstream ancillary industries. Tailings are used for underground backfill, embankments and sealants for reactive waste rock, and production of construction materials for mine use. All excess tailings are used for "soil" development. Excess process water (after use in recycling) is treated for being

Environmental effect/ Development	Social environment*	Physical environment**	Biological environment***
Exploration Prospecting – Surveys – Drilling – Sampling			
Opencast mining – Overburden stripping – Blasting – Dewatering – Crushing **Underground mining** – Methods used – Ventilation system – Dewatering **Dredging** – Floating plant – Pond formation			
Ore processing – Water supply – Washing plant – Process used – Stockpiling – Wastewater treatment			
Tailings – Tailings dam – Run-off control			
Rehabilitation – Contour shaping – Planting – Overburden use			
General – Surface infrastructure – Access roads – Energy source			

*Social Environment covers the following considerations: Public participation, Employment, Sentiment, Land values, Risks and Anxieties, Personal and social values, Historical/ Cultural, Landscape/Visual, Recreation.
** Physical Environment covers the following considerations: River regime, Erosion/Land stability, Sedimentation, Surface water, Groundwater, Agricultural soil, Foundation materials, Climate? Atmosphere, Nuisance (noise, dust, smell), Landform.
***Biological environment covers the following considerations: Forest, Shrubland, Grassland, Herbfield (Alpine), Sand, Shingle/rock, Cropland, Urban land, Lakes, Rivers, Estuaries, Intertidal, Marine, Wetlands.

Figure 3.4.2 Matrix diagram for the visualization of mining impacts
Source: UNEP Tech. Rept. No, 5, 1991, p. 83

used in fish farm ponds and crop irrigation. The TPD approach benefits the various entities in the following ways:

1. Mining company benefits from (1) increased metal recoveries and additional revenue through the retrieval and sale of non-target minerals, (2) reduced operational

expenses through the maximum utilization of tailings in backfill, preparation of tailings/concrete blocks for underground and surface constructions, and replacement of topsoil, (3) reduced mining costs by saving on the construction of tailing ponds and waste rock dumps, (4) reduced rehabilitation costs, and (5) income from productive use of post-mine land use. The mining company would also have some intangible benefits such as good public image, making it easier to get the required environmental permits.

2. The local community benefits from increased employment and income levels, long-term food security and livelihood after the closure of the mine. The improved infrastructure helps in communications, and access to wider markets. TDP benefits the environment through the elimination of waste rock dumps, drastic reduction in AMD and contamination from dust, heavy metal contamination of soil and water, etc.

As Struthers (2001) pointed out, most companies have been practicing some element or other of the TDP. What is needed is to plan and implement *all* the components of TDP for every mine, through the cooperation of the government, mining companies, local communities, and technical experts (such as, engineers and mineral economists).

Chapter 3.5

Artisanal mining

The contribution of small-scale or artisanal mining to the overall mining output in the world is estimated to be 10–16%. In general, the percentage contribution with respect to industrial minerals is higher than for metallic minerals. Estimates vary widely about the number of persons involved in artisanal mining (upto 16 million). In some countries, the contribution of artisanal mining is of considerable economic significance. For instance, Peru has about 3,000 small-scale mines which produce 100% of the antimony, 90% gold, 15% tungsten, etc. of the national production (Gocht, 1980, *Natural Resources and Development*, v. 12, pp. 7–18). About 90% of Brazil's gold production is attributable to artisanal mining.

Capital-intensive, mechanized mining is not cost effective for the exploitation of small deposits, even of high grade. Such a deposit is amenable to small-scale mining, particularly where the deposit occurs at or close to the surface, and where the mineral could be won by simple methods, such as hand-picking, panning, sluicing, etc. It therefore follows that artisanal mining may be a cost-effective option in the case of several economic minerals occurring in the soil. For instance, by the nature of its occurrence and its properties, opal can only be mined manually (in candle light!).

Small-scale mining has several advantages: it is labor intensive, can be initiated on any scale, with simple technology, at low capital cost, low consumption of energy, and short lead time, and without expensive imported equipment. It can also promote local industries. This, however, does not mean that small-scale mining is the panacea for the problems of the developing countries. Artisanal mining suffers from the following serious disadvantages: (i) it tends to be haphazard, since in most countries there is no systematic exploration activity to support small-scale mining; (ii) destructive exploitation by the gouging of rich pockets; (iii) low recovery rates; (iv) low labor productivity (about 4% of highly mechanized mines); and (v) non-extraction of valuable byproducts which are therefore irretrievably lost to the country.

In future, vegetative methods of reclamation of mined land may emerge as a significant, economically viable, and employment-generating activity.

It is possible to improve the efficiency of small-scale mining, while concomitantly reducing its deleterious consequences, by the adoption of the following innovative approaches:

1. Developing simple techniques of prospecting, which could be used by unskilled labor, e.g. use of smoky quartz as indicator of cassiterite-lepidolite pegmatites, and looking for cassiterite resistate in the soils near pegmatite, training of miners on-site about simple methods of mineral search and extraction.

   Using a portable X-ray fluorescence analyzer, it is possible to make a quick and fairly accurate on-site assay of several ore metals in the material mined or to be mined by a miner. Such an assay can serve two purposes: (i) to make the miner aware of the economic value of the material already mined by him (through a knowledge of what kind of ore metals and in what concentrations occur in the material mined by him), and (ii) to advise him as to what kind of material he should be mining in order to get greater returns.
2. Research and development to design improved methods of ore search and ore extraction relevant to small-scale mining. Placer gold is a case in point. An artisanal miner can extract gold only if it is coarse grained (say, 30 μm) and high grade (say, about 25 g/m^3). He uses the mercury amalgam method of extraction which is highly polluting. New carbon-in-pulp and carbon-in-leach technologies have several advantages: (i) they are capable of extracting fine-grained gold (about 10 μm) and at low concentrations (about 2 g/t), and (ii) they are environmentally benign. These technologies need to be adapted for small-scale operations. In extremely dry areas, pneumatic methods of gold separation have to be developed.
3. Using mobile units for preconcentration and extraction on site: Truck-mounted, diesel-powered, self-contained, ore-dressing modules are taken to the site of the artisanal mining and the ore is concentrated/extracted on site. The mobile unit can be owned and operated by a cooperative or a private company. A part (say, one-third) of the output could be collected in kind towards service charge due to the mobile unit and the royalty due to the government. As the recovery rates by the mobile unit are at least 2–3 times higher than by manual methods, the artisanal miner is still left with considerably more saleable material than he would have been able to recover on his own.
4. Through the use of the mercury amalgam method, artisanal gold mining industry in Tanzania has severely contaminated the waters and soils. About 78% of the water samples analyzed contained concentrations of mercury higher than the permissible level of 1 μg/l. Mercury levels in the mine tailings range from 1.31 to 18.7 μg/g. The Institute of Production Innovation of the University of Dar es Salaam, Tanzania, has developed a rugged, easy to handle, highly portable and locally manufactured, and inexpensive (eq. USD 50) retort which has efficiency of 99.6 to 100%. The use of such a retort is the most effective way to reduce the pollution of airborne mercury produced by the firing of the Au–Hg amalgam.

5. The "Portable" gold plant developed by Libenberg, Rundle and Storey of San Martin mining company (*Mining Magazine*, July, 97, pp. 8–10) is a veritable godsend for small-scale gold miners. The salient points of the plant are as follows: San Martin's claim encompasses two dumps around Bonda, Kenya, with 250,000 t of material, grade: 1–3 g/t. Carbon-in-pulp/carbon-in-leach technique; Capacity of the plant: 10,000 t/month; production cost: USD 150/oz. The total steel requirement (12 t) for tanks, baffles, agitator mountings, and the pumps and piping, were brought from South Africa in one container and erected on site. Dump material is reclaimed by high-pressure water. In the first leach tank, lime (5 kg/t) and sodium cyanide (0.5 kg/t) are added. The residence time in each of the absorption tanks is approximately 1.3 hr at a throughput of 10,000 t/month, and carbon concentration of 15–20 g/l. The eluate is heated in a diesel-fired cast iron burner. Gold is recovered onto steel wool cathodes. Security is ensured by having the recovery cell protected by a 220 V inner cage, and 15,000 V outer cage (powered by solar cells). Doré is 80% pure. Total power consumption: 145 kW (diesel generators). Water is pumped from the Yala River (3 km from the plant).

   The plant has been in operation for several years now and has the following advantages: environmentally-benign as no mercury is used; can be erected even in remote areas, and shifted and reassembled without much problem; can be operated with minimal expatriate assistance; and economically viable. This technology can be used in two kinds of situations: (i) as a private enterprise, for treating dumps, where they exist, and (ii) as a cooperative, by setting up the plant at a central place where a number of artisanal miners (50 to 100) operate.
6. The modular plants designed by M/S Bateman Project Holdings Limited, Boksburg, South Africa, have revolutionized the recovery efficiencies in artisanal mining of a number of minerals (such as, modular process plants for diamonds, gold, etc.).

Chapter 3.6

Ways of ameliorating the adverse consequences of the mining industry

3.6.1 REHABILITATION OF MINED LAND

Though the mining companies are required by law to submit plans and commit funds for the rehabilitation of the mined land when once the mine is closed, enforcement has not always been strict enough. It is particularly difficult in the case of artisanal miners ("here today, gone tomorrow").

Though the climate, soil and hydrological characteristics and methods of mining vary greatly in different areas, there are some common elements in the techniques of rehabilitation: (i) Removal and retention of top soil, to be respread in the area that is being rehabilitated, (ii) Reshaping the degraded areas and waste dumps in such a manner that they are stable, well drained, and suitably landscaped for the desired long-term use, (iii) Minimizing the likelihood for wind and water erosion, (iv) Deep ripping of the compacted surface, (v) Revegetating with appropriate plant species in order to control erosion, and facilitate the development of a stable ecosystem compatible with the projected long-term use.

Amelioration methods can be custom-made for a given situation, as follows (Chadwick et al., 1987): (i) Low pH (usually <5): Amelioration by liming. Acid-tolerant species may be planted; (ii) High pH (usually >8): Salt content may be removed by leaching. Salt/alkali- tolerant plants may be grown; (iii) Low nutrient status: Nitrogen deficiency may be ameliorated by nitrogenous fertilization or by growing legumes; (iv) Low moisture levels: Ridging, furrowing and mulching, etc. and growing drought-tolerant plants; (v) Soil amendment: use of other wastes, such as fly ash, slag, etc., and (vi) Planting of artificial wetlands for the treatment of acid mine drainage and polluted runoff.

3.6.2 BENEFICIAL USE OF MINING WASTES

The volume of wastes generated in the process of mining increases with increased volume of mining activities, and increased mechanization. No-waste and low-waste mining technologies can in principle bring down the volume of wastes that need to be disposed of, but there is little doubt that wastes in mining cannot be avoided altogether. The use of waste rock for back-filling, recycling, and the large-scale use of wastes for the construction of roads, buildings and other civil engineering structures are some of the ways by which the wastes can be used beneficially.

Coal mining wastes: Taking the mining industry as a whole, there is little doubt that coal mining produces the largest volume of solid wastes. Mine gangue and coal-washing tailings are being increasingly used as filling materials, additives in concrete and for agricultural purposes.

The gangue material in coal waste tips generally has a porosity of about 35%. The relatively high combustible content of the waste coupled with its high porosity, makes the waste liable for spontaneous combustion. It has been estimated that 40% of the 17,000 rock waste tips in the world, are burning. Smoke from the burning tips pollutes large areas around them. An ingenious way to reduce the porosity of the waste tips and thereby reduces their proneness for spontaneous combustion, is the addition of fly ash from the wastes of the thermal power plants. In this manner, one kind of waste is made use of to reduce the environmental harm from another kind of waste!

After strengthening, the gangue material from the coal mining industry can be used in the construction of road embankments and railway lines, landscaping of building sites, and earth dams, etc. The porosity of the gangue is reduced and the strength increased by compaction with bulldozers, and addition of pore-filling materials, such as fly ash from power plants, sand, and flotation tailings. By this process, the porosity can be reduced to 20%, and the density increased to 2.1 t/m^3.

Clays with high content of organic matter can be used to make a material called karamzite. In Belgium and France, mine gangues and coal washing tailings are made use of to fabricate commercial building materials, trademarked AGRAL. The gangue material can also be made use of to make bricks, and as aggregates for light weight concrete. For instance, the brick-works of "Lvovstrojmaterialy" in Ukraine which produces 300 million bricks a year, found that the use of 10% coal wastes has reduced the consumption of fuel by 20–25%, besides improving the quality of bricks.

The CSIR Laboratories in India (principally, the Central Building Research Institute, Roorkee, and the Regional Research Laboratory, Bhopal) have developed innovative approaches for the use of fly ash from the coal industry and red mud wastes from the aluminium industry (vide *CSIR Rural Technologies*, 1995, pp. 83–88).

Clay may be mixed with fly ash (to the extent of 10–40%) and made into bricks, which can then be fired in conventional Bull's kiln, or intermittent type kilns at a temperature of 950 to 1,050°C. The use of fly ash permits the production of 40% more additional bricks from the same quantity of soil. The clay-fly ash bricks have lower bulk density, better thermal insulation and reduced dead load on the brick masonry structure. These bricks can be used for all types of construction, where normal clay bricks are used.

In areas where good quality clay is not available, fly ash-sand-lime bricks can be made. Fly ash could be used to the extent of 70%. The bricks will have a wet compressive strength of 100–200 kg/cm and water absorption of 10 to 20%. Drying shrinkage (0.01–0.05%) and thermal conductivity are comparable to those of the clay bricks. Unlike the clay bricks, the fly ash – lime – sand bricks do not need drying.

The lime – fly ash blends can be used as stabilizers in road construction. For granular soils, 3–6% lime and 10–25% fly ash should be used. For clayey soils, 5–9% lime and 10–25% fly ash, need to be used.

Bricks can be made with red mud wastes from the aluminium industry. Red mud improves the quality of bricks made from clay-deficient soils. When fired, bricks made with red mud develop a pleasing pale brown, orange or golden yellow colour, depending upon the composition of the raw material, and firing temperature. They therefore have a good architectural value as facing bricks. The presence of 4–5% alkalis in red mud makes for good fluxing action. Consequently, the red mud bricks have better plasticity and bonding than the normal bricks. They may be fired in the usual Bull's trench kiln. Black coal flotation sludges can be dried to reduce their moisture content to 8 to 10%, and the resulting product can be burnt in the thermal power plants. Brown coal sludges are finding numerous uses in agriculture. When added to the soil, the humic acids contained in coal form organo-mineral humus and sorption complexes and becomes repositories of nutrient elements. This improves the structure, pH and fertility of the soil. In Russia, the combination of manure and high-ash coal (the so-called mineral manure) proved very successful. In Hungary, brown coal dust mixed with manure is used as a fertilizer. Coal waste can be used as bio-organic mineral fertilizer.

Acid mine effluent often contains copper which can be recovered cheaply by treating the effluent with scrap iron. Methane generated in the underground mining can be collected and used to feed the boilers.

Other kinds of mining wastes: Nepheline tailings in the production of apatite concentrates can be used in the production of glass, and as a binder for silica bricks. Wastes of chalcopyrite ore concentrates can be used for the manufacture of silicate wall and facing materials, glass, etc.

Solid wastes from mining could be used as fillers in concrete and other cement-based materials. Mine soil fill material can be effectively used for the renovation of wastewater. Red mud waste is produced when bauxite is processed to produce alumina, and is available in large quantities around the bauxite mines. It contains compounds of Al (22–37%), Fe (24–26%), Ca (2–4%), Na and Si. It has been found that red mud mixed with medium-sized sand is highly effective in removing P, BOD, suspended solids and fecal coliforms from domestic sewage (Brandes et al., 1975).

Residential and municipal wastewaters contain numerous pathogens, such as enteric viruses (which can cause meningitis and hepatitis), bacteria (which can cause typhoid and gastroenteritis), protozoans (which can cause amoebic dysentery and giardiasis), and helminthes (which can cause a number of chronic diseases such as anaemia and gastroenteritis). Size-wise, the enteric viruses are the smallest, and the helminths the largest. Considerations of size enter the picture because the larger the organism, the more readily it is trapped and retained when wastewater containing the pathogen percolates through the soil. Consequently, the greater the percentage of fines (silt- and clay-sized particles) in the soil, the greater is its capacity to retain bacteria. Besides, the charged nature of bacteria and viruses facilitates their adsorption on soil constituents. As it is difficult to detect viruses in soils and waste disposal systems, the abundance of faecal streptococci, and faecal coliforms are used as indicators of pathogenicity.

Excess amounts of NO_3^- may be toxic to infants and young animals, and both NO_3^- and P promote eutrophication of surface waters. NH_4^+ concentrations have decreased to background levels after percolating through 76 cm of soil fill. The mine soil-fill has been found to be very efficient at removing PO_4–P from the wastewater.

Wastewater could be applied at the rate of (say) 19.3 L/m/d on at least 0.76 m of mine soil-fill. Uniform distribution of effluent could be ensured by using low-pressure distribution or drip irrigation system (Harrison et al., 1999).

Tailings have been used in USA as bulk fill in highways, embankment material, as aggregate for sub-base and bituminous paving mixtures, in building bricks and blocks, and in the manufacture of low-grade glass (Collins and Miller, *Minerals and Environment*, 1979). China's largest gold producer, Shangdong Gold Group Co. Ltd, has recently commissioned a 4 million m^3 tailings brick manufacturing plant – it is expected to generate annual profits of Yu 12 million and pay back the company's investment in five years (*Mining Journal*, Aug. 18, 2000 issue). Zambia converted the abandoned open pits to fish ponds.

3.6.3 REUSE OF MINE WATER

Mine water is invariably highly acidic, besides containing undesirably high quantities of toxic metals. There is severe scarcity of drinking water in the coalfield areas of eastern India. On one hand, the water-table has gone down to 200–250 m due to mining activities, thus making the tapping of groundwater prohibitively expensive. On the other hand, there is abundance of mine water, which, however, is not potable because of its high acidity, and the high content of metals, such as iron. The Central Mining Research Institute (CMRI), Dhanbad, Bihar, 826 001, India, has developed a treatment process which is claimed to render the mine water potable (item 6.2.13, *CSIR Rural Technologies*, New Delhi, India, 1995). Filtration is done adjacent to the settling pond. Two filter beds are used to work alternatively at the time of changing the bed. A slow or rapid filtration may be employed depending upon the situation. A disinfectant is incorporated in the treatment process to destroy the pathogens.

The presence of high iron content in groundwater is objectionable because of discoloration, turbidity, bad taste and tendency to form deposits in the distribution mains. The National Environmental Engineering Research Institute, Nagpur 440 020, India, developed a simple plant to remove iron from groundwater by precipitating the iron impurity as a ferric sludge (item 6.2.4, *CSIR Rural Technologies*, 1995). The plant is to be attached to a hand pump. It has a capacity of 2500 L/d (10-hr operation) and costs about Rs. 25,000 (USD 500). The plant has three chambers. "The water from the hand pump is sprayed over an oxidation chamber. The aerated water flows over baffle plates to a flocculation chamber and then to sedimentation chamber. The water then passes through plate settlers and to the filter from where the filtered water is drawn through a tap after chlorination". The ferric sludge needs to be scoured out twice a month.

Chapter 3.7

Iron ore mine, Kiruna, Sweden – A case study

The LKAB iron ore mines in Kiruna, Sweden, are located above the Arctic Circle. Two underground mines, located at Kiruna and Malmsberget, employ sublevel caving method, to produce 30 Mt/y of iron ore. Much of the mining operation is automated – for instance, the production drilling rigs, loading machines and transport systems on the new main haulage level are remote-controlled. The magnetite ore is crushed, ground, screened and upgraded by magnetic separation techniques. The iron ore concentrate is pelletised using bentonite as a binder.

The LKAB complex demonstrates how continuous improvements in technologies could bring about high productivity, while at the same time reducing water and air pollution (Nordstrom, 2001). Mining and process water goes through large pond systems, to facilitate the removal of the sludge by sedimentation. The water is then pumped to the clarifying ponds from which the process water is recirculated. Since 1980, LKAB's atmospheric emissions have been halved, while the production of pellets has doubled. The external energy consumption in the pelletising plants which was 639 MJ (million joules)/t of pellets in 1970 has been brought down to 250 MJ/t in 2000.

The following air pollutants are produced in the process of pelletisation:

(i) Fine dust, mainly composed of iron oxides, before induration,
(ii) Sulphur dioxide, from sulphur in the fuel, and sulphides in green pellets,
(iii) Hydrogen fluoride and hydrochloric acid from apatite residues in green pellets,
(iv) Nitrogen oxides, from nitrogen in the fuel and in the atmosphere.

That the emissions in the pelletising plants in Kiruna are less than the statutory limits is evident from the following information:

Parameter	*Statutory limit (g/t of pellets)*	*Emissions in Kiruna (g/t of pellets)*
Particles	100	60
Sulphur dioxide	15	13
Hydrogen fluoride	6	2
Hydrochloric acid	6	2

Noise is generated by ventilation fans, mining and transport equipment. The maximum noise level nearest to the dwelling should not exceed 40–45 dB (A). Blasting in mines produces vibrations in the bedrock in the surrounding areas. Noise and vibrations are measured continuously to minimize the inconvenience to property owners in the neighbourhood.

Chapter 3.8

Basic research and R&D

The public abhors the mining industry because of the environmental damage it causes. So the most important goal of the basic research, and R&D has to be to minimize the adverse environmental impact of the mining industry. In the Indian context, most of the deposits of iron ore, manganese ore, bauxite etc. are located in the forest areas that are inhabited by adivasis (tribal people) who have a strong attachment to land, and worship some promontories as sacred. Bastar District in Central India, has huge deposits of iron ore, and tribals live there. The mining of the iron ore by the National Mineral Development Corporation (NMDC) displaced the tribals. At a later stage, when the Govt. of India offered to provide them compensation, the tribal leader refused it, saying, "*Land is my Mother. I will not sell my Mother*".

All mining companies are signatories to the UNEP's International Declaration of Cleaner Production. In June 2000, UNEP set up the Global Reporting Initiative (GRI) Sustainability Reporting Guidelines, to which all the mining companies agreed to adhere. The vision of UNEP (Hoskin, 2001) to achieve environmentally – sustainable mining industry in the early part of the twenty-first century, has the following components:

1. Recycling of metals should approach 100% – recycling of metals reduces disposal pressures, and results in great energy savings. The limited amount of virgin metal that may be needed should be obtained from highest-grade reserves.
2. Technological improvements: (1) Application of remote sensing and hydrospectral analysis in exploration, and monitoring of tailing impoundments, closed mines, and compliance with environmental regulations, (2) minimization of mine wastes, reducing air and water pollution to essentially zero level, and fabrication of lighter, stronger and more durable materials, and secondary recovery of useful materials

from mine wastes, (3) Remediation of abandoned mine sites to increase arable land for agricultural production.

3. One of the major sources of accidents in the mining industry is the failure of tailings' dams. This problem can be mitigated by (1) new technologies to dewater waste slurries – production of paste-consistency material from mill tailings, (2) A thorough analysis of all the design components including site selection, drainage systems, impoundments, measurements and inspections needed with respect to water balance, taking into consideration unusual conditions arising from rain, ice and snow and seismic activity.
4. Training and assistance to small-scale miners, particularly in Africa, Asia and Latin America, to improve the commodity recovery, reduce environmental damage, and improve local health and safety conditions. Phasing out of the use of mercury in artisanal gold mining.

The environmental impact of a given mineral industry depends upon the geologic setting, genesis and mode of occurrence of the mineral deposit, which in turn determines how the mineral is to be mined, and processed and how the wastes are to be disposed. This concept is illustrated with two examples.

Because of the nature of the genesis, bauxite, the ore of aluminium, is invariably a surface deposit, which is therefore mined by opencast methods. The mining of bauxite removes the vegetation, dries up the water resources and the area becomes pockmarked with pits. Aluminium dusts generated by the crushing of bauxite create health problems for the community. Huge quantities of "red mud" are generated in the process of chemical conversion of bauxite to alumina. Aluminium industry discharges large amounts of fluoride-loaded particulates which can cause dental mottling and skeletal fluorosis in human beings and animals. Aluminium plants produce cryolite mud (at the rate of 0.02 t of cryolite mud per tonne of cryolite used) which contains toxic heavy metals, such as, arsenic, cadmium, nickel, etc.

When an aluminium industry is planned, basic research and R&D need to be performed on the following activities to enable the design of a flowsheet whereby the adverse effects are minimized: (i) vegetating the mined area with appropriate fruit and timber trees, depending upon the agroclimatic setting of the area, (ii) use of dust control chemicals, to control dust dispersion, (iii) making bricks with red mud wastes – the presence of 4–5% alkalis in the red mud makes for good fluxing action. Consequently, the red mud bricks have better plasticity and bonding than normal bricks, (iv) determination of the mineralogy, microanalysis, morphology and textures of the dust particles to choose the most effective particulate control technology.

Underground methods are employed to mine vein deposits of (say) primary sulphide ores of base metals. Ore concentrates are produced from ROM through processes such as flotation. Sulphide ores are invariably polymetallic. A copper ore may contain lead, zinc, gold, cadmium, nickel, etc. The metals include high-value metals like gold, toxic metals, like cadmium and nickel. Basic research and R&D have to be performed in order to design a flowsheet which provides for maximum extraction of high-value metals, and minimum extraction of toxic metals. The adverse impact of the sulphide mining arises from the Acid Mine Drainage (AMD) from the mine and waste piles, SO_2 emissions and acid rain. These are mitigated by the use of scrubbers and through the prevention and control of AMD, and by passive treatment of AMD through natural or constructed wetlands.

References

Alexieva, T. 2002. The benefits of the risk assessment to the mining industry: pp. 295–301. *Proc. Int. Symp. on Tailings and Mine Waste '02.*

Angelos, M. & P. Niskanen 2001. Research into the rehabilitation of the Luikonlahti copper mine: pp. 21–30. *Proc. Int. Conf. on Mining and the Environment, Skellefteå, June 25–July 1, 2001.*

Archer, A.A., G.W. Luttig & I.I. Snezhko (eds.) 1987. *Man's Dependence on the Earth.* Nairobi–Paris: UNEP–UNESCO.

Aswathanarayana, U. 2003. *Mineral Resources Management and the Environment.* Rotterdam: A.A. Balkema

Ayres, B.K., M. O'Kane, D. Christensen & L. Lanteigne 2002. Construction and instrumentation of waste rock test covers at Whistle Mine, Ontario, Canada: p. 163–171. *Proc. Int. Symp. on Tailings and Mine Waste '02.*

Brandes, M., N.A. Chowdhry & W.W. Cheng 1975. Experimental study on removal of pollutants from domestic sewage by underdrained soil filters: pp. 29–36. *Proc. National Home Sewage Disposal Symp., Chicago, IL, Dec. 9–10, 1974.*

Brierley, C.L. 1990. Bioremediation of metal-contaminated surface and groundwaters. *Geomicrobiology J.* 8: 202–233.

Brierley, C.L., J.A. Brierley, and M.S. Davidson.1989. Applied microbial processes for metal recovery and removal from wastewaters. In T.J. Beveridge & R.J. Doyle (eds.) *Metal Ions and Bacteria.*

Chadwick, M.J., N.H. Highton & J.P. Palmer 1987. *Mining Projects in the Developing Countries – A Manual.* Stockholm: Beijer Inst.

Chadwick, J. 2001. Mining chemicals. *Mining Mag.*, Sep. 2001, 185(3): 120–126.

CCORE 1996. Proposed mining technologies and mitigation techniques: A detailed analysis with respect to mining of specific offshore commodities. CCORE Publ. 96-C, 15.

Davé, N.K. & A.D. Paktunc 2001. Hydrogeochemistry and mineralogy of inactive and rehabilitated pyritic uranium tailings – case histories of Stanrock and Lower Williams lake tailings sites, Elliot Lake, Ontario, Canada: pp. 127–136. *Proc. Int. Conf. on Mining and the Environment, Skellefteå, June 25–July 1, 2001.*

Davison, J. 1993. Successful acid mine drainage and heavy metal site bioremediation. In G.A. Moshiri (ed.) *Constructed wetlands for water quality improvement.* Chap. 16, Boca Raton, USA: CRC Press.

Eger, P. & K. Lapakko 1989. Use of wetlands to remove nickel and copper from mine drainage. In O.A. Hammer (ed.), *Constructed Wetlands for Wastewater Treatment – Municipal, Industrial and Agricultural*, Chap. 42e. Boca Raton, Fl., USA: Lewis Publishers.

Ellis, D.V. & J.D. Robertson 1999. Underwater placement of mine tailings: case examples and principles. In J.M. Azcue (ed.), *Environmental Impacts of Mining Activities*, Chap. 9: pp. 123–140. Berlin: Springer-Verlag.

Eriksson, N., M. Lindvall & M. Sandberg 2001. A quantitative evaluation of the effectiveness of the water cover at the Stekenjokk tailings pond in northern Sweden: Eight years of follow-up: pp. 216–227. *Proc. Int. Conf. on Mining and the Environment, Skellefteå, June 25–July 1, 2001.*

Ferrow, E.A. 2001. Application of Mössbauer spectroscopy in mining and agrochemical prospecting. pp. 228–239. *Proc. Int. Conf. on Mining and the Environment, Skellefteå, June 25–July 1, 2001.*

Förstner, U. 1999. Introduction. In J.M. Azcue (ed.), *Environmental Impacts of Mining Activities,* Chap. 1: pp. 1–3. Berlin: Springer-Verlag.

Grayson, R.L. 1999. Mine Health and Safety: Industry's march towards continuous improvement – the United States experience. In J.M. Azcue (ed.), *Environmental Impacts of Mining Activities*, Chap. 7: p. 83–99. Berlin: Springer-Verlag.

Gusek, J.A. 1995. Passive-treatment of acid rock drainage: what is the potential bottom line? *Mining Engg.* 97: 250–253.

Harrison, A-I, B.B. Reneau Jr. & C. Hagedorn, 1999. Wastewater renovation with mine-derived fill material. In J.M. Azcue (ed.) *Environmental Impact of Mining Activities.* Chap. 11, pp. 163–177. Berlin: Springer-Verlag.

Höglund, L.O. 2001. MiMi – Research to meet future demands: p. 280–291. *Proc. Int. Conf. on Mining and the Environment, Skellefteå, June 25–July 1, 2001.*

Hoskin, M.A. Wanda 2001. Environment and Mining in the 21st century. *Appendix to the Proc. Int. Conf. on Mining and the Environment, Skellefteå, June 25–July 1, 2001.*

Hutchinson, R.W., 1987, Metallogeny of Precambrian gold deposits – space and time relationships. *Econ. Geol.*, v.82, pp. 1993–2007.

Khanna, T. 1999. Mining and the environmental agenda. *Mining Mag.* 183(1): 158–163.

Klapper, H. & M. Schultze 1997. Sulfur acidic mining lakes in Germany – ways of controlling geogenic acidification. In Canadian Mine Environmental Neutral Drainage Program (MEND) (ed.), *Proc. IV Int. Conf. on Acid Rock Drainange, Vancouver, May 31–June 6, 1997*, Vol. IV: 1727–1744.

Kolbash, R.L. & T.L. Romanovski 1989. Windsor Coal Company Wetland: An overview. In O.A. Hammer (ed.), *Constructed Wetlands for Wastewater Treatment – Municipal, Industrial and Agricultural*, Chap. 42e. Boca Raton, Fl., USA: Lewis Publishers.

Lacki, M.J., J.W. Hummer & H.J. Webster 1992. Mine-drainage treatment as habitat for herptofaunal wildlife. *Environ. Management* 16(4): 513–520.

Lindström, P., K.E. Isaksson, J. Ljungberg & M. Lindvall 2001. The gold leach plant expansion at the Boliden Mill – Design considerations, baseline investigations and approval process: p. 438–445. *Proc. Int. Conf. on Mining and the Environment, Skellefteå, June 25–July 1, 2001.*

Maxwell, P. & S. Govindarajalu 1999. Do Australian companies pay too much? Reflections on the burden of meeting environmental standards in the late twentieth century. In J.M. Azcue (ed.), *Environmental Impacts of Mining Activities*, Chap. 2: pp. 7–17. Berlin: Springer-Verlag.

McNulty, T. 2001. Cyanide substitutes. *Mining Mag.* 184(5): 256–261.

MEND (Mine Environment Neutral Drainage Program) 1997. *Evaluation of techniques for preventing Acid Rock Drainage.* Final Report no. 2.35.2b. Natural Resources Canada, Ottawa.

Miller, R.O. 1999. Mining, environmental protection, and sustainable development in Indonesia. In N.J. Vig & R.S. Axelrod (eds), *The Global Environment: Institutions, Law and Policy.* Washington, D.C.: Congress Quar. Inc.

*MiMi 1998. *Prevention and control of pollution from mining waste products, State-of-the-art Report.* Compiled by P. Elander, M. Lindvall & K. Håkansson.

Moellerherm, S. & P.N. Martens 2002. Use of copper mine tailings as paste backfill material in mining operations – Approach to minimize land occupation?: p. 149–153. *Proc. Int. Symp. on Tailings and Mine Waste '02.*

Nordström, K. 2001. Environmental impacts of pelletising at LKAB: p. 604–614. *Proc. Int. Conf. on Mining and the Environment, Skellefteå, June 25–July 1, 2001.*

Pai, B.H.G. & B.V. Shenoi 1988. Occupational hazards of deep-level hard-rock mining on the Kolar Gold Fields. In S.C. Joshi & G. Bhattacharya (eds), *Mining and Environment in India*: p. 206–212. Himalayan Research Group, Nainital, U.P., India.

Robbins, G. et al. 2001. Cyanide management at Boliden, using the Inco SO2/AIR process. p. 718–728, *Proc. Int. Conf. on Mining and the Environment, Skellefteå, June 25–July 1, 2001.*

Sengupta, M. 1993. *Environmental Impacts of Mining.* Boca Raton, Fl., USA: Lewis Publishers.

Singh, B., G.B. Mathur & P.R. Landge 1995. Environmental management for the proposed Chapri – Sidheswar mine at ICC: p. 203–213. *Proc. First World Mining Congress, New Delhi, India.*

Skarzynska, K.M. & P. Michalski 1999. Environmental effects of the deposition and reuse of colliery spoils. In J.M. Azcue (ed.), *Environmental Impacts of Mining Activities*: p. 179–200. Berlin: Springer-Verlag.

Sofra, F. & D.V. Boger 2002. Planning, design and implementation strategy for thickened tailings and pastes: p. 129–137. *Proc. Int. Symp. on Tailings and Mine Waste '02.*

Stottmeister, U. et al. 1999. Strategies for the remediation of former opencast mining areas in Eastern Germany. In J.M. Azcue (ed.), *Environmental Impacts of Mining Activities*, Chap. 16: pp. 263–294. Berlin: Springer-Verlag.

Struthers, S. 2001. Total Project Development – A new approach to mining: p. 814–823. *Proc. Int. Conf. on Mining and the Environment, Skellefteå, June 25–July 1, 2001.*

UNEP, 1986. Tech. Rev., *Environmental aspects of Iron & Steel production.*

UNEP, 1991, Tech. Rept. No. 5, *Environmental aspects of selected non-ferrous metals ore mining.*

Vartanyan, G.S. (ed.) 1989. *Mining and Geoenvironment.* Paris–Nairobi: UNESCO–UNEP.

Vermuelen, N.J., E. Rust & C.R.I. Clayton 2002. Variations in composition on South Africa gold tailings dam: p. 45–52. *Proc. Int. Symp. on Tailings and Mine Waste '02.*

White, C. & G.M. Gadd 1996. A comparison of carbon/energy and complex nitrogen sources for bacterial surface reduction: potential application in bioprecipitation of toxic metals as sulphides. *J. Industrial Microbiology*, 17, 116–123.

Section 4

Energy resources management

U. Aswathanarayana (India)

This section drew the basic concepts and data from the earlier works of the author (Aswathanarayana, 1985, 2001, 2003, 2008, 2009, 2010), and then updated.

Chapter 4.1

Coal resources

4.1.1 IMPORTANCE OF COAL IN THE ENERGY ECONOMY

Coals are formed from the accumulation of vegetable debris in geologic formations ranging in age from Upper Palaeozoic to Recent. The rank of coal (peat, lignite, sub-bituminous coal, bituminous coal, semi-anthracite and anthracite, in order of increasing rank) is determined by the depth and length of burial. Thus, palaeozoic coals tend to bituminous and anthracitic, and Tertiary coals, lignitic.

Fossil fuels (coal, oil and natural gas) continue to supply much of the energy used worldwide. The world energy consumption by source in 2000, was as follows:

Oil – 35%, Coal – 24%, Natural gas – 21%, Biomass waste – 11%, Nuclear – 7%, Hydropower – 2%, Geothermal, wind and solar – less than 1%. Thus, fossil fuels account for about 80%.

Coal provides about a quarter of the global energy needs, and accounts for 40% of the world's electricity. Some countries are highly dependent upon coal for their electricity production: Poland – 93%; South Africa – 93%; Australia – 80%; China – 78%; Israel – 71%; Kazakhstan – 70%; India – 69%; USA – 50% (Source: World Coal Institute, 2007 edition).

Coal is mined in all continents, except Antarctica. Proven Coal reserves in the world (847 billion tonnes) are expected to last 118 years, while the oil and gas reserves have life times of 41 and 63 years respectively. Also, unlike coal whose occurrence is wide-spread, about two-thirds of oil and gas deposits in the world are confined to Middle East and Russia.

Countries with more than 0.1% of the global coal reserves are listed in Table 4.1.1.

It may be seen that just four countries, USA, Russia, China and India, account for about two-thirds of the coal reserves of the world.

Table 4.1.1 Proved recoverable coal reserves at end-2006 (in million tonnes)

Country	*Bituminous & anthracite (Mt)*	*Sub-bituminous & lignite (Mt)*	*Total (Mt)*	*Share (%)*
USA	111,338	135, 305	246,643	27.1
Russia	49,088	107,922	157,010	17.3
China	62,200	52,300	114,500	12.6
India	90,085	2,360	92,445	10.2
Australia	38,600	39,900	78,500	8.6
S. Africa	48,750	0	48,750	5.4
Ukraine	16,274	17,879	34,153	3.8
Kazakhstan	28,151	3,128	31,279	3.4
Poland	14,000	0	14,000	1.5
Brazil	0	10,113	10,113	1.1
Germany	183	6,556	6,739	0.7
Colombia	6,230	381	6,611	0.7
Canada	3,471	3,107	6,578	0.7
Czech Rep	2,094	3,458	5,552	0.6
Indonesia	740	4,228	4,968	0.5
Turkey	278	3,908	4,186	0.5
Greece	0	3,900	3,900	0.4
Hungary	198	3,159	3,357	0.4
Pakistan	0	3,050	3,050	0.3
Bulgaria	4	2,183	2,187	0.2
Thailand	0	1,354	1,354	0.1
N. Korea	300	300	600	0.1
New Zealand	33	538	571	0.1
Spain	200	330	530	0.1
Zimbabwe	502	0	502	0.1
Romania	22	472	494	0.1
Venezuela	479	0	479	0.1
TOTAL	478,771	430,293	909,064	100.0

Source: B.P. Statistical Review of World Energy Supplies, 2007

The coal production in the world has been steadily going up. The total global hard coal production in the world in 2010 was 6185 Mt (million tonnes). The total brown coal/lignite production in the world in 2010 is 1042. The top four hard coal producers in the world and the production in Mt are as follows: China – 3162; USA – 932; India – 538; South Africa – 255. Thus, China alone produces more than half of the coal production in the world. About 13% of hard coal is used in the iron and steel industry.

4.1.2 ENVIRONMENTAL IMPACT OF THE COAL CYCLE

4.1.2.1 General considerations

The major environmental and health impacts of the coal cycle (mining, preparation and beneficiation, transport and usage and utilization) are summarized in Table 4.1.2 (Chadwick et al., 1987, pp. 135–136).

Table 4.1.2 Major environmental and health impacts of coal cycle

Operation	*Environmental impact*	*Health impact*
Surface mining: Many variations (contour stripping, mountain top area mining, open pit), and machinery-use options (shovel-truck, dragline, continuous mine). All basically involve removal of vegetation, top soil and overburden to expose coal. Usual depth limit ~300 m.	Destruction and disruption of vegetation, natural drainage patterns and land use in the area of the mine. Erosion of cleared areas and soil and overburden dumps leading to sedimentation and pollution of water courses. Possibility of acid mine drainage. Dust created during operations causing visibility problems and loss of agricultural production. Water consumption effects in arid areas.	Noise and vibration effects from machinery. Blast effects. Potential silicosis and respiratory problems.
Deep mining: Basically two types – longwall, and board-and-pillar, but many variations and degrees of mechanization. Access to seams by vertical shaft or drift,	Production of surface spoil heaps with potential erosion effects such as sedimentation and acidification of water courses. Possibility of spontaneous combustion of spoil heaps causing air pollution and tip instability. Mine drainage adversely affecting the water quality of a large area by removing soluble minerals from aquifers and by the acidification of surface water courses. Loss of agricultural productivity over large areas caused by subsidence. Water consumption effects in arid areas.	Noise, vibration and blast effects. Pneumoconiosis and other respiratory problems from dust. Effects of mine gases. Poor working environment – high temperatures, wet conditions, inadequate light. Hard physical work. High accident rate.
Coal transport and storage: Conveyor, slurry pipeline, truck, railway, barge, ship.	Dust effects particularly during transit and at transfer points. Water pollution from disposal of untreated slurry water.	Dust effects.
Utilization: Coking, direct combustion, coal conversion.	Emissions from all processes of particulates, nitrogen and sulphur dioxides, carbon monoxide, hydrocarbons and trace elements. Disposal of liquid effluents, e.g. ammoniacal liquor from coking. All processes produce large amounts of solid waste which can pose problems of erosion, runoff, toxicity and contamination of water courses.	Emissions of noxious gases, heat and dust. Process and End-product related to occupational health risks.

The environmental impact of coal industry arises from the preparation, transportation; washing and combustion of coal, etc. (see Das & Chatterjee, 1988, for a good summary of the topic).

4.1.2.2 Preparation of coal

Good part of the global production of coal (about 6,185 Mt) undergoes some form of preparation before it is used directly (say, in a pit-mouth thermal power station)

or sold. Consumers demand a high degree of consistency in the product sold to them, and the environmental regulations need to be adhered to.

To start with, the ROM coal and the associated refuse material are analysed for their mineralogy, size distribution, hardness, calorific value, coking properties, etc. in order to determine their treatability by the main separation techniques available. ROM coal is crushed using jaw, gyratory or rolls crushers, and then screened to produce different size fractions.

"Clean" coal is lighter, because of its lower ash content. Hence coal separation is effected using the density criterion. Jigging using the water medium is by far the oldest and the simplest method of separation of clean coal. The density of the medium can be raised to the desired level by making use of water-based suspensions of sand, shale, barite, magnetite, etc. The lighter clean coal particles float to the top of the washery cell, whereas the higher density waste particles accumulate at the bottom, with middlings in between. The cycle is repeated until the needed separation is effected.

Separation of coal and waste material can also be effected using cyclones, which may make use of water or some other appropriate (heavier) medium. Occasionally, shaking tables, launders and spirals are used.

Froth flotation could be made use of to clean coal fractions with a maximum diameter of 0.5 mm. In the flotation cell, air is bubbled through the coal slurry, which contains the collector reagents. The aerophyllic coal particles rise to the surface, while the hydrophilic shale and pyrite particles sink to the bottom of the cell.

4.1.2.3 Disposal of coal mine tailings

Coal mining industry produces enormous quantities of wastes, considering that the world production of coal is 6,185 Mt in 2010. Apart from the solid wastes (shale rock, dolerite, burnt coal, etc.) produced due the mining of coal, coal preparation plants produce large quantities of coarse and fine particles, and contaminated water. The usual practice is to impound the tailings and slurry in the lagoons. As breaching of the lagoons constitute some of the most serious environmental disasters associated with mining, it is critically important that great care be taken in the design of the lagoons. The stability of, and control of seepage from, the lagoons can be ensured by keeping in mind the following design parameters: (i) Construction of the lagoon at or below the ground level, (ii) the banks should have a slope of 34° on the lagoon side, and 26.5° on the outer side, (iii) there should be a toe drain to allow the water table drawdown below the outer slope, (iv) there should be a free board (water surface to the crest of the lagoon bank) of at least one metre, (v) the inner surface of the bank should be capable of withstanding erosion due to wave action, (vi) there should be provision for drawing off the supernatant, as also rainwater due to abnormally heavy rainfall, etc. (Chadwick et al., 1987, pp. 126–127). The supernatant water from the lagoons may be either recycled, or discharged into the natural waterways after treatment.

Coal mines generally use timber for roof support. Experience shows that a coal mine with a production of (say) 400,000 tonnes per annum uses 9,000 to 12,000 m^3 of timber, which is usually obtained from the local forests. Thus, local forests tend to disappear unless tree crops are grown to provide the wood needed on a continuing basis. Dewatering of the mines may lead to significant changes in the vegetation in the mining area.

Plants are particularly susceptible to atmospheric pollution. The intensity of photosynthesis is adversely affected by pollutants such as, sulphur dioxide, carbon monoxide and hydrocarbons, which cause necrosis of leaves, inhibition of growth and early leaf fall. Eventually, the plants wither and die. Space photographs clearly show the devastation of the vegetation caused by mining in different parts of the world.

4.1.3 WASTES FROM COAL INDUSTRIES

4.1.3.1 Solid wastes

When the soil or sediment cover is removed in the process of quarrying, their filtering and attenuation capabilities would have been lost, thus exposing the groundwater to greater risks of pollution. Groundwater could also be contaminated due to some ancillary activities associated with quarrying, such as, accidental spillages of fuel oil, leakages from storage tanks or toilets for workers, or draining of water from the surrounding areas into the quarry, etc.

Solid wastes arising from the mining of coal and lignite tend to contain pyrite (FeS_2). Under oxidizing conditions, and in the presence of catalytic bacteria, such as *Thiobacillus ferrooxidans*, pyrite gets oxidized into sulphuric acid and iron sulphate. Thus, surface runoff and groundwater seepages associated with waste piles tend to be highly acidic, and corrosive. The discharge of such waters (known as Acid Mine Drainage or AMD) into streams destroys the aquatic life, and the stream water is rendered non-potable.

The solid wastes produced in mining activities which could contaminate water resources through leaching and effluent production, are listed in Table 4.1.3.

Fly ash is reactive because of its high surface area: volume ratio. Leaching of the fly ash may produce effluents containing toxic elements, such as, Mo, F, Se, B and As. Low pH leachates from fly ash may give rise to problems of iron floc formation in surface

Table 4.1.3 Solid wastes from mining

Source	*Potential characteristics of leachate/effluent*	*Rate of effluent or solid waste production*
Coal-mine drainage	High total dissolved solids. Suspended solids. Iron. Often acid. May contain high chlorides from connate water	10^5–10^7 m^3y^{-1}
Colliery waste	Leachate similar to mine drainage waters	10^5–10^7 ty^{-1} of wastes per colliery. Quantity of leachate depends on climate
Power generation (thermal)	Pulverized fuel ash. Upto 2% by weight of soluble constituents, sulphate. May contain Concentrations of Germanium and Selenium. Fly ash and flue gas scrubber sludges. Finely particulate, containing disseminated heavy metals. Sludges of low pH unless neutralized by lime addition.	10^4–10^6 ty^{-1}

Source: Laconte & Haimes, 1982

waters. When the flue gases are scrubbed, the resulting sludge will typically contain cyanide and heavy metals. Its pH will be low, unless neutralized by lime. It has been reported that mixtures of sludge, lime and fly ash will set rapidly to a load-bearing, low-permeability, solid which is not easily leachable. Two benefits accrue from this process – on one hand we will have a useful construction material, and on the other, we would have minimized the pollution risk.

An environmentally-sound, and technoeconomically viable approach to minimize the contamination potential of the wastes, such as fly ash, is to put them to some useful purpose soon after they are produced. Fly ash – clay bricks can be prepared by the incorporation of 10–40% of fly ash to the clay, and subsequent firing in conventional continuous type Bull's kiln or intermittent type kilns at a temperature of 950°C and 1050°C. With the use of fly ash, 40% more bricks can be produced from the same quantity of soil. The clay – fly ash bricks are lighter, have better thermal insulation and lower dead load than clay bricks. Incorporation of siliceous materials would make the red mud suitable for making building bricks. Its alkali content (4–5%) provides a fluxing action, resulting in good plasticity and bonding of the bricks. Red mud bricks develop a pleasing pale brown, orange or golden yellow color, depending upon the composition of the red mud and firing temperature. They have therefore good architectural value as facing bricks (source: *CSIR Rural Technologies*, New Delhi, India, 1995, pp. 82–83).

4.1.3.2 Liquid wastes

How industrial effluents arising from coal mining affect the environment is illustrated with the example of coal mining in the Damodar River basin in eastern India.

Damodar river basin in eastern India contains about 46% of the coal reserves of India. Apart from underground and opencast mines of coal, the area has numerous coal-based industries, such as, steel, chemical and fertilizer plants. Fortunately, the pyrite content of coal is not high. The water of the Damodar River is contaminated by (i) huge volumes of polluted water from underground mines – for instance, Bharat Coking Coal Limited (BCCL) mines in Jharia pump out 300 gallons of mine water daily, (ii) runoff water leaching the overburden dumps, (iii) industrial effluents from coal-based plants. Large amounts of fly ash and fine coal particles discharged by the thermal power plants and washeries settle down to the bed of the river and hinder the growth of the biota. About 25 million tonnes of coal is washed annually in the area, involving the use of about 2000 t of pine oil. Large amounts of suspended solids, oil and grease arising from the washeries are discharged into the river. The coke oven plants serving the steel industry, release highly toxic substances like phenols and cyanides. The low DO contents and high content of heavy metals in the effluents make it almost impossible for biota to survive in the river water.

The quality of the water in the upstream part of the Damodar river is fairly good with TSS in the range of 18–168 mg/l. But serious deterioration of water quality occurs when the industrial waters from Patratu thermal power plant, and the steel plant at Durgapur are discharged into the river. But people who live in the area have no option except to drink the contaminated water. A health survey conducted in 1993–94 showed high incidence of water-related diseases, such as, dysentery, diarrhea, skin infections, jaundice, typhoid, etc. (Tiwary et al., 1995).

Table 4.1.4 Summaries of the physico-chemical characteristics of the industrial effluents in the Damodar river basin

Parameter	*IS:2490*	*Coke oven plant*	*Thermal power plant*	*Coal washery*	*Steel plant*
Flow rate (m^3/hr)	–	10.8–72	81–33768	30–810	112–7245
pH	5.5–9.0	7.67–8.62	7.98–8.46	7.31–7.49	2.98–7.3
Temp. (°C)	40.0	29.2–40.0	28.0–39.0	29.1–31.2	20.0–36
TSS	100	447–636	780–20400	160–83560	334–1465
TDS	2100	486–581	206–398	304–775	16–186
DO	–	0.5–2.2	3.98–5.89	2.61–3.84	2.86–7.52
BOD	30	7.8–81.0	3.0–22.0	4.0–10.0	8.0–32.0
COD	250	208.3–331.2	359–5192	2073–63848	55–1098
Phenols	–	0.1–0.2	Nil	Nil	0.02–1.13
Cyanide	0.2	0.002–0.2	Nil	Nil	0.01–0.07
Oil & Grease	10	0.98–1.26	0.7–1.8	0.4–32.5	0.86–112.4
Fe	–	2.89–6.54	5.61–83.6	0.11–181.3	2.17–37.0
Mn	–	0.28–0.45	0.08–1.19	0.06–6.20	0.14–0.41
Cr	2.0	BDL–0.05	BDL–0.17	0.18–0.43	BDL–0.34
Pb	0.1	0.0–0.5	BDL–0.09	020–0.75	BDL–0.06
As	0.2	BDL	BDL	BDL–0.02	BDL
Cd	2.0	BDL	BDL	BDL–0.2	BDL–0.12

BDL – Below Detection Limit
All parameters are expressed in mg/l, except flow rate, and pH
Source: Tiwary et al., 1995

Table 4.1.5 Average emissions from 1000 MW coal-fired and oil-fired stations (in tonnes)

	Coal-fired station	*Oil-fired station*
Sulphur dioxide	110,000	37,000
Nitrogen oxide	27,000	25,000
Particulates	3,000	1,200
Carbon monoxide	2,000	710
Hydrocarbons	400	470
Ash	360,000	9,000

Source: El-Hinnawi, 1981

4.1.3.3 Emissions due to coal industries

Energy industries (particularly coal-fired, thermal power stations produce huge quantities of gaseous pollutants (Table 4.1.5).The global emissions of heavy elements and the contribution by the combustion of coal and oil (in terms of $10^9\,g\,y^{-1}$) are given in Table 4.1.6.

4.1.4 POWER GENERATION TECHNOLOGIES

The following account is largely drawn from "*World Energy Outlook 2007*", pp. 220–222.

Traditionally, coal-based power generation involved the combustion of pulverized or powdered coal to raise steam in the boilers. The efficiency of the process currently

Table 4.1.6 Heavy element emissions from coal and oil combustion (10^9 g y^{-1})

Element	*Total*	*Coal*	*Oil*
As	23.6	0.7	0.002
Se	1.1	0.42	0.03
Cd	7.3	–	0.002
Hg	2.4	0.0017	1.6
Pb	449	3.5	0.05

Compiled by Fergusson, 1990

ranges between 30 to 40%, depending on the quality of coal, ambient conditions and the back-end cooling employed. A number of power-generation technologies are being developed which are characterized by higher thermal efficiencies, but also lower emissions of nitrogen oxides (NOx) and sulphur dioxide (SO_2).

The most important new technologies are as follows:

Supercritical and ultra-supercritical pulverized coal combustion: Steam pressure and temperature largely determine the efficiency of the steam cycle. Most of the current plants use subcritical technologies which operate at 163 bar pressure and 538°C temperature. In the supercritical designs, the pressure used is typically 245 bars, and temperature of more than 550°C – under these conditions, water turns into steam without boiling. In the ultra-supercritical designs, the temperature used exceeds 600°C. Though such units need specialized materials for fabrication, and higher capital costs, their higher efficiencies make them cost-effective. Supercritical technologies have become the norm in OECD countries. China and India are building supercritical power stations. Japan, Germany and Denmark have commercial ultra-supercritical plants. Extensive R&D effort is ongoing to develop new kinds of materials, which allow efficiencies of more than 50%.

Circulating fluidized bed combustion (CFBC): It is possible to design CFBC plants for a variety of coals and particle sizes. As coal is burned at low temperatures, less NO_x is produced compared with conventional pulverized coal (PC) boilers. Also, the operating temperatures are low enough for the *in situ* capture of SO_2. The efficiency of CFBC plants is similar to that of PC units. CFBC units that can be operated at supercritical mode are now available commercially. Poland is currently building the first supercritical CFBC plant (460 MW). CFBCs may not be practicable at ultra-supercritical plants, as the temperature of operation may exceed 550°C.

Integrated Gasification combined – cycle (IGCC): In this system, coal is gasified under pressure with air or oxygen to produce fuel gas which is then cleaned and burned in a gas turbine to produce power. Heat is recovered from the exhaust gas from the gas turbine, and is used to generate extra power. Efficiencies of the order of 41% have been achieved, and even higher efficiencies are possible with new gas turbine models. IGCC plants have been built in Europe, USA, China and Japan. The advantages of the IGCC plants are: (i) it is possible to control emissions as the gas clean-up takes place before the combustion of fuel gas, (ii) equipment needed is not too elaborate, and (iii) solid waste is in the form of vitrified slag. When CO_2 Capture and Storage (CCS) systems become operational, it would become a plus point for IGCC plant, as CO_2 capture from the IGCC plant is technically easier than from a conventional steam plant.

Other technologies: USA and Japan are in the process of developing a number of hybrid systems, such as integrated gasification – fuel cell, which combines a fuel cell with IGCC, and could achieve efficiencies of the order of 60%. New technologies are being developed to make efficient use of low-grade coals, more efficient drying systems for high-moisture coals, ways of overcoming fouling problems in gasification and combustion, etc. In the case of CCS technologies, the purpose of R&D is to reduce costs and improve reliability.

4.1.5 CHINA – A COUNTRY CASE STUDY

China has emerged as an industrial giant, with USD 10 trillion economy in 2006, and there is little doubt that coal sector played an important role in the phenomenal growth of GDP (~11% in 2006) in the country. China accounts for more than half of the coal production in the world (3162 Mt, out of 6185 Mt). However, twenty out of thirty most polluted cities in the world are in China. The following steps that are being taken by China to improve the coal sector, are of general interest to the coal community in the world: (i) High degree of mechanization in coal mining designed to lead to high productivity per person-year (e.g. Shenua Group achieved the productivity of 30,000 t person-year), reduce hazards and fatalities, and increase recovery rate (to >45%), (ii) Dedicated rail lines, with trains capable of large pay-loads (25,000 t), (iii) Installation of Flue Gas Desulphurization (FGD) units to reduce noxious emissions, (iv) Improve energy efficiency in power generation, iron and steel, cement, fertilizer, etc. industries.

China has made major improvements in the coal-based power-generating technologies.

Table 4.1.7 Economics of coal-based power generating technology

Technologies	*Availability of technology*	*Cost ($/kW)*	*Efficiency (%)*	*Market share in China*
Subcritical		500–600	30–36%	Most of China's current generating units are of this type
Supercritical	Available now	600–900	41%*	About half of the new units on order are of this type.
Ultra-supercritical	Available now, but needs further R&D to increase efficiency	600–900	43%*	Two 1,000 MW units of this type are under operation
IGCC (Integrated Gasification Combined Cycle)	Available now, but costs are high, and needs more R&D to increase efficiency	1,100–1,400	45–55%	Twelve units are planned, and are awaiting clearance by NDRC

*Indicates current efficiency. Improved efficiencies are expected in the future

Chapter 4.2

Oil and natural gas

4.2.1 OIL

Oil has been and continues to be the dominant energy source, because of its critical role in transportation and industrial end-use sectors. The projection of Total energy consumption may be based on the basis of economic growth and price. Consumption of all fuels is increasing, but not at the same rate (Fig. 4.2.1). The use of liquid fuels is projected to increase at a higher rate than others. Increased urbanization and personal incomes, particularly in countries like China and India, have led to increase in air travel

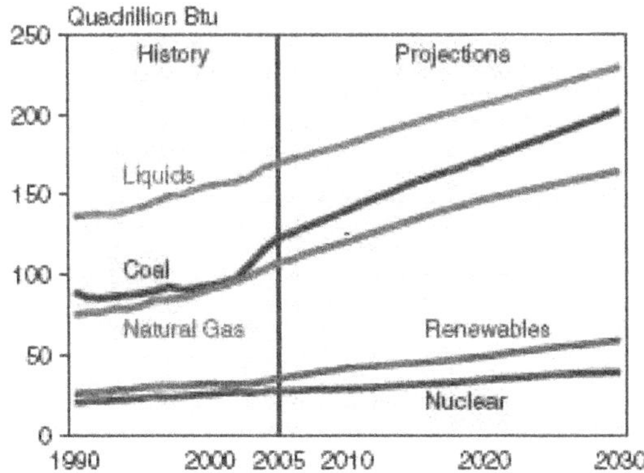

Figure 4.2.1 Historical and projected energy use of fuel type
Source: EIA, 2008

and also increase in the number of cars (In India, even middle class families can afford Tata's Nano car, which costs about USD 2,200). Transportation of fright by trucks, by air and marine vessels is increasing.

Oil and natural gas were produced by the burial and transformation of biomass over the last 200 million years. The transformation of kerogen to hydrocarbons (*catagenesis*) is a complex process which is largely dependent on the way kerogen is "cooked" (based on the temperature and duration of heating which are related to thermal gradient, depth of burial and age of sediments).

The proven oil reserves (in thousand million barrels) at the end of 2007 in the six regions of the world (Source: BP, 2008) and the percent of global reserves, are as follows:

Depth (km)	*Hydrocarbon that could be expected*	*Maturity of phase*
0–2	Gas	Immature (under-cooked)
2–4	Oil	Mature (well-cooked)
3–7	Gas	Over-cooked

Asia – Pacific: 40.8 (3.2%), North America: 69.3 (5.5%); South and Central America: 111.2 (9.0%), Africa: 117.5 (9.4%); Europe and Eurasia: 143.7 (11.6%); Middle East: 755.3 (61.0%). Thus, Middle East holds more oil than the rest of the world put together.

World oil production in Dec. 2007 has been 86 Mbd (million barrels/day).

The US Geological Survey (USGS) made use of the geologic characteristics as the primary criterion for the assessment of reserves. USGS prepared the most comprehensive assessment of oil resources in 96 countries (USGS, 2000). This assessment quantifies the resources of oil, gas and natural gas liquids (NGL) as they are known in 2000, but also have the potential to be added to the reserves within a 30-year time frame as a result of the exploration effort during 1995–2000. Masters et al. (1998) gave the resource estimates for parts of 128 geologic provinces in 96 countries (Fig. 4.2.2).

The proven oil reserves by region and country, the share of the reserves via-vis the global total, and R/P (Reserves to annual production ratio) are given in Table 4.2.1.

Evidently, oil reserves are finite, and will be exhausted sooner or later. Hubbert (1956) developed a tool ("Hubbert Curve") for forecasting future production of oil by estimating the likely peak in production in the case of any oil-producing region. According to this theory, known as Peak Oil Theory, the profile of annual production volumes in any oil-producing region, would be a bell-shaped curve ("Hubbert Curve") and the maximum production is reached when about 50% of the ultimate production volume has been extracted. When it was propounded, Hubbert's view was widely criticized for being pessimistic, but when the time came, his theory proved to be correct, in the case of USA (Fig. 4.2.3).

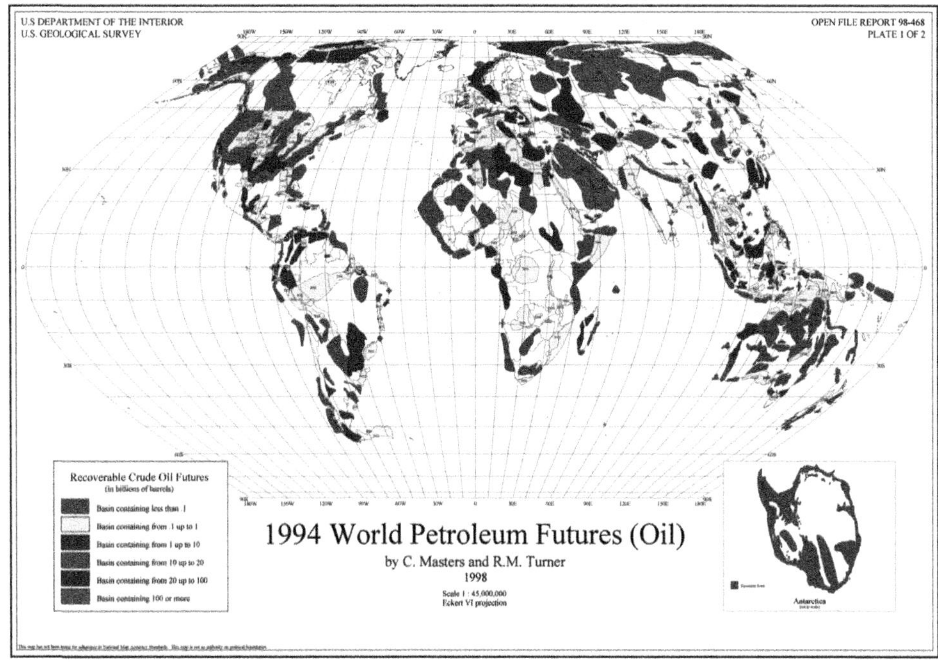

Figure 4.2.2 Geological provinces (128) in different countries (96) of the world, and the estimated petroleum resources in each of the provinces
Source: Maters et al., 1998

Table 4.2.1 Proven oil reserves (in thousand million barrels) at the end of 2007 by country and region. R/P is the reserves to annual production ratio

Country	*Oil reserves (in thousand million barrels) at the end of 2007*	*Percentage of global total (%)*	*Reserves to annual production ratio*
USA	29.4	2.40	11.7
Canada	27.7	2.20	22.9
Mexico	12.2	1.00	9.6
Total North America	**69.3**	**5.60**	**13.9**
Argentina	2.6	0.20	10.2
Brazil	12.6	1.00	18.9
Colombia	1.5	0.10	7.4
Ecuador	4.3	0.30	22.5
Peru	1.1	0.10	26.4
Trinidad & Tobago	0.8	0.10	14.1
Venezuela	87	7.00	91.3
Other South & Central America	1.3	0.1	25.2
Total South & Central America	**111.2**	**9.00**	**45.9**

(*Continued*)

Table 4.2.1 Continued

Country	*Oil reserves (in thousand million barrels) at the end of 2007*	*Percentage of global total (%)*	*Reserves to annual production ratio*
Azerbaijan	7	0.60	22.1
Denmark	1.1	0.10	9.8
Italy	0.8	0.10	17.6
Kazakhstan	39.8	3.20	73.2
Norway	8.2	0.70	8.8
Romania	0.5	0.00	12.4
Russian Federation	79.4	6.40	21.8
Turkmenistan	0.6	0.00	8.3
United Kingdom	3.6	0.30	14.3
Uzbekistan	0.6	0.00	14.3
Other Europe and Eurasia	2.1	0.20	12.8
Total Europe and Eurasia	**143.7**	**11.60**	**22.1**
Iran	138.4	11.20	86.2
Iraq	115	9.30	*
Kuwait	101.5	8.20	*
Oman	5.6	0.50	21.3
Qatar	27.4	2.20	62.8
Saudi Arabia	264.2	21.30	69.5
Syria	2.5	0.20	17.4
United Arab Emirates	97.8	7.90	91.9
Yemen	2.8	0.20	22.7
Other Middle East	0.1	0.00	10.9
Total Middle East	**755.3**	**61.00**	**82.2**
Algeria	12.3	1.00	16.8
Angola	9	0.70	17.2
Chad	0.9	0.10	17.2
Congo (Brazzaville)	1.9	0.20	23.9
Egypt	4.1	0.30	15.7
Equatorial Guinea	1.8	0.10	13.2
Gabon	2	0.20	23.8
Libya	41.5	3.30	61.5
Nigeria	36.2	2.90	42.1
Sudan	6.6	0.50	39.7
Tunisia	0.6	0.00	16.7
Other Africa	0.6	0.10	10.2
Total Africa	**117.5**	**9.50**	**31.2**
Australia	4.2	0.30	20.3
Brunei	1.2	0.10	16.9
China	15.5	1.30	11.3
India	5.5	0.40	18.7
Indonesia	4.4	0.40	12.4
Malaysia	5.4	0.40	19.4
Thailand	0.5	0.00	4.1
Vietnam	3.4	0.30	27.5
Other Asia Pacific	0.9	0.30	11
Total Asia Pacific	**40.8**	**100.0**	**14.2**
Total world	**1237.9**	**0.50**	**41.6**

Source: BP, 2008

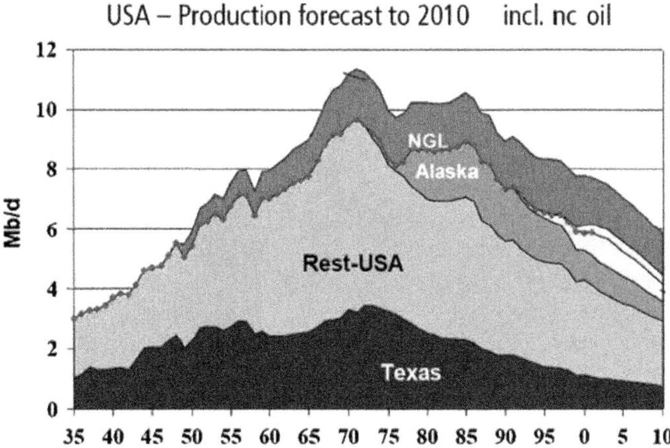

Figure 4.2.3 Forecast of oil production in the USA
Source: Hubbert, 1956

Hubbert Theory predicts that global oil production will be reached when about 50% of the world's Ultimately Recoverable Resource (URR) has been consumed. The unresolved question is how high URR is, and how close the world is to the oil peak.

According to Gabrielle, the peak has been already reached in 2010. Taking into account the present global reserves (1,238 Billion Barrels of oil) and production (82 Million Barrels of Oil per day), the exhaustion of oil is only 40 years away. The extractability, transportability and versatility of oil is unequalled by any other fuel. Considering the uneven distribution of oil resources (with Middle East accounting for about 60%), and the non-availability of equally convenient fuel, oil will continue to be a powerful geopolitical tool.

4.2.2 NATURAL GAS

Proved gas reserves (in trillion cubic metres) in 2007 are as follows in different regions, with percentage of global total: South and Central America – 7.73 (4.5%); North America – 7.78 (4.5%); Asia – Pacific – 14.46 (8.4%); Africa – 14. 58 (8.5%); Europe and Eurasia – 59.41 (34.7%); Middle East – 73.21 (42.8%); Total – 177.17. Though Middle East dominates the gas resources, is dominance is much less than in the case of oil.

USGS (2000) estimates that the world has undiscovered natural gas resources of 4133 million cubic ft.

The world production of natural gas is (3127 billion cubic metres). Important producers of natural gas (in billions of cubic metres) are: USA – 593, Russia – 583, and Iran – 200. The high consumption of natural gas in USA is because of cold winters and strong demand for gas for power generation.

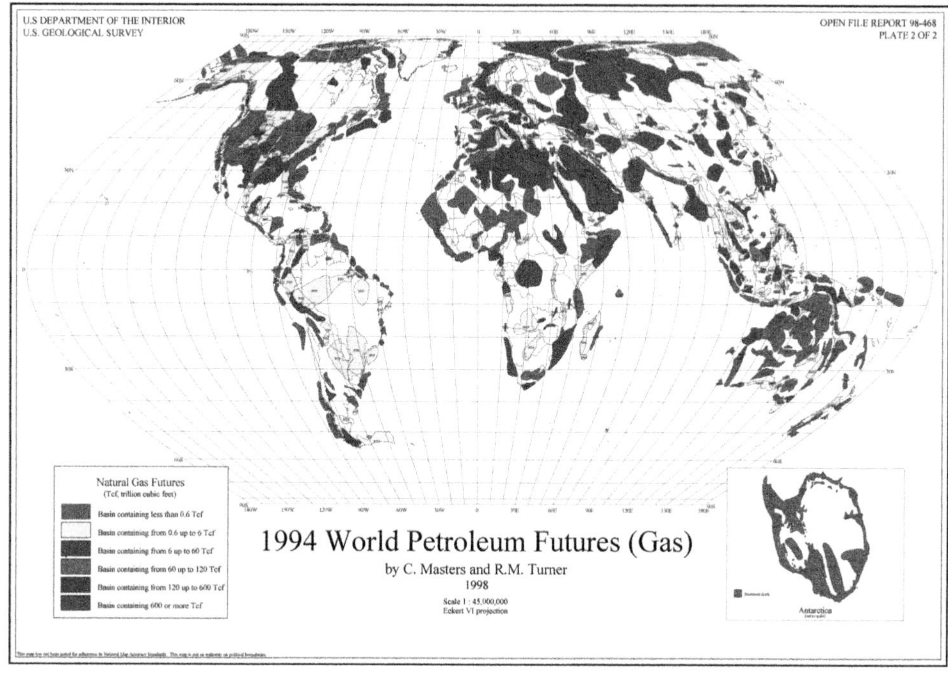

Figure 4.2.4 Estimated gas resources in different geological provinces of the world
Source: Masters et al., 1998

Masters et al. (1998) gave the estimated natural gas resources in the geological provinces of the world (Fig. 4.2.4). Detailed country-wise estimates of resources are given in Table 4.2.2 (source: BP, 2008).

Oil and natural gas resources are dwindling fast. Peak production is estimated to occur in two decades, and exhaustion is likely occur by 2050. For this reason, and also to reduce the carbon dioxide emissions, it is necessary to use oil and natural gas wisely.

There is urgent need to improve the energy efficiency, and achieve technological breakthroughs in energy utilization and management. Marginal emission reduction costs for ACT and BLUE scenarios for the global energy system are shown in Fig. 4.2.5.

A giant oil and gas field is one which has 500 millions barrels of ultimately recoverable oil or gas equivalent. There are 932 such giant oil and gas fields in the world. These giants account for 40% of the world's petroleum reserves. There are clustered in 27 regions of the world, with two largest clusters in the Arabian Gulf, and Western Siberian Basin (Fig. 4.2.6, Mann et al., 2007). A majority of the world's giant oil and gas fields occur in passive margins and rift environments.

Apart from the surface characteristics and the tectonic setting of the geologic structures that contain the hydrocarbons, an important criterion to characterize a giant oil field is the identification of the dominant geological event that facilitated the structure's ability to trap and contain oil and gas in large quantities.

Table 4.2.2 Rankings and estimates of the principal petroleum provinces (province names in the Saudi Arabian region are in bold). BCFG is Billion cubic feet of gas. Cumulative gas is the reported volume of gas that has been produced. Remaining gas is the calculated remaining reserves (the difference between known and cumulative gas. Known gas is the reported discovered volume (cumulative production plus remaining reserves). Mean undiscovered gas is the mean volume estimated. Future gas is the remaining gas plus estimated mean undiscovered gas. Gas endowment is the known gas estimated mean undiscovered gas volumes. Discovery maturity is the percentage of gas discovered (known gas) with respect to gas endowment

Rank	*Province Code*	*Province Name*	*Major Commodity*	*Cumulative Gas (BCFG)*	*Remaining Gas (BCFG)*	*Known Gas (BCFG)*	*Mean Undiscovererd Gas (BCFG)*	*Future Gas (BCFG)*	*Gas Endowment (BCFG)*	*Discovery Maturity, Gas (%)*
1	1174	West Siberian Basin	Oil	218,465	1,051,584	1,270,049	642,934	1,694,518	1,912,983	66
2	2030	Zagros Fold Belt	Oil	10,333	389,111	399,444	212,008	601,119	611,451	65
3	**2019**	**Rub Al Khali Basin**	**Oil**	**8,303**	**173,934**	**182,237**	**425,712**	**599,646**	**607,949**	**30**
4	2022	Qatar Arch	Gas	1,018	464,582	465,600	17,385	481,967	482,985	96
5	**2021**	**Greater Ghawar Uplift**	**Oil**	**26,493**	**222,057**	**248,550**	**226,969**	**449,026**	**475,519**	**52**
6	**2024**	**Mesopotamian Foredeep Basin**	**Oil**	**13,129**	**285,156**	**298,285**	**83,722**	**368,878**	**382,007**	**78**
7	1154	Amu-Darya Basin	Gas	78,513	151,830	230,343	163,650	315,480	393,993	58
8	1016	North Caspian Basin	Oil	1,745	155,161	156,906	119,051	274,212	275,957	57
9	1050	South Barents	Gas	0	70,020	70,020	160,857	230,877	230,877	30
10	7192	Niger Delta	Oil	6,232	87,579	93,811	132,716	220,295	226,526	41
11	1112	South Caspian	Oil	8,591	27,403	35,994	173,310	200,713	209,304	17
12	6098	East Venezuela Basin	Oil	22,053	106,066	128,119	93,561	199,627	221,679	58
13	4017	Vestfprd-Helgeland	Oil	0	15,662	15,662	165,201	180,863	180,863	9
14	4025	North Sea Graben	Oil	31,235	128,603	159,839	37,745	166,348	197,584	81
15	**2020**	**Interior Homocline-Central Arch**	**Oil**	**0**	**9,825**	**9,825**	**122,074**	**131,899**	**131,899**	**7**
16	3948	Northwest Shelf	Gas	1,840	54,405	56,245	64,711	119,116	120,956	47
17	2058	Grand Erg/Ahnet	Gas	4,980	109,056	114,034	8,664	117,717	122,697	93
18	**2023**	**Widyan Basin-Interior Platform**	**Oil**	**60**	**7,295**	**7,355**	**94,746**	**102,041**	**102,101**	**7**
...										
...										
132										

Source: USGS, 2000

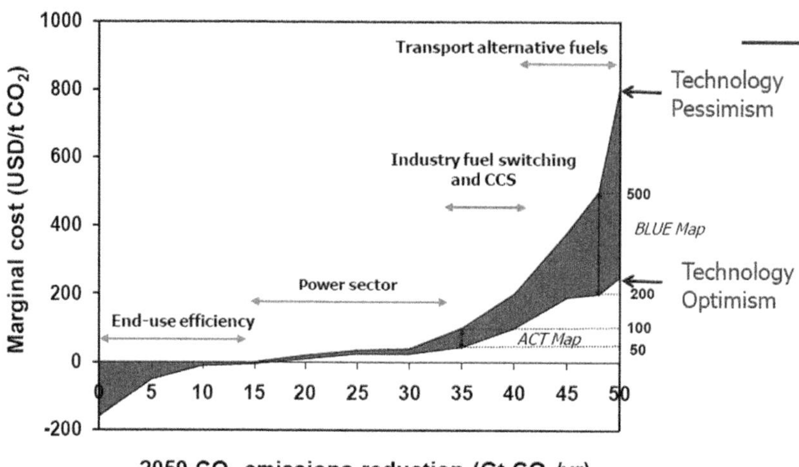

Figure 4.2.5 Marginal emission reduction costs for the ACT and BLUE scenarios for the global energy system
Source: IEA, 2008

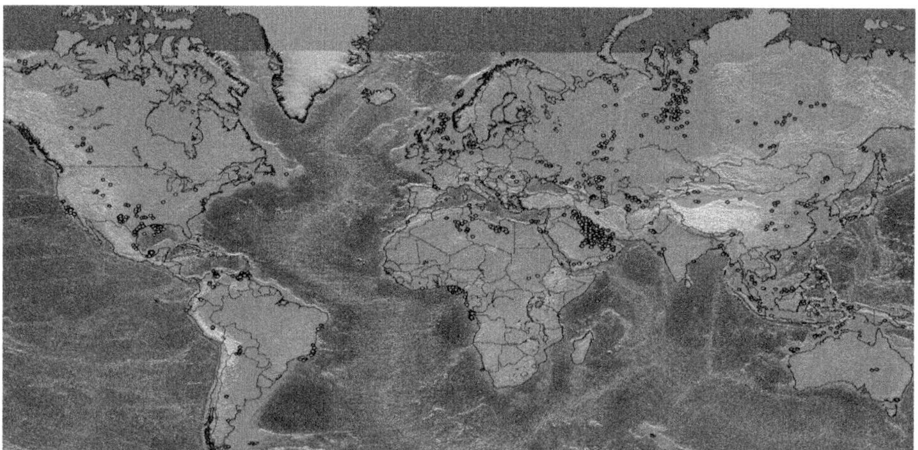

Figure 4.2.6 Distribution of giant oil fields in the world, showing two large clusters in the Arabian Gulf and Siberian Basin
Source: Mann et al., 2007

4.2.3 SHALE GAS

Shale gas is an unconventional source of natural gas produced from shale. It has been reported in 48 basins in 38 countries. Shale gas has become increasingly important in USA during the last decade. Canada, Europe, Asia, and Australia are extracting shale gas. Shale gas may provide half of natural gas in North America by 2020. One

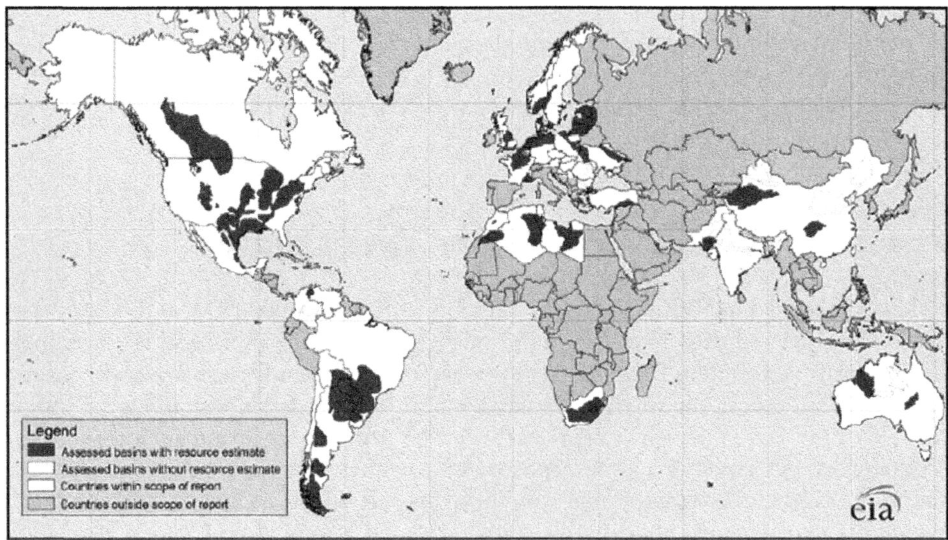

Figure 4.2.7 World shale gas resources
Source: Wikipedia

analyst expects shale gas to supply as much as half the natural gas production in North America by 2020. Shales that have potential for shale gas are rich in organic matter (0.5% to 25%). Shale intervals with high gamma radiation are most productive. When hydrocarbons present in shale are "overcooked" because of being subjected to high pressure and temperature, they get converted to shale gas.

Unlike sandstones and limestones, shales are impervious and it is not easy to extract gas from them. Two technological advances, namely horizontal drilling and and hydraulic fracturing (fracking) made it possible to extract gas from shale. Shale gas wells may involve lateral lengths of over 3000 m. to create maximum borehole surface area.

Production and use of shale gas has environmental health consequences. It is now established that shale gas use aggravates global warming, as shale gas emits methane which is a more potent green house gas than conventional gas. The footprint of shale gas is far worse than that of coal and fuel oil when viewed for the integrated 20-year period after emission. Chemicals added to water to facilitate the underground fracturing process, can some times lead to the contamination of groundwater

A 2011 study by the Massachusetts Institute of Technology concluded that "The environmental impacts of shale development are challenging but manageable." The study addressed groundwater contamination, noting "There has been concern that these fractures can also penetrate shallow freshwater zones and contaminate them with fracturing fluid, but there is no evidence that this is occurring".

The world reserves of shale gas has been estimated to be 6,622 trillion cu.ft. Unlike the case of oil and natural gas, the occurrence of shale gas is wide-spread. The most important shale gas countries are: China – 1,275 trillion cu., ft., USA – 862; Argentina – 774; Mexico – 681; South Africa – 485 (Fig. 4.2.7).

Shale gas tends to cost more to produce than gas from conventional wells, because of the expense of the massive hydraulic fracturing and horizontal drilling. This is, however, offset by the low risk of shale gas wells. The global shale market is estimated to be worth about USD 27 billion. Some say that this figure should be taken with a pinch of salt, as some companies intentionally overstate the productivity of the wells and the size of their reserves.

4.2.4 SAUDI ARABIA – A COUNTRY CASE STUDY

Oil in commercial quantities was discovered in Saudi Arabia in 1938 in the Dammam area. The clustered distribution of giant oil fields in Saudi Arabia and the neighboring countries accounts for the oil riches of the country. Saudi Arabia has the largest proven oil reserves in the world, estimated at 267 billion barrels, including 2.5 billion barrels in the Saudi- Kuwaiti neutral zone. The geologic environment of oil in the Middle East comprises of a large depositional platform along a pre-Mesozoic passive margin of Godwin. Extensive migration of hydrocarbons into traps underlying thick, regionally extensive evaporate seals, facilitated the accumulation and preservation of oil.

Saudi Arabia produces the world's largest crude oil (10.5–11 Million bbl/d). Saudi Arabia's oil production mainly from six oil fields:

1. Ghawar (onshore) – 5 million bbl/d of 34° API Arabian Light Crude. Ghawar alone produces half of Saudi Arabia's total oil production, and is the world's largest oil field.
2. Abqaiq (onshore) – produces approximately 400,000 bbl/d of Arab Extra Light Crude.
3. Najid (onshore) – produces 200,000 bbl/d of Arab Super Light.
4. Safaniya (offshore) – produces around 1 million bbl/d of Arab Heavy crude.
5. Zuluf (offshore) – produces about 500,00 bbl/d of Arab Medium Crude.
6. Marjan (offshore) – produces approximately 270,000 bbl/d of Arab Medium crude.

Saudi Arabia produces a wide range of crude oils, ranging from heavy to super light, with 65 to 70% of crudes rated as light gravity. Lighter grades generally are produced onshore, and heavy grades are from onshore. The heavy crudes are "sour" (i.e. contain sulphur). Saudi Aramco operates the world's largest crude processing facility of more than 7 million bbl/d at Abqaiq in Eastern Saudi Arabia. Saudi Arabia has seven domestic refineries which processs combined crude of 2.1 MBO/d.

Saudi Arabia has three primary oil export terminals:

1. Ras Tanura Complex, with a capacity of about 6 million bbl/d,
2. Ras al-Ju'aymah facility on the Persian Gulf, with a capacity of 3 to 3.6 million bbl/d,
3. Yanbu terminal on the Red Sea, with a capacity of 4.5 million bbl/d of crude and 2 million bbl/d for NGL and other products.

Saudi Arabia has 9,000 miles (~14,400 km) of oil pipelines. Saudi Aramco has a shipping subsidiary, Vela International Marine Ltd., which operates sixth largest fleet of oil tankers in the world, including 19 VLCCs (Very Large Crude Carriers) with a capacity of 200,000 to 320,000 DWT (Dead Weight Tonnes).

Saudi Arabia has 252 trillion scf reserves of natural gas, with production of 85 billion cubic metres of natural gas/yr.

In order to compensate for 6–8% decline rates in the existing oil fields, Saudi Arabia is planning to build 700,000 bbl/d of additional capacity.

Chapter 4.3

Nuclear fuel resources

4.3.1 INTRODUCTION

Nuclear power came into prominence for the following reasons (Ewing, 2006).

1. It is a proven technology – nuclear power currently accounts for about 16 % of the electricity production in the world.
2. Energy release from nuclear reaction per unit mass is about a million times more than chemical combustion,
3. The volume of waste produced is small (typically, 20 t per reactor per year), and is generally containable (as against wastes from coal use which get widely dispersed),
4. The radioactivity and radio toxicity of the Spent Nuclear Fuel (SNF) drop steeply in thousand years, after removal from the reactor,
5. New reactor designs and advanced fuel cycles permit the virtual elimination of "waste" radio nuclides like Np, and permit the generation of new fissile material (^{239}Pu from ^{238}U, or ^{233}U from ^{232}Th),
6. Most importantly, nuclear power plants do not emit greenhouse gases.

While the considerations listed above continue to be valid, the nuclear disaster, which took place in Fukushima, Japan, on Mar. 11, 2011, raised grave doubts among the countries whether the benefits of nuclear power are worth the cost in human misery and economic destruction when nuclear accidents happen. Germany has chosen to forego the nuclear option altogether. In France, there is a talk of reducing the country's dependence on nuclear power from the current 75% to 50% by 2025. The EPR plant under construction in Flamanville, France, has seen long delays and massive cost hikes. A recent accident in the oldest nuclear site in Marcule further undermined the image of the French nuclear industry, though the French Government took great pains to describe it as an industrial accident rather than a radiological accident. No wonder, India has reportedly informed France that it will import French reactors only after post-Fukushima certification.

Only fissile isotopes can be used directly in nuclear fission for generating power. Fertile isotopes have to be rendered fissile first, before being used to produce nuclear fission. Uranium has two isotopes – fissile isotope ^{235}U with an abundance of about 0.7%, and fertile isotope ^{238}U with an abundance of about 99.3%. Neutron absorption in ^{238}U produces the fissile plutonium isotope, ^{239}Pu. On the other hand, thorium has no fissile isotope, but consists entirely of fertile isotope ^{232}Th which has to be converted to fissile ^{233}U through neutron absorption before being used for nuclear fission. ^{233}U produces the most neutrons per neutron absorption at thermal energies (slow neutrons) and is hence superior to both ^{235}U and ^{239}Pu as nuclear fuel in the common light water thermal reactors.

Enriched uranium fuel, which is used in the reactors, is produced by enriching the ^{235}U content to 4–5%. ^{233}U does not occur in nature – it has to be "bred" (hence the term, breeder reactors). Enriched uranium (typically, 20% ^{235}U) or ^{239}Pu is needed for the start-up of such reactors. When thorium is substituted for natural uranium in a breeder reactor, as much (or more) ^{233}U is produced by neutron absorption in thorium, as is consumed in the reactor operation. Also, when once the reactor is started, no further enriched uranium is needed.

Power reactors: The fissioning of one kg of ^{235}U yields 22.66×10^6 kWh of energy, equivalent to 2750 t of bituminous coal. The fuel value of one ton of uranium is 6.8×10^{13} BTU. The very high fuel value of uranium has profound implication for siting the nuclear power plants. While coal-fired thermal power stations have to be established at pithead in order to save on transportation costs, and hydroelectric power stations have to be set up at dam site, a nuclear power station is not subject to such constraints (a 1000 MWe power station would need annually about 3.1 million tonnes of hard coal, but only about 24 tonnes of enriched uranium).

4.3.2 RESOURCE POSITION

Australia, Canada, Kazakhstan and South Africa have large resources of uranium, followed by Russia, Brazil, Namibia, Niger, Ukraine, China and Uzbekistan.

OECD (2005) has given the following prediction of the world nuclear generating capacity, in GWe: 2005 – 373; 2010 – 390; 2015 – 427; 2020 – 455; 2025 – 491, and prediction of uranium demand (in tonnes of uranium): 2005 – 66,800; 2010 – 72,000; 2015 – 79,000; 2020 – 80,900.

The following information has been drawn from the Red Book of 2007.

4.3.2.1 Thorium resources

Monazite (Ce, La) PO_4; ThO_2 upto 12%; U_3O_8: 0.1–10%, is the principal ore of thorium. Vein deposits of monazite are known, but they are economically insignificant. The most important resources of monazite are detrital, and are associated with ilmenite-bearing heavy mineral sands, which occur as coastal and inland placers.

Table 4.3.1 Proposed production methods to exploit Inferred Resources

	<USD 40/kg	<USD 80/kg	<USD 130/kg
Open-pit mining	202,100	199,300	251,900
Underground mining	296,700	692,400	767,000
In situ leaching	344,400	378,200	359,700
Heap leaching	12,700	22,300	23,900
In place leaching	1,500	24,800	24,800
Co-product/by-product	367,000	445,800	493,200
Unspecified mining method	10,200	95,600	180,100
Total	**1,203,600**	**1,858,400**	**2,130,600**

Table 4.3.2 Reasonably assured resources (t) by deposit type (totaling 7,702,700 U t)

	<USD 40/kg	<USD 80/kg	<USD 130/kg
Unconformity-related	424,100	485,200	491,600
Sandstone	347,800	537,000	999,500
Hematite-breccia complex	492,300	492,300	449,400
Quartz-pebble conglomerate	88,100	126,400	163,600
Vein	0	80,600	156,800
Intrusive	47,400	131,480	183,700
Volcanic and caldera-related	50,400	135,700	157,800
Metasomatite	121,200	291,300	304,900
Other*	162,300	221,000	284,300
Unspecified	32,800	67,800	96,700
Total	**1,766,400**	**2,598,000**	**3,338,300**

*Includes surficial, collapse breccia, phosphorite, etc.

Table 4.3.3 Inferred resources (t) by deposit type (totaling 5,102,600 U t)

	<USD 40/kg	<USD 80/kg	<USD 130/kg
Unconformity-related	148,300	152,300	158,100
Sandstone	374,800	468,100	524,400
Hematite-breccia complex	393,900	399,900	401,500
Quartz-pebble conglomerate	113,700	132,000	138,300
Vein	0	108,500	167,700
Intrusive	61,600	78,800	104,200
Volcanic and caldera-related	1,000	44,600	53,500
Metasomatite	14,800	289,200	368,800
Other*	77,800	133,900	154,400
Unspecified	17,700	51,100	59,700
Total	**1,203,600**	**1,858,400**	**2,130,600**

*Includes surficial, collapse breccia, phosphorite, etc.

India has one of the largest resources of monazite in the world – about 10.1 million tonnes which contain ~0.72 million tonnes of ThO_2 (Dhanaraju, 2005, p. 53). They are located in Chhatrapur-Gopalpur in Orissa, Bhavanapadu-Kalingapatnam- Bhimunipatnam in Andhra Pradesh, Manavalakurichi in Tamil Nadu, Chavara in Kerala, and Ratnagiri in Maharashtra.

These are all surface deposits, and can be mined easily with the least environmental degradation. Floating plants with hydraulic dredging or bucket line dredging could be employed for the dislodgement and lifting up of the heavy mineral sands. Heavy minerals, which typically constitute about 5% of the sands, are separated electromagnetically on board the ship, and the rest about 95% is dumped back at the same site. This way, the environmental impact of mining is minimized

Incidentally, the Corridor Sands deposit of heavy minerals (~17 billion tonnes of heavy mineral sands, with about 5% of the Total Heavy Minerals) in the Limpopo palaeo-delta in the Gaza province of Mozambique, has emerged as probably the largest deposit of monazite in the world (Taylor et al., 2003).

The estimated resources of thorium in the other parts of the world (USD 80/kg Th) are as follows: Australia: 452,000 t; USA: 400,000 t; Turkey: 344,000 t; Venezuela: 300,000 t; Brazil: 221,000 t; Norway: 132,000 t; Egypt: 100,000 t; Russia: 75,000 t; Greenland: 54,000 t; Canada: 44,000 t; South Africa: 18,000 t; Other countries: 33,000 t. *(source: Geoscience Australia 2006 and OECD/NEA Red Book retrospective, 2006)*

The 2005 IAEA-NEA "Red Book" gives a figure of 4.5 million tonnes of thorium reserves, but this data is neither complete globally, nor is it based on systematic exploration.

An MIT study projects that the nuclear power will grow from 350 GWe today to 1500 GWe in 50 years (Ansolabehere et al., 2003). This would need 10 Mt of uranium. The total of all resources known (~23 Mt) is twice the amount of uranium projected to be needed. A number of areas in the world have not been adequately explored. Considering the large variety of geologic settings in which uranium could occur, and taking into account the economic value of ore elements that might be co-extracted with uranium, the known resources of uranium listed in the Red Book are conservative. It can be safely said that the uranium resources are available for a few hundred years.

A 1000 MWe of nuclear power will displace 15–25% of the predicted growth of carbon emissions.

The cost of uranium was $27.74/kg in 2004, and $31.59/kg in 2005. Though the fuel costs of nuclear power ($0.018/kWh) are not much different from those of thermal power ($0.023–0.05/kWh), the capital costs of building a new nuclear plant ($2000/kWe) are much higher than for a modern gas turbine plant, which could be built as cheaply as $500/kWe (source: Macfarlane and Miller, 2007).

Nuclear power plants face problems of getting a license, political opposition to plant siting, and concern about health effects, and contamination of water resources and the Environment.

Presently, the cost of the nuclear fuel in India is roughly twice that of the global price. The price will go down when India is able to access the international uranium market through the Nuclear Suppliers Group (NSG). In the Indian context, nuclear power is likely to become competitive with thermal power at places 500 km away from coal deposits. In the case of coal-fired power stations, the fuel cost accounts for about 60% of the electricity cost, as against 15% in the case of Light Water Reactors. Consequently, nuclear power costs will increase at a lower rate than for coal power, as a consequence of inflation.

Australia, Canada, India, Kazakhstan, Russia, USA and Uzbekistan concentrated on exploration of unconformity-type, sandstone-type and haematite-breccia type uranium

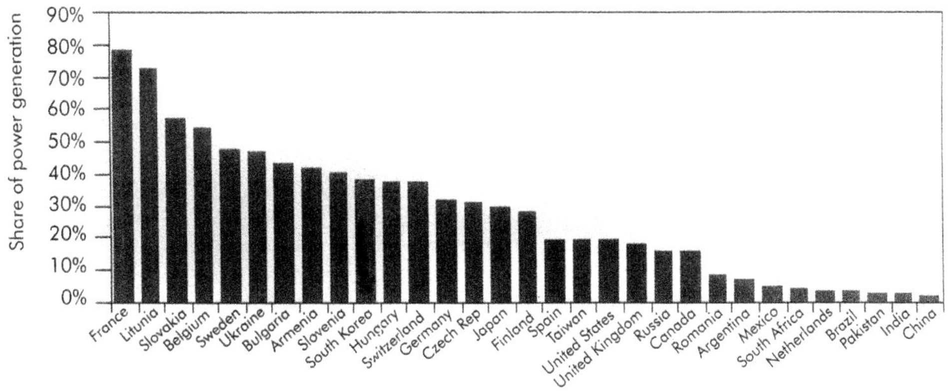

Figure 4.3.1 Nuclear share of the electricity generation by country, 2006
Source: IEA, 2008, p. 285

occurrences in 2004. Hardly any exploration work for uranium was done elsewhere in the world during the past 15 years. But as the price of uranium is picking up because of increased interest in nuclear power, exploration and mining of uranium is likely to be taken up in several countries in the future.

In Aug. 2007, there were 438 operating nuclear plants in 30 countries, with a total capacity of 372 GW. Thirtyone reactors which are under construction have a capacity of 24 GW. Nuclear power accounted for 2700 TWh in 2006, which corresponds to 16% of the global electricity production. The world has experience of 12,000 reactor-years. USA, France and Japan account for about 60% of the world's nuclear power. In 2006, France has the highest share (78%) of nuclear power in the world. Fig. 4.3.1 gives the nuclear share of electricity generation by country in 2006. The current demand for uranium is 67,000 tonnes per year.

4.3.3 COST OF NUCLEAR POWER

The direct costs of nuclear power have three components: construction costs, operation, maintenance (O&M) and fuel costs, and backend costs (towards waste management and decommissioning costs). To compare different technologies, a "levelized" cost per kWh or MWh is computed. The construction costs vary from country to country, depending upon the length and complexity of the preconstruction period, capital costs (excluding interest), construction time and the cost of capital. The capital cost may vary depending upon whether it is turnkey project, or a project built directly by the organization. The construction time depends upon the management efficiency and availability of high-quality components. For instance, Japan could build an 800 MW BWR in 41 months, whereas Finland took 72 months for the purpose. The cost of the capital which has a major impact on the construction costs, depends upon the financing scheme – for instance, the discount rate is higher for nuclear power

Table 4.3.4 Results of recent studies on the costs of nuclear power

Study	*Cost of capital (%)*	*Overnight cost kW*	*Levelized cost per MWh*
Massachusetts Institute of Technology (MIT, 2003)	11.5	USD 2000	USD 67
General Directorate for Energy and Raw Materials, France (DGEMP, 2004)	8	EUR 1280	EUR 28
Tarjanne & Luostarinen (2003)	5	EUR 1900	EUR 24
Royal Academy of Engineering (2004)	7.5	GBP 1150	GBP 23
University of Chicago (2004)	12.5	USD 1500	USD 51
Canadian Energy Research Institute (2004)	8	CAD 2347	CAD 53
Department of Trade and Industry, UK (DTI, 2007)	10	GBP 1250	GBP 38
IEA/NEA (2005)	10	USD 1089–3432	USD 30–50

Source: IEA, 2008, p. 290

plants than for fossil power plants, as the investors and stakeholders see nuclear power as a high risk venture.

The operating (O&M and fuel) costs vary greatly between the countries, depending upon the labour costs, plant size and age, inspection and insurance costs, etc. Though the spot prices of uranium went up significantly, relatively small quantities are traded in this manner. In 2003, nuclear production costs in USA were US cents 1.72/kWh. Finland and Sweden have reported production costs of EUR 0.01/kWh. In the case of France which has 58 reactors, the O&M and fuel costs are EUR 0.014/kWh.

Decommissioning and majority of waste management costs are incurred at the end of the life of a reactor. Economists say that these costs do not significantly affect the levelized costs.

Some countries have achieved very high energy-availability factors in the existing plants. Consequently, while the generating capacity during 2003–2005 went up by only 1%, the nuclear electricity production went up by 2–3%, thus bringing about cost reduction.

The cost of nuclear power in new nuclear power plants are given in Table 4.3.4.

For future nuclear power stations, the levelized costs are expected to be in the range of USD 30–50/MWh. The levelized cost is higher than this range in the case of Japan (USD 69/MWh), because of high labour costs. Also, the levelized costs for coal, gas and nuclear power are very similar in Japan, possibly because Japan imports all fuels.

In some countries (e.g. Canada, Czech Republic, and France), nuclear power is cheaper than coal power and gas power by a margin of 10%. As fossil fuels have become costlier since 2004, this makes nuclear power more cost-competitive. The cost position in the case of nuclear power becomes even more attractive when the absence of carbon emissions is taken into account in cost calculations.

The Department of Trade and Industry (DTI) of UK came up with ranges of nuclear generating costs: High cost: USD 88/MWh; Central cost: USD 76/MWh; Low-cost: USD 62/MWh. The high-cost scenario is unlikely. The low-cost scenario is comparable to the estimates of the General Directorate for Energy and Raw materials (2004). DTI (May, 2007) made a sensitivity analysis of the key parameters that determine

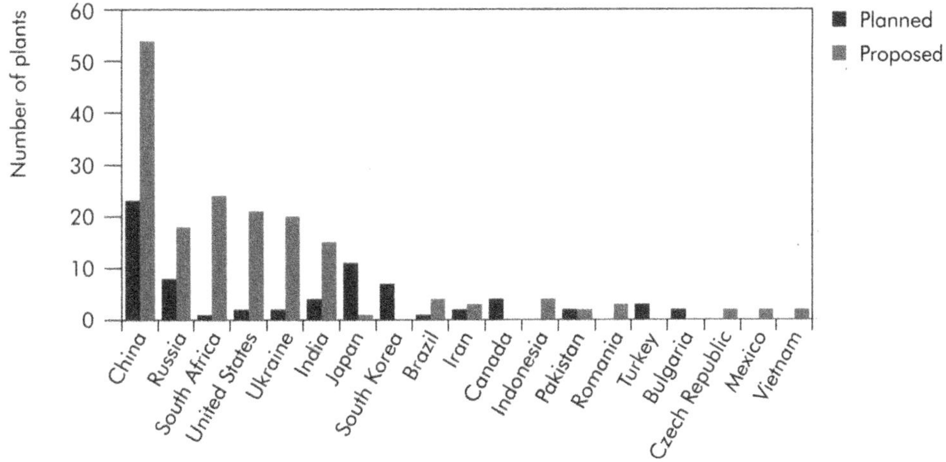

Figure 4.3.2 Plans and proposals for new power reactors
Source: IEA, 2008, p. 299

the cost of the nuclear power. The most important factors are the overnight cost and the cost of the capital. Other factors are: construction period, and O&M costs. Parameters such as Early load factor, operational lifetime, fuel cost, and waste disposal and decommissioning costs, have little impact on the levelized cost of nuclear power.

Nuclear power offers energy density and security of supply advantages. As the energy release from nuclear reaction per unit mass is about million times more than chemical combustion, it is easier to stockpile and transport uranium fuel, relative to fossil fuels. Also, countries, such as, Canada and Australia, which are politically stable, supply uranium. Disruption of fuel supplies is not expected to happen in the case of uranium supplies.

4.3.4 PROJECTED NUCLEAR POWER CAPACITY

By 2020, the following countries have plans to increase their nuclear capacity (in terms of GW): China – 40, India – 16, Russia 22. Also, Japan – 9 GW by 2015, South Korea – 12 GW by 2017, Ukraine – 16 GW by 2030. Plans and proposals for new nuclear power reactors are given Fig. 4.3.2. France built 58 new nuclear reactors during 1917 and 1933 – an average of 3.6 reactors per year. As the world economic activity is about 30 times that of France, global construction of 100 new reactors (160 GW/y, if the reactors were to be EPRs – European Pressurised water reactors) is theoretically possible. There are however constraints, not only in skills, but also manufacturing constraints. For instance, there is only one company in the world, in Japan, which has the capacity to fabricate ultra-large forging for large reactor pressure vessels, and their books are full for the next three years. Similarly, heavy-lift cranes, large diesel generators for emergency core cooling, sulphur hexafluoride for switchgear, etc. are in short supply.

Highly skilled personnel are required not only to replace the retiring workforce, but also to build and maintain new generations of sophisticated reactors.

4.3.5 NEW REACTOR DESIGNS

4.3.5.1 Pebble-bed reactors

The following account is drawn from Wikipedia and Internet sources.

Rudolf Schulten invented the pebble-bed reactor which is simple, safe and makes use of commoditized nuclear fuel. The pebbles are of the size of tennis balls (60 mm hollow spheres of pyrolitic graphite which acts a the primary neutron moderator), with a mass of 210 g. A pebble contains 9 g of uranium (thorium or plutonium or enriched uranium and possibly MOX fuels could also be used). The nuclear fuels are in the form of ceramic (usually oxides or carbides). Thus, each pebble is effectively a "mini-reactor". When enough number of spheres is piled together in a bin, criticality is attained. A 120 MWe reactor needs 380,000 pebbles. The reactor is design is such that it is inherently self-controlling. An inert gas, helium, nitrogen or carbon dioxide circulates through places between the fuel pebbles. The heated gas can run the turbine. If the heated gas has been rendered radioactive by the neutrons in the reactor, it may be made to pass through a heat exchanger where it heats another gas or generates steam. The exhaust from the turbine is hot enough for being used to warm buildings or chemical plants or even another heat engine.

In the case of conventional water-cooled, nuclear power plants, the cooling systems are complex, expensive and occupy much space because of the redundant backups. Because of the neutrons present in the reactor, the water and the impurities dissolved in it may become radioactive. Consequently, the high pressure piping on the primary side may become brittle, and needs to be regularly checked or even replaced. In contrast, inert gas like helium which serves as a coolant in the pebble bed reactor does not become radioactive because of neutrons. In the case of pebble-bed reactors, there is no piping as such (the spaces between pebbles serve as "piping"), and hence the problem of embrittlement of piping does not arise.

The pyrolitic graphite which is the main structural component of the pebbles sublimes at 4000°C. This is more than twice the design temperature of most reactors. Pyrolitic graphite slows neutrons very effectively, besides being strong and inexpensive. A pebble-bed reactor can therefore operate at a much higher temperature than the conventional light water reactors, without any problem. This permits the turbine to extract more mechanical energy from the same amount of thermal energy. This system uses less fuel per kWh, and is hence more energy efficient than the light water reactors.

Most pebble-bed reactors contain several (about seven) levels of containment. There are four containments right in the pebble itself – fission fuel in the form of a ceramic; high density, non-porous pyrolitic carbon; wrapping of silicon carbide; another wrapping of pyrolitic carbon. In addition, the reactor vessel is sealed, and the reactor is kept in a room with two-meter walls. The reactor system as a whole is enclosed in a containment building, which can withstand earthquake shocks and aircraft crashes.

Some pebble-bed designs are throttled by temperature – not control rods. This makes the design simpler, as no provision need be made for control rods.

Even if all the supporting machinery fails, a pebble-bed reactor will not melt or explode or release radioactive contaminants into the atmosphere and the environment.

The modular design of the pebble-bed design allows small reactors (100–200 MWe) to be mass-produced. This simplifies certification procedures, and installation practices. Such small reactors are needed for remote areas which cannot be economically connected to the national grid. Also, if one unit fails, another unit can be quickly substituted in its place, without waiting for repairs. Several pebble-bed reactors could be linked, if larger quantities of power (say, 1000 MWe) are needed.

Lastly, many experts believe that pebbled radioactive waste does not need immobilization – it can be directly disposed of in the geological storage.

Anil Kakodkar and Ratan Sinha of the Atomic Energy Commission of India (*Nuclear Engineering International*, May 2010 issue) came up with an innovative design of 300 MWe nuclear power reactor. As a result of its fuel mix and fuel breeding properties, the reactor will require 42% less mined uranium per unit of energy produced than a modern high burn up PWR. It uses heavy water at low pressure, thereby eliminating the maintenance of coolant pressure and drive motors, and hence less electricity will be needed. It will have double containment, passive water seal in the event of coolant accident. The fuel will be composed of 19.75% enriched uranium, and the rest, thorium. A significant fraction of the reactor power (about 39%) comes from the fission of U-233, produced by the in situ conversion of Th-230. The proposed design produces much less plutonium, and thus enhances its proliferation resistance. Since minor actinides (which have a relatively low half-life) produced will be less than for PWR, the problem of waste disposal will be correspondingly less.

4.3.6 R&D AREAS

Fourth generation reactor systems involve R&D in the following areas (International Energy Agency, 2008, p. 585):

- (i) Development of sodium-cooled, fast reactors, gas or lead-cooled fast reactors, high temperature reactors, supercritical water reactors, molten salt reactors; accelerator-driven sub-critical systems.
- (ii) Fuel cycles: Partitioning and Transmutation (P&T) and actinide recycling.
- (iii) Materials to enhance safety.
- (iv) Solutions to nuclear waste problems.
- (v) Development of internationally approved safety standards and designs.

Generation V+ reactors: Designs which are theoretically possible, but which are not being actively considered or researched at present. Though such reactors could be built with current or near term technology, they trigger little interest for reasons of economics, practicality, or safety.

Seven partners – European Union, China, India, Japan, Russia and USA – have joined together to build the International Thermonuclear Experimental Reactor (ITER) in Caderache in France. ITER will explore the techno-economic feasibility of fusion power. Fusion power is likely come into use only in the second half of the century.

4.3.7 COUNTRY CASE STUDY OF FRANCE

The particulars in this study have been drawn from Barré & Garderet (2009).

France has hardly any coal or oil resources. Most of the sites for hydropower have been developed. France even built a unique tidal power plant at La Rance. France desperately needed an alternate energy source. As is well known, radioactivity was discovered in France by Bequerel and the Curies, and France decided to tap the energy from the atom as the only recourse available. As its own reactor technology was in its infancy, France initially acquired licence for PWR from Westinghouse, and licence for BWR from GE.

After the oil shock from OPEC after the Yom Kippur war of 1973, the French Government took a momentous decision in Mar. 1974 to cancel all the planned fossil power plants and accelerate dramatically the nuclear power programmes. France made two innovative developments in reactor technology – standardization of reactor design, which facilitated training and operation teams which can move from one reactor to another, and allowed interchangeability of spare parts, and procedures, and clustering of several reactors in one site.

In 2001, France created the company, AREVA, which comprehensively covers all the activities involved in the nuclear cycle. In the Front End of the cycle, AREVA supplies uranium, and offers the conversion and enrichment services to fabricate the fuel assemblies. The Reactors and Services Division takes care of design, construction, maintenance and performance of the reactors, principally PWR and FBR. The Back End Division is concerned with recycling of the used fuel of the customers to recover reusable material (uranium and plutonium), produce MOX fuel from these recovered materials, treat and package the waste for disposal.

France with a population of 63 million, has 59 nuclear reactors (mostly PWRs), which supply about 80% of the electricity. France produces the cheapest electricity in Europe, and has the lowest per capita emissions of green house gases.

The image of nuclear power suffered grievously post-Fukushima disaster in Japan. For instance, India was planning to buy from AREVA six nos, of 1650 MWe European Pressurised Reactors (EPRs) at a cost of about 40 billion Euros. Now India is insisting of post-Fukushima certification of the reactors.

Chapter 4.4

Renewable energy resources

4.4.1 WHY RENEWABLES?

There is little doubt that the Renewables are the energy resources of the future, for the simple reason that, unlike the fossil fuels, they do not get depleted when used. Most are related to sun in some way. Sunlight produces solar energy directly. It indirectly produces hydropower (through the movement of rain water), biomass (through photosynthesis), and tidal power (through tides caused by moon and sun).

The green, renewable energy economy is fundamentally different from our 20th century economy with its overdependence on fossil fuels. The renewable fuels, such as, wind, solar, biomass or geothermal, are entirely indigenous. The fuels themselves are often free. They just need to be captured efficiently and transformed into electricity, hydrogen or clean transportation fuels. In effect, the development of renewal energy invests in people, by substituting labour for fuel. Renewable energy technologies provide an average of four to six times as many jobs for equal investment in fossil fuels. For instance, while natural gas power plant provides one job per MW during construction and ongoing operations and maintenance, equivalent investment in solar photovoltaic power technology would generate seven jobs per MW. The approval of the Renewable Electricity Standard (RES) by USA would involve the construction and maintenance of 18,500 MW/yr of wind, solar, geothermal and biomass plants, and if all the components needed for the project are manufactured in USA, this would generate 850,000 jobs.

If a society decides that climate change should be mitigated, such a policy should be reflected in the choice of technologies made by that society. It should not be forgotten that a desirable outcome, say, the development of renewable energy technologies, does not happen by itself – it should be made to happen through appropriate policy change. Policy and markets are generally in conflict, as their objectives are different. "A policy is a market intervention intended to accomplish some goal – a goal that presumably would not be met if the policy did not exist" (Komor, 2004, p. 21). The object of the public policy is to promote public good, while the object of the market is to make money as quickly as possible. On the basis of considerations of public good, a government may decide on a policy of promoting renewable energy technologies, but the market would not wish to participate in it unless there is a reasonable prospect of profiting from it.

The trick is in figuring out the "intersection" point, where the two pathways converge – i.e. whereby the market finds that it is possible to make money from an activity that the government wishes to promote. This is easier said than done. The intersection point is not a fixed point – it is a floating point. It is changing all the time in response to changes in technology and market penetration. It follows that the policy makers and private makers have to engage one another in continuous dialogue in order to arrive at the "intersection point" which is acceptable to both the sides.

The share of the total renewables in the world primary energy supply (TPES) in 2005, was 12.7%. Coal: 25.3%, Oil: 35.0%, Natural gas: 20.6%; Non-renewable waste: 0.2%, Nuclear: 6.3%, Hydro: 2.6%, Renewable combustibles and wastes: 9.9%, others: 0.57%.

Product shares in the world renewable energy supply, 2005: Renewables combustibles and waste: 78.6% (comprising liquid biomass: 1.6%, renewable municipal waste: 0.7%, solid biomass/charcoal: 75.6%, gas from biomass: 0.9%); Wind: 0.6%, hydro: 17.4%, solar/tide: 0.3%, geothermal: 3.2%.

The power of the technology innovation and market penetration can be illustrated with two recent examples. Apple's iPhone-3GS, costing just $200, sold more than one million pieces in three days after it was issued – busy people queued for hours to buy the gadget. Similarly, the demand for Tata's nano, the world's cheapest car at USD 2,000, is in millions – it has already sold 100,000 units.

As Komor (2004, p. 12) puts it succinctly, "If renewables are to succeed, they must succeed in a competitive market". Policy should be aimed at facilitating it. Energy policy of any country has to have two objectives: job generation as a way of getting out of the recession, and mitigation of climate change impacts through low-carbon technologies. Consequently, governments could consider formulating sustainable energy policy frameworks for their countries based on the following strategy:

(i) how to promote greater use of renewable energy for on-grid, large-scale electricity production – this will also help to overcome the intermittency problems of wind power and solar PV,
(ii) how to use discentives, such as carbon tax, to phase out fossil fuel use; how to use technology to reduce the carbon footprint and improve efficiency of fossil fuels, where the use of fossil fuels is unavoidable,
(iii) how to use market-based strategies, such as green certificates, and how to develop innovative technologies for the production of new kinds of fuels (e.g. algal biofuels), new ways of energy storage, and demand-side management, etc. Energy policy case histories of some countries are analyzed to delineate what works, and what does not work.

Power plants use fuel to generate electricity, which is then transmitted and distributed to the user (domestic, commercial and industrial). It is the generation part of the system that largely determines the cost of the electricity and is responsible for the environmental damage and climate change impact. It is also the part where renewable fuels can play a major role. The chapter seeks to explore the policy framework for promoting greater use of renewable energy for on-grid, large-scale electricity production.

Brief description is given of different forms of renewable energy.

4.4.2 RENEWABLE ENERGY SOURCES

4.4.2.1 Wind power

Wind energy is believed to be the most advanced of the "new" renewable energy technologies. Since 2001, wind power has been growing at a phenomenal rate of 20% to 30% per annum. Wind power (2016 GW) is expected to provide 12% of the global electricity by 2050, thereby avoiding annually 2.8 gigatonnes of emissions of CO_2 equivalent. This would need an investment of USD 3.2 trillion during 2010–2050. Atmospheric scientists are developing highly localized weather forecasts to enable the utility companies to know when to power up the wind turbines.

Wind turbines do not need gusty winds; they need only moderate but steady winds. Wind turbines start producing electricity when the wind speed reaches 18–25 km/hr (5 to 7 m/s), reaching their rated output when the wind speed reaches about 47 km/hr (13 m/s). So any area where the wind speeds are greater than about 18 km/hr (5 m/s) is suitable for generating wind electricity, and such areas are plentiful. When the wind speeds exceed 22 to 26 m/s, the turbine is shut off to avoid damage to the structure.

Availability of wind turbine is defined as the proportion of the time that it is ready for use. Operation and maintenance costs are determined by this factor. Availability varies from 97% onshore to 80–95% offshore.

Improved turbine design is aimed at extracting more energy from the wind, more of the time, and over longer period of time. Affordable materials with higher strength-to-mass ratio are needed for the purpose. More power is captured by having a larger area through which the turbine can extract energy (the swept area of the rotor), and installing the rotor at a greater height (to take advantage of the rapidly moving air).

There has been a steady growth in the size of the wind turbines (Fig. 4.4.1). A typical 2 MW wind turbine has two or three blades, each about 40 m long, and

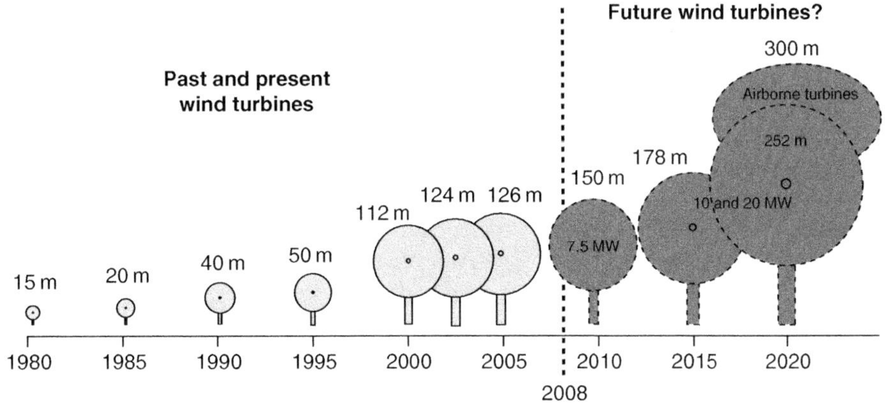

Figure 4.4.1 Growth in the size of the wind turbines
Source: Technology Roadmap: Wind power, p. 22, © OECD-IEA

made of fiberglass or composite material. The nacelle, which is the housing on the top of the tower, contains the generator and gearbox to convert the rotational energy into electricity. The tower height is ~80 m. The largest wind turbines presently in operation in the world, has a capacity of 5–6 MW each, with rotor diameter up to 126 m.

4.4.2.1.1 *Offshore wind turbines*

Till now, offshore wind turbine designs have been essentially "marinised" forms of onshore turbines. It is realized that future designs of offshore wind turbines should take into account the special characteristics of the marine environment

New designs of offshore wind turbines will have two blades rotating downwind of the tower, with a direct-drive generator. There will be no gearbox. The rotor will be 150 m in diameter. The turbine capacity could be 10 MW. It will have a self-diagnostic system, which is capable of taking care of any operational problems on its own. Such an arrangement will reduce the requirement of maintenance visits to the minimum.

4.4.2.1.2 *Further technology development*

The European Union has launched an impressive wind energy R&D initiative, code named *UpWind*, aimed at developing very large turbines (8 to 10 MW) and large wind farms of several hundred megawatt capacity. The programme would involve better understanding of wind conditions, development of materials with high strength to mass ratios, and improved control and measuring systems.

4.4.2.2 Solar power

Solar PV has many strong points – (i) PVs will work any where the sun shines – the sunnier it is, the more the electricity produced, and the lower the per kWh cost. Some electricity is produced even when it is cloudy, (ii) unlike the wind turbines which needs to have winds with speeds of more than 18 km/hr to start producing power, PVs have no such constraints – they work any where the sun shines, (iii) it is quiet, has no moving parts, can be installed easily, and can be sized at any scale, ranging from a single bulb, to powering the entire community. It has two serious drawbacks: (i) it is expensive – its levelized cost (US cents 20–40/kWh) is several times more than that of electricity from the fossil fuels (US cents 3 to 5/kWh). The PV costs are coming down all the time, but they are a long way from being competitive. (ii) It is intermittent – no power is generated during nights when there is no sun.

Solar energy can be used in the following ways: (i) Direct supply of solar heat to buildings and industrial processes – provision of heat accounts for about 40% of the global energy needs, (ii) electricity can be produced through the photovoltaic cells, or through steam turbines by the concentration of solar rays, and (iii) production of hydrogen which can be used as fuel. Among all the energy systems, solar energy is projected to grow the fastest. Between now and 2050, solar energy is expected to grow thousand-fold, to 2,319 TWh/yr in the ACT scenario, and 4,754 TWh/yr in the BLUE scenario. It is assumed that during the next ten years, there will be sustained support to the solar energy sector to enable it to become competitive. Under both ACT and BLUE scenarios, major growth is likely to occur after 2030. PV is expected to grow fast in the solar-rich OECD countries (e.g. North America) and the emerging economies of

China and India. CSP (Concentrated Solar Power) will grow strongly not only in these countries, but also in the sun belts of Africa and Latin America.

A photovoltaic cell (PV cell) is a semiconductor device that is capable of converting solar energy into direct current (DC) electrical energy. A PV cell is typically a low voltage (~0.5 V) and high current (~3 A) device. PV modules are built by combining a number of PV cells in series. A commercial module with an area of 0.4 m^2 to 1.0 m^2 can produce peak power of 50 Wp (peak Watts) to 150 Wp. By linking together appropriately large number of PV cells, it is possible to build huge power units with a capacity of tens of MWs.

Thin films

Thin film technology is rapidly emerging as a viable alternative to silicon wafer technology. A thin layer of photosensitive material is deposited on a low-cost backing, such as, glass, stainless steel or plastic. Initially, amorphous silicon (α-Si) was used, but now-a-days, Cadmium Telluride (CdTe) or Copper-Indium-Diselenide (CIS) are used instead. The efficiency of CIS gets improved when it is doped with gallium, to produce a CIGS module. The thickness of the thin film may range from 40–60 μm in the case of c-Si to less than 10 μm in the case of CdTe.

Thin films use smaller quantities of feedstock, and are amenable to automation. They can be integrated into buildings more readily and have better appearance. Their efficiencies are, however, lower than c-Si modules. Recent improvements in CIS modules have allowed them to have efficiencies of the order of 11%, which figure is comparable to the efficiency of mc-Si modules. An efficiency of 22% is projected for CIS modules by 2030. But the availability of Cd and Te may prove to be a constraint. Thin films are likely to increase their market share by 2020. After that, hybrid systems which combine crystalline and thin-film technologies, may dominate the market. These hybrid systems have the best of both the worlds- higher efficiencies of the order of 18%, lower material consumption and amenability to automation.

The module efficiencies of different PV systems are summarized in Table 4.4.1.

The consensus in the PV industry is that after 2020, the market share of c-Si PV systems will decrease, and that thin-film technology will dominate the market. Two types of Third generation PV devices are expected to come up during 2020–2030:

(i) Ultra-low cost, low to medium efficiency cells and modules, such as dye-sensitized nanocrystalline solar cells (DSC) which could attain an efficiency of 10%, if not

Table 4.4.1 Present module efficiencies for different PV technologies

	Wafer-based c-Si		*Thin Films*		
	Sc-Si	*mc-Si*	*α-Si, α-Si/mc-Si*	*CdTe*	*CIS/CIGS*
Commercial module efficiency (%)	13–15	12–14	6–8	8–10	10–11
Maximum recorded module efficiency (%)	22.7	15.3	–	10.5	12.1
Maximum recorded laboratory efficiency (%)	24.7	19.8	12.7	16.0	18.2

Source: Frankl, Manichetti and Raugei, 2008

15%, by 2030. Organic solar cells with efficiencies of the order of 2% are being developed. It is too early to speculate on their economic viability. They may figure in applications where space is not a problem.

(ii) Ultra-high efficiency cells and modules, based on advanced solid-state physics principles, such as, hot electrons, multiple quantum wells, intermediate band gap structures and nanostructures. It is difficult at this stage to predict their efficiency levels, but some experts predict that these devices may attain efficiencies of 30–60%. Most likely, several types of PV devices may coexist in 2050.

R&D needed

c-Si module technology has been successful as it is reliable, takes advantage of the electronics industry, with ready availability of feedstock. Further advances that are needed in order for PV systems to reach large production volumes and low target cost of USD 1.25/W, are summarized below (PV-TRAC, 2005; EUPVPLATF, 2007):

c-Si technology

(i) Materials: Epitaxial deposition; substitution of silver (because of cost) and lead (because of its health impact).

(ii) Equipment: Manufacturing processes (including ribbons) that use less silicon and less energy per watt. High degree of automation.

(iii) Device concepts and processes: Design of new modules which can be assembled more easily, low-cost and longer life span (25–40 years).

Thin film technologies

(i) Materials and devices: Increase of module efficiencies from the current 5% to 15% or more; development of new multi-junction structures; reduced materials consumption; alternative module concepts, such as, new substrates and encapsulation modalities; enhancement of the module life for 20 to 30 years, with less than 10% reduction in efficiency.

(ii) Processes and equipment: Techniques of ensuring uniformity of film properties over large areas, and bridge the efficiency gap between laboratory modules and large-scale industrial modules; reduction in the pay-back time of the modules from the present 1.5 years to 0.5 years.

Concentrated solar power (CSP)

General considerations

Concentrated solar power needs cheap water, cheap land and plenty of sun. The water need not be fresh water – brackish water or sea water will do.

Concentrated Solar Power (CSP) systems concentrate direct sunlight to reach high temperatures. This heat can then be used to power a steam turbine which drives a generator. CSP is best suited to areas with high direct solar radiation, the minimum requirement being 2,000 kWh/m^2.

Unlike c-Si systems in which solar energy is directly converted to electrical energy, CSP system has thermal energy as an intermediate phase. In other words, it can store heat in various forms, and release power as and when needed. CSP can be used for continuous solar-only generation. Also it can burn fuel in hybrid plants and use the traditional steam turbines to generate power.

A strong plus point in favour of CSP is that it reaches peak production (at noon) exactly at the time when the electricity demand is at its highest (say, for air-conditioners) in tropical, arid and semi-arid areas. In these areas, CSP electricity is much cheaper than PV electricity, though it is costlier than fossil fuels and wind power.

As CSP plants are invariably large (typically several hundred megawatts), they have to be linked to transmission networks. For instance, it is possible to export CSP electricity from North Africa to Europe at the cost of USD 30/MWh – this figure is les than the cost difference in solar electricity between the two regions (DLR, 2006). Barring some special situations, space should not be a constraint for CSP plants. It has been estimated that the electricity requirements of the whole of USA, could be generated with CSP plants occupying an area of a hundred square miles. CSP could be combined with conventional devices, such as, steam turbines, and it is possible to scale up CSP plants to several hundred megawatts, using well-established technologies.

Apart from electricity generation, CSP plants can be used to heat/cool buildings, to desalinize water, and to produce fuels like hydrogen. In areas where water is scarce, dry coolers may be used. In some arid countries, cogeneration of heat for desalinization and power may turn out to be an attractive proposition.

The constraint of intermittency of solar energy can be got over either by storing heat and conversion to electricity when solar energy is unavailable, or to have a fossil fuel backup that uses the same steam cycle as the CSP plant (this needs only an additional burner).

The solar flux is concentrated in three ways, with different flux concentration ratios: troughs (30–100), towers (500–1,000) and dishes (1,000–10,000) (Fig. 4.4.2).

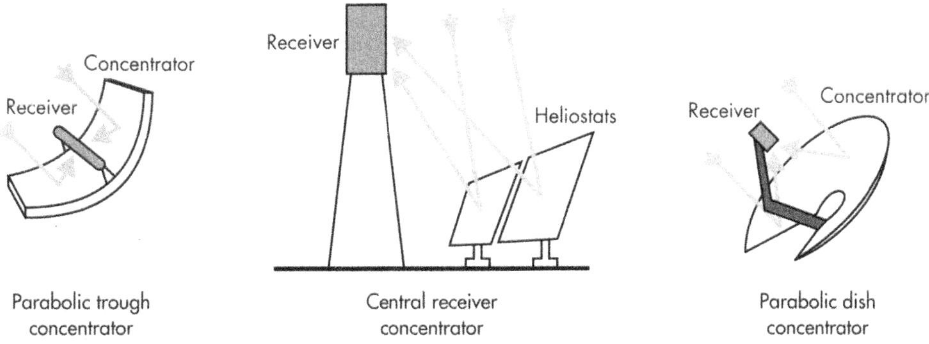

Figure 4.4.2 Troughs, Towers and Dishes
Source: Energy Technology Perspectives, 2008, p. 381

Table 4.4.2 Bioenergy conversion plant technologies

Conversion type	*Typical capacity*	*Net efficiency*	*Investment costs (in USD)*
Anaerobic digestion	<10 MW	10–15% electrical, 60–70% heat	
Landfill gas	<200 kW to 2 MW	10–15% electrical	
Combustion for heat	5–50 kW_{th} residential 1–5 MW_{th} industrial	10–20% open fires, 40–50% stoves, 70–90% furnaces	~23/kW_{th} stoves; 370–990/kW_{th} furnaces
Combustion for power	10–100 MW	20–40%	1975–3085/kW
Combustion for CHP	0.1–1 MW 1–50 MW	60–90% overall 80–100% overall	3333–4320/kW 3085–3700/kW
Cofiring with coal	5–100 MW existing. >100 MW new plant	30–40%	123–1235/kW + power station costs
Gasification for heat	50–500 kW_e	80–90%	864–980/kW_e
BIGCC* for power	5–10 MW demos 30–200 MW future	40–50% plus	4320–6170/kW 1235–2470/kW future
Gasification for CHP using gas engines	0.1–1 MW	60–80% overall	1235–3700/kW
Pyrolysis for bio-oil	10 t/hr demo	60–70%	864/kW_{th}

BIGCC – Biomass Integrated Gasification with Combined Cycle
Source: Energy Technology Perspectives, 2008, p. 312

4.4.2.3 Biomass

Biomass is biologically-produced matter, and includes agricultural and forestry residues, municipal solid wastes and industrial wastes, renewable landfill gases, etc. Biodiesel is produced from leftover food products, such as animal fats and vegetable oils and crops grown solely for energy purposes. Algae which are aquatic organisms that range from pond scum to seaweeds, have great potential for producing liquid transportation fuels (Exxon Mobil has just announced funding of USD 600 million for algal biofuels). A strain of *Escherichia coli* has been engineered to produce biodiesel fuel directly from biomass. A strong point of biomass power is that it is dispatchable – it can be turned on and off, in contrast with wind and solar power which are non-dispatchable. Biomass burning does produce particulates, but emissions of SO_x and NO_x are much less than from fossil fuel power stations.

Technology

The technoeconomic characteristics of bioenergy conversion plant technologies are given in Table 4.4.2.

Algal biofuels

Photosynthesis which is a fundamental biological process, needs sunlight, carbon dioxide and water. Cyanobacteria can convert up to 10% of the sun's energy into biomass. This rate is 5% for algae, and 1% for corn and sugar cane. The fossil fuels that we

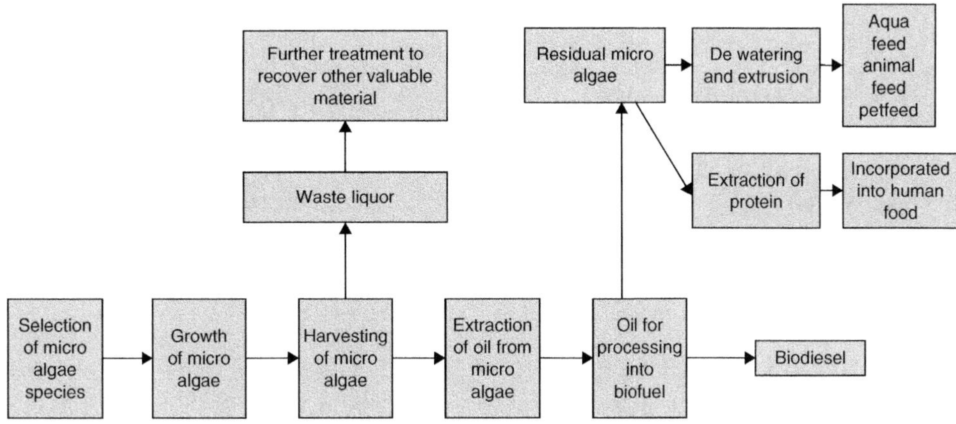

Figure 4.4.3 Detailed process of biodiesel from algae
Source: "Green Energy", p. 142

use now, were produced by cyanobacteria in the geological past. Single-cell algae are capable of producing a chemical "mix" that contains extractable fuel usable in cars and trucks. The "green" crude thus produced is chemically identical to crude oil, but it is carbon-neutral, non-toxic and sulphur-free. Sapphire Energy of California has reportedly succeeded in producing the green "crude", and expects to produce one million gallons (3.8 M. liters) of biodiesel and jet fuel per year starting from 2011.

Algae can be cultivated in tanks and ponds. As oil-producing algal strains are not fast growing, there is always a danger of faster-growing, non-oil producing algae invading such tanks (on the analogy of weeds). This can be prevented by covering a pond with a greenhouse, or using tubes which allow sunlight, and allow the circulation of nutrients and CO_2 to enhance the productivity ("bioreactors"). Closed systems are generally more expensive to build and operate, but their costs can be brought down by locating the algal plants near sources of CO_2. Firm cost information is not available in this regard.

Second generation biofuels which are expected to be produced from a range of ligno-cellulosic feedstocks through the use of thermochemical or biochemical pathways, will be superior to first generation biofuels in that they will be high yielding, they will be characterized by low emission of GHGs, and most importantly, they can be commercially produced sustainably. Second generation biofuels may initially complement first generation biofuels from grains and oil seeds, but may later supersede them (Fig. 4.4.3).

Ethanol

Ethanol (C_2H_2OH) is produced by the fermentation of sugars. Table 4.4.3 gives the ethanol yields from various raw materials.

Brazil's production of ethanol from sugar residues is the largest commercial biomass system in the world. Its impetus came from high oil prices and low sugar prices.

Table 4.4.3 Ethanol yields

Raw material	*Litres per tonne*	*Litres per hectare per year*
Sugar cane (harvested stalks)	70	400–12,000
Maize (grain)	360	250–2,000
Cassava (roots)	180	500–4,000
Sweet potato (roots)	120	1,000–4,500
Wood	160	160–4,000

Source: Boyle, 2004, p. 135

The production reached the maximum of about 15 billion litres/yr. Ethanol is blended with gasoline up to 26% (gasohol). Brazil saved about USD 40 billion in foreign exchange by the use of gasohol to run cars.

Prognosis

Bioenergy is not only the largest contributor to renewable energy presently but also has the highest technical potential among the renewable energy resources. Biomass is used inefficiently for traditional domestic cooking and space heating in developing countries. According to a projection by IEA, biomass use would increase four-fold by 2050 (150 EJ/yr or 3,604 Mtoe/yr). This would involve the delivery of 15,000 Mt of biomass to the production plants annually, with half of this coming from crop and forest residues, and the rest from purpose-grown energy crops. The projected use of bioenergy would be as follows: about 700 Mtoe/yr for transport biofuels, about 700 Mtoe/yr to generate electricity (2,450 TWh/yr), and 2,200 Mtoe to produce biofuels, heating, and cooking (including production of dimethyl ether), and industry (including process steam from CHP (Combined Heat and Power) and black liquor (used for the extraction of lignin). By co-firing biomass with coal in steam turbines of about 100 MW capacity, electricity could be produced at USD 60–80/MWh. Biofuel from algae has great potential – much research is being devoted to cultivating specific algal strains suitable for biodiesel production.

4.4.2.4 Hydropower

Presently, hydropower accounts for 90% of the renewable power generation. Some countries, notably Sweden, Switzerland and Norway, get virtually all their electricity supplies from hydropower. The total hydropower potential in the world is estimated to be 2,000 GW (Taylor, 2007). IEA estimates that by 2050, hydropower capacity could go up by 1,700 GW, producing 5,000–5,500 TWh/yr. Though the world's technically feasible hydropower is 14,000 TWh/yr, the realistic potential is 6,000 TWh/yr. Around 808 GW of hydropower is in operation or under construction. Hydropower is one of the cheapest ways of producing electricity. Most plants have been built many years ago, and their capital costs have already been amortized. The capital costs of new plants vary from USD 1,000/kW in developing countries to USD 2,400/kW in OECD countries. Generating costs vary from USD 0.02–0.06/kWh.

Table 4.4.4 Hydro potential

Region	*Technical potential (TWh/yr)*	*Annual output (TWh/yr)*	*Output as percentage of technical potential (%)*
Asia	5,093	572	11
South America	2,792	507	18
Europe	2,706	729	27
Africa	1,888	80	4.2
North America	1,668	665	40
Oceania	232	40	17
World	14,379	2,593	18

Source: Boyle, 2004, p. 153

Table 4.4.4 provides the hydro potential and output of different regions in the world.

Hydropower output (in TWh/yr) of some important countries are: Canada – 345; Brazil – 288; USA – 264; China – 231; Russia – 167; Norway: 129; Japan – 91; India – 76; Sweden – 74; France – 74; Venezuela – 61; Italy – 51; Austria – 42; Switzerland – 40; Spain – 35, etc.

There are three kinds of hydropower facilities: (i) "Storage" projects (Aswathanarayana, 2001, p. 191–195), (ii) "Pumped Storage" projects, and (iii) "Run-of-river" projects.

"Storage" projects

Large surface reservoirs are constructed to increase the availability of water during the periods of low flow or dry years. Though the construction of water reservoirs has been going on since ancient times, all the large reservoirs with a total volume of more than 50 km^3 have been constructed during the second half of the last century. According to Shiklomanov (1998), the total volume of the reservoirs of the world is about 6,000 km^3, and their total surface area is about 500,000 km^2. The volume of water in the reservoirs corresponds to about 5% of the annual rainfall on land or 15% of the annual runoff. They trap 2–5 Gt y^{-1} of sediments, which is about a quarter of the total sediment yield of the land. Thus, the reservoirs have a profound impact on the circulation of water on the land surface, discharge of sediments and nutrients to the sea. The evaporation from the surfaces of the reservoirs is considerable, and thus the reservoirs are one of the greatest users of freshwater. The map by Vörösmarty et al. (1997) shows the location of the major reservoirs in the world.

Pumped storage hydroelectricity

Pumped storage is a high capacity form of grid energy storage presently available. It can be used to flatten out load variations on the power grid, which may be linked to coal-fired plants, nuclear plants or renewable energy power plants. In the context of the back-up provided by pumped storage, these plants could continue to operate at peak efficiency (Base Load Power plants), while reducing the need for "peaking" power plants that use costly fuels. Thermal plants are less able to respond to sudden changes in the electricity demands, and may cause voltage and frequency instabilities. In contrast, pumped storage plants, like the normal hydropower plants, can respond to load changes almost instantly (less than 60 seconds). Pumped storage is expected to become important for load balancing in tandem with large capacity solar and wind mill plants.

A recent technological development is the variable speed turbines which can generate electricity in synchronization with the network frequency.

"In-river" hydroelectric projects

Small-scale hydropower projects are designed to run in-river. These are environment-friendly energy conversion options as they do not affect the river flows significantly. They can be stand-alone applications to serve rural communities or as replacements to diesel generators.

4.4.2.5 Geothermal energy

"Geothermal energy" covers both the direct use of geothermal power for space, heating, water heating and industrial processes, which are more common, and the generation of geothermal electricity, which are rarer. Geothermal electricity plants of more than 100 MW installed capacity are listed below, country-wise (MW installed capacity in 2000): USA – 2228; Philippines – 1909; Italy – 785; Mexico – 755; Indonesia – 590; Japan – 547; New Zealand – 437; Iceland – 170; El Salvador – 161; Costa Rica – 143. The total capacity of geothermal power plants in the world is 10 GW in 2007, generating 56 TWh /yr of electricity.

Geothermal energy has several advantages: (i) It is non-polluting and has no carbon footprint, (ii) It is of large magnitude – the heat stored in the earth is estimated to be about 5 billion EJ, which is 100,000 times more than the world's annual energy use, (iii) It is available all the year round, and production costs are low. There are, however, some drawbacks: (i) Air pollution may sometimes be caused by H_2S, CO_2, NH_3, Rn, etc. gases vented into the air, (ii) Low magnitude earthquakes may be triggered and land subsidences may take place due to changes in the reservoir pressure, (iii) The overall efficiency of geothermal power production (15%) is less than half of the coal-fired plants, (iv) Drilling costs are high (USD 150,000 – 250,000 per well).

Compared with wind electricity and solar PV electricity, which are intermittent, geothermal electricity can be generated round the clock, and could therefore serve as baseload electricity. This factor is reflected in the capacity factor which is defined as the actual plant output as a percentage of the maximum output of the plant operated at full capacity. Geothermal plants have a capacity factor of 90%, compared to 25 to 30% in the case of wind electricity.

Large scale geothermal plants are currently possible in high heat flow areas such as, plate boundaries, rift zones, mantle plumes and hot spots, that are found around the "Ring of fire" (Indonesia, The Philippines, Japan, New Zealand, Central America, the west coast of USA) and the rift zones (East Africa, Iceland).

4.4.2.6 Tidal power

The use of tidal energy to generate power is similar to that of hydroelectric power plants. A dam or barrage is built across a tidal bay or estuary where there is a difference of more than five metres between the high tide and low tide. Water flowing in and out of the dam runs the turbines installed along the dam or barrage, and generates electricity. Tidal plants have periods of maximum power generation every six hours. During periods of low electricity demand, extra water is pumped into the basin behind the barrage, on the analogy of pumped storage.

Apart from grid-connected electricity generation, ocean renewable energy could also be used for off-grid electricity generation in remote areas, aquaculture, desalination, production of compressed air for industrial applications, integration with other renewable energy resources, such as offshore wind power, solar PV, etc.

Tidal barrage projects are more environmentally intrusive than wave and marine current projects. The adverse environmental impact of tidal barrage projects is sought to be reduced by integrating oscillating water turbines with breakwater systems that convert water pressure into air pressure and use the compressed air to drive a Wells turbine. Such breakwaters linked projects (about 0.3 MW capacity) are being developed in Spain and Portugal. Portugal is also actively developing wave energy plants with the goal of achieving 23 MW by 2009.

Resource position

The World Energy Council has estimated the world wave power at 2 TW. The realistically recoverable ocean energy resource is put at 100 GW. The estimated wave electricity potential is 300 TWh/yr.

4.4.2.7 Deployment of Renewable Energy Technologies (RETs)

Selected characteristics and costs of common Renewable Energy Technologies (RETs) are given in Table 4.4.5. It may be noted that in general the costs of RETs are higher than conventional energy technologies which are typically around US cents 4 to 8/kWh. The position, however, is not static. The costs of many RETs are declining significantly due to technology improvements and market maturity. At the same time, the costs of some conventional energy technologies (for example, gas) are also declining. New kinds of gas deposits (such as, shale gas), new methods of mining (such as, horizontal drilling), and improvements in gas turbine technology, have brought down the costs of electricity production from gas.

An integrated strategy for the deployment of RETs

The deployment of Renewable Energy Technologies (RETs) has two concurrent goals: (i) exploit the "low-hanging" fruit of abundant RETs which are closest to market competitiveness, and (ii) developing cost-effective technologies for a low-carbon future in the long term. Instances are known whereby non-economic barriers, such as beauracratic red tape, complex administrative procedures, grid access, social acceptance of new technologies, lack of information or training, impeded the progress of RETs even when they are close to economic competitiveness with conventional technologies. High priority should be given for the removal of such impediments.

Table 4.4.5 Key characteristics and costs of Reneweable Energy Technologies

Technology	*Typical characteristics*	*Typical current investment costs*[1] *(USD/kW)*	*Typical current Energy Production costs*[2] *(USD/MWh)*
POWER GENERATION			
Hydro			
Large hydro	Plant size: 10–18,000 MW	1,000–5,500	30–120
Small hydro	Plant size: 1–10 MW	2,500–7,000	60–140
Wind			
Onshore wind	Turbine size: 1–3 MW Blade diameter: 60–100 meters	1,200–1,700	70–140
Offshore wind	Turbine size: 1.5–5 MW Blade diameter: 70–125 meters	2,200–3,000	80–120
Bioenergy[3]			
Biomass combustion for power (solid fuels)	Plant size: 10–100 MW	2,000–3,000	60–190
Municipal solid Waste (MSW) incineration	Plant size: 10–100 MW	6,500–8,500	n/a
Biomass CHP	Plant size: 0.1–1 MW (on-site) 1–50 MW (district)	3,300–4,300 (on-site) 3,100–3,700 (district)	n/a
Biogas (including landfill gas) digestion	Plant size: <200 kW–10 MW	2,300–3,900	n/a
Biomass co-firing	Plant size: 5–100 MW (existing); >100 MW (new plant)	120–1,200 +power station costs	20–50
Biomass Integrated Gasifier Combined Cycle (BIGCC)	Plant size: 5–10 MW (demonstration); 30–200 MW (future)	4,300–6,200 (demonstration) 1,200–2,500 (future)	n/a
Geothermal Power			
Hydrothermal	Plant size: 1–100 MW; Types: Binary, single and double flash, Natural steam	1,700–5,700	30–100
Enhanced geothermal system	Plant size: 5–50 MW	5,000–15,000	150–300 (projected)
Solar energy			
Solar PV	Power plants: 1–10 MW; Rooftop systems: 1–5 kWp	5,000–6,500	200–800[4]
Concentrating Solar power (CSP)	Plant size: 50–500 MW (trough), 10–20 MW (tower), 0.01–300 MW (future) (dish)	4,000–9,000 (trough)	130–230 (trough)[5]
Ocean energy			
Tidal and marine currents	Plant size: Several demonstration Projects up to 300 kW capacity; Some large scale projects under development	7,000–10,000	150–200

(Continued)

Table 4.4.5 Continued

Technology	*Typical characteristics*	*Typical current investment costs*[1] *(USD/kW)*	*Typical current Energy Production costs*[2] *(USD/MWh)*
Heating/Cooling			
Biomass heat (excluding CHP)	Size: 5–50 kW_{th} (residential)/ 1–5 MW_{th} (industrial)	120/kW_{th} (stoves); 380–1,000/kW_{th} (furnaces)	10–60
Biomass heat from CHP	Plant size: 0.1–50 MW	1,500–2,000/kW_{th}	n/a
Solar hot water/heating	Size: 2–5 m^2 (household); 20–200 m^2 (medium/multifamily); 0.5–2 MW_{th} (large/district heating); Types: evacuated tube, Flat-plate	400–1250/m^2	20–200 (household); 10–150 (medium); 10–80 (large)
Geothermal heating/ cooling	Plant capacity: 1–10 MW; types: Ground-source heat pumps, direct use, chillers	250–1450/kW_{th}	5–20
Biofuels (1st Generation)			
Ethanol	Feedstocks: sugar cane, sugar beets, corn, cassava, sorghum, wheat (and cellulose in future)	0.3- 0.3–0.6 billion per billion litres/year of production capacity for ethanol	0.25–0.3/litre gasoline equivalent (sugar); 0.4–0.5/litre gasoline equivalent (corn)
Biodiesel	Feedstocks: soy, oilseed rape, mustard seed, palm, jatropha, tallow or waste vegetable oils	0.6–0.8 billion per billion litres/ year of production capacity	0.4–0.8/litre diesel equivalent
Rural (off-grid) Energy[6]			
Micro-hydro	Plant capacity: 1–100 kW	1,000–2,000	70–200
Pico-hydro	Plant capacity: 0.1–1 kW	n/a	200–400
Biomass gasifier	Size: 20–5,000 kW	n/a	80–120
Small wind turbine	Turbine size: 3–100 kW	3,000–5,000	150–250
Household wind turbine	Turbine size: 0.1–3 kW	2,000–3,500	150–350
Village-scale Mini-grid	System size: 10–1,000 kW	n/a	250–1,000
Solar home system	System size: 20–100 W	N/a	400–600

n/a – Not applicable

1. Using a 10% discount rate. The actual global range may be wider. Wind and solar include grid connection cost.
2. Costs in 2005 or 2006.
3. Wide range. Costs of delivered biomass feedstock vary by country and region due to factors such as variations in terrain, labour costs and crop yields.
4. Typical costs 20–40 US cents/kWh for low latitudes with high solar insolation of 2,500 kWh/m^2/year. 30–50 cents/ kWh (typical of southern Europe) and 50–80 cents for higher latitudes.
5. Costs for parabolic trough plants. Costs decrease as plant size increases.
6. No infrastructure required which allows for lower costs per unit installed.

Source: " *Deploying Renewables: Principles of effective Policies*", 2008, pp. 80–83

A policy is a market intervention intended to accomplish some goal that presumably would not be met if the policy did not exist (Paul Komor, 2004). The transition to mass market integration of renewables requires some policy corrections. For instance, the price placed on carbon and other externalities need to be enhanced. It should be realized that most renewables need economic subsidies, and the removal of non-economic barriers which are impeding the deployment of RETs. The policies should be able to lead to a future energy system in which RETs should be able to compete with other energy technologies on a level playing field. When once this is achieved, RETs would need no or few incentives for market penetration, and their deployment would be accelerated by consumer demand and general market forces.

Technology-specific support schemes need to be fashioned depending upon the level of maturity of a given RET at a given time, employing a range of policy instruments, including price-based, quantity-based, R&D support and regulatory mechanisms. Apart from continued R&D support, less mature technologies which have not yet achieved economic competitiveness generally need very stable low-risk incentives, such as capital cost incentives, feed-in-tariffs (FITs) or tenders. In the case of low-cost gap technologies, such as onshore wind and biomass combustion, more market-oriented instruments such as feedin-premiums may be used. Also, TGC (Tradable Green Certificates) systems may be used innovatively, by linking technology differentiation with quota obligation either by awarding technology multiples of TGCs or by introducing technology-specific obligation (known as technology banding).

Technology banding may sometimes be necessary as a transitional phase, or it may be bypassed by the adoption of a technology-neutral TGC system. When once a given RET is competitive with other CO_2-saving alternatives and is in a position to be deployed on a large scale, the support systems for the RET may no longer be necessary now that a level playing field with other energy technologies has been achieved. The position is not static. All RETs are evolving rapidly, in response to technology improvements and market penetration. Renewable Energy policy frameworks should be so structured as to facilitate technological RD&D and market development concurrently, within and across technology families.

Chapter 4.5

Strategy for a low carbon footprint

4.5.1 CARBON EMISSIONS AND CLIMATE CHANGE

Climate Change is any significant, long-term change in the "average" weather that a region experiences. The UN Human Development Report (2007) calls the Climate Change, "the defining human development issue of our generation". The World Health Organization estimated that the health impacts of climate change caused 150,000 additional deaths in 2003, and that the proportion of the world population affected by weather disasters doubled between 1975 and 2001. This is just an inkling of the disasters that are in store for humanity if we go on increasing the emission of greenhouse gases.

Fossil fuels (coal, oil and natural gas) account for 88% of the world's commercial primary energy. Burning of the fossil fuels leads to the production of climate-relevant emissions of carbon dioxide (CO_2), methane (CH_4), nitrogen oxides (NO_x), carbon monoxide (CO) and volatile organic compounds (VOC). The extent of emissions from various sources is as follows: (i) Energy industries – about 50%, out of which CO_2 accounts for 40%, (ii) Chemical products, particularly, CFCs – about 20%, (iii) Destruction of tropical rain forests and related causes – about 15%, (iv) Agriculture and others (e.g. waste deposit sites) – about 15%.

The CO_2 concentration in the atmosphere which was about 280 ppm at the time of the Industrial Revolution in 1850s went up steadily with the increasing consumption of fossil fuels. It is now more than 387 ppm today (highest in the last 650,000 years, and higher than about 40% since the industrial revolution). The annual mean rate of growth is 2.14 ppm in 2007 (as against the rise at the rate of 1.5 ppm during the period, 1970–2000). The Developed countries which have 15% of the world's population are responsible for 50% of the CO_2 emissions.

Table 4.5.1 CO_2 concentrations and emissions

(1)	*(2)*	*(3)*	*(4)*	*(5)*
350–400	445–490	2.0–2.4	2000–2015	−50 to −85
400–440	490–535	2.4–2.8	2000– 2020	−30 to −60
440–485	535–590	2.8–3.2	2010–2030	+5 to −30
485–570	590–710	3.2–4.0	2020–2060	+10 to +60
570–660	710–855	4.0–4.9	2050–2080	+25 to +85
660–790	855–1130	4.9–6.1	2060–2090	+90 to +140

(1) CO_2 concentrations (ppm)
(2) CO_2 equivalent concentrations (ppm)
(3) Global mean temperature increase above pre-industrial equilibrium (°C)
(4) Peaking year for CO_2 emissions
(5) Global change in CO_2 emissions in 2050 (% of 2000 emission).
Source: "*World Energy Outlook 2007*", p. 206

The relationship between the increase in CO_2 concentrations (ppm) and the rise of global mean temperature above pre-industrial equilibrium (°C) is shown in Table 4.5.1.

It is expected that by about 2015, the CO_2 level will reach 400 ppm, with the concomitant rise of global warming by 2°C above the pre-industrial temperature. Limiting the global rise in temperature to 2°C provides a window of opportunity. The UN Human Development Report (2007) suggests a number of techno-socio-economic measures, such as carbon taxation, cap-and-trade programmes, reduction in emission quotas, encouraging renewable energy production through economic incentives, stringent implementation of efficiency measures in industry, building and transport sectors, support for breakthrough technologies of carbon capture and storage.

Global warming and sea level rise are going to have serious adverse consequences for agriculture, forests and ecosystems, water and the coastal zones in the country. The expected more frequent and intense heat waves are not only debilitating to humans, but are likely to spread malaria and dengue fever.

Between 2000 and 2004, 262 million people were affected by natural calamities attributable to climate change – of these 98% were in developing countries. A temperature increase of 3–4°C is projected to cause the displacement of 330 million people due to floods, malaria infection for 200–400 million people, extinction of 20–30% of all land species. In nature, life cycles of many wild plants and animals (for example, a wild flower and its pollinating insect) are seasonally synchronized. By affecting the temperature and precipitation, climate change may throw the seasonal synchronization out of gear, and thereby adversely affect the biodiversity.

The tipping point is 4°C. If the global warming goes beyond 4°C, the earth would face a catastrophe – the Amazon Rain forest may become a desert, and the sea level may rise by 80 ft. If the earth is to be saved from such an environmental catastrophe, the Developed countries need to cut their greenhouse gas emissions by 80% by the year 2050, with 20–30% cuts by the year 2030.

IPCC came up with an ambitious scenario of stabilizing the CO_2 emissions in the 445–490 ppm CO_2 – equivalent, with energy-related CO_2 emissions reduced to 23 Gt in 2030.

Table 4.5.2 World Energy Demand in the 450 Stabilisation case (Mtoe)

	2005	*2015*	*2030*	*2005: 2030**
Coal	2,892	3,213	2,559	−0.5
Oil	4,000	4,278	4,114	0.1
Gas	2,354	2,736	2,644	0.5
Nuclear	721	1,037	1,709	3.5
Hydro	251	393	568	3.3
Biomass	1,149	1,484	1,966	2.2
Other renewables	61	223	471	8.5
Total	**11,429**	**13,364**	**14,031**	**0.8**

*Average annual rate of growth
Source: "*World Energy Outlook 2007*", p. 209

In the IPCC-favoured scenario, global energy-related CO_2 emissions peak at 2012 around 30 Gt, and then decline to 23 Gt in 2030, apart from 6 Gt of CO_2 from non-energy related sources, such as, land-use changes and forestry. The figure of 23 Gt is 13% lower than the emission figure for 2005. The International Energy Agency, Paris, (2007) came up with a pathway to achieve this goal (Table 4.5.2).

The following steps need to be taken to achieve the IPCC goal: (i) Improved efficiency in fossil fuel use in industry and buildings (savings of more than 25%, relative to the Reference Scenario) (ii) Lower electricity demand through more efficient electricity use in buildings (savings of about 13%), (iii) Switching to second generation in biofuels in transport (savings of 4%), (iv) Enhanced use of renewables in the power sector (savings of 19%), (v) Increased use of nuclear power (savings of 16%), and (vi) CO_2 capture and storage (CCS) (savings of 21%). It should be emphasized that these steps are not going to be easy. Nuclear power faces many policy and regulatory hurdles, and the techno-economic viability of CCS is yet to be established. Evidently, governments all over the world have to work promptly and vigorously to make the huge investments needed for the purpose. Any delay will make it progressively more difficult to achieve the target.

4.5.2 MITIGATION OF CLIMATE CHANGE

For the mitigation of climate change, CCS (Carbon dioxide Capture and Storage) is a technology option that would allow the continued use of fossil fuels. Pulverised coal combustion (PCC) accounts for 97% of the coal-fired capacity. Supercritical steam plants and Integrated coal gasification combined cycle (IGCC) plants have been able to achieve high thermal efficiencies of 42 to 45%. Post-combustion capture of CO_2, followed by geological storage, is nearest to commercialization. CCS has considerable flexibility in technological improvement, such as the use of new absorbers. Flue gas scrubbing with amines is the most promising for plants of 500 MW or higher capacity. Transport of CO_2 is the key for CCS deployment. Pressure vessels can be used to transport CO_2 in the liquid form. Pipeline transport may be used for supercritical CO_2

above the critical point (31.1°C; 73.9 bars). Sleipner-type offshore storage of CO_2 has technical, social, political and economic advantages.

Present global nuclear share of electricity amounts to ~15% (370 GWe), and it helps reducing the global emission by ~3 giga tonnes of CO_2. Nuclear power is slated grow to 473–748 GWe by 2030. It is seen as a renewable energy, considering its vast energy potential. Current nuclear reactor technologies are based on the utilization of thermal or slow neutrons to fission low-enriched uranium (3–5%^{235}U) using light water as moderator and coolant. Most of the reactors currently operating utilize less than 1% of the fissionable content of the fuel. The rest is treated as waste in once-through fuel cycles. In a closed fuel cycle the unutilized materials are recycled. In fast breeder reactors the energy utilization of the fuel is multiplied by almost 100 times, as the reactor "breeds" more fissile material than it consumes. India has come up with the design of a low-enriched uranium – thorium fuelled, heavy water reactor (AHWR-LEU), whose fuel cycle is proliferation-resistant, and which can be deployed without any safety and security prescriptions. Such a technological innovation allows the bypassing of proliferation and security legal frameworks which are difficult to enforce.

Next Generation Green Technologies, which are in the process of development, may have a potential comparable to other renewable energies. Biomass gasification is potentially more efficient than the direct combustion of the original fuel. In the town of Güssing in Austria, a plant supplies 2 MW of electricity and 4 MW of heat, generated from wood chips, since 2003.

The global marine energy resource is estimated to be the order of 200 GW for osmotic energy; 1 TW for ocean thermal energy; 90 GW for tidal current energy; and 1–9 TW for wave energy. The total worldwide power in ocean currents has been estimated to be about 5000 GW. It is estimated that capturing just 1/1000 of the available energy from the Gulf Stream, would supply Florida with 35% of its electrical needs. On an average day, 60 million km^2 of tropical seas absorb an amount of solar radiation equal in heat content to about 250 billion barrels of oil. Ocean thermal energy conversion (OTEC) uses the temperature difference that exists between deep and shallow waters to run a heat engine. The osmotic pressure difference between fresh water and seawater is equivalent to 240 m of hydraulic head. Theoretically a stream flowing at 1 m^3/s could produce 1 MW of electricity. The worldwide fresh to seawater salinity resource is estimated at 2.6 TW.

Tidal power has potential for future electricity generation. Tides are more predictable than wind energy and solar power. Wave energy can be considered as a concentrated form of solar energy. Useful worldwide resource has been estimated at >2 TW. Low head hydropower applications use river current and tidal flows to produce energy. If the viable river and estuary turbine locations of the US are made into hydroelectric power sites it is estimated that up to 130,000 gigawatt-hours per year could be produced.

Enhanced Geothermal System (EGS) technologies "enhance" geothermal resources in hot dry rock (HDR) through hydraulic stimulation. It is reported that in the United States the total EGS resources from 3–10 km of depth is over 13,000 zetta joules. Out of this over 200 ZJ would be extractable, with the potential to increase this to over 2,000 ZJ with technology improvements.

Algae can be used to produce not only several kinds of fuel end products, but also byproducts which have wide ranging applications in chemical and pharmaceutical industries. They can be grown using land and water unsuitable for plant and food production. They are energy-efficient. They consume carbon dioxide. They can be mass produced. Algae may be cultivated in photobioreactors, and harvested using rotary screening methods. Expeller press and ultrasonic assisted extraction technologies may be used to produce energy products, such as, biodiesel, ethanol, methane, hydrogen, etc.

2050 is only 40 years away. During the next five to ten years, it is imperative that we shift to long-term trajectories while meeting the interim targets in respect of Industry, Buildings and Appliances and Transport. This would involve undertaking the required RD&D programmes, improving efficiencies, achieving increased market penetration, making appropriate investments, changing of the policies, and so on.

Chapter 4.6

Exercises

Adapted from Carol A. Dahl (2004), "*International Energy Markets*", Pennwell Corporation, USA, p. 92.

1. Calculate the electricity costs of a small village community, on the basis of the following particulars:

Costs of electricity generation

$$TC = FC + VC(Q)$$

where
TC is total cost;
FC is fixed cost;
Q is quantity of production (kWh/yr)
VC is variable cost;
VC(Q) is a variable cost which is a function of Q

P is electricity cost in US cents/kWh

$P = 75 - 4Q$
$TC = 80 + 19Q - 0.25Q^2$
$FC = 80$
$VC = 19Q - 0.25Q^2$
$AFC = FC/Q = 80/Q$
$AVC = VC/Q = 19 - 0.25$
$MC = 19 - 0.50Q$

Total cost for producing 30 units is
TC = 80 + 19 * 30–0.25 (30^2) = 425 cents = $4.25
Average fixed cost for 30 units=
AFC = 80/30 = 2.67 cents
AVC = 19 – 0.25Q = 19 – 0.25 * 30 = 11.5 cents
MC = Marginal cost of the 30th unit is = 19 – 0.5 * 30 = 4 cents

There will be economies of scale. As more electricity is produced, unit costs will fall.

2. Calculate the real cost per kWh ($\$_k$) of wind power, on the basis of the following particulars.
Turbine lasts for 20 years.
Turbine runs in full capacity at 25% of the time.
Dicount (interest) rate: 10%
Capital cost of a 2 MW wind turbine @ $1500 kW = $3,000,000/-
Installation and maintenance costs = 250,000
Total cost that needs to be recovered over 20 years = $3,250,000/-
In a year, the turbine generates 2000 kW * 0.25 * 8766 hrs = 4,383,000 units
($\$_k$) = $3,250,000/4,383,000/\Sigma_{i=0}^{20} 1/1 + 0.10)^i$) = $0.746 per kWh

3. Calculate the hydroelectricity production, given the following particulars:
P = Power created in KW,
H = Hydraulic head = 20 m
F = Flow rate = 30 m^3/s
P = 5.9 HF = 5.9 × 20 × 30 = 3540 kW

4. Adapted from US Department of Energy (2005), "*A Manual for the Economic Evaluation of Energy Efficiency and Renewable Energy Technologies*", p. 45.

TLCC = I + PVOM,

where I = Initial investment, and PVOM is the present value of all Operation and Maintenance costs.

$$PVOM = \sum_{n=1}^{N} O\&M_n/(1+d)^n$$

Compare the Total Life Cycle Cost (TLCC) of an incandescent light bulb versus fluorescent lamp.

Assume a five–year life-time of the project, and a discount rate of 12%.

Alternative A: An incandescent light bulb (75 W) costing USD 1 is used every night for 6 hours round the year. It needs to be replaced every year, and so during a five-year life-time, 5 bulbs will be required. Electricity costs USD 6 cents/kWh. The bulb is purchased at the beginning of each year, and the electricity bill is paid at the end of each year. Annual electricity consumption 164.25 kWh @ USD 6 cents/kWh, costs USD 9.86/yr. TLCC for Alternative A works out to USD 39.56.

Alternatibe B: A fluorescent lamp (40 W) costing USD 15, and has a life-time of 5 years, is used 6 h. every night round the year. It need not be replaced, as it could be used during the whole life-time of the project. Annual electricity consumption is 87.56 kWh at USD 6 cents/kWh. TLCC for Alternative B works out to USD 33.95.

The use of fluorescent lamp thus saves USD 5.61.

References

Ansolabehere, S., J. Deutsch, M. Driscoll, P.E. Gray, J.P. Holdren, P.L. Joskow, R.K. Lester, E.J. Moniz, & N.F. Todreas, 2003, *The Future of Nuclear Power*, MIT, Cambridge, Mass., USA.

Aswathanarayana, U. 1985, *Principles of Nuclear Geology. Rotterdam*: A.A. Balkema.

Aswathanarayana, U. 1995. *Geoenvironment: An Introduction*. Rotterdam: A.A. Balkema.

Aswathanarayana, U. 2001. *Water Resources Management and the Environment*. Lisse, Holland: A.A. Balkema.

Aswathanarayana, U. 2003. *Mineral Resources Management and the Environment*. Lisse, Holland: A.A. Balkema.

Aswathanarayana, U. (2008) A low-carbon, technology-driven strategy for India's energy security. *Curr. Sci.*, 94 (4), 440–441.

Aswathanarayana, U. & R.S. Divi (eds.) 2009, *Energy Portfolios*, A.K. Leiden, The Netherlands: CRC Press/Balkema.

Aswathanarayana, U., Harikrishnan, T. & K.M. Thayyib Sahini (eds.) 2010. *Green Energy: Technology, Economics and Policy*. A.K. Leiden, The Netherlands: CRC Press/Balkema.

Barre, B., & Garderet, P. 2009, *Nuclear Power in France – A success story*, pp. 305–313, *Energy Portfolios*, A.K. Leiden, The Netherlands: CRC Press.

Boyle, G. 2004. *Renewable Energy*. 2nd. Ed. Milton Keynes: The Open University.

B.P. 2008. *BP Statistical Review of World Energy*. 45 pp., http://www.bp.com.

Chadwick, M.J., N.H. Highton & J.P. Palmer. 1987. *Mining projects in the developing countries – A manual*. Stockholm: Beijer Inst.

Das, A & S.K. Chatterjee. 1988. Environmental pollution caused by coal mining and coal related industries: An appraisal. In S.C. Joshi & G. Bhattacharya (eds.) *Mining and Environment in India*, Himalayan Research Group, Nainital, U.P., India, pp. 108–120.

Dhanaraju, R. 2005. *Radioactive Minerals*. Bangalore: Geol. Soc. Ind.

DLR (2007) *Aqua-CSP. Concentrating Solar Power for Seawater desalinization*. Institute of Technical Thermodynamics, Stuttgart, Germany.

El-Hinnawi, E. Essam 1981. The environmental impacts of production and use of energy. *Nat. Res. Environ.*, Ser. 1, I–XV, 1–322, Dublin.

Energy International Administration (EIA), 2008. *International Energy Outlook, 2008*. US Dept. of Energy (DOE), Washington, D.C. 250 pp.

EUPVPLATF (European Photovoltaic Technology Platform) (2007). *Strategic Agenda*. EU PV Technology Platform, Brussels, 2007.

Ewing, R.C. 2005. The Nuclear Fuel Cycle. A role for mineralogy and Geochemistry. *Elements*, 2, 331–334.

Fergusson, J.E. 1990. The *heavy elements: Chemistry, environmental impact and health effects*. Oxford: Pergamon Press.

Frankl, P., Menichetti, E., and Raugei, M. (2008). *Technical Data, Costs and Life Cycle Inventories of PV applications*. NEEDS (New Energy Technology Externalities Developments for Sustainability) Report prepared for the European Commission (under publication).

Hubbert, M.K. 1956. *Nuclear Energy and Fossil Fuels*. API Drilling and Production Practice. Proc. Spring Meeting. San Antonio, Texas, USA. 57 pp.

International Energy Agency (Paris). 2007. *World Energy Outlook 2007*.

International Energy Agency, 2008a. *Deploying Renewables: Principles for Effective policies. Paris*

International Energy Agency. 2008b. *Energy Technology Perspectives*.

Komor, Paul. 2004. *Renewable Energy Policy*. Lincoln, NE: iUniverse.

Laconte, P. & Y.Y. Haimes (eds.) 1982. *Water Resources and Land-use Planning: A Systems Approach*. The Hague: Martinus Nijhoff Publishers.

Macfarlane, A.M., and M. Miller. 2007. Nuclear Energy and Uranium Resources. *Elements*, 3, 185–192.

Mann, P., Horn, M., and Cross, I. 2007. *Map of world's oil and gas giants*. Jackson School of Geosciences, University of Texas at Austin, USA

Masters, C.D., Root, D.H., and Turner, R.M. 1998.World of resource statistics geared for electronic access.*Oil and Gas Journal*, **95 (41)**, 98–104.

PV-TRAC (Photovoltaic Technology Research Advisory Council) (2005) *A Vision for Photovoltaic Technology*. PV-TRAC. European Commission.

Shiklomanov, I.A. 1998. *World Water Resources – A new appraisal and assessment for the 21st century*. Unesco: Paris

Taylor, B., P. Mazzoni, F. Brown, C. Kubank & L. Redemeyer. 2003. Process flow sheet development for the Corridor Sands Project. *Heavy Minerals 2003*. South African Inst. of Mining and Metallurgy, Johannesburg.

Taylor, R. 2007. *Hydropower Potentials*. International Hydropower Association.

Tiwary, R.K., J.P. Gupta, N.N. Banerjee & B.B. Dhar. 1995. Impact of coal mining activities on water and human health in the Damodar River basin: 595–603. *Proc. First World Mining Congress, New Delhi, India*.

USGS, 2000. *US Geological Survey world Petroleum Assessment 2000*. USGS Digital Data Series – DDS-60.

Vörösmarty, C. et al., 1997. The storage and aging of continental runoff in large reservoir systems in the world. *Ambio*, **26**, 210–219.

Section 5

Bio resources and biodiversity conservation

S. Balaji
Tamil Nadu Forest Academy, Coimbatore, India

Chapter 5.1

Introduction

The term biodiversity denotes the variability of all life forms on this earth. It is a vital resource which needs to be assiduously conserved for posterity, because it has in it the key to progress in medicine, agriculture, forestry, aquaculture and numerous other fields. Loss and fragmentation of natural habitats, over exploitation of plant and animal species, impact of exotics, industrial effluents, climate change and above all the greed of man are some of the causative factors for erosion of this biodiversity. This chapter discusses among other things, definition of biodiversity, global endowment of biodiversity, looming threats to its survival and importance of Biodiversity in Medicine, Agriculture and Forestry. The relationship between biodiversity and climate change, how bio resources maintain ecological services and support human livelihood is dealt with subsequently. The policy and legal instruments, the role of traditional knowledge and the need for equity in access and distribution of benefits and future portends in biodiversity conservation are discussed towards the end.

Chapter 5.2

What is biodiversity?

Biodiversity refers to the number, variety and variability of living organisms. Biological diversity came into prominence with publications that appeared in 1980. Lovejoy used the term first in 1980, essentially to denote the number of species present. Norse and McManus employed it to include two related concepts viz., genetic diversity and ecological diversity. Norse et al., in 1986 expanded this usage to refer unequivocally to biological diversity at three levels viz., genetic (within species), species (species numbers) and ecological (community) diversity. Rosen coined the constricted term "biodiversity" in the first planning conference on National Forum on Biodiversity convened in Washington D.C. in September 1986. The proceedings of that forum edited by Wilson in 1988 under the title "Biodiversity" launched the word into general use. The Global Biodiversity Assessment defines biodiversity as the total diversity and variability of the living things and of the systems of which they are a part. This encompasses the total range of variation in and variability among systems and organisms at the bioregional, landscape, ecosystem and habitat levels, at the various organism levels down to species, populations and individuals, and at the level of population and genes (UNEP, 1995). It also embraces complex sets of structural and functional relationships within and between these different levels of organizations including human action and their origin and evolution in space and time. The multifaceted nature of biodiversity is reflected in the many definitions that have been propounded so far. Jutro has recorded 14 definitions of biodiversity. The three levels of biodiversity are hierarchical in nature and overlap with one another. They are covered by three major disciplines of biology viz., ecology, genetics and taxonomy. The species is the basic unit of classification and the most practical and commonly used currency while referring to biodiversity. Species is defined as a group of similar organisms that interbreed or share a common lineage of descent. Species do not occur in isolation but exist in a very wide array of ecological groupings, which we can recognize as distinct. These together with the physical environment form the main eco-regions and biomes of the world. For practical purposes, the term biodiversity is synonymous with biological diversity as defined by Norse et al., in 1986. This is reinforced by the official definition of the Convention on Biological Diversity (CBD) ratified by 159 nations in the Earth Summit held at Rio in 1992. The broad scope of the CBD is illustrated in its Article 1, that includes "conservation of biological diversity, sustainable use of its components and fair and equitable sharing of benefits arising out of the utilization of genetic resources." Resolving biodiversity into three levels viz., genetic, species and ecosystem diversity is not without its difficulties as they are inextricably interlinked and actions at any one level will have impact on other levels of the hierarchy. However biodiversity as a unifying concept brings together people from different disciplines and interests with a common goal of understanding of conservation and wise use of biological resources. Several attempts to unite the components into a universal paradigm have been made. A DIVERSITAS Programme has been designed and launched as a multi-level and multi-scale endeavour looking at the interactions and the integration of the different disciplines involved; this programme provides an acceptable framework or *leitmotif* for exchange of information across the globe (di Castri and Younes, 1996).

5.2.1 Endemism and keystone species

A native species with unique distribution restricted to particular locality is called an endemic species. Endemism has an important role in biodiversity and presence of endemic species enhances conservation values. Endemism in plants is demonstrated at various geographical scales. A plant species can be restricted to a continent, region or even a locality. Four forms of endemism are identified in neo tropical plants viz., Island endemism, relict endemism, neo endemism and anthropogenic endemism. Island endemism is the most common form restricted to islands that are spatially separated. Relict endemism is represented by an ancient taxonomic lineage or distribution restricted in range by specialized habitats as in the case of genera such as *Mahonia* and *Rhododendron* in the Nilgiris in South India. Neo endemics are newly evolved plants, which are usually restricted to the site of origin as in the case of the species of *Nilgirianthus* in the hills of South India. Anthropogenic endemism is more a result of recent human destruction of habitats exterminating some species locally. The peninsular regions are a close second to islands in having conditions for endemism. Keystone species are those species playing a major role in maintaining ecosystem structure and integrity. The fruits of figs are a fundamental resource for primates and many frugivorous birds that they ensure the perpetuation of figs in the Bodongo forests. Yet, foresters for decades, indulged in eradicating figs for purely silvicultural reasons (Balaji, 2004). The keystone species not only determine the ecosystem function in that particular seral stage of succession where it appears, but even decide the very process of succession due to their impact on nutrient cycling. Indeed, they modulate ecosystem function, both in space and time.

Chapter 5.3

Why conserve biodiversity?

Variety is both the spice of life and basis of its survival. The importance of biological diversity to human society is hard to overstate. An estimated 40 percent of the global economy is based on biological products and processes. Poor people, especially those living in areas of low agricultural productivity, depend heavily on the genetic diversity of the environment. Forests protect the watersheds, moderates climate and act as the foster mother of agriculture. They hold the key to global food and water security. Darwin's hypothesis of natural selection and survival of the fittest will have no operational validity if our planet is not so rich in biological diversity; this provided opportunities for the domestication of plants and animals leading to the birth of agriculture about 10,000 years ago. Loss of diversity at any of the level is detrimental to the life-supporting environment of the earth and disruptive of the natural processes that are vital for biological evolution. As many of the world's diverse life forms from microbes to higher animals and plants have a direct or indirect bearing on agriculture, conservation of these is essential for sustainable agriculture (Swaminathan and Jana, 1992). Each time we take a medicine, the chances are one in two that its origin was a wild plant. The commercial value of such drugs is around US$15 billion per year in the United States and about US$40 billion worldwide.

Ecosystem stability is another compelling reason for preserving biodiversity. Nature is beautifully balanced; each little thing has its own place, its duty and special utility. Any disturbance creates a chain reaction which may not be visible for sometime. All living organisms are an integral part of the biosphere and provide invaluable services. These include the control of pests, the recycling of nutrients, the replenishment of local climate, the control of flood etc. Civilization depends on the survival of the biological world. By conserving biodiversity at the ecosystem level, not only are the constituent species preserved, but the ecosystem functions and services are also protected. These ecosystem functions include pollutant cycling, climate control, as well as non-consumptive recreation, and scientific values (Norton and Ulanowicz, 1992). Biodiversity also needs to be preserved for the aesthetic services it provides. According to Norman Myers "We can marvel at the colours of a butterfly, the grace of a giraffe, the power of an elephant and the delicate structure of a diatom. Every time a species goes extinct we are irreversibly impoverished. Protection of biodiversity also makes good economic sense." He has further stated that, "from morning coffee to evening nightcap we benefit in out daily life style from the fellow species that share out one earth home. Without knowing it, we utilize hundreds of products each day that owe their origin to wild animals and plants. Indeed our welfare is intimately tied up with the welfare of wildlife. We may proclaim that by saving the lives of wild species, we may be saving our own" (Myers, 1986).

Chapter 5.4

Global biodiversity resource

To date, 1.4 million life forms have been named and described by science. But Biological estimates suggest that there are at least five million but perhaps as many as 50 million species may be existing today (McNeely et al., 1990). The estimate has increased dramatically in recent years following research in the rain forest canopy. The tropical rain forests support millions of insect species. Our knowledge about individual groups such as fungi, nematodes, mites and bacteria is meager. Marine biodiversity, especially that of the deep sea is poorly known. New plant and animal species from all major families are still being discovered in the rainforests. During the last few years over 300 new fish species have been described from the Amazon region. New species are being discovered among Amphibians and even Primates. The diversity of all living things depends on temperature, precipitation, altitude, soil, geography and the presence of other species. Botanist Alvin Gentry estimates that about 15,000 to 20,000 unknown species of tropical flowering plants are yet to be documented. The total number of named species of organisms is calculated to be about 1.4 million. Of these about 250,000 are vascular plants and bryophytes, 44,000 vertebrate animals and 750,000 insects (Wilson, 1988). But, this biodiversity is not evenly distributed, and varies greatly across the globe as well as within regions. In this context, Nature has been very benevolent to the developing countries in the distribution of biodiversity on the earth. Thus the developing countries, located in subtropical/tropical belt, are far richer in biodiversity than the developed countries in the temperate region. The Vavilovian centres of diversity of crops and domesticated animals are also located in developing countries. The initiative for biodiversity assessment was taken by UNDP as early as 1991. At present, 192 countries and the European Union are party to the agreed text on the Convention on Biological Diversity. Global Biodiversity Assessment

(UNEP, 1995) estimates the total number of animal and plant species to be between 13 and 24 million as against 1.7 million species described so far. Each species contains up to 400,000 genes and no two members of the same species may be identical genetically. Ecosystems diversity has not even been reasonably explored as yet. There seems to be a wide gap in the knowledge at global, regional and local levels.

Chapter 5.5

Erosion of biodiversity

The rich global biodiversity is threatened with erosion on an unprecedented scale. While the rates of extinction were roughly equal to those of speciation for most of the history of life on earth, contemporary extinction rates are several times faster than those of speciation leading to erosion of Biodiversity (UNEP, 1995). We may already be losing 50 to 100 species per day (Myers, 1986). Conservation biologists caution that 25 percent of all species could become extinct during the next 20 to 30 years. During the last 200 million years about 100 species became extinct in each century due to natural evolutionary processes. At the same time evolution ushered in new life forms that more than compensated the species lost. Today the extinction rate is approximately 40,000 times higher than this background rate due to the depredations of *Homo sapiens*. The Holocene extinction rates of decline in biodiversity in this sixth mass extinction match or exceed rates of loss in the five previous mass extinction events in the fossil record. Previous mass extinctions had no palpable effect on terrestrial plants; but today for the first time an enormous proportion of terrestrial plant species which form the basis of land ecosystems are threatened (Knoll, 1984). A disappearing plant can take with it 10 to 30 dependent species such as insects, higher animals and even other plants. In 2006 many species were formally classified as rare or endangered or threatened. Moreover, scientists have estimated that several species are at risk which have not been formally recognized. About 40 percent of the 40,177 species assessed using the IUCN Red List criteria are now listed as threatened with extinction. There is a growing concern on the prospect of accelerating loss of species, populations, domesticated varieties and natural habitats such as tropical rainforests and wetlands. Recent estimates suggest that more than half the habitable surface of the planet has been significantly altered by human activity. World Conservation and Monitoring Centre (WCMC, 1992) has estimated the number of species of plants and animals extinct since 1600 as 654 and 484 respectively. One of the estimates suggests that tropical forests are being denuded

Table 5.1 Different estimates of species extinction (as quoted by Reid, 1992)

Estimate of species loss	*% of Global loss per decade*	*Method of estimation*	*Proponents*
1 million species 1975–2000	4	Extrapolation of past exponentially increasing trend	Myers (1979)
15–20% of species 1980–2000	8–11	Species area curves	Lovejoy (1980)
25% of species 1985–2015	9	Loss of half species in area likely to be deforested by 2015	Raven (1988)
2–13% of species 1990–2015	1–5	Species area curves	Reid (1992)

at the rate of 15.4 million ha per annum or approximately 1.8% of the remaining forest cover. This will undoubtedly cause the loss of innumerable species and populations of plants and animals and thus will impoverish the global species and genetic diversity. Despite their high diversity, tropical forests are fragile ecosystems and are less capable to recover from repeated human depredations than temperate forests. Tropical humid forests in general are amongst the most diverse, most productive and most threatened of the biological communities with indeed 14 of the 18 biodiversity hotspots identified by Myers representing these biomes. Two of these hot spots viz., the Eastern Himalayas and the Western Ghats occur in India. But both these hot spots are threatened. To quote Dr. M. S. Swaminathan, both are paradises of valuable genes but are fast inching towards the status of "Paradise Lost." At least 10 percent of India's recorded wild flora and possibly more of its wild fauna are on the list of threatened species many of which are on the brink of obliteration. Of the wild fauna 80 species of mammals, 47 of birds 15 of reptiles, 3 of amphibians and a large number of moths, butterflies and beetles are listed as endangered. Out of 19 species of primates, 12 are endangered. The cheetah (*Acinonyx jubatus*) and the pink headed duck (*Rhodonessa caryophyllacea*) are among the well known conspicuous species that have become extinct; but there must be many more that have been totally annihilated unrecorded either because they were not that spectacular or because we simply were not aware of their existence (Balaji, 2010).

Despite the discrepancy in the different estimates (Table 5.1) all prognoses lead to the conclusion that what is taking place is not just loss of individuals but biodiversity as a whole is endangered.

5.5.1 CAUSES FOR THE EROSION OF BIODIVERSITY

The primary cause for the erosion of diversity is the greed of man. Never before has one species influenced the environmental conditions all over the planet to such a magnitude as today. The cause of gradually disappearing biodiversity in both rich and poor countries is different; in rich countries humans destroy environment by affluence, luxury and perverting the precious expression of freedom; on the other hand in developing countries destruction of environment is due to poverty. The damage in the latter could be in the form of collecting fuel wood or questionable land reclamation in marginal areas due to scarcity of land which could be mainly due to population escalation and the gross failure to mitigate poverty. Therefore, poverty and illiteracy need eradication in most of the developing countries in tropics. Edward O. Wilson while explaining the causative factors for erosion of biodiversity, prefers the acronym **HIPPO**, standing for Habitat destruction, Invasive Alien Species, Pollution, Over population of human beings and Over exploitation of Natural resources (Wikipedia, 2011).

5.5.2 HABITAT LOSS

Habitat size and numbers of species are systematically related. Physically larger species and those living at lower latitudes or in forests or oceans are more sensitive to reduction in habitat area (Wilson, 1988). Conversion to "trivial" standardized ecosystems such as plantations following deforestation effectively destroys habitat for the more diverse species that preceded the conversion. In some countries lack of property rights or lax law/regulatory enforcement necessarily leads to biodiversity loss. Habitat destruction has played a key role in extinctions, especially related to tropical forest destruction. Deforestation and increased road-building in the Amazon rainforest are a significant concern because of increased human encroachment upon wild areas, increased resource extraction and further threats to biodiversity.

5.5.3 INVASIVE ALIEN SPECIES

That impact of Invasive Alien Species is an over-riding factor in the loss of biodiversity which is often under-estimated. The International Biodiversity Day 2009 was devoted to the theme of Invasive Alien Species. Barriers such as large rivers, seas, oceans, mountains and deserts encourage diversity by enabling independent evolution on either side of the barrier. Invasive species occur when those barriers are blurred. Without barriers such species occupy new niches, substantially reducing diversity. Repeatedly humans have helped these species circumvent these barriers, introducing them for food and other purposes. Introduction of exotic species has proved a deterrent to natural regeneration of indigenous flora. Invasive species threaten biodiversity by (1) causing disease, (2) acting as predators or parasites, (3) acting as competitors, (4) altering habitat, or (5) hybridizing with local species. The American chestnut tree once dominated the eastern deciduous forests of the USA. The accidental introduction of Asian chestnut blight fungus in the first half of 20th century through nursery trade eliminated American chestnut over 72 million ha. The loss of chestnuts was a disaster for many animals that were highly adapted to live in forests dominated by this tree species. Then moth species that could live only on chestnut trees became extinct. The fast growing exotic, water hyacinth, from South America is fast spreading over water bodies unchecked and has become serious problem to fishes and form a congenial breeding ground for mosquitoes. The predatory fish, Nile Perch was introduced into Lake Victoria in Africa as a food fish. Being a voracious predator, it eliminated over one hundred species particularly native cichlid fishes (Kondas, 2010). Not all introduced species are invasive, nor all invasive species deliberately introduced. In case of oil palms in Indonesia and Malaysia, the introduction produces substantial economic benefits, but the benefits are accompanied by costly unintended consequences. An introduced species may unintentionally injure a species that depends on the species it replaces. In Belgium, *Prunus spinosa* from Eastern Europe leafs much sooner than its West European counterparts, disrupting the feeding habits of the *Thecla betulae* butterfly which feeds on the leaves. Introducing new species often leaves endemic and other local species unable to compete with the exotic species and unable to survive (Koh et al, 2004). Endemic species can be threatened with extinction through the process of genetic pollution, i.e. uncontrolled hybridization, introgression and genetic swamping.

5.5.4 POLLUTION

Pollution of soil and water, Pesticides, trophospheric ozone, sulphur and nitrogen oxides and global warming as a result of industrialization contribute to degradation of natural ecosystems. Many organisms are very sensitive to enrichment of CO_2 concentration in the atmosphere (Jeffrey, 2004). According to the millennium ecosystem assessment 2005, fresh water ecosystems are the most threatened at present,

5.5.5 HUMAN POPULATION

The global population increased from 2 billion in 1950 to 6.6 billion now. The pressure is more in the biodiversity rich tropical forests. While the natural resources are limited, burgeoning human population coupled with fast pace of economic development in biodiversity rich areas is posing grave threat to survival of several species. Human mobilized material and energy flows rival those of nature. The human species presently uses forty percent of the annual net photosynthesis production of the planet (Vitousek et al., 1986). The consumption of two-fifths of annual net food resources of the planet by one species is clearly incompatible and at loggerheads with biological diversity and stability. Over the past 45 years, about 11% of the earth's vegetated soils have become degraded to the point that their original biotic functions were damaged and reclamation may be costly or in some cases impossible (UNESCO, 1990).

5.5.6 OVEREXPLOITATION

Overexploitation occurs when a resource is consumed at an unsustainable rate. About 25% of world fisheries are now overfished to the point where their current biomass is less than the level that maximizes their sustainable yield. Poaching is another factor responsible for the pressure on wild animals. Elephants are being ruthlessly hunted for their tusks while the tiger is being shot for its valuable skin and bones. The lion-tailed macaque is being hunted for its fur. The striped hyena is being persecuted for its proneness to predating on livestock. Pythons are killed for their ornamental skin.

The flesh of turtle, tortoise and terrapin is widely consumed. Therefore the number of these animals keeps dwindling. Moreover man-made forest fires cause decline in forest regeneration and loss of biodiversity. Satellite technology such as MODIS of USA comes quite handy in monitoring forest fires and in giving fire alarm to the field officers at the range level and helps in taking preventive and remedial measures.

5.5.7 ARRESTING BIODIVERSITY LOSS

Chapter 15.5 of Agenda 21 calls on signatories to undertake Country studies or use other methods to identify components of biological diversity important for its conservation and the sustainable use of biological resource. There are a number of international initiatives to preserve the rich tropical biodiversity such as IWOKRAMA International Centre for Rain Forest Conservation and Management in Guyana promoted by Guyana Government and the Commonwealth Secretariat. The IWOKRAMA Centre covers an area of about 4,000 km^2; half of this area is being maintained as a wilderness preserve in order to study evolution in action (IWOKRAMA, 1996). Several Nations are engaged in developing their own National Biodiversity Strategies or Action Plans. Successful biodiversity conservation of degraded forests requires a sound knowledge of the ecological process involved in restoration so that human action may facilitate nature to assert itself. There are success stories on Joint Forest Management (JFM) in many parts of India, where over 100,000 Village Forest Committees are protecting 22 million hectares of degraded forests at present. For instance, JFM in Tamil Nadu State through Tamil Nadu Afforestation Project, over the last 15 years, has helped to restore forests and rejuvenate biodiversity over 650,000 ha of degraded forests besides improving the quality of life of local people (Balaji, 2010).

Chapter 5.6

Climate change and biodiversity

It is reported that by 2100, the ecosystems will be exposed to substantial increase in the level of atmospheric CO_2, much higher than in the past 650,000 years. Global temperature will also rise, at least among the highest, as compared to those experienced in the past 740,000 years. The increase in the level of atmospheric CO_2, is a consequence of unbridled development in industries, transport and energy sector. Stern review of economics of climate change has projected an increase of CO_2 equivalent (CO_2e) of Green House Gases (GHG) from 430 parts per million (ppm) now to 550 ppm by 2035, almost double the level of CO_2e prior to the industrial revolution. This is expected to warm up the atmosphere by 2°C. The poor countries will be hit hard by the climate change. Life and property in coastal towns and island nations will be endangered. As the size of global economy is likely to grow 3 to 4 times the present level by 2035, emissions at that time will have to be brought down by 25% below the current level. Stern review has estimated that if climate change is not addressed to and mitigation and adaptive measures are initiated worldwide, it would eventually damage economic growth and cause major disruption in economic and social activity. It will increase flood risk, reduce crop yield and cause water scarcity. Many organisms are sensitive to carbon dioxide concentration in the atmosphere that may lead to disappearance of 15–40% of species. Projected sea-level rise is very likely to result in significant loss to coastal ecosystems of South-East Asia. Stability of wet lands, tidal forests with mangroves and coral reefs around Asia might be increasingly threatened. The annual of cost of achieving cuts in emission at the rate of 1–3% to stabilize CO_2e level at 500–550 ppm will be 1% of GDP, a level significant but manageable (Stern, 2006). Protection of the atmosphere and reduction of GHG emissions is an important environmental issue today. Some pro active measures have been taken by few countries to check the global warming. The Canadian House of Commons has become the first parliament in the world to pass a climate act (the Climate Change

Accountability Bill), which commits the country in reducing its GHG emissions from the present 80% to that of 1990 levels by 2050. New Zealand has pledged to become carbon-neutral.

5.6.1 ROLE OF FORESTS IN CLIMATE CHANGE MITIGATION

Forests play a critical role in combating climate change, collectively capturing and storing significant amounts of carbon that would otherwise pollute the atmosphere. Forests have four major roles in climate change; when forests are cleared, overused or degraded they also contribute about one-sixth of global carbon emissions; forest species react sensitively to a changing climate; when managed sustainably, they produce wood fuels as a benign alternative to fossil fuels; and finally, they have the potential to absorb about one-tenth of global carbon emissions projected for the first half of this century into their biomass, soils and products and store them in perpetuity. The tropical rain forests of Congo of West Africa, Amazon basin of South America and Western Ghats of India serve as earth's gigantic carbon sinks. India has rich biodiversity heritage which not only traditionally respected forests, but actually venerated the Mother Nature. Deforestation and accelerated denudation of tropical forests due to extensive lumbering, urbanization and other non-forestry activities, has diminished the capacity of forests for assimilation of CO_2 from the atmosphere. On the other hand Green House Gas emissions are increasing day by day with increased urbanization and industrialization. Increasing anthropogenic activities around the Protected Areas (PAs) and Reserved Forests (RFs) are putting tremendous pressure on the limited forest resource. The average amount of carbon per acre varies regionally and by type of forests. The National Forests of United States contain an average of 77.8 metric tons of carbon per acre (0.4 hectare). The U.S. forests offset roughly 11 percent of industrial green house gas emissions released each year in the United States, according to the USDA Forest Service. It reports that 41.4 billion metric tons of carbon is currently stored in the Nation's Forests. Due to increases in both the total area of forest land and the carbon stored per hectare, an additional 192 million metric tons of carbon is sequestered by United States forests each year. The additional carbon sequestered is the equivalent of removing almost 135 million passenger vehicles from the Nation's Highways (www.fao.org.in). It has been estimated that cessation of deforestation in India and annual afforestation of 1 million hectares, can potentially remove as much as 6 million tonnes of carbon from the atmosphere annually, and also yield 15 cubic meters of wood per hectare per year (ICFRE, 1993). According to a recent study, the global forests can absorb 27% of current levels of anthropogenic emissions. The carbon stock in Indian forests is estimated to be 6622 million tonnes. The sequestration of Indian forests is showing an annual increase of 138 million tonnes of CO_2e while that of China is showing an annual increase of 301.23 million tonnes of CO_2e (FSI, 2009). Indian Government has embarked on eight Missions to check ill effects of climate change including the Green India Mission to increase green cover by Planting trees over 5 million hectares and improving crown cover over 5 million hectares of degraded forests.

Chapter 5.7

Role of biodiversity in medicine, agriculture and forestry

5.7.1 BIODIVERSITY IN MEDICINE

Biodiversity has direct consumptive value in Agriculture, Medicine and Forestry. Biodiversity's relevance to human health is becoming an international issue, as scientific evidence builds on the global health implications of biodiversity loss. This issue is closely linked with the issue of climate change, as many of the anticipated health risks of Climate change are associated with changes in biodiversity. A significant proportion of drugs are derived, directly or indirectly, from biological sources: at least 50% of the pharmaceutical compounds on the US market are derived from plants, animals, and micro-organisms, while about 80% of the world population depends on medicines from nature used in either modern or traditional medical practice for primary health-care. Only a tiny fraction of wild species has been investigated for medical potential. Around 20,000 plant species are believed to be used for medicines in the developing world. In India the knowledge about medicinal value of plants has evolved in the form of traditional systems of medicine known as *Ayurveda*, *Siddha* and *Unani*. More than 8,000 species are used in some 10,000 drug formulations. The demand outstripping supply has put unreasonable pressure on our wild plant resources. Due to this, certain species are at risk of over exploitation and extinction (Basappanavar, 2009). The global plant based drug trade is projected at around US$62 billion with a 7% annual growth rate. An effective strategy for bio-prospecting for medicine from indigenous species will require careful floristic inventory, an understanding of traditional knowledge, habitat-species relationship, identification of suitable genotypes, evolving bio-technological tools to conserve the species and patenting of potential medicinal species. Some of the health issues influenced by biodiversity include dietary health, nutrition security and spread of infectious diseases.

5.7.1.1 Biodiversity prospecting – IN-Bio-model

For decades, ecologists and environmentalists have argued that agricultural, pharmaceutical and other commercial applications of biodiversity should help justify its conservation. In September 1991, Costa Rica's National Biodiversity Institute (INBio), a private, non-profit organization and the US-based pharmaceutical firm Merck & Co., Ltd. announced an agreement that provides US$ 1.135 million to INBio from Merck to conduct research and sampling of wild plants, insects and microorganisms in exchange for chemical extracts which Merck could screen for potential pharmaceutical applications. If any commercial products resulted, Merck agreed to pay INBio royalties which INBio will use to further its inventory and research, and to support a fund for the management of Costa Rica's National Parks (Reid et al., 1993). Since, then, a growing number of biodiversity prospecting agreements have been negotiated between industry, research institutions and government around the world.

5.7.1.2 Addressing risks

The dramatic changes taking place in the use of biodiversity in the agricultural and pharmaceutical industries are primarily the result of advances in gene transfer and biochemical screening technologies. Unregulated biodiversity prospecting can speed up the destruction of a species. In one particulars notable case, the entire adult population of the shrub *Maytenus buchnanii* – the source of the anti-cancer compound *maytansine* – was harvested (27,215 kg) in Kenya by the US National Cancer Institute for testing in its drug development programme (Oldfield, 1984). To avoid such problems, research agreements, such as those specified under the Philippines Presidential Executive Order on Biodiversity Prospecting, can specify ecological and population studies to determine limits for sample collections. All collections must be approved in accordance with these limits.

5.7.1.3 Indigenous knowledge

Local technical knowledge of Natives helps to Conserve Biodiversity. A High level of cultural diversity is dependent on high levels of biological diversity, which in turn supports it. The importance of this interdependence is being increasingly realized in traditional resource management systems in many developing countries. Indigenous people with a historical continuity of resource use practices, often possesses valuable knowledge about the behaviour of complex ecological systems in their localities. This knowledge has accumulated through a series of observations that stretch over many generations. Their practices for biodiversity conservation were built through lengthy trial and error processes. This implies an intimate relationship between the application of ecological knowledge with belief systems. Such knowledge is difficult for Western science to understand. It is vital, however, that the value of the knowledge-practice-belief complex of indigenous peoples as it relates to biodiversity management is fully recognized. Conserving this knowledge might best be accomplished through promoting the community-based resource management system of indigenous people (Gadgil

Box 1 TBGRI benefit sharing model

Tropical Botanical Garden Research Institute in Trivandrum, India, facilitated patenting of *Arogya pacha* ***Trichopus Zeylanicus*** used in indigenous medicines of Kani Tribes of Kerala. The unripe fruits of "Arogyapacha" are eaten fresh to remain healthy and agile by Kanis during their long trekking trips in the forests. Detailed chemical and pharmacological investigations showed that the leaf of the plant contained various glycol lipids and some other non-steroidal compounds with profound immmuno-enhancing properties. The fruits showed mainly anti-fatigue properties. The Tropical Botanical Gardens Research Institute (TBGRI) was successful in developing a scientifically validated and standardized herbal drug, based on the tribal knowledge. The drug was named as Jeevani and was released for commercial production in 1995 by Arya Vaidya Pharmacy, Coimbatore. While transferring the technology for production of the drug to the pharmaceutical firm, TBGRI agreed to share the license fee and royalty with the tribal community on a fifty-fifty basis (ENVIS centre, Kerala, 2009). This model ensures that the local technical knowledge of the tribes is duly recognized and rewarded. Such models are necessary to ensure equity in bio prospecting.

et al., 1993). Brazilian Kayapo Indians at the southern edge of the rain forest develop forest islands or *apete*. Beginning as small vegetation mounds, these *apete* are carefully managed and supplemented as they grow with a complex species mix of medicinal plants, palms and vines. Now *apete* fields reach peak production levels in 2–3 years, although some species, such as sweet potatoes, yams, taro, banana and papaya are productive for longer periods. Even when the primary crops disappear, old *apete* fields keep producing a range of useful products. They become heterogenous forest patches, some managed for fruit and nut trees and for attracting wildlife.

Gadgil et al. (1993) lists four kinds of indigenous conservation practices which are of particular relevance to biodiversity conservation. They include,

- Total protection to individual biological communities, including ponds, meadow, pools along river courses, often protected as sacred groves.
- All individuals of certain species of plants and animals may be afforded total protection. For example, *Ficus* trees are widely protected in may parts of the Old World. These trees are considered a keystone resource by ecologists as they are refuge for a wide range of birds, bats and primates.
- Resource harvesting is often carried out as a group effort once a year which may help groups to adjust harvesting levels. Certain vulnerable stages in the life cycle of an organism may be given special protection. For example, fruit bats are not hunted while roosting in south India which shows ecological prudence and native wisdom.
- Natural Forests around tribal villages are protected as sacred groves.

5.7.2 AGRO-BIODIVERSITY

Earth's surviving biodiversity provides resources for increasing the range of food and other products suitable for human use. Though People world over depend on 40,000 different species for their requirement of food shelter, and clothing about 80 percent of global food supply comes from just 20 species. Russian scientist, Nikolai I Pavlov estimated that about 80,000 species of edible plants have been used at one time or the other in human history, of which about 150 have been cultivated on a large scale. The native tribes obtain considerable part of their daily food requirement from wild plants and wild animals. The reservoir of genetic traits present in wild varieties and traditionally grown crops is extremely important in improving crop performance. Important crops, such as potato, banana and coffee, are often derived from only a few genetic strains. Crop Improvement over the last 250 years has been largely due to incorporating genes from wild varieties into cultivars. Plant breeding for useful traits under Green Revolution has helped to more than double crop production in the last 50 years. For instance, wheat production in India went up from 10 Million tonnes in 1960 to 86 Million tonnes in 2010-11. A broad genetic base helps recovery when the dominant cultivar is attacked by a disease or predator. On the other hand, Intensive monoculture in the past has been one of the causative factors to several agricultural disasters. The collapse of European wine industry in the late 19th century, and the Southern Corn Leaf Blight epidemic of 1970 in United States of America were due to monoculture. The Irish potato blight of the year 1846 was one of the major factors that lead to the death of one million people and the emigration of another million. It was the result of planting only two varieties of potato, which were vulnerable to the blight. When paddy was attacked by Rice Grassy Stunt Virus, out of 6,273 varieties of paddy, only one resistant variety from India was found which was subsequently used as the mother plant to get more resistant hybrids. When coffee crop in Sri Lanka and Brazil was threatened in 1970, a resistant variety was discovered in Ethiopia. The evolution of crops over the past 10,000 years has been affected by both natural and artificial selection, the latter involving isolation of stocks followed by migration and seed exchanges; the resultant hybridization and recombination of genes has undoubtedly added to the genetic diversity of crop species (Wikipedia, 2011). India is endowed with 9,500 plant species of ethno-botanical importance, including 3,900 for edible purposes used by native tribes. There are 356 major and minor crop plant species besides, 326 wild related species of crop plants. Twenty-five major and minor crop plant species are domesticated. There are over 1,500 edible species in wild including 145 species of roots and tubers, 521 species of leafy vegetables, 101 species of bulbs and flowers, 647 species of fruits and 118 species of seeds and nuts (Balaji, 2004). The germ plasma collection of agricultural crops in India is around 3,01,220 which includes variety of cereals, legumes, oilseeds, millets, minor millets, vegetables, fruits, fiber crops etc.

However, modern cultivation techniques are signaling the end of this period of diversification. The major threat to traditional varieties in the developing world is the process of agricultural modernization. In 20 years, a few modern high-yield varieties (HYVs) have replaced ancient crop varieties or "landraces" as part of international and national efforts to develop agriculture. Genetic erosion is occurring because farmers are changing their farming systems, pushed by social, economic and technical

Box 2 Impact of agricultural intensification in the Aral Sea

A tragic example of how agricultural intensification can go wrong is observed in the Aral Sea, the largest body of water between the Caspian and the Pacific. Vast new irrigation systems including the 1,100 km Karakum Canal through southern Turkmenistan, led to the draining of the Amu Darya and Syr Darya rivers, causing the Aral Sea to shrink by over half and lose more than two-thirds of its volume since 1960. Its largest city is now 100 km from the shore. The loss of biodiversity has been profound; its waters formerly supported 25 species of fish, but now support none, and its 60,000 fishermen are unemployed. The former seabed is now a salt desert swept by the prevailing north wind, which dumps between 40 and 150 million tons of sand and salt on the precious fertile land of the Amu Darya delta each year. The problems are made worse by the use of pesticides and chemical fertilizers which are considered necessary to maintain yields of cotton, but which accumulated in the sediments of the Aral Sea, and are now turned into dust and carried to the surrounding lands (Ellis, 1990 as quoted in UNEP, 1995).

forces. The pattern and rate of adoption of HYVs is very uneven across geographic regions. Greatest losses of traditional varieties have occurred in lowland valleys close to urban centres and markets. The least genetic erosion has happened in mountain areas distant from urban centres and markets. Impact of modern agriculture intensification through extension of canal irrigation in South Turkeministan has made fertile delta of Aral Sea a Salt desert and has lead to disappearance of 25 species of fishes. The detailed case study is given in Box 2. Different strategies have been developed for controlling agricultural pests and diseases through different microbes. In many cases microbial genes have been used. The endotoxin gene from *Bacillus thuringiensis* has been engineered against insects. For viral resistance, sequences of the virus itself have been introduced into the plant in order to interfere in the viral life cycle. Antimicrobial compounds from grasses, fungi (*Trichoderma, Aspergillus*), bacteria and animals (insects, rat, cow) are being used to obtain resistance against fungi and bacteria. Finally, fine tuning of endogenous, general resistance mechanisms in the crop of interest is another promising strategy for pest and disease resistance. The diversity of crop genetic resources is affected by distribution of human populations, crop improvement and extension programmes and Changes in consumption patterns in populations.

5.7.2.1 Domesticated plants and animals

The estimated accessions of domesticated agricultural crops in *ex-situ* gene banks across the globe amount to 4.6 million. Similarly the breeds of domesticated animals on file amount to 2,322 out of which 442 breeds are reported to be at risk. The details are given in the Table 5.2.

Table 5.2 Estimated accessions of domesticated Plants and Animals in *ex-situ* gene banks (Adapted from UNEP, 1995)

(a) Plants

Type of germsplasm	*Mean % landraces*	*Accessions*
Cereals	79	2,082,621
Food legumes	68	743,756
Forage legumes and grasses	N/A	452,043
Vegetables	70	374,467
Fruits	N/A	243,823
Industrial crops	65	241,826
Roots and tubers	55	186,687
Oil Crops	N/A	91,067
Sugar, beverage, spice crops	N/A	74,517
Others/unknown	N/A	117,253
Total Plants		4,608,060

(b) Domesticated Animals

Species	*Breeds on File*	*Breeds at Risk*
Ass	69	6
Buffalo	70	1
Cattle	845	120
Goat	330	39
Horse	360	99
Pig	368	69
Sheep	880	108
Total Animals	2,922	442

5.7.2.2 Aquaculture

Aquaculture, whether marine or in fresh waters, is an old tradition, but has been greatly expanded over the past 20–30 years with sophisticated hatchery technologies for mass-rearing of larvae of many species. Technical advances in cultivation systems have enabled operations to be moved progressively into deeper and less protected ocean waters. During 1989–91, average annual aquaculture production of fish and shellfish was 0.12 million tons, and the production is increasing. The largest producers are found in Asia (China, India, Indonesia Japan, Republic of Korea, Philippines, Thailand), Europe (France, Italy, Norway, Spain) and North America (USA) (WRI, 1994). The increase in aquaculture production is rapid and very promising, but industrial fish farming is not without problems and conflicts.

5.7.3 BIODIVERSITY AND FORESTRY

Forests covering 30% of the earth surface provide significant socio economic and environmental benefits. The concern for sustainable development of forests is for both tangible and intangible benefits from the forest. The concerns could be local such as fuel

Table 5.3 **Change in Land Use**: Human-induced Global conversions in selected land uses (WRI, 1994)

Cover	*Date*	*Area 10^6 km²*	*Date*	*Area 10^6 km²*	*Percentage change*
Cropland	1700	2.65	1990	14.41	+543.0
Irrigated cropland	1800	0.08	1989	2.00	+2400.0
Closed forest	pre-agricultural	46.28	1983	39.27	−15.1
Forest and woodland	pre-agricultural	61.51	1983	52.37	−14.9
Grassland/pasture	1700	68.60	1980	67.88	−1.0
Lands drained			1985	1.606	
Urban settlement			1985	2.47	
Rural settlement			1990	2.09	

wood, fodder, non-timber forest produce, soil and water conservation, and consequent agricultural development or global such as the influence of forests over climate, rainfall and biodiversity conservation. Forests provide several goods and services including eco system services essential for very survival of this biosphere. Eco system services include provision of clean air and water, maintenance of soil fertility and structure, maintenance of livable climates, pollination of crops, control of pests, provisions of genetic resources, production of goods like food and fiber and provision of cultural, spiritual and intellectual values. In the context of climate change forests can act as a carbon sink. Therefore there is urgent need to augment existing forest resources through afforestation and integrated management for conserving biological resources and enhancing human welfare (Balaji, 1997).

5.7.3.1 Deforestation

Deforestation is a dramatic alteration of habitat which leads to long-term and perhaps permanent changes in species composition. Logging for international and domestic consumption, although only one of the causes of forest destruction has proved to be of great importance because it opens up the forest to further encroachment by agricultural settlers, in itself believed to be the single greatest cause of forest destruction. According to an estimate, the closed Forest area has decreased by 15.1% (Table 5.3). Deforestation is a major issue confronted by many countries. FAO (1993) reports that between 1980 and 1990, an annual average of 15.4 million ha of tropical forests were cleared, amounting to an annual loss of about 0.8%. The total loss of tropical forests over the decade is almost equivalent to three times the size of France. Deforestation has three major effects: habitat loss; habitat fragmentation; and edge effects at the boundary zone between forested and deforested areas.

The various consequences of tropical deforestation on Biodiversity are enumerated in Box 3. Therefore it is essential to restore degraded ecosystems to conserve species. It can be achieved by erosion control, reforestation, range improvement and public participation. Planting of native fruit trees such as *Ficus*, *Zizyphus*, *Santalum album*, *Artocarpus heterophyllus* etc., may enhance biodiversity restoration. This would encourage wildlife to spread out and reduce compaction of soil (Kondas, 2010). There has been a positive trend in Asia-Pacific region in the last

Box 3 The consequence of tropical deforestation for biodiversity (after Grainger 1992)

1. Reduced diversity of species and genes
   (a) Species extinctions
   (b) Reduced capacity to breed improved crop varieties
   (c) Inability to make some plants economic crops
   (d) Threat to production of minor forest products.
2. Changes affecting local and regional ecosystems
   (a) Soil degradation
   (b) Changes in water flows from catchments
   (c) Changes in buffering of water flows by wetland forests
   (d) Increased sedimentation of rivers, reservoirs, etc.
   (e) Possible changes in rainfall characteristics.
3. Changes affecting the global ecosystem
   (a) Reduction in carbon stored in the terrestrial biota
   (b) Increase in carbon dioxide content of the atmosphere
   (c) Changes in global temperature and rainfall patterns by greenhouse Effects.

decade. As against an annual loss of over 0.7 million hectares of forests during 1990 to 2000, an annual increase of 2.3 million hectares during 2000 to 2005 and an annual increase of 0.5 million hectares between 2005 and 2010 has been achieved in this Region. This turnaround is due increased public awareness and large scale afforestation initiatives in China, India, the Philippines, Thailand, Bhutan Fiji, Vietnam and Sri Lanka. But deforestation in Australia, Indonesia and Myanmar still remains high (FAO, 2010).

5.7.3.2 Protected areas for biodiversity conservation

The strategies for conservation of biodiversity in the past included constitution of Protected Areas (PAs) such as National Parks, Wildlife Sanctuaries and Biosphere Reserves. A Protected Area is defined as a "geographically defined area", which is designated or regulated and managed to achieve specific conservation objectives. The land in the PAs has certain legal provisions, which facilitate the protection and management. Management Plans are made for the management of individual Protected Areas. The PAs have figured very prominently in biodiversity conservation efforts around the world. The total extent of the PAs across the globe increased to 12% by the year 2010. The PAs can be more effective in conserving biodiversity, if they are well distributed across different bio-geographic region and conflict with local people is minimized through participatory management through eco development (Box 4).

Box 4 Forest restoration in India through JFM

India has 69.09 million hectares under forests (21.02%). Only 37.85 m ha of land has a dense crown cover of over 40%. The per capita forest cover is abysmally low at 0.075 ha. The tree cover outside forests account for 9.28 million hectares taking the total area of Forests and Tree cover to 78.37 million hectares (23.84%). Promulgation of Forest Conservation Act, 1980 brought down diversion of forestlands substantially. The National Forest Policy 1988 emphasizes the role of forests in environmental stability and importance of people's participation in forest management. JFM is based on the premise that there are extraordinary possibilities with ordinary people. In the past two decades, over 22 m ha of forests has been brought under Joint Forest Management (JFM) and are managed through over 100,000 Village Forest Committees. The decadal increase between 1997 to 2007 is estimated to be 3.13 million hectares. Participatory forest management in recent years implemented by dedicated Foresters has not only helped in the restoration of degraded forests, but has initiated comprehensive rural development of remote forest fringe villages (Balaji, 2004; FSI, 2009).

5.7.3.3 Biodiversity action plans

Biodiversity action plans are internationally recognized proactive programmes based on ecosystems focusing on conservation of threatened species and habitats. National Biodiversity Strategy and Action Plan (NBSAP) was launched in India for stock taking of biodiversity related information, including distribution of endemic and endangered species and site-specific threats. The Plan envisages:

- Digitized mapping of forest soils and characterization of vegetation types and to determine their density including endemic flora.
- Estimation of the growing stock and to assess productivity of the habitat.
- Determine the food requirement of large herbivores in the study area and also the extent of predators that can be sustained by ungulate populations.
- Assess the impact of wild fires on biodiversity.
- Assess the potential for eco-tourism and Impact of tribal art and culture on biodiversity conservation.

Chapter 5.8

Biodiversity and biotechnology

In the Convention of Biological Diversity, biotechnology is defined as "any technological application that uses biological systems, living organisms, or derivatives thereof, to make or modify products or processes for specific use". Biotechnology, includes genetic engineering, recombinant DNA technologies and molecular technologies. The Chapter 16 of UNCED *Agenda* 21 is devoted to biotechnology. It also makes a distinction between "modern" biotechnology (DNA manipulation) and "traditional" biotechnology practiced by the plant breeders. Various bio technology applications in biodiversity conservation are given in a diagrammatic form in Figure 5.1.

Bio Technology is useful in biodiversity conservation in several ways as given below:

- Increasing the availability of food, feed and renewable raw materials.
- Improving human health.
- Enhancing protection of the environment.
- Enhancing bio safety and developing international mechanisms for co-operation.
- Establishing enabling mechanisms for the development and environmentally sound application of biotechnology.

5.8.1 BIOTECHNOLOGY FOR BIODIVERSITY ASSESSMENT

Biotechnology provides important tools for biodiversity assessment and monitoring through DNA fingerprinting. In biodiversity conservation, it allows assessment of optimal or minimal population sizes for maintaining diversity, and best practices for augmenting wild populations through transfers from other wild populations or

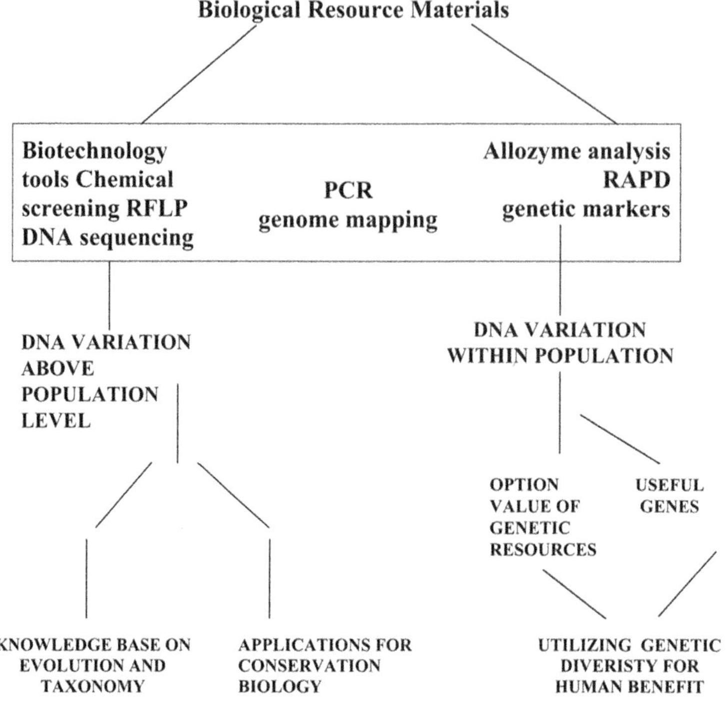

Figure 5.1 Biotechnology applications in biodiversity assessment and management

from captive breeding programmes. Biotechnology can increase the quality of *ex situ* germplasm collections. Samples characterized on the basis of DNA sequence rather than on phenotype can provide better coverage of existing diversity, and can provide a means of avoiding redundancy. Biotechnology is the entire basis for DNA libraries. It allows compact storage of large amounts of genetic information which are recoverable for conservation and utilization.

5.8.2 BIODIVERSITY UTILIZATION

Biotechnology provides important tools for biodiversity utilization for human benefit: Genetic Engineering can be used to improve domesticated varieties to increase yield, control diseases and eradicate pests. Living organisms can be used as factories for manufacturing specific products. Biotechnology can rehabilitate polluted ecosystem.

5.8.3 IMPACTS

Biotechnology can impact on biodiversity in various ways. These impacts can be direct or indirect. The outcomes can be positive or negative for human benefit and for environmental conservation. Direct impacts of biotechnology are ecological or evolutionary, and can be assessed by scientific methods. They include introgression, weediness, pathogenicity, altered nutrient cycling and a range of responses in population of other species. There are realistic ways of assessing potential impacts, to gain maximum benefits from biotechnology applications. Procedures exist in many countries for assessing the possible impact of release of genetically manipulated organisms. In general, these adopt a case-by-case approach, although generic *a priori* risk appraisal is essential. Risk reduction is achieved mainly by confinement. Good containment possibilities exist, even for microorganisms. Measures for minimizing direct impacts on native ecosystems aim to impose specified limits on gene transfer, on exposure to toxic products and on direct invasion. Past experience with release of conventionally bred varieties is often a good guide for best practice. Indirect impacts of biotechnology are predominantly socio-economic. They can be of major importance because of the way social system drive our use of biotechnology and biodiversity. Risk assessment for indirect impacts should therefore be independent of that for direct impacts. Biotechnology can increase the value of genetic resources. In absolute terms these values may be limited, but value changes may lead to indirect effects. These may include conflicts over ownership, including intellectual property; altered rates of utilization and therefore of resource conservation and environmental protection and ethical debate on the acceptable scope of the biotechnology applications themselves. While biotechnology results in increased agricultural productivity, both positive and negative impacts can be predicted. These include changed pressures for cultivating new land, replacement of traditional landraces, and marginalization of fringe production areas through price changes.

5.8.4 BIOTECHNOLOGY FOR PROSPECTING GENETIC DIVERSITY

The combination of chemical and DNA screening technologies with genetic engineering provides an extremely powerful tool for the exploitation of genetic resources. For example, medium chain fatty acids such as laurate accumulate in the seeds of tropical trees and are harvested for dietary and industrial purposes at the rate of at least a million tonnes annually. The knowledge that production of medium chain fatty acids is due to a specific protein (acyl carrier protein-BTE) has led to the cloning of the DNA of the gene coding for this protein from the Californian bay (*Umbellularia californica*). This gene can be introduced into an oilseed species and can increase the levels of laurate. In Industry, transgenic animals, plants and micro-organisms can be used as source of (Voelker et al., 1992) insulin, growth hormones, antibodies, vegetable oils; essential oils and secondary metabolites such as pharmaceuticals and bio-pesticides. Genetic Engineering can be used for enhancing efficiency of microorganisms in industrial processes, such as secondary recovery of oil from reservoirs, bioleaching, extraction of metal from low grade ores, production of industrial enzymes, antibiotics and bio plastics. Biotechnology is also useful in Agriculture and Horticulture for production of better crops, ornamentals with enhanced flavour and shelf life. In environmental management they can be used to in bioremediation of polluted environment.

5.8.5 GENETICALLY MODIFIED FOODS

Genetically modified organisms (GMOs) can be defined as organisms in which the genetic material (DNA) has been altered in a way that does not occur naturally. Genetically Modified (GM) foods are developed and marketed because there is some perceived advantage either to the producer or consumer of these foods. This is meant to translate into a product with a lower price, greater benefit (in terms of durability or nutritional value or both). Insect resistance is achieved by incorporating into the food plant the gene for toxin production from the bacterium *Bacillus thuringiensis* (BT). Virus resistance is achieved through the introduction of a gene from certain viruses which cause disease in plants The safety assessment of GM foods generally investigates their direct health effects (toxicity), tendencies to provoke allergic reaction, the stability of the inserted gene and any unintended effects. No allergic effects have been found relative to GM foods currently on the market. Gene transfer from GM foods to cells of the body or to bacteria in the gastrointestinal tract could adversely affect human health. This would be particularly relevant if antibiotic resistance genes, used in creating GMOs, were to be transferred. Although the probability of transfer is low, the use of technology without antibiotic resistance genes has been encouraged by a recent FAO/WHO expert panel. The movement of genes from GM plants into conventional crops or related species in the wild (referred to as "out crossing") as well as the mixing of crops derived from conventional seeds with those grown using GM crops, may have an indirect effect on food safety and food security. This risk is real, as was shown when traces of a maize type which was only approved for feed use appeared in maize products for human consumption in the United States of America. Several countries have adopted strategies to reduce mixing, including a clear separation of the fields within which GM crops and conventional crops are grown. The Cartagena Protocol on Bio safety (CPB), regulates trans boundary movements of living modified organisms (LMOs) (www.fao.org).

5.8.6 ENVIRONMENTAL BIOTECHNOLOGY

Chemical contamination of our environment has been a highly visible and undesirable outcome of industrialization. Environmental clean-up through biological technologies, especially microbial ones, are often sought because they are usually less expensive than mechanical ones, Bioremediation can often be carried out *in situ*, eliminating or greatly reducing costs for soil excavation and movement; the biological catalyst multiplies as the expense of the pollutant, thus naturally enhancing the remediation rate. A consortium of bacteria is used in bioremediation of polluted lakes across the world. Polluted Lake Superior in USA was cleaned in this way. Ooty lake in India was cleaned using similar technology in the year 2004–05.

Box 5 Bioremediation of environmental pollutants

Chlorinated solvents. These include chemicals such as tetrachloroethene (PCE), trichloroethene (TCE), trichloroethene (TCA), and carbon tetrachloride (CCl_4). These chemicals are water-soluble, and cause serious groundwater contamination problem.

Chlorinated aromatics. This group includes the polychlorinated biphenyls (PCBs), chlorinated dioxins, tetrachlorophenol (PCP) and chlorinated benzenes. The members of this group with higher molecular weights are strongly sorbed to soil and are not biodegradable.

Explosives and other nitroaromatics. This group includes TNT. RDX, and some industrially used chemicals such as nirophenol. The reduction of the nitro-group does not occur under Natural condition but occurs readily under anaerobic conditions.

Polynuclear aromatic hydrocarbons (PAHs). This class includes with multiple aromatic rings, such as phenantherene, pyrene and benzo(a) phrene. These chemical are strongly sorbed to soil, some are carcinogenic, and those with higher molecular weights are very difficult to biodegrade. Co-metabolism is the only known way to degrade those with higher molecular weights.

Components of gasoline such as benzene, toluene, ethylbenzene and the zylenes (BTEX compounds). These chemicals are easily degraded by aerobic microorganisms and, except for benzene, by some anaerobes. The chemicals are common groundwater pollutants as they are water-soluble and are amenable for bioremediation through Bio venting and air sparging technologies.

Metals. Heavy metals are perhaps the most difficult class of pollutant chemical for which bioremediation may be useful. Some metals can be removed from waste streams by absorption of the metal onto the microbial biomass, and some metals can be oxidized or reduced by microbes.

Source: UNEP, 1995

5.8.7 PRAGMATIC USE OF BIOTECHNOLOGY

Biotechnology has great potential to generate benefits, both environmental and socioeconomic. It can enhance our understanding of biodiversity and our capacity to conserve, manage and utilize it. It can improve our production systems and the diversity of the amenities we desire. In general, however, its impacts on biodiversity may not be different in nature, but only in degree, from those of our traditional or conventional practices. These impacts may be direct ecological and evolutionary outcomes of biotechnology applications, or indirect, resulting from altered socioeconomic circumstances. The Convention on Biological Diversity calls for appropriate safeguards to ensure safe applications of biotechnology. We have a growing capacity to assess likely impacts, and to make informed decisions both for conservation of biodiversity and for equitable sharing of the benefits of biotechnology.

Chapter 5.9

Economics and policy of biodiversity management

5.9.1 ECONOMICS AND POLICY

Economics provides a useful perspective for understanding human relationships with biodiversity. Many industrial materials are derived directly from biological sources. These include building materials, fibers, dyes, rubber and oil. Biodiversity is also important to the security of resources such as water, timber, paper, fiber, and food. Biodiversity loss is a significant risk factor in business development and a threat to long term economic sustainability. Loss of biodiversity results in the loss of Natural capital that supplies ecosystem goods and services. The economic value of 17 ecosystem services for Earth's biosphere Is estimated to be US$ 33 trillion (3.3×10^{13}) per year (Costanza et al., 1997). Biodiversity enriches leisure activities such as hiking, bird watching or natural history study. Biodiversity inspires musicians, painters, sculptors, writers and other artists. Many cultures view themselves as an integral part of the natural world which requires them to respect other living organisms. Indian culture emphasizes people to live in harmony with nature. Popular activities such as gardening, fish keeping and specimen collection strongly depend on biodiversity. Philosophically it could be argued that biodiversity has intrinsic aesthetic and spiritual value to mankind. This idea can be used as a counterweight to the notion that tropical forests and other ecological realms are only worthy of conservation because of the services they provide.

Biodiversity supports many ecosystem services that are often not readily visible. It plays a part in regulating the chemistry of our atmosphere and water supply. Biodiversity is directly involved in water purification, recycling of nutrients and providing fertile soils. Experiments with controlled environments have shown that humans cannot easily build ecosystems to support human needs; for example insect pollination cannot be mimicked, and that activity alone represents tens of billions of dollars in ecosystem services per year to humankind. But ecosystem services are not valued or taken into national accounting. Biodiversity, is a public good that is provided to everyone, rather like Law and Order and defence. It does not confer any property right. Market economics, left to itself, typically under-provide public goods. Thus property rights work well for bread as a private good, but much less well for a public good such as genetic variation in wheat types. The full social benefits of tigers, rhinos, portfolios of germplasm, marine resources of a global commons, and so forth, are public goods beyond appropriation by markets, even when market value is fully enhanced by all the devices of the law. Therefore, new policies are required which help enable public goods to be managed for the benefit of society.

5.9.2 TANGIBLE AND INTANGIBLE USES OF BIODIVERSITY

- The direct use value of biological resources derives from their direct role in consumption or production. Agriculture, Forestry and Fisheries are all based on the direct use value of foods and fibres.
- The value of the mix of biological resources, biodiversity, is an indirect use value. It derives from the role of the mix of species in (a) supporting one or more organisms (the value of habitat) and (b) supporting the provision of the economically interesting range of ecological services (the value of ecosystem functions). The value of ecosystem functions is sometimes referred to as a secondary value.
- The non-use or passive use values of biological resources derive from: altruism towards friends, relatives or others who may be users (vicarious use value); altruism towards future generations of users (bequeath value); and altruism toward non-human users (existence value).
- The social value of an ecosystem exceeds its private or market value as full complexity and coverage of the underpinning “life-support” function of healthy ecosystems are currently not precisely known in scientific terms.
- The continued functioning of a healthy ecosystem is more than the sum of its individual components.

Estimation of the total value of wetlands, forests, etc. involves complex problems to do with system boundaries, scale, thresholds and value component aggregation. Given the inevitable uncertainties involved, a precautionary approach based on maximum ecosystem conservation is a high priority. A suitable comprehensive and long-term view of use value – one that protects ecosystem services by protecting the health and integrity of system over the long term – is probably sufficient to realize the case for more biodiversity conservation and will carry with it aesthetic and intrinsic moral value as well.

5.9.3 CONSERVATION STRATEGY

The Convention on Biological Diversity 1992 provides an overall legal framework for addressing biodiversity management, supplemented by the Cartagena Protocol on

Bio safety. A number of other treaties that address some aspects of biodiversity management are the International Convention for the Regulation of Whaling 1946, the Convention on International Trade in Endangered Species of Wild Fauna and Flora (CITES) 1973, the International Tropical Timber Agreement 1994. The broad scope of the Biodiversity Convention is illustrated in its Article 1, that includes "conservation of biological diversity, sustainable use of its components and fair and equitable sharing of benefits arising out of the utilization of genetic resources including by appropriate access to genetic resources and by appropriate transfer of relevant technologies taking into account all rights over those resources and technologies and by appropriate funding." The **Nagoya Protocol** on access to genetic resources and the fair and equitable sharing of benefits arising from their utilization and thereby contributing to the conservation of biological diversity and the sustainable use of its components was adopted by the Conference of the Parties (COP-10) to the Convention on Biological Diversity at its tenth meeting on 29th October 2010 in Nagoya, Japan. The Nagoya Protocol will be open for signature by parties to the convention from 2 February 2011 until 1 February 2012 at the United Nations headquarters in New York.

Chapter 5.10

Future prospects

Virtually every decision people make can have an influence on biodiversity, either positive or negative. Loss of biodiversity at the hands of people will continue and may even accelerate, with increasing pressures from population expansion and faster pace of economic development. Therefore, there is need to manage these bio resources more effectively and adopt national and regional polices in consonance with tenets of Convention on Biological Diversity. Although it is by no means, clear whether poverty, with its pressures to survive, or affluence, with its pressures to consume, ultimately leads to greater loss of biodiversity, it is obvious that the rural poor cannot conserve their biological resources if this is in conflict with their immediate survival needs. However, biodiversity is most likely to prosper in areas with the least human interference, such as Protected Areas. Tropical islands can provide a preview of the environmental situation that is likely to become more prevalent on the world's continents in the future. These islands typically have high population densities, exhibit highly fragmented landscapes, and have already experienced significant extinction events. Humans have endeavoured rather successfully, to acquire their growing biomass needs from intensifying the productivity of a small number of domesticated species through plantations and aquaculture. These replace large tracts of diverse natural ecosystems with species-poor systems supported by high levels of technological inputs; they also promote extensive use of pesticides resulting in more widespread negative impacts on biodiversity. While bio productivity can be enhanced through managed plantations to meet our biomass requirement, it is imperative to maintain biodiversity rich areas as Protected Areas. In order to integrate production objectives without sacrificing environmental concerns, Foresters in pluralistic setting, must be highly innovative and able to build consensus (Balaji, 1997). In this regard, countries across the globe recently adopted Aichi targets in Nagoya, Japan.

5.10.1 THE STRATEGIC PLAN – AICHI TARGETS 2011–2020

The Conference of the Parties (COP) established under the CBD at its sixth session in 2002 set global targets to reduce the loss of biodiversity by 2010. These include:

- Conservation of biodiversity at the level of ecosystems, species and genes;
- Addressing risks such as invasive alien species, global warming and developments that threaten the natural environment;
- Maintaining the rights of the aboriginal people and protecting their traditional knowledge;
- Ensuring equal and equitable distribution of profits from the use of genetic resources.

The vast majority of nations have fallen far short of the Convention on Biological Diversity's (CBD's) 2010 target to reduce the rate of loss of biodiversity. This prompted the COP-10 held in Nagoya in Japan in October 2010 to develop a new Plan of Action for conservation of biodiversity and for better access and benefit-sharing of genetic resources. It formulated 20 "SMART" (Specific, Measurable, Ambitious, Realistic, and Time-bound) targets for 2020. The ecosystem services framework of Nagoya has four main consequences for target setting. First, what and how much biodiversity should be targeted for conservation depends on what services are important to maintain and with what reliability. Second, the temporal and spatial scale of targets should be based on the changing temporal and spatial distribution, and risk profiles of ecosystem services. Third, target development and implementation should include all agencies involved with management of biodiversity and the ecosystem services they support. Fourth, interdependence among ecosystem services, the benefits they provide, and the value placed on those benefits implies that targets must be conditional (Perrings, C. et al., 2010). In this regard Japan has assured to provide two billion dollars over three years to help developing countries save their ecosystems.

The Strategic Plan 2011–2020 (Aichi Target) is aimed achieving a world, Living in Harmony with Nature. The mission was to take effective urgent action to halt loss of Biodiversity in order to ensure that by 2020 ecosystems are resilient and continue to provide essential services. The 20 targets are as follows:

1. All people are made aware of the values of biodiversity and the steps they can take to conserve and use it sustainably.
2. The values of biodiversity are integrated into national accounts, national and local development, and poverty reduction strategies.
3. Incentives including subsidies harmful to biodiversity are eliminated, phased out, or reformed in order to minimize negative impacts.
4. Governments, business, and stakeholders take steps to achieve or have implemented plans for sustainable production
5. The rate of loss and degradation, and fragmentation, of natural habitats including forests is at least halved and brought close to zero.
6. Overfishing is ended, destructive fishing practices are eliminated, and all fisheries are managed sustainably.

7. Areas under agriculture, aquaculture, and forestry are managed sustainably, ensuring conservation of biodiversity.
8. Pollution, including from excess nutrients, be brought to levels that are not detrimental to ecosystem function and biodiversity.
9. Invasive Alien Species are identified, prioritized, and controlled or eradicated.
10. The multiple pressures on coral reefs and other vulnerable ecosystems affected by climate change are minimized.
11. At least 20% of terrestrial, inland-water, and of coastal and marine areas are conserved.
12. The extinction and decline of known threatened species is prevented.
13. The loss of genetic diversity of cultivated plants and domestic farm animals in agricultural ecosystems and of wild relatives is halted.
14. Ecosystems that provide essential services and contribute to health, livelihoods, and well-being, are safeguarded.
15. Ecosystem resilience and the contribution of biodiversity to carbon stocks is enhanced, through conservation and restoration.
16. Access to genetic resources is enhanced, and benefits are shared.
17. Each party has developed, adopted, and implemented, an effective, participatory, and updated National Biodiversity Strategy and Action Plan.
18. Have *sui generis*/legal systems in place to protect traditional knowledge, innovations, and practices relevant to biodiversity.
19. Knowledge, the science base and technologies relating to biodiversity, its values, functioning, status, and trends, are improved.
20. Capacity of human resources and financing for implementing the Convention is increased tenfold.

5.10.2 SCOPE FOR FUTURE RESEARCH

The present level of understanding of biodiversity is inadequate as already discussed. Therefore there is need for research on Biodiversity assessment and monitoring. Similarly Research on livelihood improvement through biodiversity prospecting is also essential. Biotechnology offers scope for understanding relationship amongst species through molecular taxonomy. Biotechnology research is essential for improved understanding of this relationship besides utilizing the knowledge about species for their sustainable benefit of the humanity. Geographical Information System offers substantial scope for future research in forest inventory, biodiversity assessment and monitoring. Forestry research on *in situ* and *ex-situ* biodiversity conservation is essential to unravel the mystery of living things and ecosystems. Biotechnological research for prospecting biodiversity should take into consideration safeguards required in genetic manipulation.

Chapter 5.11

Conclusion: Living in harmony with nature

Our life on this earth is ecologically intertwined. Sustainable development denotes improving the quality of life within the carrying capacity of the supporting ecosystem. Sustainable development and effective rehabilitation and management of ecosystem with adequate concern for biodiversity represent two sides of the same coin. The conservation of our rich biodiversity heritage portends better quality of life for the present as well as future generations. The biodiversity outlook of the last decade indicates that biodiversity is declining at all levels and on geographical scales. However, targeted response options such as the expansion of protected areas, resource management and pollution prevention can reverse this trend for specific habitats and species. For instance, protected area coverage has doubled over the past 20 years and terrestrial Protected Area net work now covers 12 percent of the earth's surface. Tropical Forest cover in several countries in Asia and Pacific is showing signs of recovery in the last decade with an increase of 0.5 million hectares every year between 2005 to 2010. Water quality in Europe, North America and Latin America has improved since 1980. The National River Conservation Plan and the National Lake Conservation Plan of the Union Ministry of Environment and Forests, Government of India over the last two decades aim to improve water quality in rivers and lakes albeit with limited success. But the active participation of all stakeholders including local bodies, industries and the public at large alone can bring glory back to rivers and lakes. The experience in other developed countries indicates that such transformation is not impossible. **Biodiversity conservation is essential for the sustenance and wellbeing of the present and future generations. As John Sawhill said, "In the end our society will be defined not by what we create but what we refuse to destroy."** The global agreements such as CBD provide a framework for sustainable future for mankind. But success of these global agreements depends on the commitment and action at the national and local level to live in harmony with nature and conserve biodiversity for the posterity.

Chapter 5.12

Sample exercises

1. Define Biodiversity.
2. What are the ecological services provided by Biodiversity? Explain with reference to your neighbourhood.
3. Why tropical countries are endowed with rich Biodiversity?
4. What is the extent of erosion of Biodiversity? How will you halt this process in your country's forests?
5. What is HIPPO? How does it relate to Biodiversity loss?
6. What are Invasive Alien Species? What Invasive Species are common in your region? How can they be checked?
7. What is Climate Change? How is it related to Biodiversity?
8. What is the role Biodiversity Conservation in Medicine? How do you propose to recognize Local Technical Knowledge in your country?
9. How will you protect Agro Biodiversity in your country for sustained food security?
10. Can aquaculture be practiced without affecting the local environment?
11. What are the functions of Protected Areas available in your country?
12. Explain the devastation caused by forest fires in your country. What measures would you suggest tocombat Forest fires?
13. What biotechnological tools will you use in protecting your country's environment?
14. How Genetically Modified foods are regulated in your country?
15. What are Aichi Targets for 2020? Explain what steps are required to achieve the targets in your country?

References

Balaji, S., 1997. The Technical aspects of Sustainable forestry and Rural Development in a Pluralistic Institutional Environment In: *Pluralism and Sustainable Forestry and Rural Development*, FAO, Rome.

Balaji, S., 2004. '*Joint Forest Management for Biodiversity Enhancement*' p. 216, Bishen Singh Mahendrapal Singh, Dehradun.

Balaji, S., 2010. Biodiversity Challenges Ahead, *The Hindu*, Chennai May 27, 2010.

Basappanavar, C.H., 2009. *Biodiversity conservation an antidote to Climate Change*. Vanasuma Prakashana, Bengaluru. p. 240.

Costanza, R. et al. 1997. The value of the World's Ecosystem Services and Natural capital *Nature*, 387 (6630): 253–260.

di Castri, F. & T. Younes (eds.) 1996. *Biodiversity, Science and Development Towards a New Partnership*. International Union of Biological Sciences, Paris. p. 646.

ENVIS Centre, 2009. Kerala State Council for Science, Technology and Environment (KSCSTE), Government of Kerala, Thiruvananthapuram, 695004.

FAO. 1993. Forest Resources Assessment, 1990: Tropical Countries. FAO Forestry Paper 112, FAO, Rome.

FAO. 2010. Tiger paper, vol. xxxvii, No. 5, July–September, 2010, FAO, Regional Office, Bangkok.

FSI. 2009. *India – State of Forest Report-2009*, Forest Survey of India, Dehradun.

Gadgil, M., Berkes, F. and Folke, C. 1993. Indigenous knowledge for biodiversity conservation. *Ambio* 22: 151–156.

Grainger, A. 1992. *Controlling Tropical Deforestation*. Earthscan, London

Jeffrey, K. McKee 2004. *Sparing Nature: The Conflict Between Human Population Growth and Earth's Biodiversity*. Rutgers University Press. p. 108. ISBN 9780813535586. Retrieved 28 June 2011.

ICFRE, 1993. *Annual Report of Indian Council of Forestry Research and Education*, Dehradun. p. 315.

IWOKRAMA, 1996. *International Rain Forest Programme*, Operational Plan 1996–2000, Working Document, George Town, Guyana.

Koh, L. P., Dunn, R. R., Sodhi, N. S., Colwell, R. K., Proctor, H. C. & Smith V. 2004. *"Species Co extinctions and the Biodiversity Crisis"* (PDF). Science **305** (5690): 1632–4.

Knoll, A. 1984. Patterns of extinction in the fossil record of vascular plants. In; *Extinction* (ed. M. Nitecki). Chicago Press, Chicago.

Kondas, S. 2010 *Management of Biological Diversity of Indian Forests.* Future Graphic, Chennai. 500 pp.

McNeely, J., Miller, K., Reid, W., Mittermeir, R. & Werber, T. 1990. *Conserving the World's Biodiversity*, IUCN Gland Switzerland WRI, WWF-US and the World Bank, Washington, DC, USA.

Myers, N., 1986. Tropical deforestation and a mega extinction spasm. In: *Conservation Biology* (ed. M.E. Soule), Sinauer Associates, Sunderland, USA.

Norton B.G. & Ulanowicz, R.E. 1992. 'Scale and Biodiversity Policy: A hierarchical approach' *Ambio* **213**: 244–249.

Oldfield, M.L. 1984. *The value of Conserving Genetic Resources.* US Department of the Interior. National Park Service, Washington, DC.

Perrings, C. (and 17 others) (2010), *Ecosystem Services for 2020. Science*, **330** 15 October 2010. www.sciencemag.org

Reid, W.V. 1992. How many species will there be? In: Whitmore, T.C. and Sayer, JA. (eds), *Tropical Deforestation and species Extinction* 55–74. Chapman and Hall, New York.

Reid, W.V., Laird, S.A., Meyer, C.A., Gamez, R., Sittenfeld, A., Janzen, D., Gollin, M.A. and Juma, C. (eds) 1993, *Biodiversity Prospecting; Using genetic resources for sustainable development.* World Resources Institute, Washington, DC.

Swaminathan, M. S. and S. Jana, 1992. *Biodiversity implications for global food security*, MacMillan India Ltd., Madras.

Stern, N. 2006. *Stern review on the Economics of Climate Change*, Executive Summary, Cambridge University, UK.

UNEP, 1995. *Global Biodiversity assessment*, U.K. edited by Heywood, V. H. and R. T. Watson. University of Cambridge Press, Cambridge CB 2 IRP., p. 1140.

UNESCO, 1990. State of the environment in Asia and the Pacific. *United Nations Economic and Social Commission for Asia and Pacific, Bangkok, Thailand.* p. 331.

Vitousek, P. M., P. Ehrlich, A. Ehrilich and P. Matson, 1986. Human appropriation of the products of photosynthesis. *Bioscience.*, **36**: 368–373.

Volker, T., Worrell, A., Anderson, L., Blehaum, J., Fan, C., Hawkins, D., Radke, S. and Davies, H. 1992. Fatty acid biosynthesis redirected to medium chains in transgenic oil seed plants. *Science*, 257, pp. 72–74.

WCMC. 1992. *Global Biodiversity – Status of the Earth's Living Resources* (ed.) Brian Groombridge, Chapman and Hall. London U.K. p. 594.

Wikipedia, 2011. Web source on Biodiversity conservation.

Wilson, E. O. 1988. The Current State of Biological Diversity. In: *Biodiversity* (ed. E.O. Wilson and Peter, F.M.). National Academy Press, Washington. D.C.

WRI 1994. *World Resources 1994–95.* Oxford University Press, New York.

www.fao.org.in; on Genetically Modified Crops.

Section 6

Disaster management*

U. Aswathanarayana (India)

---

*This Section draws the basic concepts and data from Chap. 9 of the author's works, "*Geoenvironment: An Introduction*" (1995), and Chap. 36 of "*The Indian Ocean Tsunami*" (2006).

Chapter 6.1

Hazardous events (natural, mixed and technological)

All lands on earth are subject to natural hazards, such as, floods, droughts, landslides, earthquakes, typhoons, volcanism, etc. The hazard itself cannot be prevented, but through an understanding of the land conditions which are prone to a given hazard and the processes which could culminate in damage to life and property, it is possible to minimize the damage to life and property, through preparedness for a particular eventuality. For instance, Bangladesh and the province of Florida in USA, are highly susceptible to hurricanes. But the number of deaths per million of population in Florida due to hurricanes is much less than in Bangladesh, because of the prevalence of more efficient systems of preparedness and crisis management.

Natural Hazard (H) is probability of occurrence, within a specified period of time and within a given area, of potentially damaging phenomenon (e.g. landslide). Vulnerability (V) is the degree of loss (or damage) resulting from H, of a given magnitude, expressed on a scale from 0 (no damage) to 1 (total loss). Specific risk (R_s) is the expected degree of loss due to particular, H. Elements at risk (E) are the population, properties, and economic activities, including public services at risk in a given area. Total risk (R_t) is the expected number of lives lost, persons injured, damage to property, or decline of economic activity due to a particular, H (Carrara, 1984).

Thus,

$$R_s = H \times V; \quad R_t = E \times R_s = E \times H \times V$$

The general model of management response to a hazard event is given in Fig. 6.1.1.

A geological hazard is a geological condition, process or a potential event that poses a threat to the health, safety and welfare of a group of citizens, or the functions or economy of a community or a larger governmental entity (US Geological Survey, 1977). The spectrum of hazards is quite large. Hazards may arise from natural or technological causes. Hazardous conditions set the stage for future harm. It is therefore necessary to recognize how certain conditions of land when acted upon by a particular process, can culminate in perpetuating a particular kind of damage (Tables 6.1.1 and 6.1.2).

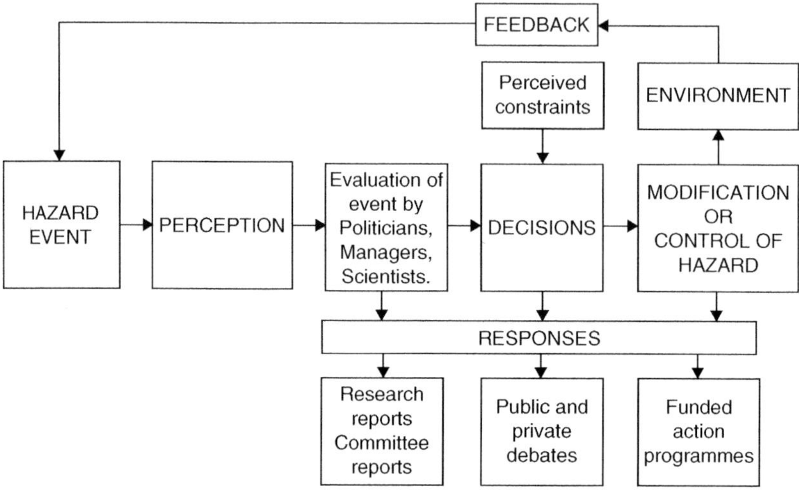

Figure 6.1.1 General model of management response to a hazard event
Source: Kasperson, 1969

Table 6.1.1 Hazardous events

Type of hazard	*Events*
Natural: Purely natural, and not induced by, and controllable by, humans	Large earthquakes, volcanic eruptions. Subsidence in karst areas, large river floods, coastal floods, landslides
Mixed: Natural in character, but influenced by human action	Small earthquakes in regions of below-ground disposal of liquid wastes; landslides in developed areas; river floods in developed areas; coastal floods on lakes
Technological: Generated by human action	Subsidence in the area of sub-surface mines; burial of toxic wastes; failure of a dam or a nuclear power plant.

Source: Lundgren, 1986

Table 6.1.2 Culminating events that could arise from hazardous conditions

Land condition	*Process*	*Culminating event*
Fault	Crustal deformation	Faulting and earthquakes
Volcano	Magma emplacement	Volcanic eruptions
Metastable slope/old landslide	Gradual change in groundwater regime at depth	Landslides
Carbonate, gypsum or salt rock	Dissolution of rock	Collapse (sink hole formation)
Layers of water-saturated well-sorted sand	None	Liquefaction
Flood plain	Storm generation	River flood
Coastal plain	Storm generation	Storm surge or flooding (hurricane)
Abandoned mines	Gradual dissolution. Fracturing, ground fires	Subsidence

Source: Lundgren, 1986

Chapter 6.2

Vulnerability to hazardous events

6.2.1 EARTHQUAKES

It is now well recognized that the earthquakes are concentrated along the margins of plates, and are caused when the plates converge, diverge or slide past each other. Such earthquakes occur on faults, some of which may be deep-seated.

Analysis of damage due to earthquakes has shown that the extent of ground-shaking and resultant damage are critically dependent upon the nature and properties of rocks and soils at or near the ground surface. Many countries prepared seismic risk maps on the basis of the past records (seismicity) in an area, and the structure and composition of rocks and soils.

Earthquakes could also occur due to anthropogenic activities, such as the creation of large water reservoirs, mining, and forceful injection of fluids into pores and cracks in crustal rocks. There is no unequivocal correlation between the height of the dam, the volume of water impounded and the magnitude of the largest earthquake. It appears that the crucial factor is the hydraulic continuity between the body of the impounded water and the deeper groundwater (facilitated by deep-seated faults?). "If there is no water in the rocks, there would be no tectonic earthquakes" (Bolt, 1993). The reservoir-induced Koyna earthquake of Dec. 10, 1967, killed about 200 people, and caused severe damage to Koyna Nagar Township in western India. Earthquakes of magnitude equal to or greater than 5 seem to occur when the rate of loading in the reservoir exceeded 13 m per week (Gupta, 1993).

Animals appear to be sense the low-magnitude precursors of the earthquakes which the human beings cannot sense (e.g. snakes coming out of their holes). Using animal behaviour as an indicator of an imminent earthquake is, however, not easy.

The damage due to an earthquake is not confined to the consequences of the ground vibration. Often the secondary effects, such as landslides, floods, fires and tsunamis, have caused more damage than the earthquake itself.

It has been estimated that on an average about 10,000 people are killed by earthquakes – the 1976 earthquake in Hopei in China resulted in the death of 650,000 people.

6.2.2 TSUNAMIS

The vulnerability of a coastal site to tsunami damage is dependent on the slope and morphology of the coast. In order for an earthquake to generate tsunami, the magnitude should be ≥ 7.9, the nature of the faulting should be either thrust or normal, and the earthquake should be shallow enough to cause vertical uplift of the ocean floor.

The Indian Ocean Tsunami which had a magnitude Mw 9.3, energy of 1.1×10^{17} Nm, occurred on Dec. 26, 2004, at a depth of 20–30 km, close to the Indonesian forearc. The earthquake rupture had a maximum length of 1,200 km along the interface between the Indo-Australian and the Burmese plates. There was a 20 m displacement along the fault plane, and the seafloor thrusted up several metres. The earthquake appeared to have produced a recognizable pole shift, and a small change in the length of the day.

The tsunami affected 12 countries, from Indonesia to Somalia, and killed 176,260 people. The tsunami destroyed billions of dollars worth of property. There is a scientific explanation for the linear path of the tsunami – why it affected Sri Lanka and Thailand coasts, but not Bangladesh and Orissa coasts. Singh (*Nature*, 2005) showed that a consequence of the existence of a lithosphere-scale boundary around the Simuelue Island which continues upto the east of Nicobar Island, the Dec. 26, 2004 earthquake rupture which seem to have been initiated west of the boundary, did not cross the boundary to the east, but got propagated northwards upto the Andaman Islands.

Animals appear to sense the approaching tsunami wave. In the case of the Dec. 26 Indian Ocean Tsunami, elephants carrying tourists along a beach in Thailand, suddenly started running inland. Such a behaviour was perplexing the mahouts, but it saved many lives.

Evidently, the elephants sensed the tsunami well before it hit the coast.

The design of a cost-effective strategy for the warning and preparedness for tsunamis has to take into account two attributes of tsunami, namely, the genesis and impact. Tsunami is triggered by submarine earthquakes, volcanism and submarine landslides, etc. and hence it can be conveniently dovetailed with the existing administrative structures for earthquake disaster management. A tsunami is akin to a tidal wave in its impact on the coasts, and can hence be treated as an add-on to the tidal wave warning system.

6.2.3 VOLCANIC HAZARDS

There is a close relationship in the distribution of earthquakes and the volcanoes, for the simple reason that both occur along the plate margins, and are related to plate movements. But the magnitude of hazard due to the volcanoes is much less than due to earthquakes. As against about a million earthquakes that take place in a year, there are only 760 active volcanoes. The manifestations of the volcanic hazards are: lava flows, dome eruptions, ejected materials, nuées ardentes (a fluidized cloud of hot ash dust and gas generated by the eruption of Mt. Pelee in 1902 killed 30,000 people, and destroyed the town of St. Pierre on Martinique Islands), poisonous gases (the emissions of poisonous gases from Lake Nyos in Cameroons, West Africa, on Aug. 21, 1986, killed about 1,700 people, and 3,000 cattle by asphyxiation), volcanic mudflows or lahars (a lahar in Armero, Colombia, killed 22,000 people in 1985).

6.2.4 SLOPE FAILURES, LANDSLIDES AND SUBSIDENCE

Vulnerability to hazards such as landslides can be evaluated from neotectonic activity. Cooke and Doornkamp (1990, p. 353) compiled a list of geomorphological indicators of netectonic activity.

Direct: (1) Emerged coral reefs, (2) Displacement of dated beaches, (3) Deformed shorelines, (4) Offset in coastline configuration, (5) Distortion of river terraces, (6) Sedimentation of alluvial fans, (7) Deformation of alluvial fans, (8) Displacement of dated terraces, (9) Changes in lake depth, (10) offset of glacial features, (11) Warping of planation surfaces, (12) Displacement of synthetic structures, (13) Fractured cave structures, (14) Fault scarps, (15) Spurs and facets, (16) Shutter ridges, (17) Separation of river terraces, (18) Separation of river terraces, (19) River reversals, and (20) Displacement of man-made structures.

Indirect: (1) Response to stream channels, (2) Downstream changes in river sinuosity, (3) River capture, (4) Rates of sedimentation, (5) Fluvioglacial gravel deposition, (6) Formation of lakes. River terraces and alluvial fans have been extensively studied in China to delineate the neotectonic activity. Kali Gandaki gravel deposits in central west Nepal have been used to decipher neotectonic activity

Slope failure, leading to subsidence and landslides, could occur due to natural reasons (e.g. earthquakes or river cutting) or due to anthropogenic activities (e.g. construction of high-rise buildings, dams, road, forest clearance, etc.)

In 1985, the Nevada del Ruiz volcano in Colombia (South America) became active. The heat melted the ice covering the mountain summit, and while flowing down, the melt water picked up the ash on the slopes of the mountain. The resulting torrent of mud (lahar) came roaring down the mountain, and buried the town of Armero killing about 22,000 people.

How a civil engineering project could trigger unpredictable, disastrous consequences could be illustrated with the example of the Vaiont Dam in Italy. The water level of the reservoir was lowered too rapidly. This has left the perched water table within unsupported reservoir slide slopes and created instability in the slope. A huge quantity of rock mass slid into the reservoir. The landslide generated a giant wave in the reservoir which burst the dam. The resulting flood drowned over 2,000 people in the village downstream.

Removal of forest cover from steep slopes almost always leads to slope failure and landslides, as it happened in the foot-hills of the Himalayas in the Bhagirathi and Alakananda Valleys in the Ganga system. Disastrous landslides occurred in Rio de Janeiro, Brazil, when houses were constructed on steep slopes after cutting down trees.

Chapter 6.3

Marine hazards

K.S.R. Murthy
National Institute of Oceanography (N.I.O.), Regional Centre, Lawsons Bay, Visakhapatnam, India

6.3.1 INTRODUCTION

Plate tectonics provides the framework for interpreting the history and character of the continental margins. Plate movements and the basic difference in the density of oceanic and continental crustal units initiate the structural pattern of continental margins and result in a tectonic classification of coastlines as active (Pacific, leading edge) or passive (Atlantic, trailing edge) or transform margins, each of which have certain fundamental characteristics (Kious and Telling, 1996; Mead, 2005; Morelock et al., 2006, Murthy, et al., 2011).

6.3.2 TYPES OF MARINE HAZARDS

Continental margins and adjacent oceanic basins are affected by a variety of marine hazards, which can be broadly categorized as:

- Natural hazards
- Man-made hazards

Some of the important forms of natural hazards are:

- Earthquakes/volcanoes/tsunamis/landslides
- Cyclones, hurricanes and tornadoes
- Sea level rise due to global climatic changes

Earthquakes, volcanoes, tsunamis have been dealt with in Chapter 6.2. In this section we mainly deal with the other natural hazards like cyclones, hurricanes and tornadoes and sea level rise due to global climatic changes. Under the man-made hazards, we discuss mainly about oil spills, oil blow-outs and marine pollution due to land-based activities.

6.3.3 NATURAL HAZARDS

6.3.3.1 Cyclones, hurricanes and tornadoes

A cyclone is a low pressure system. A tropical depression is a tropical cyclone with maximum sustained winds of 51 kmph or less, where as a tropical storm is a tropical cyclone with maximum sustained winds between 51 and 117 kmph. A hurricane is a tropical cyclone with maximum sustained winds of 118 kmph or more. A tornado is a violently rotating column of air that extends from the base of a cumulonimbus cloud and touches the ground. They move with phenomenal speeds, touching 480 kmph on land before dissipating. In Indian Ocean regions, a cyclone is also referred as a hurricane. In fact, typhoons, hurricanes and tropical cyclones are three different region specific names for the same kind of storm system (www.wiresmash.com/amazing/cyclone-vs-hurricane/-).

The average wind speed of most tornadoes ranges from 64–176 kmph, though some of the most powerful ones attain wind speeds in excess of 480 kmph! There are three different types of tornadoes, which are land spouts, multiple vortex tornadoes and waterspouts. Tornadoes mostly occur in USA in spring season as it is a time of transition in temperatures.

Hurricanes or Tornadoes are extremely powerful over the sea, causing tides and torrential rains around, but weaken and die out as they more over land, causing major damage in coastal areas (http://www.buzzle.com/articles/typhoon-vs-hurricane-vs- tornado.html). Both hurricanes and typhoons are vortex based systems. The central part or the eye of a typhoon could be as big as 370 kilometers! The hurricane season for the North Atlantic Ocean starts from 1st of June and continues up till November 30. Hurricane activity is most commonly observed in late summer, due to the striking difference between sea temperatures and seasonal basin patterns. September is considered the most active, season for hurricanes worldwide.

A storm surge is a rise of ocean water associated with tropical cyclones and strong extra-tropical cyclones (http://en.wikipedia.org/wiki/Storm_surge). Storm surges are caused primarily by high winds pushing on the ocean's surface. The wind causes the water to pile up higher than the ordinary sea level. Low pressure at the center of a weather system also has a small secondary effect, as can the bathymetry of the body of water. It is this combined effect of low pressure and persistent wind over a shallow water body which is the most common cause of storm surge flooding problems. Storm surge is measured as water height above predicted astronomical tide level. In areas where there is a significant difference between low tide and high tide, storm surges are particularly damaging when they occur at the time of a high tide. In these cases, this increases the difficulty of predicting the magnitude of a storm surge since it requires weather forecasts to be accurate to within a few hours. Factors that determine the surge heights for tropical cyclones include the speed, intensity, size of the radius of maximum winds (RMW), radius of the wind fields, angle of the cyclone track relative to the coastline, the physical characteristics of the coastline and the bathymetry of the water offshore.

Monitoring and mitigation mechanism

Warning systems are in place worldwide for predicting and monitoring the storm surges due to extreme weather events like cyclones, hurricanes, tornadoes or tsunamis. The National Hurricane Center in the US, for example, forecasts storm surge using the SLOSH model, which stands for Sea, Lake and Overland Surges from Hurricanes. SLOSH inputs include the central pressure of a tropical cyclone, storm size, the cyclone's forward motion, its track, and maximum sustained winds. Local topography, bay and river orientation, depth of the sea bottom, astronomical tides, as well as other physical features are taken into account, in a predefined grid referred to as a SLOSH basin.

Construction of dams and floodgates (storm surge barriers) is one way of reducing the impact of storm surge. They are open and allow free passage but close when the land is under threat of a storm. Creation of housing communities at the edges of wetlands with floating structures, restrained in position by vertical pylons is another way of protection. Such wetlands can then be used to accommodate runoff and surges without causing damage to the structures while also protecting conventional structures at somewhat higher low-lying elevations, provided that dikes prevent major surge intrusion (http://en.wikipedia.org/wiki/Storm_surge).

6.3.3.2 Recent events

6.3.3.2.1 *Orissa super cyclone, east coast of India, 1999*

A well-marked low pressure area that formed over Gulf of Thailand on 24 October 1999 intensified rapidly over the Andaman Sea by 26 October and further intensified into a very severe cyclonic storm by 27 October. Moving in a northwesterly direction, it intensified into a "super cyclonic storm" over a hundred kilometers southeast of the port-town of Paradip by 28th October. It crossed the coast at Paradip on 29 October 1999. The system was practically stationary for 30 h after landfall and a record rainfall of 530 mm was observed at Paradip on 30 October. The intensity of the storm increased from T4.5 to T7.0 from 28 to 29 October (i.e. 2.5 T/day), which is much higher than the usually recognized rapid rate of 1.5 T/day observed over the Atlantic Ocean 2. The lowest central pressure of 912 hPa observed in this super cyclone was the minimum so far for any tropical cyclone in the Bay of Bengal (Sadhuram, 2004).

The super cyclone in Orissa in October 29, 1999, is perhaps the most destructive natural calamity in India in this century with a wind speed of nearly 300 kms and incessant rains that lasted for about 48 hours with a total downpour between 447 mm and 995 mm (www.actionaidindia.org/emr_ori_cyclone.htm). The tidal waves from the sea reached nearly to 10 meters height and inundated up to 15 kms inland. There was massive damage to houses, vegetation, livelihood and the environment. Over 15 million people in the 12 districts were affected. Almost 20,000 persons were killed. The total estimated damages were INR 3968 crores (eq. ~USD 800 million) (Source: Government of Orissa).

6.3.3.2.2 *Hurricane Katrina, U.S. Atlantic coast, 2005*

Hurricane Katrina in the 2005 Atlantic hurricane season was the costliest natural disaster, as well as one of the five deadliest hurricanes, in the history of the United

States. Among recorded Atlantic hurricanes, it was the sixth strongest overall. At least 1,836 people died in the actual hurricane and in the subsequent floods.

Hurricane Katrina initially formed as a Tropical Depression over the southeastern Bahamas on August 23, 2005 as a result of an interaction of a tropical wave and the remains of Tropical Depression Ten (http://en.wikipedia.org/wiki/Hurricane_Katrina). The system was upgraded to tropical storm status on the morning of August 24 and was given the name Katrina. The tropical storm continued to move towards Florida, and became a hurricane only two hours before it made landfall between Hallandale Beach and Aventura on the morning of August 25. The storm weakened over land, but it regained hurricane status about one hour after entering the Gulf of Mexico. The storm rapidly intensified after entering the Gulf, growing from a Category 3 hurricane to a Category 5 hurricane in just nine hours. This rapid growth was due to the storm's movement over the "unusually warm" waters of the Loop Current, which increased wind speeds.

On Monday, August 29, Katrina made its second landfall as a Category 3 hurricane with winds of 125 mph (205 km/h) near Buras-Triumph, Louisiana. At landfall, hurricane-force winds extended outward 120 miles (190 km) from the center and the storm's central pressure was 920 mbar (27 in Hg). After moving over southeastern Louisiana and Breton Sound, it made its third landfall near the Louisiana/Mississippi border with 120 mph (195 km/h) sustained winds, still at Category 3 intensity. Katrina maintained strength well into Mississippi, finally losing hurricane strength more than 150 miles (240 km) inland near Meridian, Mississippi. It was downgraded to a tropical depression near Clarksville, Tennessee, but its remnants were last distinguishable in the eastern Great Lakes region on August 31, when it was absorbed by a frontal boundary. The resulting extra tropical storm moved rapidly to the northeast and affected eastern Canada (http://en.wikipedia.org/wiki/Hurricane_Katrina).

6.3.3.3 Global climate and sea level rise

Global sea level changes and the Earth's climate are closely linked. The Earth's climate has warmed about 1°C (1.8°F) during the last 100 years. The global climate has warmed up following the end of a recent cold period known as the "Little Ice Age" in the 19th century. As a result, sea level has been rising about 1 to 2 millimeters per year due to the reduction in volume of ice caps and mountain glaciers in addition to the thermal expansion of ocean water. If present trends continue, including an increase in global temperatures caused by increased greenhouse-gas emissions, many of the world's mountain glaciers will disappear. Most of the current global land ice mass is located in the Antarctic and Greenland ice sheets (Table 6.3.1). Complete melting of these ice sheets could lead to a sea-level rise of about 80 meters, whereas melting of all other glaciers could lead to a sea-level rise of only one-half meter.

6.3.3.3.1 *Glacial-interglacial cycles*

During cold-climate intervals, known as glacial epochs or ice ages, sea level falls because water is evaporated from the oceans and stored on the continents as large ice sheets and expanded ice caps, ice fields, and mountain glaciers. Global sea level was about 125 meters below today's sea level at the last glacial maximum about 20,000 years ago (Fairbanks, 1989). As the climate warmed, sea level rose because

Table 6.3.1 Estimated potential maximum sealevel rise from the total melting of present-day glaciers

Location	*Volume (cubic km)*	*Potential sea level rise (m)*
1. East Antarctic ice sheet	26,039,200	64.80
2. West Antarctic ice sheet	3,262,000	8.06
3. Antarctic peninsula	227,100	0.46
4. Greenland	2,620,000	6.55
5. All other ice caps, ice fields and glaciers	180,000	0.45
Total	32,328,300	80.32

Williams and Hall, 1993, http://pubs.usgs.gov/factsheet/fs50-98/

the melting North American, Eurasian, South American, Greenland, and Antarctic ice sheets returned their stored water to the world's oceans. During the warmest intervals, called interglacial epochs, sea level is at its highest. Today we are living in the most recent interglacial, an interval that started about 10,000 years ago and is called the Holocene Epoch by geologists (http://pubs.usgs.gov/fs/fs2-00/).

Sea levels during several previous interglacials were about 3 to as much as 20 meters higher than current sea level. The evidence comes from two different but complementary types of studies. One line of evidence is provided by old shoreline features (Please see section 6.3.3.4). Wave-cut terraces and beach deposits from regions as separate as the Caribbean and the North Slope of Alaska suggest higher sea levels during past interglacial times. A second line of evidence comes from sediments cored from below the existing Greenland and West Antarctic ice sheets. The fossils and chemical signals in the sediment cores indicate that both major ice sheets were greatly reduced from their current size or even completely melted one or more times in the recent geologic past. The precise timing and details of past sea-level history are still being debated, but there is clear evidence for past sea levels significantly higher than current sea level.

6.3.3.3.2 *Potential sea-level changes*

If Earth's climate continues to warm, then the volume of present-day ice sheets will decrease. Melting of the current Greenland ice sheet would result in a sea-level rise of about 6.5 meters; melting of the West Antarctic ice sheet would result in a sea-level rise of about 8 meters (Table 6.3.1). The West Antarctic ice sheet is especially vulnerable, because much of it is grounded below sea level. Small changes in global sea level or a rise in ocean temperatures could cause a breakup of the two buttressing ice shelves (Ronne/Filchner and Ross). The resulting surge of the West Antarctic ice sheet would lead to a rapid rise in global sea level (http://pubs.usgs.gov/fs/fs2-00/).

Reduction of the West Antarctic and Greenland ice sheets similar to past reductions would cause sea level to rise 10 or more meters. A sea-level rise of 10 meters would flood about 25 percent of the U.S. population, with the major impact being mostly on the people and infrastructures in the Gulf and East Coast States.

6.3.3.4 Sea level rise – a case study from east coast of India

Geophysical data (mainly bathymetry and high-resolution seismic records) over the continental shelf regions help in delineating the relict (or paleo) strandlines, which in

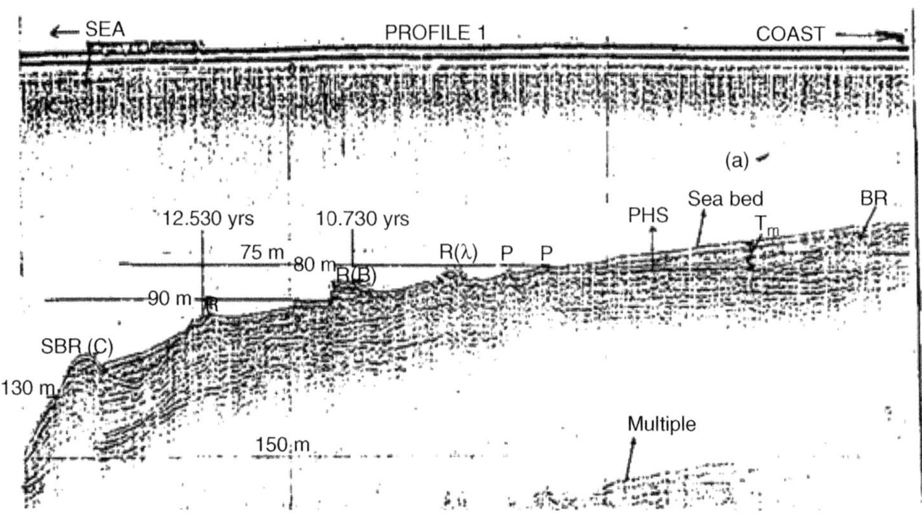

Figure 6.3.1 High resolution seismic reflection data over the continental shelf of Visakhapatnam, east coast of India indicating relict features associated with paleo-shore lines at 80, 90 and 130 m water depth (Murthy, 1989; Mohana Rao et al., 2000). The sediment samples collected at 80 and 90 m water depth were subjected to radio-carbon dating (ages indicated at the top of these features)

turn could give us a clue about the sea level history. Bathymetry and high resolution seismic data (Fig. 6.3.1) collected over the inner and outer shelf of Visakhapatnam (30 to 150 m water depth), east coast of India, revealed linear trends of relict strandlines between 30 to 130 m water depths. Sediment samples were collected over the relict features (hard rock outcrops) associated with these strandlines. Radio-carbon dating of these sediments helped in understanding the history of sea level changes during the period of late Quaternary (Mohana Rao et al., 2000). Based on these observations, a sea level curve for the period from the Last Glacial Maximum (LGM) to the present (18,000 years to the present) was drawn for the coast of Visakhapatnam (Fig. 6.3.2).

Another section of high resolution seismic data over the Visakhapatnam shelf, east coast of India (Fig. 6.3.3) revealed a faulted and rugged subsurface layer (marked as Lower Unconformity) approximately 100 m below the present shelf. Though sediment samples could not be collected over this buried layer, it was estimated that this unconformity might represent a paleo-shelf corresponding to 1.5 kya, based on the sedimentation rates in this part of the coast and the ages determined from the surface sediment samples (Murthy, 1989, Mohana Rao et al., 2000).

6.3.4 MAN-MADE HAZARDS

Both natural and man-made hazards cause significant changes in the physical, chemical and biological processes and have considerable impact on the coastal and offshore environment. In the previous section, we have discussed some of the natural hazards, their impact and the monitoring and mitigation mechanism. However, the more serious

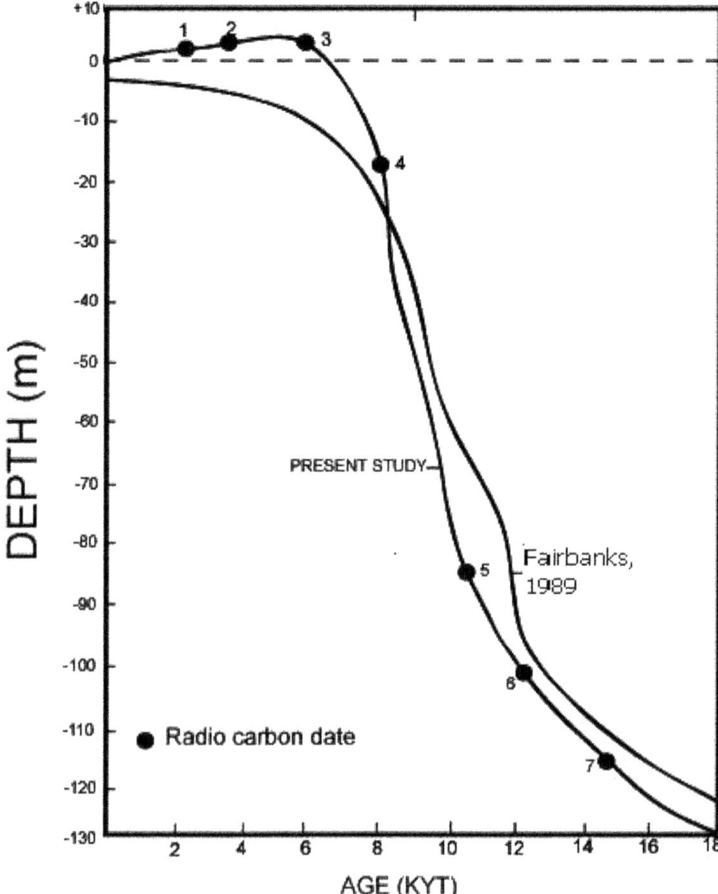

Figure 6.3.2 Tentative sea level curve (with thick circles) since Last Glacial Maximum (LGM) for the coast of Visakhapatnam, India (Mohana Rao et al., 2000). Continuous line is the global eustatic sea level curve derived by Fairbanks (1989) for LGM to present

concern at present is the impact of man-made hazards on the coastal and marine environment, which are responsible for much of the marine pollution and the consequent drastic changes in global climate.

Important man-made hazards include:

- Oil spills and blow-outs
- Pollution due to coastal activities such as, toxic releases from shore based industries, domestic sewage, ship breaking, dredging, aquaculture and other constructional activities
- Illegal exploitation of marine resources, particularly the fragile and endangered species like whales, turtles, coral reefs, etc.
- Marine pirates.

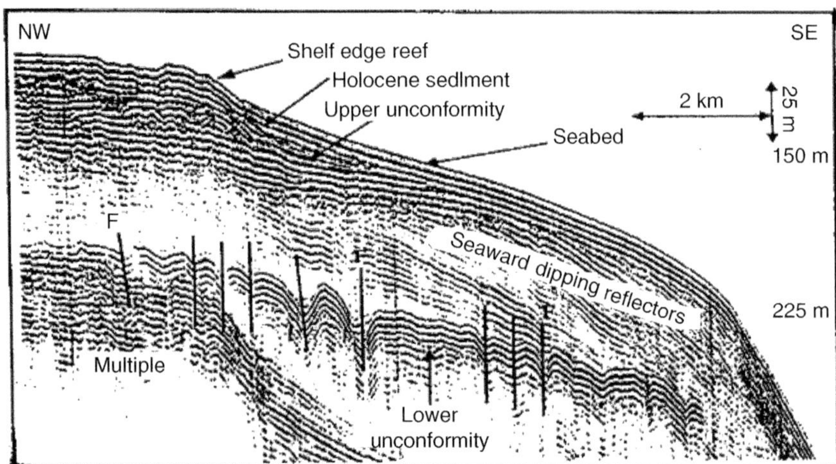

Figure 6.3.3 High resolution seismic record over the continental shelf of Visakhapatnam, east coast of India (Murthy, 1989; Mohana Rao et al., 2000). The faulted layer approximately 100 m below the present seabed indicates another paleo-shoreline with an estimated age of about 1.5 kya (Late Pleistocene)

6.3.4.1 Oil spills and blow-outs

Oil spills and blow-outs are some of the major hazards in the marine environment. The spills might be either due to leakages/blow-ups during an offshore oil drilling process, leakages from ship traffic or due to collisions of ships, leading to spillage.

6.3.4.1.1 *Effects on marine environment*

Once an oil or oily compound is discharged into water, it is spread on the surface by winds and currents, forming a thin layer. On the surface of seas in tropical or temperate zones, oils can be polymerized gradually by biodegradation and eventually form dense particles, which sink. Concentration of oily compounds in water is an important indicator of water quality, particularly in recreational water areas.

6.3.4.1.2 *Effects on human beings*

Polluted marine environment may result in contaminating fish and other living resources from the sea. Marine and coastal ecology including aquatic fauna and flora, a large number of species of bacteria, phytoplankton, zooplankton, benthonic organisms, coral, seaweed, shellfish, fish and other aquatic biota, terrestrial flora such as mangroves and wetlands will be affected. It will have a drastic effect on the living resource potential of the sea. Marine pollution is also considered as the major cause for Global Climate changes, which in turn affect human health and habitation.

6.3.4.1.3 *Effects on rivers, estuaries and channels*

The ocean process like currents, waves, tides, cyclones, storm surges have a great impact on fresh and salt water exchanges. Oil pollution in marine environment may spread to rivers, estuaries and other tidal creeks connected to sea. It might severely affect the river and estuarine environment. Particular damage may be caused to coastal habitat, and crops adjacent to such fresh water environment. Mangrove and other coastal ecosystems might be severely affected due to disasters like oil pollution. Drinking water in coastal environment might be contaminated.

6.3.4.1.4 *Mitigation measures*

Proper contingency plans and a prompt reporting system are the keys to prevention of oil blow-outs/oil spills. Appropriate regulations on ship discharges and provision of reception facilities are indispensable for proper control of oil pollution from ships. Detection of oil spills is also important for regulating ship discharges. To handle the accidental spills, recovery vessels, oil fences, and treatment chemicals should be prepared/deployed with a view to minimizing dispersal. Regular monitoring through Environmental Impact Assessment (EIA) studies is essential.

6.3.4.2 Oil spills – recent events

6.3.4.2.1 *BP oil disaster*

The Deepwater Horizon oil spill (also referred to as the BP oil spill, the Gulf of Mexico oil spill, the BP oil disaster, or the Macondo blowout) is an oil spill in the Gulf of Mexico which flowed for three months in 2010 (http://en.wikipedia.org/wiki/Deepwater_Horizon_oil_spill). It is the largest accidental marine oil spill in the history of the petroleum industry. The spill stemmed from a sea-floor oil gusher that resulted from the April 20, 2010, explosion of *Deepwater Horizon*, which drilled on the BP-operated Macondo Prospect. The explosion killed 11 men working on the platform and injured 17 others. On July 15, 2010, the leak was stopped by capping the gushing wellhead, after it had released about 4.9 million barrels (780,000 m^3) of crude oil. An estimated 53,000 barrels per day (8,400 m^3/d) escaped from the well just before it was capped. It is believed that the daily flow rate diminished over time, starting at about 62,000 barrels per day (9,900 m^3/d) and decreasing as the reservoir of hydrocarbons feeding the gusher was gradually depleted. On September 19, 2010, the relief well process was successfully completed, and the federal government declared the well "effectively dead".

The spill caused extensive damage to marine and wildlife habitats and to the Gulf's fishing and tourism industries. In late November 2010, 4,200 square miles (11,000 km^2) of the Gulf were re-closed to shrimping after tar balls were found in shrimpers' nets. The amount of Louisiana shoreline affected by oil grew from 287 miles (462 km) in July to 320 miles (510 km) in late November 2010. In January 2011, an oil spill commissioner reported that tar balls continue to wash up, oil sheen trails are seen in the wake of fishing boats, wetlands marsh grass remains fouled and dying, and that crude oil lies offshore in deep water and in fine silts and sands onshore. A research team found oil on the bottom of the seafloor in late February 2011 that did not seem to be degrading. Skimmer ships, floating containment booms, anchored barriers, sand-filled barricades along shorelines, and dispersants were used in an attempt to protect hundreds of kilometres of beaches, wetlands, and estuaries from the spreading oil. Scientists have also reported immense underwater plumes of dissolved oil not visible at the surface as well as an 80-square-mile (210 km^2) "kill zone" surrounding the blown well. By mid-June, 2011, there was a total of 9,474 days of oil-related closings, advisories and notices on Gulf coast beaches. Clean-up crews are still working to remove oil that continues to wash ashore in Alabama, Louisiana, Florida and Mississippi. Four beaches in Louisiana remain closed.

6.3.4.2.2 *The Exxon Valdez oil spill*

This oil spill occurred in Prince William Sound, Alaska, on March 24, 1989, when the Exxon Valdez, an oil tanker bound for Long Beach, California, struck Prince William Sound's Bligh Reef and spilled 260,000 to 750,000 barrels (41,000 to 119,000 m^3) of crude oil (http://en.wikipedia.org/wiki/Exxon_Valdez_oil_spill). It is considered to be one of the most devastating human-caused environmental disasters. The Valdez spill was the largest ever in U.S. waters until the 2010 Deepwater Horizon oil spill, in terms of volume released. However, Prince William Sound's remote location, accessible only by helicopter, plane, and boat, made government and industry response efforts difficult and severely taxed existing plans for response. The region is a habitat for salmon, sea otters, seals and seabirds. The oil, originally extracted at the Prudhoe Bay oil field, eventually covered 1,300 miles (2,100 km) of coastline, and 11,000 square miles (28,000 km^2) of ocean.

Beginning three days after the vessel grounded, a storm pushed large quantities of fresh oil on to the rocky shores of many of the beaches in the Knight Island chain. The ship was carrying approximately 55 million US gallons (210,000 m^3) of oil and it was estimated that 11 to 32 million US gallons (42,000 to 120,000 m^3) of oil were spilled into the Prince William Sound.

6.3.4.3 Chemical and liquefied gas leakages

The major damage due to chemical and liquefied gas leakages is for the marine environment, if it happens at mid sea and if the leakage is from the source to the sea.

6.3.4.3.1 *Effects on marine environment*

This might have severe impact on the global environment. The marine habitat like fauna and flora will be severely affected. It might lead to bottom sediment contamination. Since oceans are interconnected, large chemical and gas leakages will have adverse effect on the global environment and climate change.

6.3.4.3.2 *Effects on human beings*

If chemical and gas leakage happens during the loading and unloading operations, it might have immediate impact on human health and safety. It will not only affect the crew handling the operations, but also the habitat in the neighborhood. Large scale chemical and gas leakages may lead to an epidemic quickly spreading over a vast area, like for example the Bhopal Gas episode on land.

6.3.4.3.3 *Effects on rivers, estuaries and channels*

Rivers, estuaries, tidal creeks will be only affected if they are connected to either ship's traffic or loading and unloading operations. As already mentioned earlier, since the coastal ecology is connected to the sea, any disaster caused at sea is quite likely to spread to coastal environment due to ocean and coastal processes like waves, tides and currents.

6.3.4.3.4 *Mitigation measures*

Risk Analysis studies must be undertaken for handling hazardous cargo. Risk Analysis is the process of identifying the probability of occurrence of an accident and its consequence. Risk Analysis involves identification of hazards and the associated risks, if any, involved in these operations. Recognition of all the possible hazards and analysis of the associated risks is an important first step to improve the safety. Risk analysis is a tool to determine the consequence of operational failures (e.g. failure of pipeline carrying hazardous liquid, oil spill, fire, etc.). It enables ship's crew to determine the action that needs to be taken to improve safety of operations and deal with the probable effects of an incident in the area.

6.3.4.4 Marine pollution due to land based activities

Oil spills, blow-outs, accidents at sea and other such natural hazards contribute only a fraction of the total marine pollution. Bulk of marine pollution comes from land and coast-based activities. The coastal and near shore areas are favorable locations for a variety of industries like ports and harbors, thermal and nuclear power plants, pharmaceutical (bulk-drug) units, oil refineries, etc. both for navigational advantage and for easy disposal of treated effluents. These activities involve layout of submarine pipelines, construction of off shore break-waters/jetties/platforms/drilling rigs and considerable amount of dredging of the coastal/inner shelf areas. These developments also result in the release of industrial/domestic effluents into the sea. The fragile marine environment therefore becomes vulnerable to:

- Changes in coastal morphology (mainly coastal erosion)
- Damage to the coastal and marine vegetation and habitat
- Degradation in biodiversity due to over exploitation

Continuous monitoring of coastal and marine environment before, during and after any developmental activity is therefore very essential. These studies involve measurement of data on physical (currents, waves, tides), chemical (water quality, suspended sediments), biological (coastal and marine biota), geological and geophysical (sea bed topography, sub-surface sediments, bottom sediment contamination), processes. Site specific data on Low Tide Level (LTL), High Tide Level (HTL) along with the Coastal Regulation Zone (CRZ) guidelines are also essential to determine the suitability of the site for any constructional activities in the coastal zone. Sustainable development of marine environment requires strict follow-up of environmental guidelines stipulated by concerned authorities.

6.3.4.5 Environmental Impact Assessment (EIA)

Environmental Impact Assessment (EIA) is a planning tool generally accepted as an integral component in sound decision making. EIA is to give the environment its due place in the decision-making process by clearly evaluating the environmental consequences of the proposed activity before action is taken. Early identification and characterization of critical environmental impacts allow the public and the government to form a view about the environmental acceptability of a proposed developmental project and what conditions should apply to mitigate or reduce those risks and impacts.

ACKNOWLEDGEMENTS

The author is grateful to Dr. S.R. Shetye, Director, N.I.O., Goa for permission to publish this work. He is also grateful to C.S.I.R., New Delhi, for the award of the CSIR Emeritus Scheme under which this work was carried out. This is N.I.O., contribution No. 5095.

REFERENCES

Fairbanks, R.G., 1989. A 17,000-year glacio-eustatic sea level record; influence of glacial melting rates on the Younger Dryas event and deep-ocean circulation: *Nature*, v. 342, no. 6250, pp. 637–642.

Kious, W.J. and Tilling R.I. 1996. *This Dynamic Earth, The Story of Plate Tectonics*. 108 pp. [USGS Online Edition,US Government Printing Office, Washington DC]

Mead G.A. 2005. Ocean basins and continental margins. [Lecture Notes, University of Florida]

Mohana Rao, Reddy, N.P.C., Prem Kumar, M.K., Raju, Y.S.N., Venkateswarlu, K. and Murthy, K.S.R. (2000). Signatures of Late Quaternary Sea-level Changes and Neo-tectonic activity over Visakhapatnam-Gopalpur Shelf, East Coast of India, *Proc. Int.Quat. Seminar of INQUA Shoreline, Indian Ocean Sub-Commission*, pp. 116–124.

Morelock, J., Ramirez, W., Balsam, W. and Hubbard, D. 2006. *Continental margins, marine geology*. Details available at www.geology.uprm.edu/Morelock/marg.htm

Murthy, K.S.R. 1989. Seismic stratigraphy of Ongole-Paradip continental shelf, east coast of India, *Ind. Jour. Earth Sciences*, v16, pp. 47–58.

Murthy, K.S.R., Subrahmanyam, A.S. and Subrahmanyam, V. 2011. *Tectonics of the Eastern Continental Margin of India*. New Delhi: The Energy and Resources Institute (TERI), New Delhi, (in Press), pp. 200.

Sadhuram, Y., 2004. Record decrease of sea surface temperature following the passage of a super cyclone over the Bay of Bengal, *Current Science*, v. 86, NO. 3, 383–384.

Williams, R.S., and Hall, D.K., 1993. Glaciers, in Chapter on the cryo-sphere, in Gurney, R.J., Foster, J.L., and Parkinson, C.L., eds., *Atlas of Earth observations related to global change*: Cambridge, U.K., Cambridge University Press, pp. 401–422.

Chapter 6.4

Nuclear energy accidents*

There have been three major nuclear energy accidents till date: The Three-Mile Island accident USA (1979), and Chernobyl Reactor Accident, Ukraine (1986) were caused by technical failure, while the Fukushima – Daiichi (Japan) nuclear accident on Mar. 11, 2011, was triggered by earthquake and tsunami. The severity of nuclear accident is rated 0 to 7 on the International Nuclear Event Scale (INES). The Three-mile Island has been rated at 5, and Chernobyl and Fukushima are rated at 7, the highest (in the early stages of Fukushima accident, it was rated 4, then 6, and finally 7).

6.4.1 THE THREE MILE ISLAND (TMI) ACCIDENT

The TML accident began on Mar. 28, 1979 (Wednesday) in Dauphin County, near Harrisburg, Pennsylvania, USA, when TMI-2, one of the two pressurized water reactors, suffered a partial meltdown. The accident resulted in the release of 43 kCi of radioactive krypton (1.59 PBq) and less than 10 Ci (740 GBq) of ^{131}Iodine. About 25,000 people who lived within 8 km. radius of TMI, were alerted.

Though no one was killed or injured because of the TMI accident, the consequences of the accident were catastrophic for the nuclear power industry in USA. The popular feeling against nuclear power was so intense that there was virtual cessation of reactor construction for decades.

*Particulars cited in this chapter have been drawn from the author's own works. "*Geo-environment: An Introduction*" (1995), "*Energy Portfolios*" (2009), Wikipedia and press reports.

6.4.2 CHERNOBYL REACTOR ACCIDENT

On April 26, 1986 (Saturday), there was an explosion in one of the four, graphite-moderated, water-cooled nuclear reactors at Chernobyl, 128 km from Kiev in Ukraine (former Soviet Union). Chernobyl is on the banks of the Pripyat River which flows into the Dnieper. The region is relatively flat with gentle slopes. It is the most serious nuclear accident in the world to date, and gave a foretaste of the horrors of nuclear war. Basically, Chernobyl was the result of steam explosion. Fuel heated up very rapidly, and this caused all the coolant water to vaporize instantly, causing an explosion. Some fuel did melt, but that does not account for the magnitude of the disaster.

About 50 MCi (million curies) or 2×10^{18} Bq of radioactive fission products and noble gases (including ^{123}Xenon) which correspond to about 5% of the total fission products inventory of the reactor core, escaped. About half of this amount relates to radionuclides which figure in the food chain. Radioactivity got released in two distinct plumes – one on the first day (12 MCi), and another on the seventh day. Containment was effected by the eleventh day, and the emissions ceased after that. Meteorological conditions determined the pattern of dispersion of the Chernobyl radioactive cloud.

The Chernobyl radioactive plume rose to a height of 3 km before it started spreading horizontally. About 50% of the emission of condensable products fell in an area of about 60 km radius around the accident site. The rest of the emissions fell in an area of 10 million km^2 in Europe. The fallout was controlled by rainfall, other things being equal.

It may be noted that the meteorological conditions rather than the distance from the site of the accident, determined the extent of fallout. This explains as to why Sweden received >50 times more deposition than the neighbouring Denmark.

People were exposed to Chernobyl radiation in four ways: (1) Exposure to external radiation from the radioactive isotopes in the cloud as it passed overhead, (2) Exposure to internal radiation from inhaled radioactive isotopes, (3) Exposure to external radiation from the radioactive material deposited on the ground during the transit of the cloud, and (4) Exposure to radiation from radioactive isotopes ingested in food (Barnaby, 1986). The radioactive isotopes in the Chernobyl cloud which are of the greatest importance to human health are the isotopes of iodine and caesium (Table 3.5.7) which got widely deposited in Europe.

6.4.2.1 ^{131}Iodine and caesium isotopes

About 7.3 MCi of ^{131}I was released during the accident. ^{131}I has a half-life of 8 days. It gets concentrated in the milk of the cows when they ingest contaminated grass. When people drink the milk of such cows, ^{131}I gets concentrated in the thyroid, and could cause thyroid cancer. Two isotopes of caesium are involved. ^{134}Cs with a half-life of 2.05 years, and ^{137}Cs with a half-life if 30 years. About 0.5 MCi of ^{134}Cs and 1.0 MCi of ^{137}Cs were released in the accident. Caesium ingested by animals from the vegetation ends up in meat. Vegetables, fruits and dairy products are also contaminated by radioactive caesium. It has been reported that when bentonite clay is included in the feed of the sheep, their meat remained uncontaminated, possibly because radioactive caesium got bound by bentonite and excreted.

Korobova et al. (1998) studied the mobility of ^{137}Cs and ^{90}Sr in soils, and their transfer in the soil-plant system in the Novozybbkov District affected by the Chernobyl accident. They found a sharp contrast in the behaviour of ^{137}Cs which is strongly fixed in the soil (to the extent of 40–93%), whereas 70–90% of ^{90}Sr is present in water-soluble, exchangeable, and weak-acid-soluble forms. Vertical migration of radionuclides, which is detected to a depth of 30–40 cm, is most pronounced in local depressions with organic and gley soils. In woodlands, most of the ^{137}Cs is fixed in the plant litter and upper mineral soil layer. In the case of floodplain grasslands, radionuclides are associated with soils having fine texture. The uptake of radionuclides by plants decreases in the following order: legumes > herbs > grasses. A high accumulation of ^{137}Cs in potato tubers grown in sandy, podzolic soils, has been noted. Korobova et al. (1998) recommend that people living in the areas within the zone of contamination exceeding 15 Ci km^{-2}, should avoid eating local forest products, and cattle should not be allowed to graze on wet floodplain meadows.

The Chernobyl fall-out contaminated milk and meat products all over Europe. Extensive surveillance networks were set up to monitor the radioactivity of fresh milk, lamb meat, etc. For instance, the Government of U.K. prescribed the following Derived Emergency Reference Levels (DERL's) for milk: 2,000 Bq l^{-1} for ^{131}I, 2,000 Bq l^{-1} for ^{137}Cs, and 3,100 Bq l^{-1} for ^{134}Cs, totaling 7,100 Bq l^{-1} of milk. Some countries have adopted more stringent measures: any food-stuff with >1,000 Bq l^{-1} of radioactivity was declared unacceptable for human consumption.

It has been estimated that about 75 million people in erstwhile USSR have been exposed to Chernobyl radiation, with the "average" individual receiving about 33 mSv (millisieverts) or 33 REM over his life time. The range for the EEC countries was about 8 mSv. There was, however, considerable variation in the radiation exposure among the European countries, depending upon the pattern of movement of Chernobyl cloud. For instance, the average effective dose in Italy was eight times more than that in the UK. This is so because about half of the total radiation was received by the Italians, about a quarter by the Greeks, and about 20% by West Germans, among the EEC countries. Poland and Austria were exposed to dosages closer to 8 mSv.

The claims of deaths of tens of thousands of people in the Chernobyl accident are highly exaggerated. One estimate puts the number of fatalities as 210. According to Chernobyl Forum (2006), 28 emergency workers died due to exposure to very high radiation dose, 15 patients died from thyroid cancer, and there may be 4000 cancer deaths from among 600,000 people who received high dose of radiation. It has been estimated that there might be 10–15 million cancer death in the affected population in the erstwhile Soviet Union during the next 70 years. Estimates of the total number of cancers producible in Europe (excluding erstwhile Soviet Union) due to Chernobyl accident vary from 4,600 to 76,000. This large range in the estimate arises from the fact that the projected number of cancer deaths attributable to 1,000 man-Sv has been estimated by various authorities to have a range of 24 to 360. The carcinogenic effects of exposure to large doses of radiation are poorly known for the simple reason that, barring the survivors of Hiroshima and Nagasaki, there is no large group of people exposed to such high radiation doses.

The Chernobyl accident had serious effects on biota as well. For instance, it led to enhanced concentrations of radionuclides in the air above Sweden, which were then washed into lakes and streams in Sweden. Petersen et al. (1986) reported that as of August, 1986, increased radioactivity was found at all trophic levels of the fresh water ecosystem, from algae to top carnivore. Aquatic biota was exposed to 250–1,000 Bq l^{-1} of caesium isotopes, i.e. 25 times higher than normal and was still rising.

Figure 6.4.1 Image of the massive (~15 m) tsunami wave engulfing Fukushima-Daiichi nuclear power plant in Japan, on March 11, 2011 at 15.27 hrs JST: Courtesy: TEPCO

6.4.3 FUKUSHIMA – DAIICHI REACTOR ACCIDENT

The magnitude 9.0 Mw, Tohoku-oki earthquake and accompanying tsunami hit Japan on Mar. 11, 2011, with 23,769 people reported dead or missing and displacing 125,000 people. Among the numerous human interest stories of stoicism, discipline and kindness of the Japanese that this disaster brought to light, the following episode stands out as a monument to the eternal glory of mother's love to her child. Volunteers searching the earthquake debris dug out the body of a woman and found her dead. As they were leaving, somebody noticed that the woman's body was hugging a cloth bundle. The volunteers found a three-month old child in the bundle. A doctor was summoned, and he found the child to be alive and healthy with no scratches. Evidently, the mother protected the child with her body. A cellphone which was found near the child, carried the following, deeply touching message of the mother: "*Dear child, you will live. All your life you will remember how much I loved you*".

How the earthquake and tsunami wave damaged Fukushima reactor system

The Fukushima nuclear disaster involved a series of equipment failures, nuclear meltdowns and release of radioactive material. It is more complex than the Chernobyl disaster as multiple reactors and spent fuel pools are involved (later studies show that the spent fuel pools were unaffected).

The reactor system at Fukushima (coordinates: 37°25′17″N; 141°1′57″E) in eastern Japan coast has the combined power of 4.7 gigawatts. It is one of the largest nuclear power stations in the world. It consists of six, light water, boiling water reactors (BWRs). It was designed by GE and constructed and run by Tokyo Electric Power Company (TEPCO). Except for unit 3, which has uranium oxide/MOX (Mixed Oxide) fuel, the other five reactors had uranium oxide fuel.

The 9.0 magnitude Tohoku earthquake with its epicentre near the island of Honshu, hit Fukushima at **1446** hrs JST on Mar. 11, 2011 (Friday). Its ground acceleration (about 5.50 m/s^2) was above the design parameter (4.52 m/s^2). When the earthquake occurred, reactors in units 1, 2 and 3 were operating, but the reactors in the units 4, 5 and 6 were in cold shut down for periodic inspection (unit 4 fuel has been removed from the core vessel). Units 1, 2 and 3 underwent automatic shutdown (scram) when the earthquake occurred. The most important damage that the earthquake inflicted on Fukushima is the destruction of power supply to the reactors. Due to this total station blackout (due to failure of power supply within and without), core cooling system had to be maintained by diesel generators. Emergency generators started up to run the control electronics and water pumps needed to cool reactors.

When the 14–15 m tsunami wave triggered by the earthquake engulfed Fukushima at 1527 hrs, i.e. 41 mins after the earthquake, it topped the 5.7 m sea wall, and flooded the basement of the Turbine buildings, thereby disabling the emergency diesel generators located there (except for one generator in Unit 6). The tsunami destroyed the power lines and the connection to the electrical grid was gone. It was no longer possible to cool the reactors, which in consequence started to overheat. The flooding due to the tsunami wave hindered external assistance.

Under the circumstances, reactors 1, 2 and 3 experienced full **meltdown**. On Day 2 (i.e. 12 March) at 15.10 hrs, the upper 75% of the core of unit 1 appears to have melted and slumped in the lower part of the core. When the supply of 80,000 litres of fresh water ran out at 14.53 hrs on 12 Mar., seawater was injected into the system for purposes of cooling at 20.20 hrs JST, on the orders of the Japanese Government.

There was confusion in the communication between the Government and TEPCO about the injection of seawater. Seawater injection continued till 25 Mar.

Due to high pressures inside the reactors, the water injection into the reactors failed. Moreover the reactor pressure became very high, posing a risk of reactor breach. To bring pressure down, steam venting was resorted to.

Cooling is essential for the operation of the reactors. Zircaloy (zirconium alloy) is used to fabricate the internal components and fuel assembly of the reactors (for instance, 32.7 t of zirconium was used in unit 1). It is inert at the normal operating temperature of approximately 300°C. But when heated to above 500°C, as has happened in Fukushima due to failure of cooling, and in the presence of steam, zircaloy undergoes an exothermic reaction, and produces hydrogen.

The accumulation and explosion of hydrogen had distrous consequences. Hydrogen explosions destroyed the upper cladding of the buildings housing Reactors 1, 3, and 4, and damaged the secondary containment structure of Reactor 2. As Unit 4 was not having any fuel in the core, the explosion is assumed to have been caused due to the flowing in of hydrogen discharged by venting at unit 3. Fires broke out.

TEPCO injected nitrogen into the containment vessel to reduce the likelihood of further hydrogen explosions.

When the remnants of its reactor core fell to the bottom of its damaged reactor vessel, Unit 1 and probably other melted-down reactors in the complex, continued to leak cooling water months after the initial events. Flooding of the basement by radioactive water prevented access to the basement where repairs were needed.

Nuclear fuel rods that have reached cold shutdown temperatures, are customarily cooled for several years in a spent fuel pool. This is necessary to reduce the production of decay heat to a level when they can be safely transferred to dry storage casks. Such fuel rods which are stored in pools in each reactor building, began to overheat due to loss of cooling as circulation was interrupted. Reactors 1 to 4 remained inoperable because of flooding, fires and explosions. On 17 March, one generator in Unit 6 was started, and grid power was restored to the plant on March 20. On May 5, workers entered the reactor building for the first time since the accident on Mar. 11.

Boron can effectively absorb neutrons. On 25 June and the following day boric acid dissolved in 90 tons of water was pumped into the spent fuel pool of Reactor 3 to control the neutron flux. There were complications. TEPCO found that the water in the pool was strongly alkaline (pH of 11.2), possibly because of the calcium hydroxide leached from the concrete debris from the March hydrogen explosion of the reactor building. The alkaline water could corrode the aluminium racks holding the spent fuel rods. If the fuel assemblies would fall, the system may become critical once again. Traces of xenon found in some of the plants on Nov. 4 has been attributed to spontaneous fission, not to re-criticality.

Dispersal of radiation from Fukushima

The maximum radiation detected at Fukushima (800 mSv) is less than 0.4% of that released at Chernobyl (200,000 mSv).

As a consequence of change in the wind direction, the radioactive plume from Fukushima that initially moved towards the ocean, swerved towards NW towards land. It was too weak to have an impact in California, USA. Pressure venting of the PCVs caused the maximum discharge of radioactivity.

The level of radioactive iodine in the sea off Fukushima went up steadily and reached 4,385 times the legal limit. This evidently was caused by radiation leaking continuously into the sea. Ships proceeding to Yokohoma were alerted about it.

Radioactive iodine (I-131) has a short half-life of 8 days, and the radioactity due to this isotope reverts to normal quickly. IAEA monitoring unit found that soil radioactivity in Iitate, 40 km NE of Fukushima (i.e. outside the 20 km Exclusion zone) was twice the IAEA prescribed limit. Such radioactivity in the soil may have long-term health effects.

Health effects

Nobody died in the Fukushima accident. There were 37 cases of physical injuries and a few cases of radiation burns.Two workers received 2,000 to 6,000 mSv of radiation at their ankles because of standing in radioactive water in unit 3 on Mar. 25, but they were not life-threatening.

Fearing radioactivity releases, the Government evacuated all the citizens in a 20 km radius around Fukushima.

In areas of northern Japan 30–50 km from the plant, levels of Cesium-134 and 137 and Iodine-131 in food were higher than the threshold of 500 bql/kg (a mushroom from Shiitake recorded 8,850 bql/kg). Food grown in the area was banned from sale. The public was advised not to use tap water to prepare food for infants. Low levels of plutonium contamination detected in the soil at two sites in the plant, were attributed to previous atmospheric nuclear weapns tests.

The Japanese Government announced that the plant will be decommissioned once the crisis is over.

In April 2011, the US Department of Energy, gave an assessment of the radiation risk.

The radiation exposure may be 20 mSv/yr (2 rems/yr) in some areas upto 50 km from the plant. There may be incidence of one cancer case in 500 young adults, and the incidence may be higher (4 out of 100) in some pockets.

As a part of the massive decontamination effort, experts from IAEA are cooperating with the Japanese specialists to test for thyroid abnormalities in about 360,000 persons who are under the age of 18 when the Fukushima reactor system got damaged. No one has died in the reactor accident, and the plant is now relatively stable. But still very few are willing to go back to their own homes in the 30 km zone near Fukushima. If any thyroid abnormalities are detected, children will be tested every two years until they reach 20 yrs age.

Global reaction

The Fukushima disaster dealt a great blow to the nuclear industry globally. TEPCO did not exactly cover itself with glory by its handling of the accident. TEPCO has been found to be secretive in communicating to the public about the happenings in the reactor system, and also not acting on the suggestions made by GE, IAEA and others to improve the design to be able to withstand damage by more powerful earthquake and tsunami. In Japan itself, there is much uneasiness about continued dependence on nuclear power, Germany took a policy decision to phase out nuclear power. India which planned to buy from France six units of 1,650 MWe EPRs for Euros 40 billion, to be established in Jaitapur, western India, has postponed the decision, pending post-Fukushima certification. Even in France, where nuclear power is a holy cow, there is a move to reduce the dependence on nuclear power from the current 75% to 50% in due course. China and several other countries are waiting for the results of the ongoing Fukushima investigations to modify the design of the reactors.

A case for continuance of nuclear power

With better technology and safer reactor design, it should be possible for nuclear power to play an important role in energy generation globally. That technological advances during the last 25 years have a profound impact in improving the safety of reactors should be evident from the following comparative study of Chernobyl versus Fukushima.

Description	*Chernobyl*	*Fukushima*
Electricity output	1 GWe	4.7 GWe
Amount of fuel in the reactors	180 t	1600 t
No. of reactors involved the accident	1	4
Maximum radiation detected	200,000 mSv	800 mSv
Direct deaths due to accident	57	0

Source: Kalam & Singh, The Hindu, Nov. 6, 2011

Lessons from Fukushima (Pers. Comm. from Dr. T.Harikrishnan of IAEA, Vienna)

1. Emergency generators should be installed at high elevations or in water-tight chambers,
2. If a cooling system is intended to operate without power, make sure that all of its parts can be manipulated without power,
3. Keep power trucks on or very close to the power plant,
4. Install independent and secure battery systems to power crucial instruments during emergencies,
5. Ensure that catalytic hydrogen recombiners (power-free devices that turn dangerous hydrogen back into steam) are positioned at the tops of the reactor buildings where the gas would most likely collect,
6. Install power-free filters on vent lines to remove radioactive materials and allow for venting that would not harm nearby residents.

Chapter 6.5

Integrated disaster preparedness

6.5.1 DUAL USE TECHNOLOGIES AND PRACTICES

Resiliency has been defined as the "capability of a system to maintain its function and structure in the face of internal and external danger ...". Allenby and Fisk (2005) recommend the development and implementation of dual-use technologies that provide substantial economic benefits in addition to resiliency even if no negative events occur. The developed countries are already highly urbanized, and the developing countries in Asia, Africa and Latin America are getting rapidly urbanized. It has been estimated that by 2030, 60% of the world's population will live in cities. Consequently, information-dense urban structures are coming into vogue. Instead of rigid systems (e.g. land phone), more fluid and responsive, network-centric organizational patterns are emerging (e.g. Internet phones). A network-centric society is more equitable (e.g. employment of handicapped persons, housewives), more productive (less commuting) and less fragile. The urban systems have necessitated the development of tools that aggregate and display complex urban systems data. The availability of such a system can serve as a training tool for disaster management, besides being useful in the coordination of disaster management (Allenby and Fisk, 2005).

A good example of dual-use technology is "Mission 2007: Every Village a Knowledge Centre", launched by the Government of India. The Mission consists of an ambitious plan for establishing Information and Communication Technology based Village Knowledge Centers in each of the 100,000 villages in India. At least two persons, one woman and one man, will be trained to run each of these centers, which will serve a variety of developmental purposes, such as, health, drinking water, agriculture, hazards, guaranteed employment for 100 days for one member of the families which are below the poverty line (one USD per day), etc. The broad-band connectivity of the Village Knowledge Centre allows these centres to be used effectively in all hazard preparedness, public education, public warning, etc. campaigns.

6.5.2 RESILIENCY LINKED TO SOCIAL-ECOLOGICAL SYSTEMS

Resilience to coastal disasters is related to the capacity of linked social–ecological systems to absorb recurrent disturbances, such as hurricanes, tidal waves, tsunami, etc. so as to retain essential structures, processes and feed-backs. Just as a healthy man is less vulnerable to disease, and more capable of recuperation after he is affected by a disease, the healthier an ecosystem is the more capable it is of regenerating itself and continuing to deliver resources and ecosystem services that are essential for human livelihoods and societal development (Adger et al., 2005).

In coastal Asia, the resilience of the ecosystems have been degraded by activities, such as, deforestation of mangroves for intensive shrimp farming, overfishing, coral mining, land clearing, etc. Consequently, the adverse consequences of the tsunami got accentuated in some places in Sri Lanka and Tamil Nadu coast. Similar considerations hold good for social systems. Thanks to the inherited knowledge of tsunamis, and institutional preparedness for disasters, fishing communities in Simeulue Island, west of Sumatra, close to the epicenter of the December 26, earthquake, survived the tsunami better than the fishing communities (say) in theTamil Nadu coast which have no such knowledge or tradition. The Cayman Islands in the Carribean which suffered three devastating hurricanes during 1988–2000, has put in place specific rules and governance of hurricane risk, dedicated organizations, establishment of early warning systems, promotion of self-mobilization in civil society and private companies, public infrastructure, coral reef management, etc. This effort paid off when the Cayman Islands was hit by Hurricane Evan in 2004 (Adger et al., 2005).

Mangroves of Pichavaram along the Tamil Nadu coast saved the lives and property of the people living in seven hamlets. Some mangroves on the coast were uprooted, but beyond that, there was hardly any damage. Apparently, the velocity of the tsunami wave got drastically reduced due to friction with mangrove forest, and the tsunami water got dispersed into the creeks and canals that crisscross the mangrove forest. A fisherman said, "we saved the mangroves, and the mangroves saved us" (MSSRF note).

Similar observation has been made in Sri Lanka about the protection afforded by coastal ecosystems, such as mangroves and corals, against the tsunami.

6.5.3 RISK MANAGEMENT THROUGH SECURITIZATION

World Bank studies have shown that natural catastrophes have a strong adverse impact on the economic development of the low-income countries, and can offset poverty reduction activities. Raising taxes by a government is never popular, and people who will never be affected by a tsunami, will be totally unwilling to pay taxes to provide tsunami protection. The main purpose of the tsunami risk management is to minimize disaster losses in hazard-prone areas. These management alternatives may be structural (such as, seawalls, groins, bioshields) and nonstructural (such as, Catastrophe Bonds or cat bonds). Such actions are taken before or in advance of the event

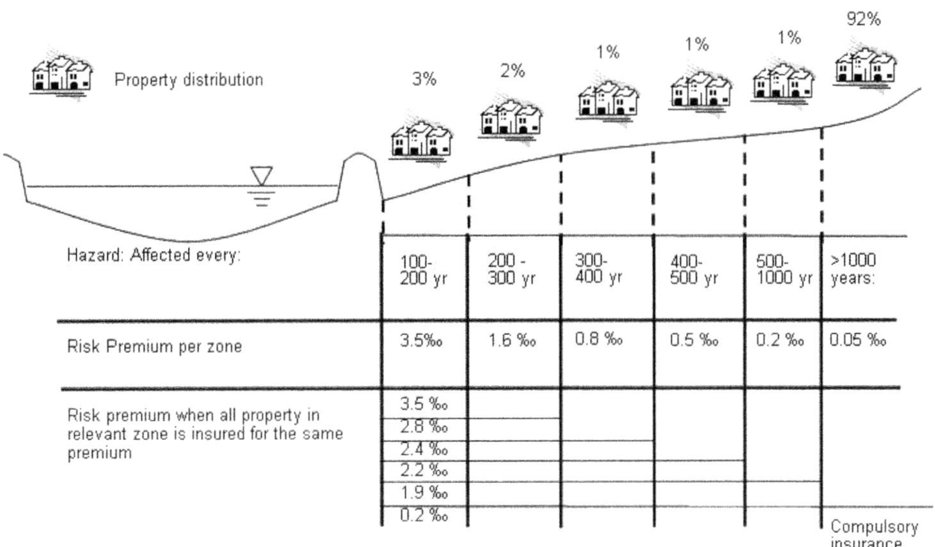

Figure 6.5.1 Risk-adequate premiums as a function of the size of the risk community
Source: Menzinger & Brauner, 2002

(*ex ante*), and are, therefore, proactive measures. A comprehensive risk management program is usually a combination of structural, nonstructural and reactive actions. Reactive actions are those taken once the event has occurred, such as evacuation and relief.

Pizarro and Lall (2005) gave a detailed account of the use of innovative insurance systems, such as cat bonds, for unpredictable, disastrous events. Figure 6.5.1 shows how the risk premium amounts are determined depending upon the function of the size of the risk community. Stipple (1998) gave a lucid account of the methodology of securitization, through the use of catastrophe or cat bonds. The institutional structure for a catastrophe bond is shown in Figure 6.5.2. To date, cat bonds have been issued for earthquakes in California and Japan, hurricanes in the coast of the USA, typhoons in Japan, and windstorms in Europe. Presently, the volume of cat bonds is about USD 6 billion. Mills (2005) proposed an insurance mechanism, based on the globalization of risk, to cover the disasters arising from climate change. A similar system may be developed to take care of the any kind of risk.

Insurance Linked Securities (ILS) are the most common form of cat bonds. A "Special Purpose Financing Vehicle" (SPFV), or "Special Purpose Reinsurance Vehicle" (SPRV) is created for the purpose. There are two kinds of bonds, namely, "principal at risk" or "interest at risk". If a catastrophe occurs, the investors will lose their principal (the invested capital), or part of it, or only the interest, depending upon the type of cat bond purchased.

It has been estimated that out of the global economic costs of USD 1,400 billion due to weather-related disasters during the period 1980–2004, only USD 340 billion were insured. Climate change is exacerbating the economic losses. Some forward-looking

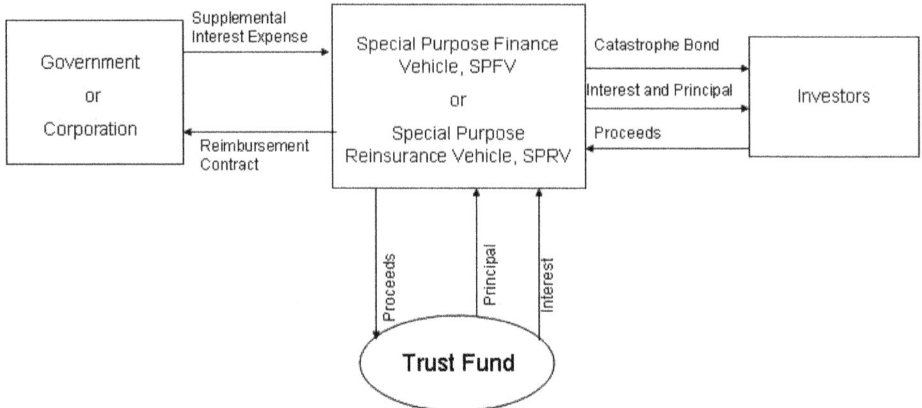

Figure 6.5.2 Institutional structure for a catastrophe bond (Cat bond)
Source: Stipple, 1998

insurance companies (like Swiss Re) are attempting to reduce their losses by promoting the protection of mangroves, reefs and wetlands that buffer storm surge and wave risks, and funding the development of energy-efficient technologies. This is akin to life insurance companies promoting practices to increase life expectancy.

In the aftermath of December 26 tsunami, the developing countries which were affected by it, sought financial assistance of the donor countries. Linnerooth-Bayer, Mechler and Pflug (2005) suggest novel insurance instruments to be adopted by the donor community for assisting the poor countries *before* the disasters happen. Under this arrangement, the donor community transfers the catastrophe risks to global financial markets. The proposed donor-supported risk transfer programme would on one hand reduce the disaster aid demands from the donor countries, while freeing the poor countries from the vagaries of post-disaster assistance, on the other. Recent advances in computer modeling have made it possible to better estimate and price low-probability extreme events for which there is limited historical data. The proposed scheme seeks to complement post-disaster humanitarian aid, with pre-disaster programmes of preparedness and risk transfer.

6.5.4 MONITORING AND WARNING SYSTEMS

Technologies exist for monitoring and warning of natural hazards, such as earthquakes, tsunamis, volcanism, landslides.

Seismograph networks like those established along the San Andreas fault in California, monitor the onset of seismic activity on the base of the precursors. Though it is still not possible when precisely a major earthquake is going to occur, the time band during which it is likely to occur can be predicted.

The upwelling of the magma chamber leading to a volcanic eruption can be monitored in a number of ways: remote sensing, tiltmeter study of the ground movement, earthquake activity, gas geochemistry, and hydrology. It is now possible to predict when the volcanic eruption is likely to occur, and what kind of eruption it is likely to be.

Unlike the Pacific Ocean where there are numerous sites from which tsunamis can be triggered, the Indian Ocean has only two belts of possible tsunamigenic sites – Sumatra–Andaman fault belt wherefrom the December 26, tsunami got originated, and the Makran fault off Pakistan. The Indian Ocean Tsunami Warning and Mitigation System (IOTWS) which is being developed, involves the participation of Australia, India, Indonesia, Iran, Malaysia, Pakistan and Thailand, and envisages the eventual installation of about 20 DART (Deep Ocean Assessment and Reporting for Tsunamis) systems in such a way that there are no gaps. The protocols for data formats, transmission, analysis and storage, etc. are being standardized.

A DART system comprises of pressure-sensitive sensors located on the ocean floor (which detects the greater pressure of a passing tsunami). This information is passed on to sondes mounted on floating buoys, which are linked to satellites and monitored continuously. India would also be putting in place 20–25 automated sea-level gauges. In future, it should be possible to identify the location and magnitude of the earthquake in less than 3 min of its occurrence, and provide tsunami warning in less than 30 min. Countries in the Indian Ocean region are in the process of developing their own national detection networks, their own risk assessment and preparedness plans and their own public education and awareness campaigns.

According to Eddie Bernard of NOAA (*EOS*, 4 January 2005), the accuracy of detection of the currently available systems is 0.5 cm and the numerical models for tsunami forecasting have an accuracy of about 80%. Bernard cautions that while smaller tsunamis generally behave in a linear way, the behavior of larger tsunamis tends to be nonlinear. Monte Carlo calculations may be useful for this purpose.

As is well known, the height of the tsunami waves increase sharply as it approaches the coast because of the sea becoming shallower near the coast. In situations where the sea floor has a gentle slope, it is possible that high-frequency shore-based radars which are used to monitor storm surges on the basis of the measurement of surface currents and wave characteristics, could provide the advance warnings of tsunami. It has been reported (Roder et al., 2005) that strong electrical signals corresponding to the M9.3 Great Sumatra Earthquake of December 26, which took place at 0058:50.7 UTC, were picked up by an electrostatic sensor in Tuscany, Italy, almost immediately after the event, as the electric signals travel at the speed of light. The p-waves from the same earthquake reached the station 740 s later. This suggests the possibility of increasing the hazard alert window.

6.5.5 SCIENCE-BASED AND PEOPLE-BASED HAZARD PREPAREDNESS SYSTEMS

The pre-hazard, during hazard and post-hazard preparedness and mitigation systems have both science-based and people-based components which need to be integrated (Table 6.5.1). Data from different spectral bands of the earth observation satellites, communication satellites and satellite web are useful in addressing all phases of the hazard.

Table 6.5.1 Science-based and people-based preparedness systems

	Science-based preparedness and mitigation systems	*People-based preparedness and mitigation systems*
Pre-hazard	1. Earthquake precursors may have electromagnetic, thermal, geotechnical, geobiological, geochemical, etc. signatures. R&D is needed to identify these signatures which have the potentiality to lead to a longer warning time (1–2 days?) for an earthquake and the resulting tsunami. 2. Network of detection systems on the seabed, on the sea surface and on land. 3. Location of the possible sites of the earthquakes, volcanism, landslides, etc. 4. Probability studies of temporal and spatial distribution and possible magnitudes, characteristics, movement patterns of potential tsunami, based on ocean depth, coastal bathymetry, land use and land cover of the coast, etc. War games-type simulations, involving Monte Carlo calculations.	1. Public education, 2. Longer warning time, 3. Warning communication systems, 4. Evacuation drills, 5. Protective bioshields (mangroves, salt-tolerant trees) which are ecologically-sustainable and employment-generating, 6. Knowledge-based coastal biovillages, 7. Hazard zoning, 8. Securitization through catastrophe bonds.
During hazard	1. The detection of (say) M7 submarine earthquake should automatically activate the tsunami – warning system, which will project, based on previously made desk studies, when, where and how the tsunami is going to hit. 2. Monitoring of the tsunami with GOOS satellites.	1. Activation of warning communication systems, 2. Evacuation drills, 3. Help to children, women and old people, 4. Safety of animals.
Post-hazard	1. Damage assessment by remote sensing, involving the nature and extent of damage, how and when it occurred, prioritization based on the severity of damage. 2. Long-term planning involving the identification of rehabilitation sites, coastal zone regeneration, identification of sensitive coastal environments, etc.	Public–private partnership through an integrated psychological, ecological, Agronomic and livelihood rehabilitation.

6.5.6 RISK COMMUNICATION

A significant component of preparedness is public education. Every schoolboy in Japan knows that a sudden recession of the sea is an indication of an imminent tsunami attack and that he should run inland as fast as possible, alerting others while doing so. That this kind of public consciousness is absent in countries like India is evidenced by the fact that in some areas both children and adults were washed away when they rushed to collect fish and crabs lying on the exposed seabed. Thus, public education at all levels (school, community, college, etc.) constitutes a critically important part of preparedness.

Tsunami hazard zoning, along the lines of earthquake zoning, will be useful not only for regulating civil constructions, but also for designing warning communications systems. Evacuation drills need to be performed periodically to ensure that warning communications and evacuation protocols are viable.

6.5.7 REHABILITATION MEASURES

An integrated plan for psychological, ecological, agronomic and livelihood generation in the affected areas is best effected through public–private partnership. A major challenge would be the restoration of the ecosystem and biodiversity, involving the desalinization of agricultural land, water wells, etc. Remote sensing is useful for damage assessment, covering the nature and extent of the damage, how and when it occurred, and prioritization based on the severity of damage, and for long-term planning involving the identification of rehabilitation sites, coastal zone regeneration and identification of sensitive coastal environments.

Chapter 6.6

Basic research and R&D

The total damage caused by a tsunami is a function of two kinds of biophysical factors. First is the cause, mode of generation, location, and geological environment in which a tsunami is generated; this would determine the how and with what energy and velocity the wave train would move. Second is the bathymetry and geomorphology of the coast concerned. A variety of fundamental studies need to be made to understand the tsunami-causing processes in the Indian Ocean region:

(i) Identification and location of the possible causes and potential sites where tsunamis could be triggered. The stress map of the region, downloadable from http://www.world.stress.map.org, could be a good starting point.

(ii) Location of sites which are prone to submarine landslides, such as submarine canyons with high slopes and thick sediment cover.

(iii) Probability studies of the temporal and spatial distribution and possible magnitudes, characteristics, dynamics, movement patterns, etc., of the potential tsunami.

(iv) Coastal bathymetry and the shoaling effect. When the tsunami wave is moving through the deep ocean, it may have a wavelength of hundreds of kilometers but a wave height of a few tens of centimeters. When such a wave approaches the coast, it slows down depending upon the depth of the water, and increases greatly in height (to 10 m or more), without losing energy.

(v) Land cover and land use of the coast. A rocky coast or a coast with extensive mangrove stands will suffer less damage than a sandy coast without mangroves.

(vi) Modeling and simulation studies involving the above parameters. Three-dimensional modeling of the propagation of the tsunami is horrendously difficult, as the wavelength of the tsunami depends upon the water depth along the propagation path, which may not be known with sufficient accuracy. As regards the coastal bathymetry, almost all countries zealously guard detailed information about this on grounds of security.

References

Adger, W.N. et al. (2005). Social-ecological resilience to coastal disasters. *Science*, **309**, 1036–1039.

Allenby, B. and Fisk, J. (2005). Toward inherently secure and resilient societies. *Science*, **309**, 1034–1036.

Aswathanarayana, U. (2005). Preparedness and mitigation systems for asian tsunami-type hazards. *EOS*, **86**(11), 2005.

Bolt, B.A. (1993) *Earthquakes*. New York: Freeman.

Carrara, A. (1984). Landslide hazard mapping – aims and methods. In Flageolet, J. C. (eds.) *Movements de Terrains*. Series Documents du BRGM, 83, 141–151.

Cooke, R.U. and I.C. Doornkamp (1990). *Geomorphology in Environmental Management*. Oxford: Clarendon Press.

Gupta, H.K. (1992). *Reservoir-induced earthquakes*. Amsterdam: Elsevier

Kasperson, R.E. (1969). Environmental stress and the municipal political system. The Brockton water crisis of 1981-66. In R.E. Kasperson & I.V. Minghi (eds.) *The structure of political geopgraphy*. 481–496. London: Univ. of London Press

Linnerooth-Bayer, J., Mechler, R. and Pflug, G. (2005). Refocussing disaster aid. *Science*, **309**, 1044–1046, 2005.

Lundgren, L. (1986). *Environmental Geology*. New Jersey: Prentice-Hall.

Menzinger, I. and Brauner, C. (2002). *Floods are insurable* Swiss Re, Zurich.

Mills, E. (2005). Insurance in a climate of change. *Science*, **309**, 1040–1043, 2005.

Pizarro, G. and Lall, U. (2005). Climate drivers, stream flow forecasting and flood risk management, In: U. Aswathanarayana (ed.), *Advances in Water Science Methodologies*, Taylor & Francis, UK, 134–155.

Roder, H., et al. (2005). Great Sumatra earthquake registers on electrostatic sensor. *EOS*, **86**(45), 2005.

Singh, S.C. (2005). Sumatra earthquake research indicates why rupture propagated northwards. *EOS*, **86**(48), 2005.

Stipple, I. (1998). *Securitizing the Risks of Climate Change*. IIASA, Laxenburg, Austria.

Section 7

Overview and integration

U. Aswathanarayana (India)

The purpose of the volume is to provide employable knowledge and skills in managing natural resources (water, soils, minerals, and ecosystem) in economically-viable, ecologically-sustainable, socially-equitable and employment-generating ways. Nature has a certain amount of resilience. Living in harmony with nature means that we take care not to push nature beyond the resilience limits.

SECTION 1 DEALS WITH GENERAL ISSUES WHICH ARE COMMON TO ALL NATURAL RESOURCES

The benefit of living in harmony with nature is illustrated with an example. The fishermen of Pichavaram, Tamilnadu, India, protect the mangroves. The mangroves saved them from the fury of the Indian Ocean Tsunami of Dec. 2004. The root exudates of the mangroves enriches water with nutrients, and thereby help in augmenting the fisheries, and the mangroves protect the coastal waters from salinization. As against this, wrong practices caused by the ignorance of the importance of symbiosis between mangroves and man led to horrendous degradation of the environment in Quelimane, Mozambique. The uncontrolled cutting down of mangrove trees for fuel wood, timber and coffins, led to extensive, *irreversible* salinization of coastal waters and soils.

The economics of natural resources management can be understood in terms of "virtual" natural resources. It is generally difficult and prohibitively expensive to transfer large quantities of water, soils, and low unit cost minerls and rocks (such as limestones) to some other places where they are required. It is, however, possible to share the *benefits* of water, instead of physically sharing the water. "Virtual" water is the amount of water needed to produce goods – for instance, 1000 L (one m^3) of water is needed to produce 1 kg of grain, and 13,000 L of water is needed to produce

1 kg of beef. Thus, a country which exports one tonne of grain is in effect exporting $1000\,m^3$ of "virtual" Water.

Globalisation

Globalization describes a process by which regional economies, societies and cultures have become integrated. The process is driven by a combination of economic, technological, sociocultural, political and biological factors. Economic globalisation is the most important part of globalisation, whereby national economies get integrated into international economy through trade, foreign direct investment, capital flows, migration and spread of technology.

Two cases of impact of globalisation may be mentioned. Specialists of Indian origin dominate the Silicon Valley, California, which is a haven of high-tech enterprises. Some publishers get their books typeset in one country, printed in another country, bound in a third country, and released in the fourth country.

An important consequence of globalisation is the modernisation and industrialisation of previously agrarian countries like India and China. These countries are gaining more knowledge and wealth. As more and more people move from Below-Poverty-Line to middle class status, they consume superior food, purchase more cars (Tata's Nano car costing about USD 2200/- is the cheapest car in the world, and is highly affordable), consume more fuel and more consumer goods. This necessitated a massive development of natural resources, including mineral resources within these countries. Besides, these countries are also making large investments in developing mineral deposits in Africa and elsewhere.

Mineral resources are finite, and when once used, are gone forever. Technology has a great potential to address the depletion problem. Since globalization connects the world, more scientists from around the world would be able to work to find solutions to a global problem. For instance, since oil production is peaking, and oil plays a critical role in the world economy, attempts are being made to substitute natural gas in its place.

Innovation

Technology development goes hand in hand with the innovation process. The framework conditions necessary for the successful prosecution of the innovation are macroeconomic stability, education and skills development, favourable business climate, protection of Intellectual Property (IP) rights, etc. Innovation process is not necessarily linear, and may not proceed smoothly – there can be many impediments enroute. RD&D is only part of the Innovation scheme – it needs to be adapted depending upon the feedback from the markets and technology users.

Innovation Chain has five phases: Basic Research → Research and Development → Demonstration → Deployment → Commercialisation (diffusion).

Governments and private sector have roles to play in all the five phases of the Innovation Chain. Generally, governments are expected to play a greater role in the early part of the chain (such as, basic science), though some large industrial houses (such as, SONY, GE) have extensive basic research programmes. Down the line, private sector alone would be involved in the last phase, i.e. commercialization. Roles overlap in the case of Applied R&D, Demonstration and Deployment. Sometimes there may be difficult technical problems that markets fail to address. Through specifying technology standards and participation in full-scale "in the field" demonstration projects, governments may induce private companies to achieve higher technological performance.

There are a number of ways in which governments could help in navigating the "Valley of Death", the transition from publicly-funded demonstration to commercial: economic incentives like tax credits, production subsidies, and guaranteed procurement, and knowledge incentives, such as, codification and diffusion of generated technical knowledge. Public – private research consortia can play a valuable role in technology transfer and commercialization. In some countries, technology parks are established to facilitate technology transfer. In these parks, governments give support to individuals or groups of scientists and technologists to perform basic research and applied R&D. When something viable emerges from this effort, the same group is helped to commercialize the invention. Governments can also create demand to new technologies through the promulgation of regulatory requirements. This would induce the supply side to respond to new regulations.

SECTION 2 DEALS WITH WATER RESOURCES MANAGEMENT

Roughly half of the book is devoted to water which is an important component of living in harmony with nature. The Chinese ideogram for "political order" is based on the character for "water", symbolizing that those who control water, control people.

"Water is many things in one: a basic life-need and right; an amenity; a cleaning agent; a social good (e.g. for firefighting, hospital use, use in schools and public institutions, etc.); a requirement for economic activity (agriculture, industry, commerce); a means of transportation; an occasional manifestation as floods; a part of our social, political and cultural life; and a substance that plays a role in rituals" (Iyer, Chap. 2.3). Every human being on earth is a stakeholder in water resources management. And so, for that matter, is every plant and animal, domesticated and wild – only they do not have a constituency. We need to protect other living things, not for any altruistic reasons, but in our own self-interest. We should realize that our well-being is inseparable from the well-being of the ecosystem. If frogs are dying, we would surely be next in line.

The many different perceptions of water is reflected in the many perspectives on water, notably, Engineering perspective, Economic perspective, Ecological perspective, Rights perspective, Social Justice perspective, Water Quality perspective, etc. These perspectives need to be integrated and harmonized, with priority given to ecological and social justice perspectives. For instance, drinking water is a human right, but not irrigation water. Ecological imperatives should override all other considerations. Sustainability demands that the pursuit of economic growth should be commensurate with its ecological footprint.

Suppose there is a cubic metre of water of a particular quality at a particular place and time in a watershed. The watershed has some hydrological and surface attributes, population, livestock, climate, evapotranspiration, fertilizer use, etc. characteristics. Using linear programming methodologies, it is possible to optimize the apportioning of water among the competing uses, taking into account both competitive criteria (i.e. which kind of allocation would yield the maximum profits for a unit water use, say, electronics industry) and social criteria (for instance, drinking water is given the highest priority), while keeping the ecological impact as an overriding constraint in all situations.

Holistic water resources management based on the hydrological cycle

There are four kinds of waters – Rainwater, Surface Water, Groundwater and Soil Water (Soil Moisture). We have no control over precipitation – it has to be treated as a given.

Surface water and groundwater form a continuum – most rivers start as groundwater springs. Human interventions are possible to apportion the precipitation between surface water and groundwater. The position about soil water is more complex.

When rainfall occurs, a part of it infiltrates into the soil and becomes the soil moisture (*green water resource*). This is the water that is available to plants through root uptake, and constitutes the main water source of the rainfed agriculture. The surface runoff, which may be stored in dams, lakes and aquifers, is called the *blue water resource.* It is the main source of water for irrigated agriculture. In the course of the hydrological cycle, green water which constitutes about 65% of the global precipitation, flows back as vapour to the atmosphere (evaporation, interception and transpiration), whereas blue water which amounts to about 35% of the global precipitation, flows as liquid recharging groundwater, and flowing in rivers to lakes, and ultimately to the ocean. The total green water flow from croplands globally is around 6800 km^3/yr which corresponds to around 6% of global precipitation. It has been estimated that 5000 km^3/yr of green water flow originates from rainfed agriculture, and the rest 1800 km^3/yr from irrigated agriculture.

The distinction between irrigated and rainfed agriculture does not actually have any conceptual basis in hydrology. Green water in rainfed agriculture originates from naturally infiltrated rainfall, whereas irrigated agriculture involves supplementation with blue water. Increased water use in one sector (say, agriculture) will affect water availability in other sectors, such as direct human use (water supply) and ecosystem use (terrestrial and aquatic ecosystems). Experience has shown that rainwater harvesting and building of small-scale water storage systems have hardly any adverse impacts on downstream flows. Investing in water management in rainfed agriculture has several beneficial effects, such as reduction in land degradation, improvement in water availability (to enable more food to be produced) and water quality downstream.

The plant water availability in the root zone and hence the crop productivity, can be enhanced by mulch practices, drip irrigation techniques and crop management to enhance canopy cover.

If steps are taken to minimize non-productive green water losses, more green water will be available for crop production. This would lead to higher yields for the same amount of green water use. Expansion of irrigation in effect involves augmenting of green water.

Green water can also be augmented by *in-situ* rainwater harvesting, before the rainfall generates runoff. This would reduce the blue water source.

Saline water, municipal and industrial wastewater can also be used in irrigation if combined with proper management. If not properly controlled, irrigation with contaminated water may introduce pathogens and toxic chemicals in vegetables and crops.

Some kinds of water management involve only demand management. For instance, mulching and growing of larger canopy plants would result in lower soil evaporation, thereby enhancing the productivity of green water.

When *ex-situ* water harvesting expands irrigation, the amount of water available to sustain the downstream ecosystems and downstream industrial, domestic use, will be reduced. Trade-offs with other land uses, such forestry and pasture will have impacts on biodiversity.

Yields are very low in rainfed agriculture. If investments in *in-situ* or *ex-situ* water harvesting in rainfed agriculture could increase yields from 1 t/ha to 2 t/ha, it would amount to increase in crop productivity in terms of water consumption from approximately 3500 m^3 of water per ton of grain to less than 2000 m^3/ton. Water collected *ex-situ* can also be used to raise a cash crop in winter. As the atmospheric demand would be low at that time, this would result in higher water productivity.

If flood water can be captured upstream and used in agriculture, there will be less runoff and less problems with erosion. When agriculture is developed in degraded lands with low infiltration capacity, more groundwater is formed and there will be reduced floods and erosion during heavy storms.

Conjunctive use of water resources

The purpose of the conjunctive use of water resources is simple: surface water surplus during the wet periods is temporarily stored in groundwater reservoirs, to be withdrawn for use during the dry periods. Conjunctive use is particularly necessary in the monsoon regions. During the rainy period lasting for a few months, the rivers are full and there is more water than could be used. During the dry period when water is desperately needed, the rivers are dry or have very little flow. Conjunctive use of water corrects the mismatch between supply and demand of water, and is a sensible strategy in monsoon climates.

Water resources endowments of countries

The total global runoff is more than adequate to meet the demands of water for the humankind for many decades to come. But the catch in it is that the distribution of freshwater resources in the world is extremely uneven. Anthropogenic activities tend to degrade the freshwater resources, thus complicating the problem even further. Under the circumstances, the transfer of waters between one country to another and between different parts of a country are unavoidable, and every effort should be made to minimize the problem.

Though some rivers like the Danube and the Rhine in Europe, Nile in Egypt, Colorado in USA and Kaveri (Cauvery) in India, are intensively used, most of the rivers in the world are used only to a limited extent. Despite enormous techno-socio-economic problems, massive transfer of waters (e.g. Three Gorges dam on River Yangtze in China which transfers about 70 $km^3\ yr^{-1}$ of water from south to north) has to be undertaken in some situations. It should be emphasized that high dams are not the only means of bringing about water transfers – a number of other techniques are available (e.g. groundwater recharge).

Pathways of pollution

Geoenvironment comprises rocks, soils, fluids, gases and organisms. It is linked to and influenced by, climate, terrain, and vegetal cover. Human activities affect the geological, physical, chemical and biochemical processes taking place in rocks, soils, hydrological systems and associated media. Surface water and groundwater are the components of the geoenvironment most affected by man. Protection of surface water from depletion and pollution (chemical, organic, thermal, mechanical, etc. contamination) constitute the most important task of the community, for the simple reason, when once the water (particularly groundwater) gets badly polluted, it is horrendously difficult, sometimes even impossible, to detoxify it. Microorganisms affect the hydrochemical processes and water quality. For instance, thionine bacteria oxidize hydrogen sulphide to sulphuric acid, and render the water more corrosive. Atmospheric pollution can penetrate the hydrosphere and biosphere, and can bring about undesirable changes in climate, soils, vegetation and water quality.

Pollutants may enter the surface water directly, and the groundwater through fractures/pores in rocks and soils. Dissolved pollutants in the surface water move much faster than the dissolved constituents in the groundwater. The hydraulic connection between the groundwater and the surface water has a profound implication – for instance, steps taken to protect the groundwater quality by reducing infiltration into the groundwater may lead to greater runoff, and consequent effect of river flow and quality.

Groundwater protection

Source protection (or wellhead protection as it is called in USA) is an essential part of groundwater protection. The objective of demarcating source protection areas or zones, is to protect the groundwater from two kinds of pollutants:

- Contaminants which decay with time – the longer they reside in the subsurface, the greater will be their attenuation, and the less will be their ability to contaminate the groundwater,
- Non-degradable contaminants, whose adverse impact on the groundwater can be mitigated through flowpath-dependent dilution.

Three source protection zones are prescribed according to the current British practice:

1. Zone III (catchment area): Catchment of the source, whose area is demarcated on the basis of water balance considerations, and whose geometry, on groundwater flowpath considerations,
2. Zone I (Microbiological protection): This is defined by the mean 50-day saturated zone travel time, based on pathogen decay criteria. The pathogens (say, in human

faeces) have persistence times ranging from weeks to months, and the length of travel of the biological pollutants from the source is of the order of 20–30 m. To be on the safe side, a zone of 50 m radius is recommended, as high-storage aquifers are prone to fissure flow.

3. Zone II (Outer Protection): The purpose of this zone to provide delay, dilution and attenuation of slowly-degrading contaminants, and to allow gradational land-use controls between zones I and III. The extent of this zone is arbitrarily set at 400 day saturated travel time.

Sequential use of water

Water is a mobile resource. Unlike gasoline which when once used is gone forever, water can be used and reused many times (e.g. municipal water supply → waste water → bio-ponds → irrigation water → groundwater recharge → withdrawal from wells, etc.). The kind of changes that water undergoes in terms of its quality, quantity, location and timing because of a given use, would determine its value for the succeeding uses.

Since many combinations of use and reuse of water are possible simultaneously or in sequence, it is critically important to know how the various uses of water combine and interact in space and time. Consideration has therefore to be given to return flow and reuse. A given physical unit of water can be used over and over again if water is used fast and returns large quantities of clean or well-treated water. Even though the value addition for each use of water is small, a pattern of multiple use can generate large values. The location and timing of water use becomes important in the system context of water use. The system gains more value when the waste-generating users are located as far downstream as possible.

Paradigm of global water resources management

There is little doubt that in the twenty-first century, water issues are going to emerge as the most serious problem facing humankind. The situation can be addressed by taking the following steps.

1. Protection of water quality. Groundwater is sought to be protected through an understanding of its vulnerability.
2. Drastic decrease in specific water consumption, particularly in irrigation and industry. In 1995, the agriculture sector accounted for 67% of the total water withdrawal and 86% of its consumption. This is not sustainable. In future, the efficiency of irrigation has to be improved greatly (through measures such as drip irrigation and cutting down of conveyance losses), so that with lesser total withdrawal (about 60%), more food could be grown.
3. Ways and means of reducing the water consumption in industries by extensive recycling.
4. Complete cessation of the practice of discharging waste water into the hydrological systems (such as, rivers and lakes). Every cubic metre of contaminated water discharged into rivers and lakes spoils up to 8–10 m^3 of good water. This is a monstrous situation, and every effort should be made to stop this practice.

5. Harvesting of precipitation and making a more efficient use of runoff: rainwater harvesting from rooftops, harvesting of surface runoff, groundwater recharge, etc.
6. Greater use of salt and brackish waters.
7. Influencing the precipitation-forming processes.
8. Use of water stored in lakes, underground aquifers and glaciers: Conjunctive use of water resources, and the design and management of groundwater reservoirs.

Agrometeorological advisories

Agrometerological advisories at various levels (district, field units, individual farmer) are valuable in optimising water use and alerting the farmer of extreme weather events (heavy rainfall and drought) to enable him to be prepared for them. The advisories seek to minimize the risk to the farmer in crop and livestock production. Farmer-specific agrometeorological advisories coupled with water-saving techniques like drip irrigation will help the farmer to achieve higher yields while using less water. India has a variety of climate and cropping patterns. The country has been divided into 127 Agro-climatic zones based on Thornthwaite Moisture Availability Index (MAI), soil type, thermal regimes and water balance. Water resources in each agroclimatic zone should be used most judiciously in the context of crop water requirements of the zone.

Climate change is projected to lead to increase in the variability of rainfall and decrease the natural storage provided by snowpack and glaciers, such as in the Himalaya that feed the rice-wheat belt of the Indo-Gangetic Plains. Through simulation studies, it is possible to adapt the advisories to meet the emerging challenges of climate change, zone-wise.

Remote sensing in water resources management

About a dozen satellite sensors, such as AVHRR, MODIS, NDVI, etc. are now available to measure land surface variables such as, Leaf Area Index, Soil Moisture, Surface temperature, Surface air temperature, and Precipitation. The spatial resolution may vary from 8 to 50 km, and temporal repeat may vary from bi-weekly to daily. Modeling is a powerful tool for the estimation of water storage and fluxes and when coupled with data assimilation and satellite remote sensing offers a very strong framework for hydrological analysis.

The usefulness of the satellite observations in agriculture and water resources management is obvious. Data and modeling software systems are now available to seamlessly integrate data from satellite, aircraft and ground sensors at various scales, ranging from individual farms to regions. For instance, Leaf Area Index (LAI) derived from satellite data can be used to compute water use and irrigation requirements to maintain crops at optimal levels of water requirement. By integrating leaf area, soils data, daily weather and weekly weather forecasts, it is possible to estimate the spatially varying varying requirements within a farm.

Hydrodynamics of an urban river The hydrodynamics of an Urban river – Case study of Musi river, Hyderabad, India, seeks to delineate the dynamics of interplay between rainfall, water storage reservoirs, groundwater recharge, and water use patterns in respect of the Musi river, Hyderabad, India, as an aid in developing future strategies of planning of water use.

Large-scale groundwater withdrawal in the upper Musi basin for irrigation has resulted in the reduction of inflows into the Himayatsagar and Osmansagar reservoirs which supply drinking water to the Hyderabad city. Climate change may not be the reason for the reduction of the inflows, as the short-term (1985–2004) temperature data show a slight decrease in temperature. It is found that the premonsoon groundwater level has greater influence on the recharge, rather than the total amount of rainfall itself. It appears that whatever the rainfall that occurs in the upper catchment is simply percolating down to recharge the depleted groundwater table, to meet the soil moisture deficiency and to meet the storage of water in smaller or bigger water conservation structures such as tanks, percolation ponds, check dams, etc., thereby leaving little or no runoff that could reach the downstream.

SECTION 3 IS DEVOTED TO MINERAL RESOURCES MANAGEMENT

This section draws attention to control technologies to minimize the adverse impact of mining on environment and make beneficial use of mine wastes.

Presently, the horrendous consequences of mining are evident everywhere. The landscape in some countries (e.g. USA, Zambia, PNG) is pockmarked with gigantic pits. The mass of the mine tailings produced worldwide (18 billion m^3/year) is of the same order as the quantity of sediment discharge into the oceans. As progressively lower grades are worked, the mass of the mine tailings is expected to double in the next 20–30 years. Vast areas are strewn with rock fragments, and in some areas, Acid Mine Drainage has rendered the soil and water so acidic that not a blade of grass grows there. Whole towns (e.g. eastern India) had to be abandoned due to subsidence caused by underground coal mining. Mine workers are exposed to a number of physical, chemical, biological and mental hazards, and mining is ranked as number one among the industries in the average annual rate of traumatic fatalities.

The iron ore mine of Kiruna, Sweden, is an outstanding example of how technology can be applied to minimize environmental pollution, reduce energy consumption and improve productivity. The Kiruna mine's production of 23 Mt/y of ore is drawn, trammed and dumped by semi-autonomous LHDs.

Emerging technological needs

The specialized and exacting requirements of modern industries led to profound changes in the ways metals are detected, extracted, alloyed and used. New applications for metals are being found all the time, e.g. use of germanium in semiconductors, use of cerium in high temperature superconductors, development of zircalloys in nuclear industry, titanium alloys in aerospace industry, metal glasses, etc. On the other hand, some traditional metals (e.g. Fe and Cu) are being substituted by plastics, fiberglass, ceramics, etc., thus increasing the demand for industrial minerals. The non-metallic minerals are being increasingly used as insulating material, fillers, glasses, and construction material. The ever-increasing need for more fertilizers (due to the need to grow more food for the increasing population of the world) will greatly increase the consumption of fertilizer raw materials, like apatite, potash feldspar, etc. Thus, the demand for a given mineral depends upon technology and markets.

Control technologies for the mitigation of adverse environmental impacts of mining

Acid Mine Drainage (AMD) causes the greatest adverse impact. It is mitigated in the following ways: (i) Changing the chemical properties of the waste (such as, separation of pyrite or addition of a buffering substance, such as lime) or physical properties of the waste (such as, compaction to reduce porosity and permeability). This is expensive. (ii) Flooding of the waste, such that the water table is established above the disposed waste, thereby limiting the transport of oxygen or air into the waste – this is by far the most cost-effective and efficient option, where it possible to implement, and (iii) dry covering of the waste.

Tailings disposal

The most serious problem facing the mining industry presently is the enormous mass of the mine tailings (about 18 billion m^3/y). Failure of the tailing ponds constitute the most serious environmental disasters in the world.

Thickened tailings and paste technologies are being increasingly used to design the tailings disposal sites to suit the surrounding environment. Paste production process is suitable for solids volume of 40 to 55%. Equipment is commercially available (e.g. GL&V) to produce paste consistency material from mill tailings, for purposes of back-fill. Tailings are introduced with a flocculent into a feedwell. A mechanical rake and helix concentrates the solid particles by inductive circulation in a compression zone while preventing ratholing. The paste-like material can be withdrawn from the bottom, and a clear liquid overflows into the launder at the top of the tank. The paste consistency material can be stored indefinitely.

Paste backfill has two main advantages: (1) the cavity is filled to the extent of 85%, and (2) it makes use of the tailings from the processing plant, thus reducing the need for surface disposal of the tailings. Thus, the use of backfill techniques has the advantage of minimizing the use of land on the surface, but the disadvantage of higher operating costs because of energy consumption.

Underwater placement of mine tailings The deep seabed emerged as a possible site for the placement of mine tailings. In the case of the Misima Gold and Silver mine, Papua New Guinea, the tailings were placed in a 1000 to 1500 m deep near shore basin in an area of tropical, open coast with coral reefs. The tailings were dispersed to ensure a slow rate of deposition so that the organisms are able to cope with it.

Dust control technologies

Dust is a problem in almost all mineral industries, though the degree of severity of the problem varies from industry to industry. Four types of dust control techniques are used in the mineral industries, including the iron and steel industry: Mechanical dust catchers, Electrostatic precipitators, Filter media, and High-energy scrubbers.

Aluminium industry discharges huge amounts of fluoride-loaded particulates which can cause dental mottling and skeletal fluorosis in human beings and animals. Aluminium plants produce cryolite mud (at the rate of 0.02 t of cryolite mud per tonne of cryolite used) which contains toxic heavy metals, such as, arsenic, cadmium, nickel, etc.

A. Bernatsky in his book, *Tree Ecology and Preservation,* strongly advocates the use of tree belts around industries to reduce particulate pollution, and noise. One ha of spruce can collect about 32 t of dust from the atmosphere, one ha of pine 36.4 t, and one ha of beech, 63 t.

Cognis has developed a number of new surfactants to provide for improved dust control on mine haulage roads while being compatible with solvent extraction and leaching processes. EnviroWet DC-100 is highly biodegradable, and has superior wetting properties relative to the traditional surfactants based on linear alkyl benzene sulphonate and similar compounds.

Low-waste technologies

The idea of low-waste technology originated with water – that it is better not to pollute the water during the manufacturing process rather than clean it up afterwards. Low-waste technologies are those that are the least environmentally-degrading, involving pollutants (dust, gas, odour), nuisance (noise, vibration), with least consumption of energy and the use of raw materials.

Recycling of scrap

Steel production in USA involves the use of 64% of scrap. Each tonne of steel scrap recycled saves 1.1 t of iron ore, 0.6 t of coal and 54 kgs of limestone, apart from savings in energy. Aluminium scrap (mostly in the form of cans), scrap of copper and lead (lead batteries) are extensively recycled.

Treatment of wastewater

There are a number of ways of treating the large quantities of wastewater produced in the iron and steel industry, namely, recycling, removal of suspended solids, oil, and organic toxic pollutants, etc. These are applicable to other mineral industries as well.

Filtration is a highly reliable method of wastewater treatment. It is used to remove suspended solids, oil and grease and toxic metals from steel industry wastewaters. It has a number of advantages – low initial and operating costs, small land requirement, no need to add flocculant chemicals which add to the discharge stream, and low solids concentrations in the effluent, etc.

Subsidence

Mining involves the extraction of large quantities of rocks, liquids and gases from the depths of the earth, and therefore causes damage not only on the surface but also to depths of hundreds and thousands of metres. Subsidence may be controlled by not carrying out mining to a specified depth – in other words, a specified cover will be left intact throughout the life of the mine. The Room-and-Pillar stopes will be supported to 1.5 m long bolts at 1.2 m spacing. The mined out stopes will be promptly backfilled with the sand fraction of tailings from the concentrator plant. The slurry will have 70% solids by weight.

Noise and vibration

Noise and vibration not only create health hazards, but also cause damage to structures.

Blasting can generate both dust and gaseous contaminants. The adverse consequences of blasting can be controlled in the following ways: (1) wait for some time before entering the area affected by the blast, (2) wetting down with water before blasting, and (3) ventilation. It is necessary to mention that respirators for particles protect against dust particles only, but not against gaseous emissions, which require gas masks. Planting of dense tree belt has been suggested as a way to reduce noise. It has been reported in the literature (A. Bernetzky) that a tree barrier of 250 m depth can achieve a reduction of 40 dB. In 1980, ILO has issued guidelines about protecting the workers from noise and vibration.

Planning for mine closure

Instances are known of the mine owners just abandoning the mines when the ore runs out. It is critically important that mine closure programme should be incorporated into any mining proposal right at the outset. Proper closure of the mine is absolutely essential, particularly if the mine wastes happen to be acid producing. The leachates from them can play havoc with the waters, soils and biota of the area for many decades, if not centuries.

Health hazards of the mining industry

Mining is undoubtedly the most hazardous industrial occupation.

There are two kinds of health impacts associated with mining: immediate impacts such as accidents, and accumulative and progressive impacts such as stress and pneumoconiosis. Opencast mining is generally less hazardous than underground mining. Industrialised countries tend to use highly automated mining systems, which not only employ lesser number of workers (who have to be highly skilled), but also have the effect of drastically reducing the hazards to which they are exposed. Developing countries cannot afford such high-tech mining systems, so much so that mining accidents are a common occurrence in developing countries such as China and India.

An examination of the statistical data (for USA, 1995) in regard to fatalities, nonfatal days lost (NFDL), total accident incident rate, and severity measurements (SM) for underground and surface mines by sector, leads to the following conclusions: (1) Among all the mining activities for various minerals, the most hazardous is the underground mining of coal, (2) The underground mining of coal, metal and nonmetal has higher severity measure than the corresponding figures for surface mining for the same minerals, (3) Surface mining of stone has a greater SM than underground mining of stone. On the basis of such analyses, the Mine Safety and Health Administration (MSHA) targets sectors, mines and jobs to enforce the regulations.

By making various improvements in mining technology and mining practices, mining fatalities have been drastically brought down in most countries. USA brought down the mining fatalities from 47 in 2000, and to 23 in 2010. The coal mining fatalities in China which used to be as high as 6,000 per annum, has been brought down to 3215 in 2008. In the case of India, which is the third largest producer of coal, the coal mining fatalities came down from 200 in 2000 to 35 in 2005.

Health hazards due to dusts

Dust is the cause of the many of the cumulative health hazards in the mineral industries.

The main sources of dust in the mining operations are:

Point sources: (1) Ore and waste loading points in trucks, railroad cars, etc. (2) Ore chutes in the haulage systems (bin, conveyors), (3) Screens in outdoor crushing plants, (4) Exhaust from dedusting installations, and (5) Dryer chimneys.

Dispersed sources: (1) Waste dumps, (2) Ore stockpiles, (3) Haul roads, (4) Tailings disposal.

There are three *E*'s of mitigation of health hazards of the mining industry: Education, Engineering and Enforcement. The goal of the mining industry should be to ensure that the workers could work their entire career without incurring death, disability or serious injury.

Total Project Development – A visionary approach

The Total Project Development (TPD) is a new holistic approach to mining. Under this approach, a mining project is developed as a part of much wider, multi-activity regional development. All the material extracted by a mining company is put to productive use. Waste rock, mine tailings, excess mine water, etc. are used as raw material for a variety of downstream ancillary industries. Tailings are used for underground backfill, embankments and sealants for reactive waste rock, and production of construction materials for mine use. All excess tailings are used for "soil" development. Excess process water (after use in recycling) is treated for being used in fish farm ponds and crop irrigation.

Artisanal mining

The contribution of small-scale or artisanal mining to the overall mining output in the world is estimated to be 10–16%. In general, the percentage contribution with respect to industrial minerals is higher than for metallic minerals. Estimates vary widely about the number of persons involved in artisanal mining (upto 16 million). In some countries, the contribution of artisanal mining is of considerable economic significance. For instance, Peru has about 3,000 small-scale mines which produce 100% of the antimony, 90% gold, 15% tungsten, etc. of the national production About 90% of Brazil's gold production is attributable to artisanal mining.

Artisanal mining may be a cost-effective option in the case of several economic minerals occurring in the soil. For instance, by the nature of its occurrence and its properties, opal can only be mined manually (in candle light!). Small-scale mining has several advantages: it is labor intensive, can be initiated on any scale, with simple technology, at low capital cost, low consumption of energy, and short lead time, and without expensive imported equipment. It can also promote local industries. This, however, does not mean that small-scale mining is the panacea for the problems of the developing countries. Artisanal mining suffers from the following serious disadvantages: (i) it tends to be haphazard, since in most countries there is no systematic exploration activity to support small-scale mining; (ii) destructive exploitation by the gouging of rich pockets; (iii) low recovery rates; (iv) low labor productivity (about 4% of highly mechanized mines); and (v) non-extraction of valuable byproducts which are therefore irretrievably lost to the country.

It is possible to improve the efficiency of small-scale mining, while concomitantly reducing its deleterious consequences. For instance, the "Portable" gold plant developed by Libenberg, Rundle and Storey of San Martin mining company is a veritable godsend for small-scale gold miners. It can extract 1–3 g/t. of gold through Carbon-in-pulp/carbon-in-leach technique.

Ways of ameliorating the adverse consequences of the mining industry

Rehabilitation of mined land

Though the climate, soil and hydrological characteristics and methods of mining vary greatly in different areas, there are some common elements in the techniques of rehabilitation (i) Removal and retention of top soil, to be respread in the area that is being rehabilitated, (ii) Reshaping the degraded areas and waste dumps in such a manner that they are stable, well drained, and suitably landscaped for the desired long-term use, (iii) Minimizing the likelihood for wind and water erosion, (iv) Deep ripping of the compacted surface, (v) Revegetating with appropriate plant species in order to control erosion, and facilitate the development of a stable ecosystem compatible with the projected long-term use.

Beneficial use of mining wastes

Taking the mining industry as a whole, there is little doubt that coal mining produces the largest volume of solid wastes. Mine gangue and coal-washing tailings are being increasingly used as filling materials, additives in concrete and for agricultural purposes. Clay may be mixed with fly ash (to the extent of 10–40%) and made into bricks, which can then be fired in conventional Bull's kiln, or intermittent type kilns at a temperature of 950 to 1050°C. The use of fly ash permits the production of 40% more additional bricks from the same quantity of soil. The clay-fly ash bricks have lower bulk density, better thermal insulation and reduced dead load on the brick masonry structure. These bricks can be used for all types of construction, where normal clay bricks are used.

Bricks can be made with red mud wastes from the aluminium industry. Red mud improves the quality of bricks made from clay-deficient soils. When fired, bricks made with red mud develop a pleasing pale brown, orange or golden yellow colour, depending upon the composition of the raw material, and firing temperature. They therefore have a good architectural value as facing bricks. The presence of 4–5% alkalis in red mud makes for good fluxing action. Consequently, the red mud bricks have better plasticity and bonding than the normal bricks. They may be fired in the usual Bull's trench kiln.

Acid mine effluent often contains copper which can be recovered cheaply by treating the effluent with scrap iron. Methane generated in the underground mining can be collected and used to feed the boilers.

Solid wastes from mining could be used as fillers in concrete and other cement-based materials. Mine soil fill material can be effectively used for the renovation of wastewater. Red mud waste is produced when bauxite is processed to produce alumina, and is available in large quantities around the bauxite mines. It contains compounds of Al (22–37%), Fe (24–26%), Ca (2–4%), Na and Si. It has been found that red mud mixed with medium-sized sand is highly effective in removing P, BOD, suspended solids and fecal coliforms from domestic sewage.

China's largest gold producer, Shangdong Gold Group Co. Ltd, has recently commissioned a 4 million m^3 tailings brick manufacturing plant – it is expected to generate annual profits of Yu 12 million and pay back the company's investment in five years. Zambia converted the abandoned open pits to fish ponds.

Reuse of mine water

Mine water is invariably highly acidic, besides containing undesirably high quantities of toxic metals. The presence of high iron content in groundwater is objectionable because of discoloration, turbidity, bad taste and tendency to form deposits in the distribution mains. The National Environmental Engineering Research Institute, Nagpur 440 020, India, developed a simple plant to remove iron from groundwater by precipitating the iron impurity as a ferric sludge (item 6.2.4, *CSIR Rural Technologies*, 1995). The plant is to be attached to a hand pump. It has a capacity of 2500 L/d (10-hr operation) and costs about Rs. 25,000 (USD 500). The plant has three chambers. "The water from the hand pump is sprayed over an oxidation chamber. The aerated water flows over baffle plates to a flocculation chamber and then to sedimentation chamber. The water then passes through plate settlers and to the filter from where the filtered water is drawn through a tap after chlorination". The ferric sludge needs to be scoured out twice a month.

SECTION 4 DEALS WITH ENERGY RESOURCES MANAGEMENT

The section covers coal, oil and natural gas, shale gas, nuclear fuels, Renewable Energy Resources, and Ways of achieving a low-carbon economy.

The case of coal reminds us of the "rose-and-thorn" syndrome. For instance, countries like China, India, USA and Poland cannot avoid using coal for generating energy. The mining, preparation, transport and combustion of coal severely pollute the environment (not only in the countries concerned, but also in the neighboring countries), and causes climate change on a global-scale. This chapter draws attention to technologies for maximizing the efficient use of coal and minimizing the environmental impact and carbon emissions.

The Coal Cycle has serious environmental and health impacts. Technologies are available to minimize the health and environmental impact of dusts which are produced in the process of mining, preparation and transport of coal, and emissions which are generated in the process of combustion of coal, disposal of coal mine tailings and control of burning of waste tips, etc. Also, systematic application of proven technologies would result in improved rehabilitation of mined land, and in limiting the adverse effects of subsidence, flooding, explosions, roof collapse, etc.

A number of power-generation technologies are being developed which are characterized by higher thermal efficiencies, and also lower emissions of NO_x and SO_2. These are: Supercritical and ultra-supercritical pulverized coal combustion (which can attain efficiencies of about 50%), Circulating Fluidized Bed Combustion (CFBC) (which allow for the *in situ* capture of SO_2), Integrated Gasification Combined Cycle (IGCC) from which CO_2 can be captured and stored, and hybrid systems, such as fuel cell-IGCC. Australia has succeeded in using a liquid for capturing 85% of Post-Combustion CO_2 (PCC). This break-through will benefit the future and present coal-fired power station.

Oil and natural gas were produced by the burial and transformation of biomass over the last 200 million years. The transformation of kerogen to hydrocarbons (*catagenesis*) is a complex process which is largely dependent on the way kerogen is "cooked" (based on the temperature and duration of heating which are related to thermal gradient, depth of burial and age of sediments).

The proven oil reserves in the world are about 1240 thousand million barrels. About 40% of the oil resources are in the form of 932 giant oil gas fields (a giant oil and gas field is one which has 500 millions barrels of ultimately recoverable oil or gas equivalent). They are clustered in 27 regions of the world, with two largest clusters in the Arabian Gulf, and Western Siberian Basin. A majority of the world's giant oil and gas fields occur in passive margins and rift environments. Middle East holds 61% of the proven oil resources of the world. World oil production in Dec. 2007 has been 86 Mbd (million barrels/day).

Evidently, oil reserves are finite, and will be exhausted sooner or later. Hubbert (1956) developed a tool ("Hubbert Curve") for forecasting future production of oil by estimating the likely peak in production in the case of any oil-producing region. According to this theory, known as Peak Oil Theory, the profile of annual production volumes in any oil-producing region, would be a bell-shaped curve ("Hubbert Curve") and the maximum production is reached when about 50% of the ultimate production volume has been extracted. When it was propounded, Hubbert's view was widely criticized for being pessimistic, but when the time came, his theory proved to be correct, in the case of USA. Hubbert Theory predicts that global oil production will be reached when about 50% of the world's Ultimately Recoverable Resource (URR) has been consumed. The unresolved question is how high URR is, and how close the world is to the oil peak.

Proved gas reserves in 2007 are about 172 trillion cubic metres. The world production of natural gas is 3127 billion cubic metres). Important producers of natural gas (in billions of cubic metres) are: USA – 593, Russia – 583, and Iran – 200. The high consumption of natural gas in USA is because of cold winters and strong demand for gas for power generation.

Oil and natural gas resources are dwindling fast. Peak production is estimated to occur in two decades, and exhaustion is likely occur by 2050. For this reason, and also to reduce the carbon dioxide emissions, it is necessary to use oil and natural gas wisely.

There is urgent need to improve the energy efficiency, and achieve technological breakthroughs in energy utilization and management.

Shale gas has emerged as a major element in the energy picture. When hydrocarbons present in shale are "overcooked" because of being subjected to high pressure and temperature, they get converted to shale gas. Two technological advances, namely horizontal drilling and and hydraulic fracturing (fracking) made it possible to extract gas from shale. Shale gas wells may involve lateral lengths of over 3000 m. to create maximum borehole surface area.

Production and use of shale gas has environmental health consequences. It is now established that shale gas use aggravates global warming, as shale gas emits methane which is a more potent green house gas than conventional gas. Chemicals added to water to facilitate the underground fracturing process, can some times lead to the contamination of groundwater

The world reserves of shale gas has been estimated to be 6622 trillion cu. ft. Unlike the case of oil and natural gas, the occurrence of shale gas is wide-spread. The most important shale gas countries are: China – 1275 trillion cu. ft., USA – 862; Argentina – 774; Mexico – 681; South Africa – 485.

Shale gas tends to cost more to produce than gas from conventional wells, because of the expense of the massive hydraulic fracturing and horizontal drilling. This is, however, offset by the low risk of shale gas wells.

Country case study of Saudi Arabia: Oil in commercial quantities was discovered in Saudi Arabia in 1938 in the Dammam area. Saudi Arabia has the largest proven oil reserves in the world, estimated at 267 billion barrels, including 2.5 billion barrels in the Saudi- Kuwaiti neutral zone. The geologic environment of oil in the Middle East comprises of a large depositional platform along a pre-Mesozoic passive margin of Gondwana. Extensive migration of hydrocarbons into traps underlying thick, regionally extensive evaporate seals, facilitated the accumulation and preservation of oil. Saudi Arabia produces the world's largest amount of crude oil (10.5–11 Million bbl/d). Saudi Arabia's oil production mainly from six oil fields, the principal oil field being Ghawar (onshore) which produces 5 million bbl/d of 34° API Arabian Light Crude. Saudi Arabia produces a wide range of crude oils, ranging from heavy to super light, with 65 to 70% of crudes rated as light gravity. Lighter grades generally are produced onshore, and heavy grades are from onshore. The heavy crudes are "sour" (i.e. contain sulphur). Saudi Aramco operates the World's largest crude processing facility of more than 7 million bbl/d at Abqaiq in Eastern Saudi Arabia. Saudi Arabia has seven domestic refineries which processs combined crude of 2.1 MBO/d.

Saudi Arabia has three primary oil export terminals:

1. Ras Tanura Complex, with a capacity of about 6 million bbl/d,
2. Ras al-Ju'aymah facility on the Persian Gulf, with a capacity of 3 to 3.6 million bbl/d,
3. Yanbu terminal on the Red Sea, with a capacity of 4.5 million bbl/d of crude and 2 million bbl/d for NGL and other products.

Saudi Arabia has 252 trillion scf reserves of natural gas, with production of 85 billion cubic metres of natural gas/yr.

Nuclear Fuel Resources: Energy from the atom has several merits. Nuclear power reactors can be set up anywhere, need only a small quantity of fuel, and has no green house gas emissions. For instance, a 1000 MWe power station would annually need

3.1 million tonnes of hard coal, but only 24 t of enriched uranium. New reactor designs and advanced fuel cycles are being designed in which meltdown and explosion are not possible, and waste generation is minimized. Public mind is, however, exercised over possibility of reactor accidents (e.g. Chernobyl and Fukushima) and proliferation issues.

Australia, Canada, Kazakhstan and South Africa have large reserves of uranium. The uranium resources (~23 Mt) of the world will suffice upto 2100, if not longer. Australia, USA, Turkey, Brazil, India and Mozambique have large reserves of thorium, mostly in the form of monazite in detrital heavy mineral sands. The world thorium resources are estimated at 4.5 Mt of thorium.

Though the fuel cost of nuclear power (USD 0.018/kWh) is not much different from thermal power (USD 0.023–0.05/kWh), the capital cost of building a nuclear power plant (USD 2000/KW) is much higher than a modern gas turbine plant (USD 500/kW).

New reactor designs and advanced fuel cycles make it possible to avoid melt-down and release of radioactive gases, and minimize the nuclear waste. For instance, pebble bed reactors are inherently self-regulating. They can of lower capacity, say, 100–200 MW. Heat transfer is effected through ciculating inert gases (say, helium or nitrogen) through spaces between the pebbles. This kind of reactor will not undergo meltdown or release radioactive gases. Since no water cooling is involved, the kind of problems that Fukushima faced, will not happen.

France with a population of 63 million, has 59 reactors, mostly PWRs., which supply 78% of electricity in the country. France has the cheapest electricity in Europe and has the lowest emission per capita of green house gases. The Fukushima disaster dealt a grievous blow to the future of nuclear power not only in France but all over the world.

Renewable Energy Technologies (RETs)

Renewable Energy Technologies have three outstanding merits: most of them do not get depleted when used, they provide 4 to 5 times more employment than fossil fuel resources, and their use emits only a small quantity of green house gases.

Wind power (2016 GW) is expected to provide 12% of the global electricity by 2050, thereby avoiding annually 2.8 gigatonnes of emissions of CO_2 equivalent. Wind power sector would need an investment of USD 3.2 trillion during 2010–2050. The lifecycle cost is projected to be USD 70–130/MWh for onshore wind, and USD 110–131/MWh for offshore wind. The life cycle costs of wind energy are sought to be reduced through resource studies, technology (e.g. larger rotors, greater heights, deep water foundations for offshore turbines), supply chains, and mitigation of environmental impacts.

Solar PV will work wherever the sun shines. Its levelized cost (US cents 20–40/kWh) is several times more than electricity from fossil fuels (US cents 3 to 5/kWh). Solar energy is expected to grow thousand-fold between now and 2050. Technical advances in thin-film production and "building-integrated PVs" (BIPV) as well as massive application are bringing down the costs rapidly, however.

Second-generation biofuels, produced by enzymatic hydrolysis of cellulosic feed-stock, and gasification of a variety of biomass material, have a great future. Single-cell algae are being used to produce a chemical "mix" that is chemically identical to petroleum crude, which is also carbon neutral and sulphur-free. Small (~100 kW) power units, which burn biomass wastes like paddy husk, are very useful to villages, which are not connected to grid.

Presently hydropower accounts for 90% of the renewable power generation in the world. Though hydropower is the cheapest way to produce electricity, it has become controversial because of human and ecosystem problems of large dams. Pumped storage is the highest capacity form of energy storage.

Geothermal energy is confined to areas of high heat flow. It is non-polluting and can be generated round the clock. High temperature geothermal sources can be used to generate electricity.

The use of tidal energy to generate power is similar to that of hydroelectric plant. The estimated global potential of wave electricity is 300 TWh/yr. The 740-m long Rance Barrage in France which produces 480 GWh of electricity, is one of the few operating tidal energy plants in the world. Considerable R&D effort is needed to ensure the commercial viability of ocean energy.

The deployment of Renewable Energy Technologies (RETs) has two concurrent goals: (i) exploit the "low-hanging" fruit of abundant RETs which are closest to market competitiveness, and (ii) developing cost-effective ways for a low-carbon future in the long term. Highest priority should be given for the removal of non-economic barriers.

Strategy for low-carbon footprint

IPCC recommends the stabilization of CO_2 levels at 450 ppm, in order to limit the temperature rise to 2 degrees Celsius, above the pre-industrial levels. A number of steps need to be taken to reduce the carbon footprint to the minimum.

In order to achieve a low-carbon, sustainable future, it is necessary for the governments to embark on two pathways simultaneously: Firstly, through appropriate regulatory mechanisms, governments should strive to improve the efficiency of today's vehicles and for the deployment of transition technologies such as plug-in hybrids. Secondly, RD&D should be promoted for the development and deployment of long-term technologies, such as, biofuels, electric and fuel cell vehicles. Governments should make investments in infrastructure such as efficient, affordable and dependable public transportation (as in Singapore), and providing incentives for making rail travel preferable to air travel for journeys around 600 kms. International cooperation is essential to reach these goals.

Climate Change is a classical case of "negative externality" – economic actors (say, a coal-fired thermal power station and the user of electricity) impose costs on others, without paying a price for their actions. Two approaches have been attempted to limit the negative externality – pollution tax and cap-and-trade. Acid rain is caused by the emissions of sulphur dioxide from power plants. It was controlled by the government prescribing compliant effluents, and taxing the power plants that were emitting beyond the permissible limits. Pollution tax is a disincentive. A company avoids pollution tax by reducing its pollution to be within compliant limits. Recently, US EPA formally declared CO_2 as a pollutant attracting pollution tax. Pollution tax in respect of CO_2 is vigorously opposed by the coal industry. In the case of cap-and-trade, the government issues a limited number of licenses to pollute. Companies which need to pollute more need to buy licenses from those which have pollution to spare. The incentive here is

Mass extinctions took place at the time of Eocene thermal maximum 55 years ago when the earth's temperature rose by 11 degrees Fahrenheit. Martin Weitzman argues that our policy analysis should be based on the nonnegligible probability of utter disaster that is taking shape, forget about the uncertainty of Climate models and difference of opinion of ways of mitigating the adverse consequences of the climate change.

SECTION 5 DEALS WITH BIORESOURCES AND BIODIVERSITY CONSERVATION

In this section, Balaji deals with an issue that is central to the survival of mankind. Biodiversity is defined as the total diversity and variability of the living things and of the systems of which they are a part. This encompasses the total range of variation in and variability among systems and organisms at the bioregional, landscape, ecosystem and habitat levels, at the various organism levels down to species, populations and individuals, and at the level of population and genes. It also embraces complex sets of structural and functional relationships within and between these different levels of organizations including human action and their origin and evolution in space and time.

The importance of biological diversity to human society is hard to overstate. An estimated 40 per cent of the global economy is based on biological products and processes. Poor people, especially those living in areas of low agricultural productivity, depend heavily on the genetic diversity of the environment. Forests protect the watersheds, moderates climate and act as foster mother for agriculture. They hold the key to global food and water security.

To date, 1.4 million life forms have been named and described by science. But biological estimates suggest that there may be at least five million, but perhaps as many as 50 million species may be existing today. The estimate has increased dramatically in recent years following research in the rain forest canopy. The tropical rain forests support millions of insect species. Our knowledge about individual groups such as fungi, nematodes, mites and bacteria is meager.

The rich Global Biodiversity is threatened with erosion on an unprecedented scale. While the rates of extinction were roughly equal to those of speciation for most of the history of life on earth, contemporary extinction rates are several times faster than those of speciation leading to erosion of Biodiversity. Tropical humid forests in general are amongst the most diverse, most productive and most threatened of the biological communities with indeed 14 of the 18-biodiversity hotspots representing these biomes.

The erosion of biodiversity is caused by habitat loss, invasive alien species, pollution due to industrialization, population growth coupled with fast pace of economic development, and overexploitation of resources, etc. Efforts are underway in SE Asia to reverse this trend.

Stern review has estimated that if climate change is not addressed and mitigation and adaptive measures are not initiated worldwide, it would eventually damage economic growth and cause major disruption in economic and social activity. Climate Change will increase flood risk, reduce crop yield and cause water scarcity. Many organisms are sensitive to carbon dioxide concentration in the atmosphere and that may lead to disappearance of 15–40% of species. Projected sea-level rise is very likely to result in significant loss to coastal ecosystems of South-East Asia. Forests play a critical role in combating climate change, collectively capturing and storing significant amounts of carbon that would otherwise pollute the atmosphere. Forests neutralize about 29% of the global carbon emissions.

Local technical knowledge of Natives helps to conserve Biodiversity. Indigenous people with a historical continuity of resource use practices, often possesses valuable knowledge about the behaviour of complex ecological systems in their localities. Though people world over depend on 40000 different species for their requirement of food, shelter, and clothing, about 80 percent of global' food supply comes from just 20 species. Crop improvement over the last 250 years has been largely due to incorporating genes from wild varieties into cultivars. Plant breeding for useful traits under Green Revolution has helped to more than double crop production in the last 50 years. Impact of modern agriculture intensification through extension of canal irrigation in South Turkeministan has made fertile delta of Aral Sea a salt desert and has lead to disappearance of 25 species of fishes.

Different strategies have been developed for controlling agricultural pests and diseases through different microbes. In many cases microbial genes have been used. The endotoxin gene from *Bacillus thuringiensis* has been engineered against insects. For viral resistance, sequences of the virus itself have been introduced into the plant in order to interfere in the viral life cycle. Antimicrobial compounds from grasses, fungi (*Trichoderma, Aspergillus*), bacteria and animals (insects, rat, cow) are being used to obtain resistance against fungi and bacteria.

Biotechnology is defined as 'any technological application that uses biological systems, living organisms, or derivatives thereof, to make or modify products or processes for specific use'. The Chapter 16 of UNCED *Agenda* 21 recommends a set of programmes for Biotechnology applications in Biodiversity Conservation: Increasing the availability of food, feed and renewable raw materials, Improving human health. Enhancing protection of the Environment, Enhancing bio-safety and developing international mechanisms for co-operation, and Establishing enabling mechanisms for the development and environmentally sound application of biotechnology.

Biotechnology provides important tools for biodiversity utilization for human benefit: Genetic Engineering can be used to improve domesticated varieties to increase yield, control diseases and eradicate pests. Living organisms can be used as factories for manufacturing specific products. Biotechnology can rehabilitate polluted ecosystem. The combination of chemical and DNA screening technologies with genetic engineering provides an extremely powerful tool for the exploitation of genetic resources. For example, medium chain fatty acids such as laurate accumulate in the seeds of tropical trees and are harvested for dietary and industrial purposes at the rate of at least a million tonnes annually. The knowledge that production of medium chain fatty acids is due to a specific protein (acyl carrier protein-BTE) has led to the cloning of the DNA of the gene coding for this protein from the Californian bay (*Umbellularia californica*). This gene can be introduced into an oilseed species and can increase the levels of laurate. In Industry, transgenic animals, plants and micro-organisms can be used as source of insulin, growth hormones, antibodies, vegetable oils; essential oils and secondary metabolites such as pharmaceuticals and bio-pesticides. Genetic Engineering can be used for enhancing the efficiency of microorganisms in industrial processes, such as secondary recovery of oil from reservoirs, bioleaching, extraction of metals from low-grade ores, production of industrial enzymes, antibiotics and bio plastics. Biotechnology is also useful in Agriculture and Horticulture for production of better crops, and ornamentals with enhanced flavour and shelf life. In environmental management they can be used to in bioremediation of polluted environment.

Genetically Modified (GM) foods are developed and marketed because there is some perceived advantage either to the producer or consumer of these foods. This is meant to translate into a product with a lower price, greater benefit (in terms of durability or nutritional value or both). Insect resistance is achieved by incorporating into the food plant the gene for toxin production from the bacterium *Bacillus thuringiensis* (BT). Virus resistance is achieved through the introduction of a gene from certain viruses which cause disease in plants. The safety assessment of GM foods generally investigates their direct health effects (toxicity), tendencies to provoke allergic reaction, the stability of the inserted gene and any unintended effects.

Chemical contamination of our environment has been a highly visible and undesirable outcome of industrialization. Environmental clean-up through biological technologies, especially microbial ones, are often sought because they are usually less expensive than mechanical ones, Bioremediation can often be carried out *in situ*, eliminating or greatly reducing costs for soil excavation and movement; the biological catalyst multiplies as the expense of the pollutant, thus naturally enhancing the remediation rate. A consortium of bacteria is used in bioremediation of polluted lakes across the world. Polluted Lake Superior in USA was cleaned in this way.

The Conference of Parties (COP-10) held in Nagoya in Japan in October 2010 developed a new Plan of Action for conservation of biodiversity and for better access and benefit sharing of genetic resources. It formulated 20 "SMART" (Specific, Measurable, Ambitious, Realistic, and Time-bound) targets for 2020. The ecosystem services framework of Nagoya has four main consequences for target setting. First, what and how much biodiversity should be targeted for conservation depends on what services are important to maintain and with what reliability. Second, the temporal and spatial scale of targets should be based on the changing temporal and spatial distribution, and risk profiles of ecosystem services. Third, target development and implementation should include all agencies involved with management of biodiversity and the ecosystem services they support. Fourth, interdependence among ecosystem services, the benefits they provide, and the value placed on those benefits implies that targets must be conditional.

The Strategic Plan 2011–2020 (Aichi Target) is aimed achieving a world Living in Harmony with Nature. The mission was to take effective urgent action to halt loss of Biodiversity in order to ensure that by 2020 ecosystems are resilient and continue to provide essential services.

SECTION 6 DEALS WITH DISASTER PREPAREDNESS SYSTEMS

All lands on earth are subject to natural hazards, such as, floods, droughts, landslides, earthquakes, typhoons, volcanism, etc. The hazard itself cannot be prevented, but through an understanding of the land conditions which are prone to a given hazard and the processes which could culminate in damage to life and property, it is possible to minimize the damage to through preparedness for a particular eventuality.

The hazard may be purely *natural*, i.e. not induced by, and controllable by, humans (e.g. earthquakes, tsunamis); or *Mixed*: Natural in character, but influenced by human action (e.g. floods, droughts); or *Technological*: Generated by human action (e.g. reservoir-induced seismicity).

Natural marine hazards

Hurricanes or Tornadoes are extremely powerful over the sea, causing tides and torrential rains around, but weaken and die out as they more over land, causing major damage in coastal regions. Warning systems are in place worldwide for predicting and monitoring the storm surges due to extreme weather events like cyclones, hurricanes, tornadoes or tsunamis. The National Hurricane Center in the US, for example, forecasts storm surge using the SLOSH model, which stands for Sea, Lake and Overland Surges from Hurricanes. SLOSH inputs include the central pressure of a tropical cyclone, storm size, the cyclone's forward motion, its track, and maximum sustained winds. Local topography, bay and river orientation, depth of the sea bottom, astronomical tides, as well as other physical features are taken into account, in a predefined grid referred to as a SLOSH basin. Construction of dams and *floodgates* (storm surge barriers) is one way of reducing the impact of storm surge. They are open and allow free passage but close when the land is under threat of a storm. The *Orissa Super cyclone, east coast of India, 1999*, killed about 20,000 persons, and caused damage worth ~ USD 800,000. The *Hurricane Katrina, U. S. Atlantic coast*, 2005, killed about 2000 people.

Man-made hazards

Man-made hazards include blow-outs such as, the BP oil spill in the *Gulf of Mexico* which flowed for three months in 2010. It is the largest accidental marine oil spill in the history of the petroleum industry. It caused an immense destruction of the Gulf Coast marine environment. An oil spill occurred in Prince William Sound, Alaska, on March 24, 1989, when the *Exxon Valdez*, an *oil tanker* bound for *Long Beach*, California, struck *Prince William Sound*'s *Bligh Reef* and *spilled* 260,000 to 750,000 barrels (41,000 to 119,000 m^3) of *crude oil*. It is considered to be one of the most devastating human-caused *environmental disasters*.

There have been three major nuclear energy accidents till date: The Three-Mile Island accident USA (1979), and Chernobyl Reactor Accident, Ukraine (1986) were caused by technical failure, while the Fukushima – Daiichi (Japan) nuclear accident on Mar. 11, 2011, was triggered by earthquake and tsunami. The severity of nuclear accident is rated 0 to 7 on the International Nuclear Event Scale (INES). The Three-mile Island has been rated at 5, and Chernobyl and Fukushima are rated at 7, the highest (in the early stages of Fukushima accident, it was rated 4, then 5 and 6, and finally 7).

The TMI accident began on Mar. 28, 1979 (Wednesday) in Dauphin County, near Harrisburg, Pennsylvania, USA, when TMI-2, one of the two pressurized water reactors, suffered a partial meltdown. The accident resulted in the release of 43 kCi of radioactive krypton (1.59 PBq) and less than 10 Ci (740 GBq) of ^{131}Iodine. About 25,000 people who lived within 8 km. radius of TMI, were alerted. No one was killed or injured because of the TMI accident.

Chernobyl accident: On April 26, 1986 (Saturday), there was an explosion in one of the four, graphite-moderated, water-cooled nuclear reactors at Chernobyl, 128 km from Kiev in Ukraine (former Soviet Union). Basically, Chernobyl was the result of steam explosion. Fuel heated up very rapidly, and this caused all the coolant water to vaporize instantly, causing an explosion. Some fuel did melt, but that does not account for the magnitude of the disaster.

About 50 MCi (million curies) or 2×10^{18} Bq of radioactive fission products and noble gases (including ^{123}Xenon) which correspond to about 5% of the total fission products inventory of the reactor core, escaped. About half of this amount relates to radionuclides which figure in the food chain. Radioactivity got released in two distinct plumes – one on the first day (12 MCi), and another on the seventh day. Containment was effected by the eleventh day, and the emissions ceased after that. Meteorological conditions determined the pattern of dispersion of the Chernobyl radioactive cloud.

The Chernobyl radioactive plume rose to a height of 3 km before it started spreading horizontally. About 50% of the emission of condensable products fell in an area of about 60 km radius around the accident site. The rest of the emissions fell in an area of 10 million km^2 in Europe. The fallout was controlled by rainfall, other things being equal.

It may be noted that the meteorological conditions rather than the distance from the site of the accident, determined the extent of fallout. This explains as to why Sweden received > 50 times more deposition than the neighbouring Denmark.

The Chernobyl fall-out contaminated milk and meat products all over Europe. Extensive surveillance networks were set up to monitor the radioactivity of fresh milk, lamb meat, etc. For instance, the Government of U.K. prescribed the following Derived Emergency Reference Levels (DERL's) for milk: 2,000 Bq l^{-1} for ^{131}I, 2,000 Bq l^{-1} for ^{137}Cs, and 3,100 Bq l^{-1} for ^{134}Cs, totaling 7,100 Bq l^{-1} of milk. Some countries have adopted more stringent measures: any food-stuff with >1,000 Bq l^{-1} of radioactivity was declared unacceptable for human consumption.

The claims of deaths of tens of thousands of people in the Chernobyl accident are highly exaggerated. According to Chernobyl Forum (2006), 28 emergency workers died due to exposure to very high radiation dose, 15 patients died from thyroid cancer, and there may be 4,000 cancer deaths from among 600,000 people who received high dose of radiation. The Chernobyl accident had serious effects on biota as well. For instance, it led to enhanced concentrations of radionuclides in the air above Sweden, which were then washed into lakes and streams in Sweden.

The **Fukushima** nuclear disaster involved a series of equipment failures, nuclear meltdowns and release of radioactive material. It is more complex than the Chernobyl disaster as multiple reactors and spent fuel pools are involved. The reactor system at Fukushima (coordinates: 37°25′17″N; 141°1′57″E) in eastern Japan coast has the combined power of 4.7 gigawatts. It is one of the largest nuclear power stations in the world. It consists of six, light water, boiling water reactors (BWRs). It was designed by GE and constructed and run by Tokyo Electric Power Company (TEPCO). Except for unit 3, which has uranium oxide/MOX (Mixed Oxide) fuel, the other five reactors had uranium oxide oxide fuel.

The 9.0 magnitude Tohoku earthquake with its epicentre near the island of Honshu, hit Fukushima at 14.46 hrs JST on Mar. 11, 2011 (Friday). Its ground accelerations were above the design tolerances. When the earthquake occurred, reactors in units 1, 2 and 3 were operating, but the reactors in the units 4, 5 and 6 were in cold shut down for periodic inspection. Units 1, 2 and 3 underwent automatic shutdown (scram) when the earthquake occurred. Emergency generators started up to run the control electronics and water pumps needed to cool the reactors.

When the 14–15 m tsunami wave triggered by the earthquake engulfed Fukushima at 15.27 hrs i.e. 41 mins after the earthquake, it topped the 5.7 m sea wall, and flooded the basement of the Turbine buildings, thereby disabling the emergency diesel generators located there. Importantly, the earthquake destroyed the power supply to the reactor. Because of this, core cooling system had to be maintained by diesel generators. The flooding due to the tsunami wave hindered external assistance.

Under the circumstances, reactors 1, 2 and 3 experienced full meltdown. On Day 2 (i.e. 12 March) at 15.10 hrs, the upper 75% of the core of unit 1 appears to have melted and slumped in the lower part of the core. When the fresh water supply ran out at 14.53 hrs on 12 Mar, seawater was injected into the system for purposes of cooling. The seawater injection continued till 25 Mar.

Zircaloy which is used in the fabrication of the internal components and fuel assembly, underwent exothermic reaction and produced hydogen which exploded, with disastrous consequences.

On 5 May, workers entered the reactor building for the first time.

That technological innovations in reactor design and operation during the last 25 years since Chernobyl had a profound effect on the safety of the reactors should be evident from the fact that though the Fukushima reactor system is much larger than Chernobyl, the maximum radiation level in Fukushima (800 mSv) is only 0.4% of that of Chernobyl (200,000 mSv), and there were no deaths in Fukushima due to radiological causes.

Massive decontamination efforts are going on the 50 km radius around Fukushima. About 360,000 Children below the age of 18, are being screened for thyroid abnormalities. Apart from body radiation, food and soil are being monitored.

Fukushima had a strong adverse effect on the image of the nuclear power. Germany took a policy decision to phase out nuclear power. India which planned to buy from France six units of 1,650 MWe EPRs for Euros 40 billion, to be established in Jaitapur, western India, is seeking post-Fukushima certification. China and several other countries are waiting for the results of the ongoing Fukushima investigations to modify the design of the reactors.

Integrated disaster preparedness systems

The development and implementation of dual-use technologies help to provide substantial economic benefits in addition to resiliency even if no negative events occur. The urban systems have necessitated the development of tools that aggregate and display complex urban systems data. The availability of such a system can serve as a training tool for disaster management, besides being useful in the coordination of disaster management.

Resilience to coastal disasters is related to the capacity of linked social–ecological systems to absorb recurrent disturbances, such as hurricanes, tidal waves, tsunami, etc. so as to retain essential structures, processes and feed-backs. Just as a healthy man is less vulnerable to disease, and more capable of recuperation after he is affected by a disease, the healthier an ecosystem is the more capable it is of regenerating itself and continuing to deliver resources and ecosystem services that are essential for human livelihoods and societal development (e.g. mangroves).

The main purpose of the disaster risk management is to minimize disaster losses in hazard-prone areas. These management alternatives may be structural (such as, seawalls, groins, bioshields) and nonstructural (such as, Catastrophe Bonds or cat bonds). Such actions are taken before or in advance of the event (*ex ante*), and are, therefore, proactive measures.

Insurance Linked Securities (ILS) are the most common form of cat bonds. A "Special Purpose Financing Vehicle" (SPFV), or "Special Purpose Reinsurance Vehicle" (SPRV) is created for the purpose. There are two kinds of bonds, namely, "principal at risk" or "interest at risk". If a catastrophe occurs, the investors will lose their principal (the invested capital), or part of it, or only the interest, depending upon the type of cat bond purchased.

Monitoring and warning systems

Technologies exist for monitoring and warning of natural hazards, such as earthquakes, tsunamis, volcanism, landslides. Seismograph networks like those established along the San Andreas fault in California, monitor the onset of seismic activity on the base of the precursors. Though it is still not possible when precisely a major earthquake is going to occur, the time band during which it is likely to occur can be predicted. The upwelling of the magma chamber leading to a volcanic eruption can be monitored in a number of ways: remote sensing, tiltmeter study of the ground movement, earthquake activity, gas geochemistry, and hydrology. It is now possible to predict when the volcanic eruption is likely to occur, and what kind of eruption it is likely to be. A DART system comprises of pressure-sensitive sensors located on the ocean floor (which detects the greater pressure of a passing tsunami). This information is passed on to sondes mounted on floating buoys, which are linked to satellites and monitored continuously.

Science-based and people-based hazard preparedness systems

The pre-hazard, during hazard and post-hazard preparedness and mitigation systems have both science-based and people-based components which need to be integrated. Data from different spectral bands of the earth observation satellites, communication satellites and satellite web are useful in addressing all phases of the hazard.

Risk communication

A significant component of preparedness is public education. Every schoolboy in Japan knows that a sudden recession of the sea is an indication of an imminent tsunami attack and that he should run inland as fast as possible, alerting others while doing so. That this kind of public consciousness is absent in countries like India is evidenced by the fact that in some areas both children and adults were washed away when they rushed to collect fish and crabs lying on the exposed seabed. Thus, public education at all levels (school, community, college, etc.) constitutes a critically important part of preparedness.

Rehabilitation measures

An integrated plan for psychological, ecological, agronomic and livelihood generation in the affected areas is best effected through public–private partnership. A major challenge would be the restoration of the ecosystem and biodiversity, involving the desalinization of agricultural land, water wells, etc. Remote sensing is useful for damage assessment, covering the nature and extent of the damage, how and when it occurred, and prioritization based on the severity of damage, and for long-term planning involving the identification of rehabilitation sites, coastal zone regeneration and identification of sensitive coastal environments.

Author index

Subject index